S0-BLS-415

Essays in Contemporary Chemistry

From Molecular Structure towards Biology

Essays in Contemporary Chemistry

From Molecular Structure towards Biology

Gerhard Quinkert, M. Volkan Kisakürek (Eds.)

Verlag Helvetica Chimica Acta · Zürich

Weinheim · New York · Chichester
Brisbane · Singapore · Toronto

Prof. Gerhard Quinkert
Institut für Organische Chemie
Johann Wolfgang Goethe-Universität
Marie Curie-Str. 11
D-60439 Frankfurt am Main

Dr. M. Volkan Kisakürek
Verlag Helvetica Chimica Acta
Hofwiesenstrasse 26
Postfach
CH-8042 Zürich

Published jointly by
VHCA, Verlag Helvetica Chimica Acta, Zürich (Switzerland)
WILEY-VCH, Weinheim (Federal Republic of Germany)

Editorial Directors: Dr. M. Volkan Kisakürek, Tomaso Vasella
Production Manager: Birgit Grosse, Norbert Wolz

Cover Design: Bettina Bank

Library of Congress Card No. applied for.

A CIP catalogue record for this book is available from the British Library.

Die Deutsche Bibliothek – CIP-Cataloguing-in-Publication-Data

A catalogue record for this publication is available from Die Deutsche Bibliothek

ISBN 3-906390-28-4

Printing: Konrad Triltsch, Print und Digitale Medien, D-97199 Ochsenfurt-Hohestadt
Printed in Germany

For

Albert Eschenmoser

Preface

Those who seek to find a common denominator for the three main periods in the life's work of *Albert Eschenmoser* do not need to look for long before coming upon *Origins of Molecules of Life*. The first clues are to be found as early as 1951, in his Ph.D. thesis from ETH Zürich, which puts forward some thought experiments, involving cation-initiated cyclizations of acyclic polyenes into cyclic isomers or their functionalized derivatives, as tools for constitutional elucidation of monoterpenes and sesquiterpenes. This proposal provided a virtual synthetic strategy – affording potential cyclization products – to supplement the tried and tested analytical strategy of obtaining simple aromatic hydrocarbons by dehydrogenative degradation (cadalene and eudalene from farnesol, for example). In this way, it was possible to identify a connection between the constitution of the acyclic sesquiterpene farnesol (or farnesene) and the final constitutional formulae of the cyclic sesquiterpenes β-carophyllene, humulene, clovene, cedrene, or lanceol. In analogous fashion, it is possible to derive an entire set of cyclic monoterpenes from geraniol, cyclic diterpenes from geranylgeraniol, and cyclic triterpenes (including lanosterol) from squalene.

If these collected sets of examples at first served only to provide constitutional formulae for products that appeared probable in terms of the reaction-mechanism rules for cation-initiated cyclizations of the acyclic terpenes mentioned, the ever more pressing question of whether the general working hypothesis used in constitutional investigations in the terpenoid area might not also be applicable for mapping their biological syntheses (with enzymes) was not to be put off for long. Especially as it could be shown that, with the aid of rules relating to the stereospecific courses of cation-initiated cyclizations recognized in the field of chemical reactivity, the configurations of potential cyclization products were predictable, with high degrees of stereoselection, at least when the respective configuration of the acyclic reactants (with regard to its C=C bonds) and its proper conformation (with regard to the folding of the polyene chain) were assumed to be known, and a nonstop process without cationic intermediates was postulated for the normative cyclization mechanism (without enzymes). In the particular case of the cyclization of squalene to lanosterol, while predictions by the virtual synthesis strategy were well in advance of experimental evidence of this biological synthesis pathway – a fact that justifiably attracted great attention in the scientific community – study of nonenzymatic, biomimetic polyene cyclizations geared towards total synthesis of triterpenoids or steroids at first lagged behind possible expectations.

It has sometimes been asked why the actual initiator of this synthetic strategy, of using cation-initiated polyene cyclizations for determination of the identities and origins of terpenoids and steroids, did not himself develop this reaction type further for goals in total synthesis, as was to take place later at Stanford over more than two decades. Well, experimental investigations into the course of acid-initiated cyclizations of specially synthesized polyenes were certainly performed in the *Eschenmoser* laboratory. They were carried out, though, with a view towards derivation of the normative chemistry of polyene cyclizations underlying the enzymatic processes. It cannot be doubted that enzymes participating in biological cyclizations restrict the conformational space of particularly suited substrates to the advantage of optimal conformational folding, and assist the controlled cyclization process through the so-called template effect. It, furthermore, should not be ruled out that, thanks to electronic effects acting in very precisely defined local regions, they may manifest as reaction-accelerating and product-determining, even when the overall cyclization is not concerted.

*

The total synthesis of vitamin B_{12}, a drama of the highest order in which well over a hundred doctoral or postdoctoral reseachers on both sides of the Atlantic had been involved, is unique in several ways. First, there is the exceedingly complex structure of the target molecule and the distinct way in which this was worked out. Since vitamin B_{12} may be degraded into cobyric acid, also a naturally occurring product and one from which it had been possible to reconstruct the vitamin, the actual target molecule of the synthesis was thus cobyric acid. In each case – both vitamin B_{12} and cobyric acid – the structure was determined by X-ray crystallography. Since chemical degradation of cobyric acid had not taken place, this molecule occupied an isolated position in chemical space, with no close-lying *islands* from which some easy route to cobyric acid might have been feasible. Whereas chemical degradation would traditionally open up an entire chemical landscape, it was now necessary to chart the nearer and more distant environment of the target molecule with the aid of chemical synthesis. Definite planning of synthetic routes became harder. Alertness and readiness to react flexibly in the face of unforeseen difficulties was called for. That this did indeed actually happen in the case of the total synthesis of vitamin B_{12} was the result of a number of events.

The first surprise was provided by the fact that the two heroes of the vitamin B_{12} saga, *A. Eschenmoser* and *R. B. Woodward*, joined forces. The Harvard group dedicated itself to the more challenging A–D half, the ETH group to the B–C component. After the C–D link had been established with the aid of the sulfide-contraction invented during the course of the synthesis in Zürich, the A–B macrocyclization took place at a ligand that, with the aid of complexation with cobalt, it had been possible to fix in the quasi-cyclic conformation.

The *Eschenmoser* sulfide contraction is an invention that, in the successive conjoining of the heterocyclic five-membered-ring moieties, has proved itself an important advance in synthetic technology. Meanwhile, the fact that the synthesis of complex target molecules after the widespread use of X-ray analysis for molecular structure determination has developed into the primary source of new scientific discoveries in organic reactivity is attested to above all else by the *Woodward-Hoffmann rules for preservation of orbital symmetry*. Their serendipitous discovery was the consequence of an unexpected stereoselectivity observed in the preparation of the A–D component of the vitamin B_{12} synthesis in Cambridge. Working with *Roald Hoffmann*, *Woodward* developed a set of ideas vastly surpassing a mere explanation of the single observation that started it all. The essence of this new concept, which permanently changed the face of organic chemistry, was that, to understand a chemical reaction, to apply it in a controlled manner, and to be able to predict the result with greater probability than before, it is important to take account of *preservation of the bonding character of all the electrons involved in a reaction*.

It is not without irony that, with the aid of the deepened understanding of reactions achieved by *Woodward*, it was *Eschenmoser* who, applying the *Woodward-Hoffmann* ideas, discovered an A–B–C–D strategy to synthesize cobyric acid. The key reaction was a most remarkable photochemical A–D macrocyclization of a *seco*corrinoid metal complex. This new synthetic approach proved to be superior not only on paper to the earlier, and still pursued A–D–C–B strategy. In addition to this, it was found in Zürich that all the heterocyclic five-membered rings of the A–B–C–D molecule could be prepared from either one or the other enantiomer of an easily accessible racemic mixture of basic building blocks. The new approach outshone the old one in aesthetic quality and elegance. In the competitive cooperation of *Woodward* and *Eschenmoser*, the weights had shifted. The former could enjoy the satisfaction of having vastly exceeded the common synthesis goal and achieved a deepened understanding of molecular reactivity. The latter took the opportunity, by studying the reaction behavior of specially synthesized model compounds, to compare the *chemical synthesis* of vitamin B_{12} with the *biological synthesis*, which was the object of study at that time in a number of laboratories. The question under debate was how the suspected A–D cyclization came to be carried out in nature.

The chemical synthesis of cobyric acid was directly based on the A–B macrocyclization. The already mentioned shift in strategy from A–B macrocyclization to A–D macrocyclization simplified the synthesis considerably. The final construction of the 19-membered ring nucleotide loop of B_{12} was thought to require differentiation of ring D throughout the synthesis. It later became evident that this complicating factor was unnecessary. Had the regioselectivity of the nucleotidation been known earlier, the synthesis of vitamin B_{12} would have been even simpler.

A posteriori knowledge of the reaction potentials of participating molecules obtained in the course of the synthetic undertaking and *a priori conjecture* regarding the reaction potential of arguable alternative structures resulted in a synthesis design developed with the aid of *Ockham*'s razor ('*to get the most with the least*'). With the biological synthesis of vitamin B_{12}, belonging in *Eschenmoser*'s words among the most adventurous seen in the field of biosynthesis of natural products of low molecular weight, the situation was quite different. It is important not to lose sight of *Francis Crick*'s warning concerning biology, that '*while Ockham's razor is a useful tool in the physical sciences, it can be a very dangerous implement in biology. Biologists must constantly keep in mind that what they see was not designed, but rather evolved*'.

*

Well-founded opinion holds that today's DNA-RNA-protein world, with DNAs serving as informational and proteins as catalytic components, emerged out of an RNA world (without protein enzymes). According to *Walter Gilbert*, who coined the term, '*the concept of an RNA world is a hypothesis about the origin of life based on the view that the most critical event is the emergence of a self-replicating molecule, a molecule that can copy itself and mutate and, hence, evolve to more efficient copying*'. In this RNA world, RNA molecules functioned both as information stores and as catalysts (ribozymes). As might be expected of witnesses from an earlier stage of evolution, they were less reliable than DNAs as information stores and less effective than proteins as reaction mediators. Those who find the leap from the monomeric components of RNA (ribose and nucleobases) to oligonucleotides excessively wide are able to find more freedom for *evolutionary tinkering* in the hypothesis of the existence of a pre-RNA world. In such a world, *Darwinian* evolution taking place at the molecular level might enable the transition from chemistry to biology to take place in small steps. While ribozymes, relics from that ancient RNA world, attest to the emergence of the DNA-RNA-protein world from the RNA world, no corresponding remains bearing witness to the emergence of the RNA world from the hypothetical pre-RNA-world are known. Needless to say, chemists are presented here with a unique chance to design a variety of potential RNA precursors with the aid of chemical reasoning, then to synthesize a few (or more) of them by chemical methods, and lastly to carry out preliminary screening for their capability for informational base-pairing according to the *Watson-Crick* model.

In a broadly defined research project, *Albert Eschenmoser* and his co-workers at the ETH-Zürich and the *Skaggs Institute for Chemical Biology*, La

Jolla, have been engaged since the mid-1970s in a search for a potential precursor type with a structure simpler than that of RNA. Numerous oligonucleotides have been synthesized, with different sugars taking the place of ribose in the sugar-phosphate backbone of RNA. The ribose analogs taken into consideration are proposals obtained from a cascade of questions intended for a systematic search of nucleic acid space. *Why pentose and not hexose? Why ribose and not another pentose? Why ribofuranose and not ribopyranose?* The question '*Why phosphates and not sulfates or orthosilicates?*' has also been put and, with the aid of a wealth of known details from the literature, answered by *Frank Westheimer* in his classic 1987 paper.

Why questions call for *because answers*. They are clearly permissable for events that have been designed. Are they suitable for processes in evolution, too? According to *Manfred Eigen* the answer is '*yes*'. *Eigen*, on the basis of mathematical models and experimental studies of biological material, has shown that *Darwin*'s grand vision of evolution by natural selection can be elaborated further. According to his view, *selection is driven by an internal feedback mechanism that searches for the best route to optimal performance. It does not work blindly and gives the appearance of goal-directedness.*

What has the *Eschenmoser* group achieved so far to bridge the gap between the simplest organic molecules readily formed under prebiotic conditions and the self-constituted building blocks necessary to make up informational macromolecules? Firstly, it has solved the ribose problem, secondly it has set up the basis for a systematic conformational analysis of nucleic acids, and, thirdly, it has synthesized a candidate for RNA precursor.

The Ribose Problem. The observation that the aldomerization of formaldehyde in aqueous alkaline solution results in an extremely complex mixture of sugars (*formose*), which contains only a very small proportion of racemic ribose, does not in itself rule out the *formose reaction* as a prebiotic pathway to ribose, but does leave a number of questions unanswered. If, however, glycolaldehyde – the key substance involved in the *formose reaction* – is replaced with glycolaldehyde phosphate, the situation changes. Base-catalyzed aldomerization of glycolaldehyde phosphate in the presence of a half-equivalent of formaldehyde gives a relatively simple mixture of tetrose- and pentose-diphosphates, and hexose-triphosphates, with racemic ribose-2,4-diphosphate as the major component. In the presence of layered hydroxides such as hydrocalcite, the reaction between glycolaldehyde phosphate and glyceraldehyde-2-phosphate smoothly furnishes the ribose derivative in question. This result considerably alleviated earlier doubts concerning prebiotic formation of ribose.

The Conformation of the Nucleic Acid Backbone. The saturated six-membered ring is conformationally more rigid and clearly defined than the corresponding five-membered ring. This is also true for nucleic acid analogs in which the ribose-phosphate backbone of RNA, possessing tetrahydrofuran rings, is replaced by a sugar-phosphate backbone incorporating tetrahydropyran rings. Two nonnatural pyranosyl-oligonucleotides, homo-DNA and p-RNA, were synthesized in Zürich and used as demonstration objects for systematic conformational analysis. The former oligonucleotide was derived from β-D-2′,3′-dideoxyglucose and composed of (6′ → 4′)-hexopyranosyl repeating units, while the latter was derived from β-D-ribose and consisted of (4′ → 2′)-pentopyranosyl repeating units. In both cases, systematic conformational analysis reduces to only one single strand out of totals of 486 or 162 formally possible conformations, respectively, with minimal strain and possessing the capability for *Watson-Crick* base pairing. Both homo-DNA and p-RNA will pair up in double helices but are not able to form duplexes with RNA.

RNA Precursor Candidates. A lack of capability for cross-pairing would be expected to rule out exchange of information between oligonucleotides of some earlier evolutionary step and those of the subsequent one, and so the chances of p-RNA having been the genetic material that preceded RNA are weakened. Systematic screening of the base-pairing properties of potential natural, sugar-based nucleic acid congeners has been extended by *Eschenmoser* from hexopyranosyl oligonucleotides through pentopyranosyl and pentofuranosyl counterparts to tetrofuranosyl oligonucleotides. TNAs, derived from α-L-threose and composed of (3′ → 2′)-tetrofuranosyl repeating units, have been synthesized in La Jolla. In a prebiotic world, tetrose-sugar derivatives ought to be produced readily, and pairs of complementary TNAs have been found experimentally to form stable *Watson-Crick* double helices. Moreover, TNAs cross-pair efficiently with complementary RNAs (and DNAs), and so the TNA type is deemed a candidate for an RNA precursor type.

To conceive that TNAs might have arisen by self-assembly and, together with other archaic nucleic acid types, would have existed in a dynamic variant population is one thing. It is a different matter to construct a detailed picture of experimentally verifiable means through which a genetic system might emerge out of autocatalytic self-replication (without involvement of protein enzymes) of informational oligonucleotides. *Albert Eschenmoser* has given some thought to this in two publications recently appearing in the journal *Science*. He pursues some of *Eigen*'s ideas, seeing the critical selection factor as being in the base-pairing, still operative after the evolving system has left thermodynamic equilibrium and entered into a nonequilibrium state, in which the participating molecules replicate, mutate, and hence evolve. Future experiments will decide whether and under what conditions this is the case.

*

The scenarios outlined above demonstrate the broad nature of problems that, over the last fifty years, have justifiably been viewed as solvable with the aid of chemical synthesis. They portray the capabilities of synthetic chemistry, and of how it freely adds to the chemical community as a whole. The choice of the problems and the style of their solutions, though, are individual matters, to be accredited to particular protagonists. Careful study of the contributions of *Albert Eschenmoser*, to whom this volume is dedicated, may warmly be recommended to the next generation of scientists. There are few with interests so broadly disseminated and with such profound insight.

Frankfurt am Main, June 2001 *Gerhard Quinkert*

Contents

Prologue

The Gold-Mine Parable [1])

Organic chemistry is a nineteenth-century term; it is arguable whether, in the twenty-first century, it will possess still more than historical significance. Even today, the name *organic chemistry* is a straightjacket constraining everyone who may be said to be an organic chemist. For *Berzelius*, organic chemistry meant the chemistry of animal and vegetable materials, and since *Gmelin* (1848) organic chemistry has been by definition the chemisty of carbon compounds.

By this definition, then, what a magnificent piece of organic chemistry is, for example, the constitutional formula of the so-called A-protein gene of the MS2 bacteriophage [2], published in 1975. Here, we are concerned with a molecule classifiable to that special field of chemistry that concerns itself particularly with the structural elucidation and structural transformations of those organic compounds that occur in nature: organic natural-products chemistry. So, the theme and purpose of natural-products chemistry is the determination of the molecular structures of substances occurring in living nature and of their interconversions. Isolation, structural determination, and investigation of the reactivity of these substances *in vitro* make up its bedrock, while its goal is to understand the transformations of substances taking place *in vivo*, using the structure model terminology of organic chemistry.

The irony of the situation will not have escaped my respected listeners. Of those organic natural products that exist on this Earth, the most important for modern natural science, the most fundamental to life, and, indeed, the most interesting to the organic chemist are the domain of scientists who do not call themselves organic natural-products chemists. These substances are isolated and processed in laboratories that are not institutes of organic chemistry, and the exciting discoveries concerning these substances are reported

---

[1]) ***Editorial Note:*** *Albert Eschenmoser*, to whom eleven authors present a collection of essays on the occasion of his 75th birthday, should have the first word here. In a talk, *'Über Organische Chemie'*, that he gave at the 75th anniversary of the *Schweizerische Chemische Gesellschaft* in 1976 he used an analogy that has remained in the memories of the participants at the ceremony as the *'Gold-Mine Parable'*. The journal *Chimia* published the manuscript section containing his introductory and concluding remarks in 1993 [1]. The printed text has been used as the basis for the English version, produced by Dr. *Andrew Beard*.

in journals that are not called *Journal of Natural Products Chemistry*, not to mention *Journal of Organic Chemistry.* As far as the chemistry of compounds produced in living nature is concerned, today's organic natural-products chemistry sees itself as relegated to the realm of low-molecular-mass substances. To the natural products chemist who particularly insists on being purely an organic chemist, entire classes of molecules such as the biopolymers are taboo; there remains only the (not at all insignificant) task of grubbing all the finest details and subtleties of the chemistry of low-molecular-mass compounds out of the depths. The deer grazing in the untamed wilderness has evolved into a mole.

How did it come to this? The mandate granted at the start of the nineteenth century to the 'carbon chemists', through the definition of the term 'organic chemistry', was to emerge as one of the most significant, all-encompassing, and difficult research mandates in all of natural science. It was not possible that *one* type of chemist, nor *one* science alone, could carry this mandate forward in the twentieth century; an entire generation of daughter sciences had to grow up around it, and this new generation, together with the daughters of classical biology, makes up the modern-day community of 'molecular life sciences'. The original definition of organic chemistry – even though it persists in so many textbooks and is still volunteered in so many lectures (my own included) – nowadays is a historical relic. Today it is quite plain and simply misleading, in that its claim does not remotely match reality. The term 'organic chemistry' could freely be put away in the category of 'history of chemistry' – and with that I could essentially close this lecture.

Historical relic? The term 'organic chemistry', maybe, but certainly not organic chemistry. Because this mother science – although naturally not the youngest – is *despite* or even *because* of its numerous blossoming daughters (biological, physical, technical) still spry. Mind you, an organic chemist asked the question *'what is organic chemistry today, then?'* would do best by quoting *St. Augustine*, *'If you* don't *ask me, then I know, but if you ask me, I don't'*. Organic chemistry is, I suppose, what is taught by the teachers of organic chemistry under this title. In passing, science does not care about academic labels; the livelier and broader an area of knowledge is, the quicker it always escapes from academic efforts to pigeonhole it definitively by goal, content, and methodology. What a simple demonstration of the inferiority of the ideological!

It has always been the case, but is nowadays more than ever, that organic chemistry finds itself suspended between two poles that may be marked out by two provocative quotations. One comes from *Immanuel Kant: 'A natural science is a science inasmuch as it is mathematical'*. The other is from *Louis Pasteur: 'Piteous are scientists who have only clear ideas in their heads'* (coming from *Pasteur*, this sentence cannot be misconstrued). It is the spirit of *Kant* that normally besets the organic chemist, as that of *Pasteur* presumably does the physical chemist. But the ghost of *Pasteur* time after time leads organic chemists in the direction in which lie the true and original sources of discovery and inspiration in organic chemistry: the molecules of living nature. The *Pasteur*ian disposition towards the complex, towards the initially qualitative, towards the biological, has always been one of the most decisive impulses in the development of organic chemistry. However, among those organic chemists for whom *Kant* dominated over *Pasteur*, schisms between organic chemistry and biological chemistry would take place at times; and it was the spirit of *Pasteur* that opened the floodgates to all that today is biochemistry and molecular biology, or 'natural-products chemistry beyond organic chemistry'. The same spirit was to bring one *Emil Fischer* to his sugars, amino acids, peptides, and purines, one *Leopold Ruzicka* to his steroid hormones and pentacyclic triterpenes, and, in more recent times, one *Gobind Khorana* to his polynucleotides. These three singled out names may suffice to illustrate the development that has taken place: the works of an *Emil Fischer* and a *Leopold Ruzicka* did not in their time merely represent the pinnacle of organic natural-products chemistry; they simultaneously marked the most advanced frontier of knowledge concerning the chemistry of life. When, in our time, the organic chemist *Khorana* sets out, starting with organic synthetic methods, on the long road to synthetic polynucleotides, then he is no longer just pushing back his own people's frontier, but he is also, so to say, journeying into foreign territory. In his own land, this act is suspect; but outside, he is met by enthusiastic, like-minded individuals; there, with his findings, he arrives straight at the furthest frontier, that of molecular biology.

That research into organic nature requires a whole mosaic, as it were, of chemical sciences, has long been a truism. Each of these sciences has, by its own measures or resolution, with its own definition of its goals, and its own methods, to push deep into the depths of its own territory, in order to be a fruitful part of the whole. The organic chemist may catch him or herself agreeing with *George Orwell*, that *'all animals are equal, but some are more equal than others'*, since, as may be argued, whatever substances and phenomena the 'biochemistries' might one day uncover, the description of their molecular essence, the chemical structure, and the chemical reactivity of these substances will ultimately have to transpire in the terms and formulae of organic chemistry. This notion is – if also partially correct – unfruitful and beside the point. It is equivalent to the belief of a physical chemist, pointing out that all organic reactions obey thermodynamics and will also ultimately be describable by quantum mechanics, and that, therefore, physical chemistry is the more fundamental and important science. More fundamental? In some sense, yes, but therefore more important? A senseless question.

In today's research in organic chemistry, mining is no longer done open-cast; the times when the gold could be found in noble form lying on the surface of the ground are gone. At great depths – like for some mines in South Africa, 2000 m underground – long, complicated, and convoluted branching galleries are extended, using the most modern (physical) methods, so as to be able to follow the extremely narrow gold-bearing seams, which twist about both horizontally and vertically. On the surface, naturally, industrious bustle reigns, so as to process the excavated ore and extract the metal in as quantitative a manner as possible. Work in the processing plant is less onerous and more popular than work below in the galleries. The miners often return late, workworn, and battered. Now and then, though, when they have once more seen the yellow metal glistering out from the freshly broken ore, they have happy faces, and then, time and again, there are young people who also want to descend to the galleries. Above, there is naturally also an engineers' office, and there are based the geologists, who chart the course of the gold-bearing seams precisely and assess them geologically. They – so they say – know the fundamentals of the geology of the mining area perfectly. The management, however, pressing for economy, pressurizes them time and again with the question of how can the detailed and sometimes so abruptly shifting course and extent of the gold-bearing seams be understood and predicted. How is it that the miners are always right when they claim that, as yet, none of these 'office people' have reliably been able to predict exactly where they would have to dig to reach the particularly rich rock chambers, and, by the way, it has always been like this: the big finds have always effectively been made by them, the miners, because they had just been standing below the galleries, looked carefully at the rock while they were drilling, and otherwise just followed their instincts. Mining, they add, using a surprising foreign word, is simply an 'experimental science'.

It would have been a worthy exercise for this talk to concern itself with establishing how, in modern organic chemistry, experiment and theory act in concert, so as, on one hand, to show how progress, as ever, comes from experimentation, but also how fruitful the paradigm shift in the 1960s was, when the world of quantum-mechanical terminology finally became assimilated into the practice of organic chemistry. The speaker, however, has capitulated before this task, since he is – by his own description – too much of a mole and, furthermore, knows too little about geology. In place of what would be desirable, I would like quite simply to let a series of works from more recent organic chemistry parade before us. Burdened with the lecture title, already provisionally adjusted at the beginning, I must emphasize that these works have been selected according to a particular point of view, namely, that of preparative organic-natural products chemistry, and that, within this outlook, they moreover belong to those that lie close to my own interests. All those among the audience who look over organic chemistry from different perspectives, I must now ask for collegiate tolerance.

The beginning is simple. It has to be the first preliminary communication from *Woodward* and *Hoffmann*, in 1965. This work is recognized as inaugurating a new era in organic-reaction theory; it was to trigger off a development that must be seen as on a par with the introduction of the classical structure concept (1860), the tetrahedral model (1874) , the octet rule (1915), and conformational analysis (1950). As a contribution to the question of where the founts of discoveries in organic chemical research lie today, it is rewarding to reflect briefly on the special circumstances of the origins of this development.

While working out a subproblem in vitamin B_{12} synthesis, *R. B. Woodward* ran into a puzzle concerning reactivity. The theoretical analysis of this was the starting point for the formulation of the rules named after him and *R. Hoffmann*. Given its practical and personal settings, the development is probably too unique to be singled out as exemplary for the function and significance of natural-product synthesis research. Nonetheless, it illustrates – albeit in an extreme manner – the potential of natural-products chemistry for discovery and stimulus in organic chemistry. Above all, it shows that what theory states can only achieve its true potential in the arena of experimental chemistry, and that it needs a comprehensive and qualified perspective over the empirical world of organic reactions to recognize the consequences of the theory for chemistry. The research field of organic natural-products synthesis requires and provides knowledge in exceptional breadth, and so it is particularly fitting that it was the protagonist of modern natural-products synthesis who succeeded, with his and *R. Hoffmann*'s rules, in bringing about the final breakthrough of quantum-mechanical structure and reaction models into the praxis of organic chemistry.

The rules of organic chemistry are ordering principles, creating order where chemists had previously believed only disorder was to be seen. At all times, chemists have all too easily become accustomed to coming to terms with an apparent *de facto* lack of order; time and again, pioneers have proven that, beneath the surface, a form of order does indeed prevail. The *Woodward-Hoffmann* rules here are star witnesses[2]).

---

[2]) ***Editorial Note:*** In the further course of the lecture, highlights from the field of mechanistic and preparative organic chemistry were introduced and commented.They referred to the *Woodward-Hoffmann* rules, *Delongchamp*'s stereoelectronic-control rules, the *Bürgi-Dunitz* trajectories in carbonyl addition reactions, the non-occurrence of front-side attack in S_N2 reactions, *Arigoni*'s 'recent' synthesis of chiral acetic acid, the challenge of an erythromycin synthesis, *Gerlach*'s method of macrolactonization, *Seebach*'s 'Umpolung', *Merrifield*'s solid-support synthesis, phase-transfer catalysis, *Gerlach*'s nonactin synthesis, enantioselective catalysis by L-proline in aldolizations, *Pedersen*'s crown ethers, syntheses of corrins, hydroporphyrins, and vitamin B_{12}.

REFERENCES

[1] A. Eschenmoser, *Chimia* **1993**, *47*, 148.
[2] W. Fiers, R. Contreras, F. Duerinck, G. Haegmean, J. Merregaert, W. Min Jou, A. Raeymakers, G. Volckaert, M. Ysebaert, J. Van de Kerckhove, F. Nolf, M. Van Montagu, *Nature* **1975**, *256*. 273.

Looking Backwards, Glancing Sideways: Half a Century of Chemical Crystallography

by **Jack D. Dunitz**

Organic Chemistry Laboratory, ETH-Zentrum, CH-8092 Zurich, Switzerland

The past is a foreign country: they do things differently there.
L. P. Hartley (1895–1972), *The Go-Between*

It is a good job that science progresses as fast as it does because it gives us older scientists something to write about. It gives us the opportunity to describe how it was in the vanished world that existed when we were young. We did things differently then. I know because I have lived through more than half a century of X-ray crystallography, during which it has transformed itself beyond anyone's wildest dreams and thereby also transformed chemistry and molecular biology in then unimaginable ways. The present was unpredictable, and the past is viewed through the distorting lens of the present. I hope I do not distort it too much.

I am not old enough to have been there in the truly pioneering period of X-ray analysis but when I started, *Max von Laue*, *Paul Peter Ewald*, *Lawrence Bragg* were still very much alive, and their brilliant followers, *John Desmond Bernal*, *Dorothy Hodgkin*, *Kathleen Lonsdale*, *J. Monteath Robertson* in the U.K., *Linus Pauling*, *Ralph Wyckoff* in the U.S.A., *Johannes Martin Bijvoet* in the Netherlands, were in their prime. *Max Perutz* was busy with problems that most of his contemporaries regarded as insoluble; *Francis Crick* and *Jim Watson* had not yet been heard of. Who could have guessed that things would progress so far that, by the end of the century, the structure analysis of medium-to-large organic molecules would have become routine, and that structure analyses of many classes of proteins would become commonplace – one or two in each weekly issue of *Nature* or *Science*? Quite likely there were a few optimists who could look forward to such fantastic possibilities – as I recall, *Bernal* was one – but I, certainly, was not among them. It has been a marvelous experience for me to follow these developments and even to share in them a little.

1. How It Was

When I started my apprenticeship in Glasgow with *J. Monteath Robertson*, crystal structure analysis of organic compounds was based mainly on the interpretation of visually estimated intensities of a few hundred X-ray reflections from the crystal, recorded on sets of photographic film. It was a difficult, highly specialized, and long drawn out business. The days of arguing purely from cell dimensions alone were past. With a few notable exceptions (such as penicillin), a successful analysis was possible only when fairly reliable information was available about the approximate arrangement of the atoms in the molecule. This was before direct methods had been developed, and most structures were solved by a trial-and-error procedure: one postulated a model, a molecular arrangement consistent with the available chemical and crystallographic information. On the basis of this model, one calculated the structure factors (related to the relative intensities) of a few chosen reflections and checked whether the results were in qualitative agreement with the observed pattern. If the agreement was good, then one calculated an electron-density map by *Fourier* synthesis (usually a two-dimensional projection down the shortest unit-cell direction), adjusted the parameters of the trial model accordingly, recalculated the structure factors, checked whether any signs of *Fourier* terms had changed (we were more or less limited to centrosymmetric projections), and repeated the process. If the agreement was bad, and this was a matter of judgment, then one started again with a new trial model. Occasionally, the structure to be solved contained a heavy atom, which considerably simplified the task of guessing a suitable trial model.

The calculations were done by hand. For the structure factor calculation, one needed tables of sines and cosines and the ability to multiply a few numbers together for each atom in the proposed structure and sum the results. The *Fourier* calculations were more formidable: even for a relatively small undertaking, a two-dimensional projection based on 100 reflections, each reflection is associated with a *Fourier* term which has to be evaluated at the points of a grid, say 30 by 30, and the 100 results then added together, a process involving 90000 multiplication and addition operations. The work could be shortened with the help of *Beevers-Lipson* strips or *Robertson* templates (does anyone still remember what they were?), but with only simple adding machines at hand, plus the strips or templates, the calculations were still agonizingly time-consuming, and the results were probably riddled with numerical errors. The electron-density contour maps were drawn on paper with a sharp pencil and the atomic centers estimated by eye from the contour curves (*Fig. 1*). The accuracy of the bond distances derived by such methods depended on the sharpness of one's pencil. It was all hard work, it took a long time to get anywhere, but what a thrill it was when the outlines of a molecule began to be visible in the *Fourier* map! The molecules seen in this way had a satisfying impression of definiteness about them. They were revealed to correspond to objects of definite size and shape, in contrast to the intellectual constructions invented to explain the results of chemical reactivity. It was hard work but satisfying, and besides, one was expected to solve only one or two structures in the course of a normal doctoral research project. Was it better than nowadays? No. Was it worse? No. It was just different, and those of us who survived look back on it as a heroic age.

As in other heroic ages, setbacks were many and victories were few. While most of the X-ray analyses in the early period were concerned with molecules of known structural formula, a remarkable exception was the 1923 analysis of hexamethylenetetramine, $C_6H_{12}N_4$ [1]. This was possible because the symmetry of the crystals required that the four N-atoms occur at vertices of a regular tetrahedron and the six C-atoms at vertices of a regular octahedron. By the time I was beginning my studies, the structure of perhaps a hundred crystals of organic compounds had been established. They were mostly planar molecules, such as aromatic hydrocarbons. We were familiar with nearly all of these structures: how they were solved, and whether they had any interesting features. Among the main achievements from this period that come to mind are the accurate molecular dimensions of naphthalene and anthracene [2] in Glasgow and of a few simple amino acids and peptides at the California Institute of Technology [3], where the use of punched cards and tabulating machines was being introduced to ease the calculation burden. Doubtless, the results were not quite as accurate as claimed at the time but they helped to put the structures of organic molecules on a quantitative, metrical basis. The results of the Glasgow school provided benchmarks for testing results of quantum chemical model calculations, and the Caltech work provided the structural basis for *Pauling*'s α-helical and β-sheet motifs of protein structure. At Oxford, the molecular structure and shape of penicillin [4], cholesterol [5] and calciferol [6] were established by *Dorothy Hodgkin* and her collaborators, and, in another great achievement, the structure of strychnine was settled in two independent analyses of heavy-atom derivatives [7]. My own first analyses were of crystals of oxalic acid dihydrate, acetylene dicarboxylic acid dihydrate (*Fig. 1*) and of the corresponding diacetylene derivative [8]. It had to do with hydrogen bonding. Fifty years later, when I gained the impression that the vocabulary of supramolecular chemistry and crystal engineering had run ahead of the concepts, I rewrote these early papers in more modern parlance [9].

The journal *Acta Crystallographica* was founded in 1948 by the *International Union of Crystallography*, and the first two volumes make interesting reading. Volume 1 (1948) contains results of nine organic crystal structures, all flat molecules and all derived from two-dimensional projections but including two structures where three-dimensional data were used to calculate sections through the electron density. The presence of purine and pyrimidine structures in this short list shows that the crystallographers were already well aware of the tautomery problem in such molecules. Besides many papers on various technical improvements in the methods of X-ray analysis, Volume 1 includes also, remarkably, one on crystals of tomato bushy-stunt virus [10].

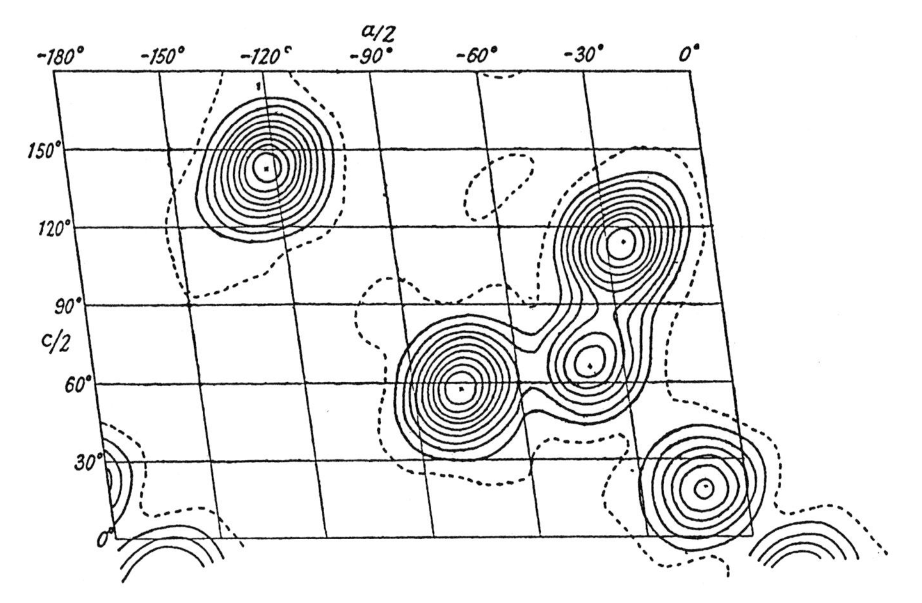

Fig. 1. *Electron density of acetylenedicarboxylic acid dihydrate projected down the short crystal axis*. Each contour line represents a density increment of approximately one electron per $Å^3$, the one-electron line being dotted (from [8]).

Volume 2 (1949) contains results of eleven organic crystal structures, all flat molecules except one, and again all the structures were derived from two-dimensional projections but now including four where three-dimensional data were used to calculate sections through the electron density. The list includes the three-dimensional analysis of naphthalene [2] and an interesting study of three different colored polymorphs of *N*-picryl-*p*-iodoaniline ([11], in French). Among the non-structural papers are again several harbingers of direct methods and a most useful tabulation of atomic scattering factors [12]. This volume also contains a remarkable paper by *Carl Hermann* on symmetry in higher dimensional space ([13], in German). As I recall, *Hermann* told me that this work was done to pass away the time when he was imprisoned in Nazi Germany during World War II.

At this point, I make a brief excursion about one of my own analyses in that period: the structure of the centrosymmetric isomer of 1,2,3,4-tetraphenylcyclobutane [14], the exception to the flat molecule structures in Volume 2. I managed to complete this work during my post-doctoral stay in *Dorothy Hodgkin*'s laboratory in Oxford. The initial analysis was based on trial-and-error methods leading to two-dimensional projections from which the atomic positions could be determined with the accuracy typical of those times (*Fig. 2*). From these projections, the bond distances in the Ph groups appeared to be normal but those in the cyclobutane ring appeared to be too long. According to the recently developed 'bent bond' model [15], bonds in small carbocyclic rings were expected to be shorter than 1.54 Å, the standard C–C bond distance in aliphatic compounds. In agreement with this expectation, the C–C distances in cyclopropane and spiropentane were known from gas-phase electron diffraction to be shorter than 1.54 Å. However, the distances I was finding in the cyclobutane ring were about 1.58 Å, distinctly longer than the standard. The reliability of this result was certainly open to challenge because of the problems of resolving the positions of individual atoms in poorly resolved projections. Following discussions with Professor *Charles Coulson* and his student *Bill Moffitt*, who were then developing the bent-bond model by the kind of quantum mechanical calculations possible at that time, I decided to undertake the arduous task of collecting three-dimensional intensity data and calculating the relevant sections of the electron density distribution. I estimated relative intensities by eye for more than 1000 reflections and carried through the necessary *Fourier* series calculations by hand. The results confirmed that the bonds in the four-membered ring were longer than normal. It was this result that led to my embarking on a further post-doctoral fellowship at Caltech. Were the long bonds an intrinsic property of the cyclobutane ring? Or were they in some way dependent on the presence of the four Ph substituents? When I discussed this problem with *Verner Schomaker* during his visit to Oxford in the early summer of 1948, we decided that it called for a gas-phase electron diffraction study of cyclobutane itself. The results, published four years later [16], showed clearly that the cyclobutane bonds were long and that, moreover, contrary to what had been assumed until then, the four-membered ring in cyclobutane itself was not planar but buckled (D_{2d} rather than D_{4h} symmetry). The reason for the striking difference between the C–C bond distances in cyclopropane and cyclobutane is that the former molecule shows no non-bonded 1,3-interactions, whereas the latter shows the strongest possible interactions of this type, which are strongly repulsive.

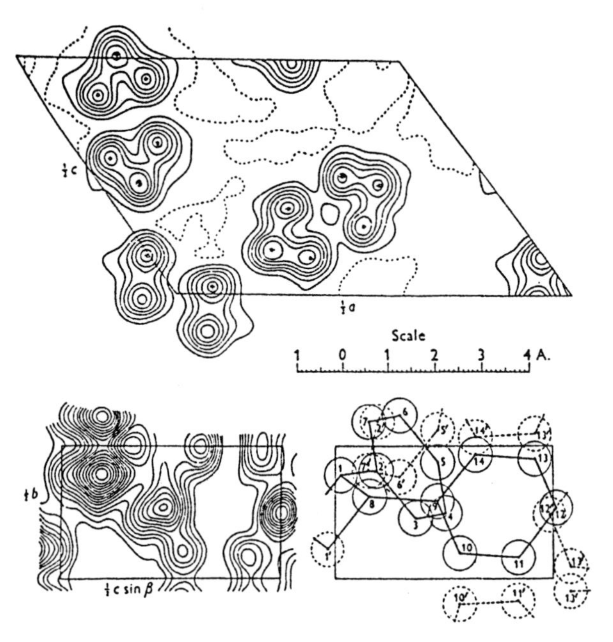

Fig. 2. *Electron density of the centrosymmetrical isomer of 1,2,3,4-tetraphenylcyclobutane projected down the 5.77 Å* b *axis* (top) *and 17.02 Å* a *axis* (bottom), *showing the asymmetric unit in both cases* (from [14]). In both maps, contours are drawn at intervals of approximately one electron per $Å^3$. The interpretation of the bottom map is indicated.

2. How It Is Today

Visual estimates of reflection intensities have long been replaced, first by point-by-point diffractometers and now increasingly by area detectors, devices which make it possible to measure hundreds of reflections simultaneously rather than one at a time. The trial-and-error method has long been overtaken by so-called direct methods, in which the missing phase angle information is derived directly from relationships among the observed intensities. Premonitions of direct methods are already present in Volume 1 of *Acta Crystallographica* but their general applicability had to wait for advances in the power and availability of electronic computers to carry out the protracted calculations that are called for. Even the first trivial step, the preparation of a list of triplets of strong reflections related as $\mathbf{h}_1$, $\mathbf{h}_2$, $\mathbf{h}_1+\mathbf{h}_2$ in three-dimensional reciprocal space, was prohibitive without computer assistance. I know from experience; I tried it in 1951 for a set of three-dimensional data collected from a calciferol derivative, and gave up after a couple of months. The same calculation today would take a fraction of a second. Similarly, we knew about least-squares analysis, but it was only with improvement in computer power that least-squares algorithms for refining atomic positions and 'thermal' parameters gradually became standard procedures. Again, it was greater computer power that enabled automated three- and four-circle diffractometers to collect hundreds, if not thousands, of reflection intensities per day. Think of calculating all the angles required by complicated trigonometry and cranking the circles by hand into the correct positions! Similarly, the task of indexing the hundreds of reflections recorded by an area detector and measuring their intensities would be out of the question without the aid of highly sophisticated hardware and software. I read recently that computer speed doubles about every eighteen months. In 50 years, that gives an improvement of about 2^{33}; what now takes a second would then have needed more than the age of the universe.

And that is roughly where we are at present from the technical point of view. We can look forward to new developments, but most of the ones I can think of are essentially improvements in existing methods rather than anything radically new: on the experimental side, more powerful radiation sources, making it possible to obtain diffraction patterns from very tiny crystals; better area detectors. On the computational side, we can expect further applications of maximum entropy methods and more routine structure determinations of medium-sized organic molecules from analysis of powder-diffraction patterns. With the exception of the last, the main thrust of these will be to overcome some of the present limitations in the area of biomolecular structure analysis, but I do not expect them to change small-molecule crystallography in any radical way. As compared to serial diffractometry, the use of area detectors should make it easier to detect and study incommensurate and disordered structures, but the principal effect will be to produce ordinary crystal structures still more automatically and more rapidly than at present; in other words, they will lead to a still more rapid accumulation of information about crystal and molecular structure. Whether this will lead to a corresponding increase in knowledge is another matter which I shall discuss later.

3. What Have We Achieved?

3.1. Molecular Structure

There is no need for me to emphasize here that the preoccupation with molecular structure is at the heart of chemistry. By the mid-1950s, X-ray analysis was being used not only to 'see' the details of molecules of more or less known structure and shape, but also with increasing success to determine the molecular structures of complex natural products of unknown constitution. This task had been regarded as one of the principal undertakings of organic chemistry, in the fulfillment of which much basic knowledge about the relationships between molecular structure and chemical reactivity had been accumulated over the years. The intrusion of X-ray analysis into natural product chemistry may even have been regarded by some chemists at the time as a kind of threat to one of their traditional activities. Such an attitude was of course short-sighted. Freedom from the task of structure proof meant freedom from the restriction that a synthesis of a given target molecule had to proceed by steps of known reaction type. In any case, it became apparent that the successful synthesis of a target molecule was not always a rigorous proof of its structure. An unexpected and unrecognized rearrangement could occur in one of the degradation steps and precisely the reverse rearrangement could happen in the synthesis [17]. In such a case, the target compound would be synthesized but its assumed structure, seemingly confirmed by synthesis, would be wrong. Fortunately, such cases are extremely rare. As might have been anticipated, the new freedom had the effect of unleashing tremendous new energies in chemistry, as was expressed in no uncertain terms by *Derek Barton* in 1973 [18]:

> *'I became convinced that the solution of structural problem in organic chemistry is in most cases much more quickly done by X-ray crystallography than it is by organic chemistry. This represented a complete change in the activities of organic chemists, because always in the past we had spent half our time on degradative and half on synthetic work. But in the early 1960's everybody realised that the degradative work was no longer going to be needed. We were not going to discover new reactions, new arrangements, new chemical phenomena by chemical degradation. We would have to discover them instead by synthesis. This has not been to the disadvantage of organic chemistry at all'.*

Today, X-ray analysis is called in not only as a big gun, to solve the difficult problems of natural product chemistry, but almost routinely, for example, even to check the identity of a reaction intermediate in a multi-step synthesis. Most major chemistry departments now have their own X-ray analysis service facilities. NMR Spectroscopy may be of comparable importance and has the advantage that it does not need the presence of a crystalline sample of the compound to be studied, but where the evidence is equivocal crystal structure analysis still provides the most clear-cut decision. In one step it can answer questions of constitution, configuration, and conformation, besides providing metrical information about interatomic distances and angles.

The vast majority of crystal structures determined today satisfy stringent quality standards. Nevertheless, it should be stressed that the nominal precision of the resulting atomic positions and derived geometric parameters, estimated by least-squares refinement methods, is quite unrealistic. The standard deviations estimated in this way merely reflect how well the least-squares model fits the observations, but it takes no account of systematic errors in the observations or inadequacies of the model (see *Sect. 2.4*). The estimated standard deviations should be doubled at least. It must also be admitted that some modern analyses are sub-standard and in some the structural information provided is even wrong. In earlier times, a published structure with suspicious features would have been scrutinized by experts, but, nowadays, in spite of checking programs, erroneous structures are more likely to pass undetected into the chemical literature. Many otherwise competent reviewers of papers submitted to the chemical journals do not know the first thing about the crystallographic aspects of a problem and are incapable of judging the reliability of the results, which may then pass unchecked into the storehouse of structural information. Unfortunately, technically advanced state-of-the-art hardware and software are no substitutes for expertise. They are quite capable of producing wrong results, varying all the way from the trivially wrong in some detail to absolute nonsense. Vociferous nonsense that claims attention is soon detected and is ultimately fairly harmless, but unpretentious nonsense can easily pass undetected for ever. It merely pollutes the storehouse of structural information.

3.2. Molecular Chirality

A unique contribution of X-ray crystallography has been the determination of absolute configuration. By the early years of 20th century (I mean the early 1900s), it had become possible to relate the configurations of hundreds of optically active compounds among each another, *i.e.*, to establish their configurations relative to some reference compound. *Emil Fischer* took this as (+)-glyceraldehyde, which was arbitrarily assigned configuration **I** and represented by projection formula **II**. Within this convention, (+)-tartaric acid was known to be represented by **III** and the naturally occurring amino acids by **IV**. However, there was no way to decide whether (+)-glyceraldehyde actually corresponded to structure **I** or to the mirror image. Once this could be settled, stereochemistry could be placed on an absolute footing, but until mid-century there seemed no way to answer the question. Indeed, when I was a student, we were told that it was impossible to answer this question. As far as X-ray diffraction was concerned, *Friedel*'s Law stood in the way. *Friedel*'s Law stated that the X-ray diffraction pattern of a crystal is centrosymmetric, whether the crystal structure itself is centrosymmetric or not. This law depends on the assumption that phase differences between waves scattered by different atoms depend only on path differences, that is, any intrinsic phase change connected with the scattering event is the same for all atoms. However, this assumption is not quite true. For non-centrosymmetric structures, there is a slight difference in intensity between reflections from opposite faces of the crystal. Such differences had been used to determine the sense of polarity of zinc sulfide crystals [19], and, in mid-century, *Johannes Martin Bijvoet* realized that the same principle could be utilized to provide a bridge between macroscopic and molecular chirality. *Bijvoet* then used anomalous scattering of X-rays to show that the absolute structure assigned by *Fischer* was indeed correct [20]. It was not necessary to rewrite all the formulas in the textbooks! In the meantime, the absolute structure of thousands of chiral and polar crystals have been determined by the anomalous scattering method, and for many years the *Prelog-Ingold-Cahn* (*CIP*) system [21] has been used to specify the sense of chirality at tetrahedral centers. What seemed an insoluble problem has become routine.

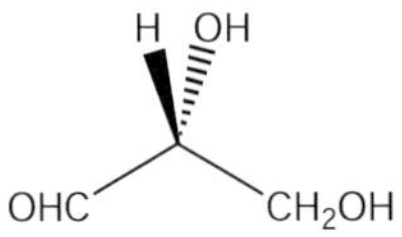

I 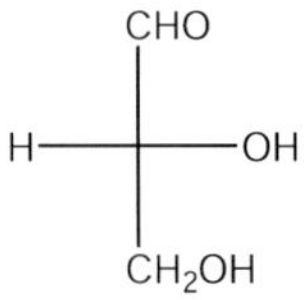 **II** 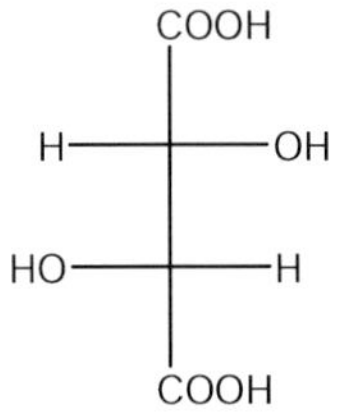 **III** 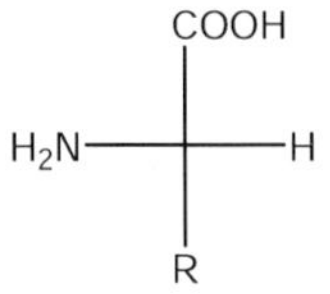 **IV**

3.3. Atomic Motion in Solids

Once least-squares methods came into general use it became standard practice to refine not only atomic positional parameters but also the anisotropic 'thermal parameters' or displacement parameters (ADPs), as they are now called [22]. These quantities are calculated routinely for thousands of crystal structures each year, but they do not always get the attention they merit. It is true that much of the ADP information is of poor quality, but it is also true that ADPs from reasonably careful routine analyses based on modern point-by-point or area diffractometer measurements can yield physically significant information about atomic motions in solids. We may tend to think of crystal structures as static, but in reality the molecules undergo translational and rotational vibrations about their equilibrium positions and orientations, as well as internal motions. *Cruickshank* taught us in 1956 how analysis of ADPs can yield information about the molecular rigid-body motion [23], and many improvements and modifications have been introduced since then. In particular, various computer programs are available to estimate the amplitudes of simple postulated types of internal molecular motion (*e.g.*, torsional motions of atomic groupings about specified axes), besides the overall rigid-body motion, from analysis of ADPs [24]. Caution may be called for in interpreting results of such calculations because of possible correlations among the parameters describing the motions. Nevertheless, in work with aspirations to high accuracy in the metrical details of molecular structure, such calculations need to be made and the results analyzed in terms of the postulated motions. This is because standard X-ray analysis locates the centroids of atomic distributions that are undergoing vibrations, and separations computed from these positions cannot be interpreted directly as interatomic distances. In general, bond distances calculated directly from the X-ray positions tend to be slightly shorter than the actual distances by an amount that depends on the details of the rotational motions.

As the amplitudes of motion are temperature dependent, multi-temperature measurements can be very useful in assessing the physical significance of results derived from such analyses. In particular, since different kinds of motion show different kinds of temperature dependence, some of the ambiguities inherent in the analysis of single-temperature data may be resolved [25]. As the technical possibilities for carrying out accurate diffraction measurements at high and low temperatures come into general use, more attention should be given to the interpretation of ADPs and the physical significance of the results, otherwise information about atomic motions in solids will be lost. For some time now, ADPs, although routinely calculated in structure refinement programs, have tended to be relegated to the Supplementary Information section of journal papers and are seldom published. They are, however, the basis for the usual 'thermal ellipsoid' pictures of molecules, and even there, visual inspection of the ellipsoids can be helpful in judging the quality of the crystal structure analysis. Unfortunately, there is at present no facility for depositing, collecting, and storing the numerical information. Much of it is destined for oblivion, and that is a pity.

3.4. Experimental Charge Density Distributions [26]

Electron density maps have been used for decades to give images of molecules in crystals (*e.g.*, example, *Figs. 1* and *2*). It has long been realized that such maps might also tell us something about the 'nature of the chemical bond'. It is fortunate that the electron density in a molecular or ionic crystal is closely similar to the superposition of the densities of the separated atoms, placed at the positions they occupy in the crystal, for it is this similarity that made it possible in the first place to use standard, spherically symmetrical scattering factors in solving crystal structures and in refining them by least-squares methods. In fact, when crystallographers take pride in their low *R* factors they pay tribute to the goodness of the pro-crystal approximation as well as to the accuracy of their measurements.

The difference $\Delta\varrho(\mathbf{X}) = \varrho(\mathbf{X}) - \varrho_{\mathrm{M}}(\mathbf{X})$ between the actual density and the pro-molecule density is known as the deformation density and can be interpreted as the electron density reorganization that occurs when a collection of independent, isolated, spherically symmetric atoms is combined to form a molecule in a crystal. Since $\Delta\varrho$ is only a very small fraction of total ϱ in the region of the atoms, it is very susceptible to experimental error in the X-ray measurements and to inadequacies in the model, namely errors in the assumed atomic positions, atomic scattering factors, and ADPs. In one approximation, a deformation density map is obtained by direct subtraction of the two densities. The density map obtained in this way is smeared by vibrational motion of the atoms, but its peaks and troughs can often be interpreted in terms of some model of chemical bonding, *e.g.*, peaks between bonded atoms being identified with 'bonding density' and so on. A difference density map for tetrafluoroterephthalodinitrile [27] is shown in *Fig. 3*.

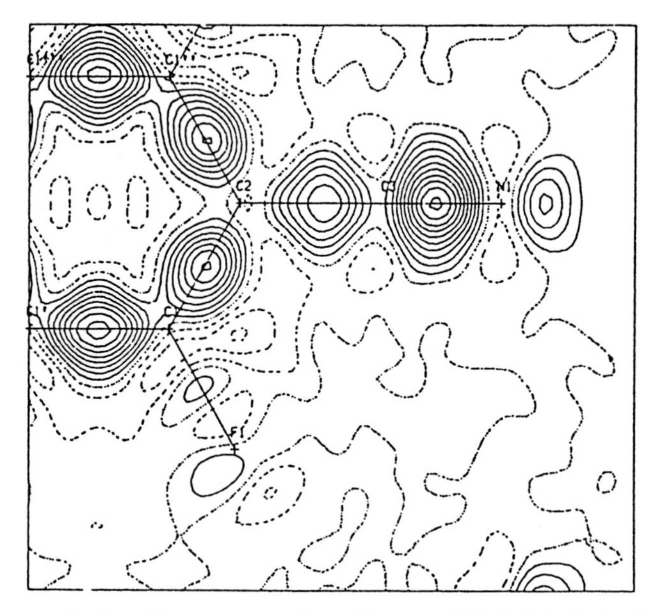

Fig. 3. *Electron density difference map of tetrafluoroterephthalodinitrile in the molecular plane* (from [27]). Contour lines are drawn at intervals of 0.075 electrons per Å^3, positive contours full lines, negative contours dashed, zero contour dotted. Note the weak density in the C–F bond.

Alternatively, $\Delta\varrho$ can be expressed in parametric form as the sum of suitably designed functions, *e.g.*, a set of multipoles, each multiplied by a radial function and centered at an atomic position. The *Fourier* coefficients of the various functions are then added to the free-atom form factors with variable population parameters, which are refined, together with the atomic positional coordinates and ADPs in one giant least-squares analysis. The density map obtained in this way is sometimes known as a static deformation map. In contrast to the difference map, it represents the charge density reorganization on going from the vibrationless pro-molecule to the vibrationless molecule in the crystal. Static deformation maps for tetrafluoroterephthalodinitrile [28], based on the same experimental data as above, are shown in *Fig. 4*. The density shown in the upper map does not satisfy the *Hellmann-Feynman* theorem: the lower one does, *i.e.*, the electric field is constrained

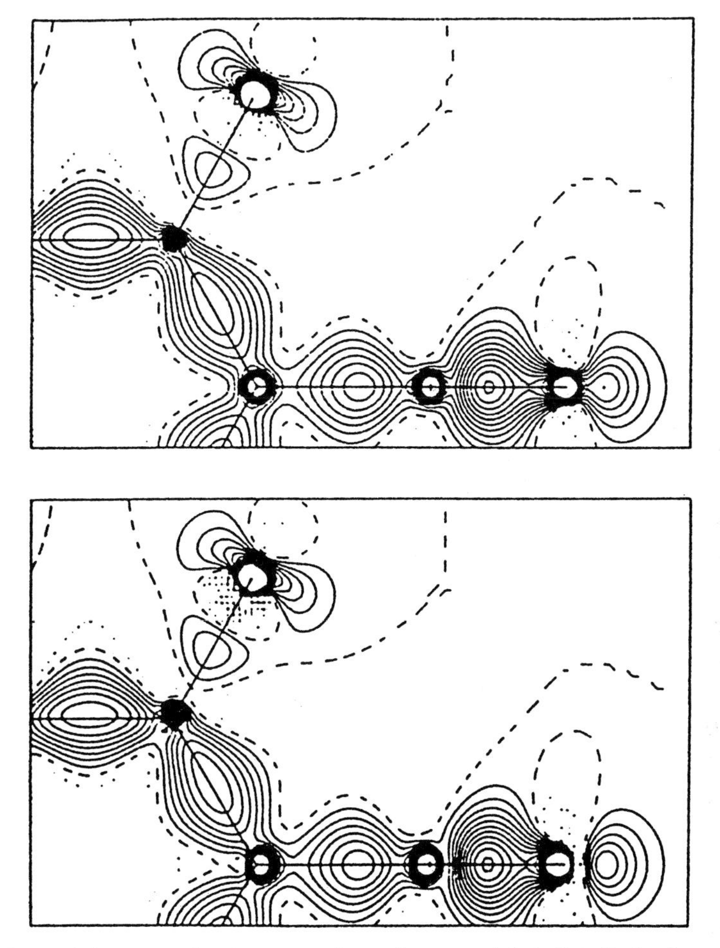

Fig. 4. *Static deformation density maps of tetrafluoroterephthalodinitrile in the molecular plane* (from [28]). The upper map is unconstrained, the lower one is constrained to satisfy the *Hellmann-Feynman* theorem (see text). Note the slight dipolar deformation at the atomic positions in the constrained map.

to be zero at all nuclear positions. The two maps look practically the same, but, on close inspection, the lower one shows sharp dipoles at the atomic positions.

By adding the static deformation density to the pro-molecule density one obtains the experimental charge density of the molecule in the crystal. In recent years, a fruitful mutual interaction has been established between *Bader*'s interpretation of bonding in terms of the Laplacian of the electron density at its topological critical points [29] and experimental static charge density distributions [26]. To avoid complications due to residual vibrational smearing, the crystal data must refer to the lowest practically attainable temperature.

3.5. Biomolecular Crystallography

Perhaps the greatest change of all has been in the area of biomolecular crystallography. When I started my career, it hardly existed. Cheerful optimists such as *Bernal* persistently predicted that the structure of proteins would be unraveled in five years time but provided no clear indication of how this was to be achieved, and *Max Perutz* persevered with attempts to interpret the *Patterson* function of haemoglobin crystals. Tangible results were wanting and it was certainly not obvious to me how the technical difficulties could ever be surmounted. Indeed, for most of us, protein crystallography was regarded as a wildly visionary target. We were poor judges of the potentialities of our methods. Today, protein crystallography has become by far the major part of our discipline, and it is still growing in power and in importance, attracting ever more funding and researchers each year, and producing ever greater wonders. Crystallographic meetings nowadays are overwhelmingly biomolecular crystallographic meetings, and the trend is likely to continue.

From its idealistic but shaky beginnings, biomolecular crystallography has developed into a more or less standard method producing hundreds of structures annually, three or four every week as far as one might judge from perusal of *Science* and *Nature*. I must admit that it is very difficult for me as a non-specialist to read most of these papers. They require considerable background in the biochemical and physiological aspects of the problem – I suppose that is true in many areas of science today. There is no question that biomolecular crystallography has provided unparalleled insights (the word is here used literally) into enzymic active sites and catalytic mechanisms, antibody structure and specificity, DNA-protein recognition phenomena, light-harvesting assembly systems and other targets that were once believed to be far beyond the range of experimental structural study. Each year stretches the limits of X-ray analysis still further; in 1999 we saw the structure of the ribosome with fascinating insights into the mechanism of the protein making machinery. There seems to be no end.

There may be no end but there was a beginning. It happened around mid-century when, within the scope of a few years, three momentous discoveries were made. The first was the construction of the α-helix and β-sheet structures as models for stable secondary structures in proteins; the second was the *Watson-Crick* model structure for DNA; and the third was *Perutz*'s discovery that heavy atoms, such as mercury, could be introduced into protein crystals without destroying the crystalline structure, thus making it possible to obtain information about the missing phases. At this point I indulge in some personal reminiscences, which some of my readers may choose to skip, having heard or read of them already.

4. Interlude: Personal Reminiscences

At mid-century I was a postdoc at Caltech. It must have been towards the end of 1950 when I attended *Pauling*'s lecture when he first publicly announced his stable H-bonded model structures for polypeptide chains. *Pauling* had a keen sense for drama. On the table in front of him stood bulky columnar objects shrouded in cloth, which naturally excited the curiosity of those in the packed auditorium. Only after describing in detail the structural principles behind the models did he turn to the table and unveil the molecular models with an icastic gesture. There were the two structures, the three-residue and the five-residue spirals, later dubbed the α- and γ-helices! I was immediately converted, a believer right from the start. As I recall, I sat beside *Max Delbrück*, who made no secret of his disapproval of *Pauling*'s manner of presentation and asked if I thought there was anything in these models. I believe I may have slightly disappointed him when I told him that the models were based on sound structural principles and were very likely to represent important building blocks of actual proteins.

While my own work at Caltech had nothing to do with protein structure, *Pauling* used to talk to me occasionally about his models and what one could learn from them. In his lecture, he had talked about spirals. One day I told him that for me the word 'spiral' referred to a curve in a plane. As his polypeptide coils were three-dimensional, I suggested they were better described as 'helices'. *Pauling*'s erudition did not stop at the natural sciences. He answered, quite correctly, that the words 'spiral' and 'helix' are practically synonymous and can be used almost interchangeably, but he thanked me for my suggestion because, on consideration, he had decided that he preferred 'helix'. Perhaps he felt that by calling his structure a helix there would be less risk of confusion with the various other models that had been proposed earlier. In the 1950 short preliminary communication [30], *Pauling* and *Corey* wrote exclusively about spirals, but in the series of papers published the following year [31] the spiral had already given way to the helix. After that there was no going back. A few years later we had the DNA double helix, not the DNA double spiral. The formulation of the α-helix was the first and is still one of the greatest triumphs of speculative model building in molecular biology, the forerunner of the untold investment in computer-assisted molecular modeling in present day research. I am pleased that I helped to give it its name.

The following summer I returned to Oxford. Before long I had a lively connection with the crystallographers in Cambridge, which brought me every few months into exciting though inconclusive discussions with *Francis Crick* over pub lunches. We argued, for example, about what would be the most favorable space group to determine the crystal structure of a protein. My preference was triclinic $P1$, with one molecule per unit cell, because every peak in the *Patterson* function would correspond to an intramolecular vector, while *Francis* was in favor of a cubic space group, because at least twelve molecules in the unit cell would be related by rotational symmetry. Whether a definite answer to this question has been provided in the meantime is unknown to me, but probably *Francis* was right. He often was. I was aware that *Crick* and *Watson* were trying to deduce the structure of DNA by model building but did not give them much chance of success, especially since they had no diffraction data to test their models. As I recall, their mood used to oscillate erratically between enthusiastic optimism and downcast pessimism. In late 1952, during a stroll with *Watson* in Oxford, I advised him to abandon the project and get down to a more promising project. From time to time I reported on the DNA model building work to my Oxford friends and discussion partners, *Leslie Orgel* and *Sydney Brenner*. In March 1953, when *Crick* telephoned to ask me to come to look at their marvelous new DNA model, all three of us traveled together with *Dorothy Hodgkin* to Cambridge. We knew enough about the problem to recognize almost immediately that the proposed DNA model must be correct in its essential features, and that it also offered the structural basis for genetic information transfer. Did we realize that we were present at the dawn of a new age? Did we feel: *'At this place and on this day a new epoch in the history of the world begins, and we shall be able to say that we were present at its beginnings?'* (These are *Goethe*'s words, written on September 20, 1792, the occasion being the defeat of the Prussian army by the ragged French militia at Valmy.) Speaking for myself, the answer is no. I must have a very limited imagination.

It is instructive that *Watson* and *Crick* built their double helix model to fit the very limited information derived from the sparse diffraction pattern of the non-crystalline B-form of DNA, which *Rosalind Franklin* had obtained under high-humidity conditions. The diffraction pattern of the crystalline A-form was much more complex and not directly interpretable. This was the pattern that *Rosalind Franklin* set out to decipher. When *Crick* saw her 1952 report giving cell dimensions and space group *C*2 of the A-form, he realized immediately that the molecule must be a double-stranded helix of *ca.* 20 Å in diameter with the individual strands running in opposite directions. The B-pattern then told him that there were 10 residues per 34 Å repeat distance along the helix axis – almost nothing else, but that was enough. I doubt whether *Crick* and *Watson* could have derived their model from the more detailed A-pattern which *Franklin* was trying to interpret. It contained more information but it was in more cryptic form. Sometimes it pays not to have too much information. It can muddy the picture and obscure the essential elements of a problem.

I also happened to be present at the conference on protein structure organized by *Pauling* at Caltech in September 1953. It was here that *Perutz* first announced that he could diffuse a heavy-atom derivative into haemoglobin crystals without altering the molecular arrangement in the crystal, leading to the same overall diffraction pattern but with small intensity changes in the reflections. The crystallographers who were present realized that this was a breakthrough. It meant that the structure of crystalline proteins could be solved – in principle at least. In practice, one needs the intensities from the native protein and from at least two isomorphous heavy-atom derivatives. That is how nearly all protein structures are solved today. With tunable synchrotron radiation, one can get by with a single derivative.

At this point the reader may well ask: if you were present at all these exciting moments at the dawn of biomolecular crystallography and realized how important they were, why did you not yourself become engaged in that branch of the subject, which was obviously destined for a glorious future? The answer is that I knew my own limitations. A research target for which one needs long-term persistence and endurance was not for me. Today, anyone can enter this field. All one needs are a few skillful young collaborators, some medium-expensive equipment, a computer and some more or less standard software, plus, of course, interesting crystalline biomolecular material and the financial backing to pay for all these necessities. In the 1950's, even after the achievements mentioned above, there were barely half a dozen laboratories in the world working on protein structure: *Perutz* and *Kendrew* in Cambridge working on haemoglobin and myoglobin, *Dorothy Hodgkin* in Oxford on insulin, *David Harker* in Brooklyn on ribonuclease, and perhaps one or two others. They were all outstanding scientists, but the projects on which they were engaged seemed quixotic at the time. There were still obvious difficulties that seemed insurmountable to many of us. For example, how on earth could they hope to carry out the enormously extended calculations that were necessary? As if by magic, the computing power kept increasing to keep pace with the demand.

Even when I went to the Royal Institution in 1956 to join the group of young scientists being assembled there by *Sir Lawrence Bragg*, I told the old master that I was not enthusiastic about the idea of concentrating exclusively on protein crystallography. In any case, it was a time when there were not enough suitable crystalline proteins problems to go around. It was agreed that I should work on other problems, for which I am not sorry. There have been enough interesting problems to keep me busy.

Thus I would plead not guilty to a charge of dereliction of duty, although possibly guilty to the lesser one of wasting my time in amusing but ultimately trivial pastimes. One can hardly be blamed for failing to keep up to the highest standards set by one's great predecessors and genial colleagues. If one does deserve censure, then surely only for failing to meet the standards imposed by one's own limitations. In one of his books, *Martin Buber* tells us that, towards the end of his days, *Rabbi Sussja* of Hanipol said: *'In the world to come I shall not be asked why I was not Moses; I shall be asked why I was not Sussja'.*

5. Chemical Crystallography

5.1. Crystal Packing and Polymorphism

The focus of interest for many crystallographers has shifted over the years from the molecular to the intermolecular level of organization. When I began my work, the structures of ionic crystals were reasonably well understood in terms of a few simple rules (*e.g.*, *Pauling*'s Rules [32]). For organic crystal structures, there were no obvious regular features, apart from the H-bond, whose importance as a structure directing element had been recognized at an early stage. One problem was that there were not many organic crystal structures from which to draw general conclusions. And another was that the positions of H-atoms could not be accurately determined by X-ray analysis. These atoms were, therefore, often simply omitted from packing diagrams, giving the false impression that there were large empty gaps between the molecules in crystals. It was *Kitaigorodskii* with his theory of close packing of molecules in crystals [33] who paved the way for future developments. Today, especially with the advent of supramolecular chemistry, there is a lively and increasing interest in the study of weak (noncovalent) interactions in organic crystals. Indeed, a crystal can be viewed as a giant supermolecule, held together by just the same kinds of noncovalent bonding interactions as are responsible for molecular recognition and complexation at all levels. The crystallization process is an impressive display of supramolecular self-assembly, involving specific molecular recognition at an amazing level of precision. Moreover, since the properties of materials depend not only on the structure of the molecules of which they are composed but also on the way these molecules are arranged, an understanding of intermolecular interactions is an essential basis for any attempts in the direction of crystal engineering and design. Another factor is the recent revival of interest in polymorphism. Polymorphs may differ greatly with respect to properties, such as color, hardness, solubility, crystal habit, chemical stability, and so on, properties that can be of vital importance in the pharmaceutical and other industries. Besides, the crystallization process is not easy to control, and even reproducibilty of an experiment to produce a given polymorph under given conditions is sometimes problematic [34]. From several directions, therefore, there is now considerable interest in the question: given the molecular formula of an organic compound, can we predict in what form it will crystallize under given conditions, and what will be its solid-state properties?

The outlook at present does not seem too promising. Results of a recent workshop suggest that, even for quite simple molecules, computed lattice energies based on various types of atom-atom potentials lead to several packing arrangements within a small energy range [35]. Generally, the observed crystal structure is among these, and it is quite possible that some of the others correspond to polymorphs. There are problems about the choice of atom-atom potentials; some workers use potentials derived from results of quantum mechanical calculations while others adopt an outspokenly empirical approach. However, regardless of the choice of potentials, the general consensus seems to be that, while it is not too difficult to predict possible crystal structures for a given compound, it is much harder to say which of these is likely to be actually obtainable under any defined conditions. Crystal design can be successful where there is strong, highly directional bonding, as in metal coordination and H-bonding, but where the intermolecular attractions are weak and non-directional, as with dispersion (*van der Waals*) interactions, there are just too many possible crystal structures with almost the same packing energy. For benzene, for example, with its highly symmetrical and nearly rigid molecular structure, recent work gave 30 crystal structures with one molecule in the asymmetric unit within a 10 jK mol^{-1} enthalpy range at zero pressure; there were 20 structures within the same range at a pressure of 30 kbar [36], where crystal enthalpy correlates strongly with inverse crystal volume (*i.e.*, with density). Even at zero pressure, calculated lattice energy correlates well with crystal density and hence with packing coefficient (defined as the ratio of molecular volume to available volume). The narrow range of lattice energy corresponds to a very narrow range of packing coefficient for these calculated benzene structures.

Indeed, it is found that in general the range of packing coefficient for small-to-medium-sized organic molecules is quite narrow. Since *Kepler*'s time it has been known that closest packing of identical spheres corresponds to a packing coefficient of 0.74, and since *Kitaigorodskii*'s work on molecular packing in crystals [33] we know that the close packing principle applies also to organic crystal structures. Molecules in crystals tend to be surrounded by twelve to fourteen neighboring molecules, the same as in typical close-packed metals, and packing coefficients in molecular crystals vary only within a range of about 0.65 to 0.80, not too different from the value for close-packed spheres. With a packing coefficient below about 0.6, substances are liquid, and below about 0.5 the attractive forces are no longer strong enough to hold the molecules together in a condensed state – the substance vaporizes. It seems remarkable that although organic molecules have very different shapes and sizes, and only a handful of very simple ones can be even remotely described as being nearly spherical, they fill space about as efficiently as spheres.

For molecular crystals, where intermolecular forces are weak, entropic factors cannot be ignored in assessing the relative thermodynamic stabilities of possible polymorphs at different temperatures. By suitable lattice dynamical calculations, based again on atom-atom potentials, the entropy contribution from lattice vibrations can be estimated reasonably well. However, even if we could compute packing energies and free enthalpies of possible structures with complete confidence, it is by no means sure that the thermodynamically stable structure will actually be formed under given conditions, because the crystallization process is under kinetic control. There are many hints that the formation of viable nuclei is the rate-determining step, but this step is poorly understood. In any case, small nuclei are almost certainly highly imperfect, and it is unlikely that they can be modeled simply as small versions of a perfectly periodic crystal. Computer models may also be quite unrealistic; in a model crystallite containing 1000 molecules, almost half the molecules are on the surface. Additional complications arise for molecules with conformational freedom for there is no reason why the conformer present in the thermodynamically stable crystal form should be the most stable conformer in solution. Thus, formation of the most stable crystal modification may be hampered by a low concentration of the particular conformer required. There are clearly many difficulties in crystal structure prediction, and it is not too clear at present how they are to be overcome.

There are obvious similarities between the crystal structure prediction problem and the protein folding prediction problem. Both problems involve unsolved questions regarding the choice of force field, the existence of many almost equi-energetic minima in a multi-dimensional energy space, and the relative importance of thermodynamic and kinetic factors, including possible nucleation steps. Both involve higher-order organization. There are also, of course, obvious differences – the crystal is built by repetition of a single fundamental building block (or at most only a few building blocks, as in salts, solvates, and co-crystals), whereas the protein is built from 20 building blocks in a presumably known sequence along the polypeptide chain. I find it hard to judge which problem is the more difficult, but in view of the vast collection of post-genome sequences of proteins of unknown structure and function that are now being assembled in the proteome project, it is understandable that the folding problem attracts much more attention. Indeed, it is likely to be one of the main challenges of the next decade or two.

As far as crystal growth is concerned, studies of the influence of 'tailor-made' impurity additives on differential face development has led to important insights [37]. An unexpected bonus has been an independent confirmation of the correctness of absolute configurations determined by anomalous dispersion methods [38]. Tailor-made additives can also be used to inhibit formation of the thermodynamically stable polymorph and thus lead to formation of a metastable one [39].

5.2. Solid-State Chemistry

When I was young, solid-state phase changes and chemical reactions were regarded more as a nuisance than as an area worthy of serious study and attention. It was the topochemical approach to solid-state chemical reactions, pioneered by *Gerhard Schmidt*, that transformed the subject for me and for many others. The textbook example is the photochemical dimerization of (*E*)-cinnamic acids; in solution such compounds yield mixtures of the various possible stereoisomeric products, but irradiation of a particular crystal leads to a single product, or to no reaction, depending on the crystal structure [40]. Thus, one can determine the relative positions of the atoms before the reaction and after it, and hence deduce the metrical relationships that needed to be satisfied for reaction to proceed. In the meantime we have learned that not all chemical reactions in solids are topochemical. Some proceed not in the ordered bulk of the crystal but at defects, on the surface, or at other irregularities.

In spite of the topochemical principle, the details of solid-state reactions may be difficult to understand. When we think of chemical reactions in solution or in the gas phase, we normally focus attention on the fate of a single molecule and its interaction with one or two immediate neighbors. This kind of simplification is generally not possible when we deal with phase transitions or solid-solid chemical reactions in which phase separation occurs. Even when overall crystal orientation is maintained between initial and final states, such transformations can involve a highly complex series of cooperative processes, in which local order is first lost and then recrystallized out of chaos. With all the similarities between solid-solid and normal chemical reactions, perhaps we may have to give up the idea of understanding the detailed mechanism of the former, except in a few very simple cases. We are dealing with many-body processes, and as with other many-body processes, a molecular dynamics simulation may be the most we can expect to achieve. The subject is now of general interest and will undoubtedly continue to be a major topic of chemical crystallography.

Just as for crystallization from the melt or from solution, nucleation may be rate-limiting in solid-solid structural transformations, including phase transitions. In a crystal, nucleation of a new phase may depend on the presence of suitable defects, such as micro-cavities and other surface irregularities between different crystal domains. Depending on the nature of such defects, nuclei of the new phase may be formed at slightly different temperatures and grow at different rates. In a sense, a defect can act as a catalyst for a structural transformation, and there are many kinds of defect, but, here again, we encounter limits to our present-day knowledge. A better understanding of crystal nucleation seems essential for further progress, but it is not clear how this is to be attained.

5.3. Structure Correlation

Much useful information in crystal structure studies is often passed over by the authors themselves and has to be actively sought and recovered, either from the original publications or from databases. From the very early days, isolated bits of information were collected and used to construct tables of standard distances for various types of covalent bonds and of *van der Waals* radii for intermolecular contact distances, as collected in *Pauling*'s influential book [41] and elsewhere, as well as standard dimensions for structural fragments that are important in biomolecular crystallography, such as the peptide moiety and the nucleic acid bases.

As time passed and more structural information became available, it was realized that, at least for some chemical groupings, the structural parameters are not constant but vary from one structure to another and often in a systematic manner. For example, as one bond in a structural fragment might tend to become longer, a neighboring bond might tend to become shorter. Expressed in the framework of a many-dimensional parameter space, the various copies of the structural fragment correspond not to a single point but to a distribution in which the individual parameters are often found to be highly correlated in a way characteristic of the fragment. The natural interpretation of such distributions is in terms of many-dimensional energy surfaces, where the observed distribution of experimental points can be assumed to delineate the low energy regions. Indeed, it is tempting to put such structure-energy relationships on a quantitative basis in terms of a *Boltzmann* distribution, but there are problems about the temperature to be used in the $\exp(-\Delta E/RT)$ expression [42]. In any case, as the population densities of 'observed' points cover only a rather small range, the corresponding energy differences are also small.

Such studies have led to a better understanding of the structural flexibility of various chemical groupings and, in some cases, to a description of chemical reaction pathways and mechanisms. For example, the geometry of the peptide moiety, as observed in many crystal structures of amides, differs from that of an isolated molecule in the gas phase; the pattern of changes is consistent with an increase in C–N and an increase in C′=O double bond character as intermolecular N–H···O=C H-bonding becomes stronger (incipient protonation of the C=O group) [43].

In his 1968 review of donor-acceptor interactions, *Bent* suggested that *'certain kinds of attractive intermolecular interactions may be viewed as incipient valence shell expansions and often as the first stage of bimolecular displacement reactions'* [44]. In particular he drew attention to the correlation between the two I···I distances observed in the linear I_3^- anion (*Fig. 5,a*) and suggested that the curve *'may be presumed to show, approximately, the changes that occur in the distances between nearest neighbors in the linear exchange reaction* $I^1 + I^2I^3 \rightarrow I^1I^2 + I^3$'. Similarly, the curves shown in *Fig. 5,b* and *c*, may be taken to depict corresponding changes in interatomic distance for the thiathiophene valence isomerization reaction (**V**) and for proton transfer between a pair of O-atoms.

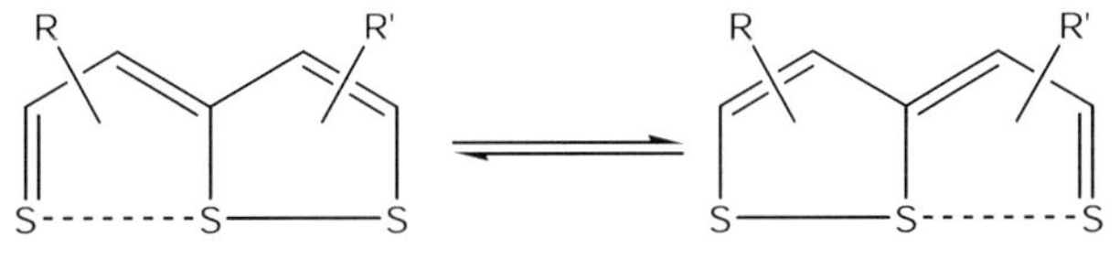

V

The correlation curves shown in *Fig. 5* are not just smooth curves drawn to fit the experimental points as well as possible. They have a special analytical form derived from a simple model of chemical bonding, based on *Pauling*'s relationship [45] for the bond number n of a fractional bond of distance d;

$$d(n) - d(l) = \Delta d = -c \ln n$$

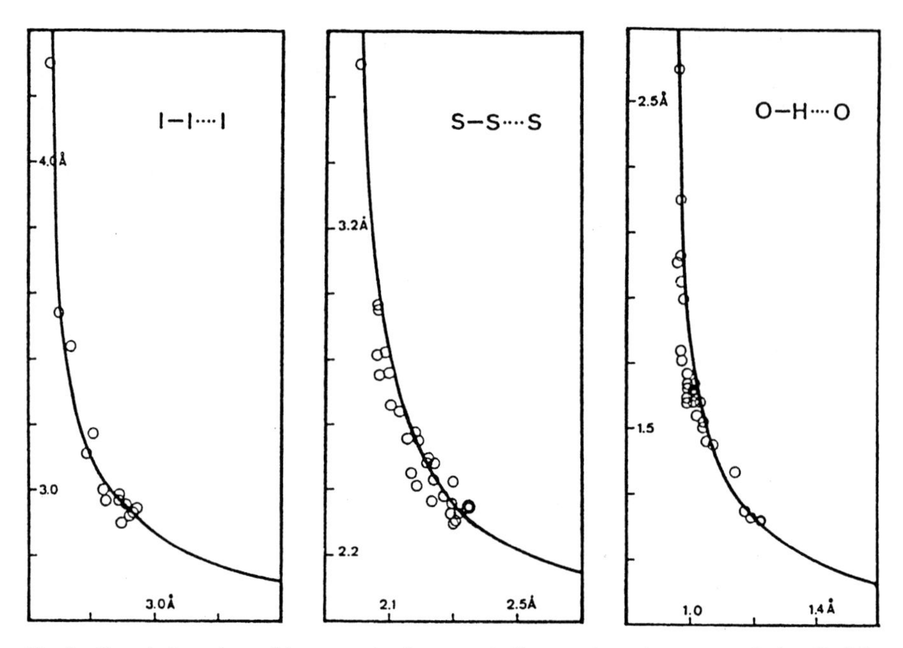

Fig. 5. *Correlation plots of interatomic distances in linear triatomic systems. Left:* triiodide anions, *center:* thiathiophthenes, and *right:* O–H···O H-bonds. Adapted from H.-B. Bürgi, *Angew. Chem., Int. Ed.* **1975**, *14*, 460.

and the assumption that the sum of the two bond numbers equals unity for all pairs of related distances (bond number conservation);

$$\exp(-\Delta d_1/c) + \exp(-\Delta d_2/c) = 1$$

The value of c must be determined separately for each system but it does not vary much. The *Pauling* relationship can be derived from the properties of the *Morse* function [46]. Bond number conservation seems to apply quite generally to reaction paths involving bond breaking and bond formation in a system of three collinear atoms, and thus to S_N2-type reactions at tetrahedrally coordinated Cd, Sn, Si, Ge, Al, and B [47] although not at C.

The structure correlation method has been applied to derive information about a variety of other reaction paths, for example, for weakening and ultimate fission of one bond of a tetrahedral MX_4 molecule to give a planar MX_3 species (S_N1-reaction type) [48] and nucleophilic addition at carbonyl C-atom [49]. From analysis of the conformations of $Ph_3P{=}O$ fragments [50], the stereoisomerization path could be mapped and identified as corresponding to a 'two-ring flip' mechanism [51].

In an extensive series of systematic observations, *Jones* and *Kirby* were able to correlate structural changes with kinetic data [52]; in a series of tetrahydropyran-2-one acetal (**VI**) with different substituents X, as the rate of the spontaneous cleavage of the exocyclic C–O bond increases (by fifteen orders of magnitude!), the length of the exocyclic C–O bond increases (from 1.41 to 1.48 Å), and the length of the endocyclic C–O bond decreases (from 1.42 to 1.38 Å), thus underlining the extraordinary sensitivity of reaction rate to small structural changes in the ground state of the reacting species [53].

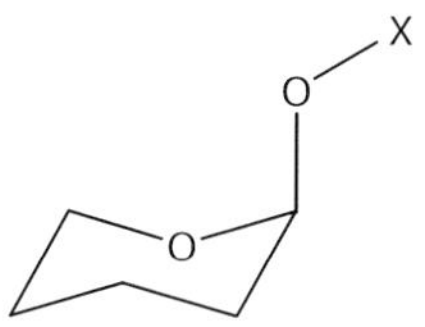

VI

Although transition states, by their very nature, cannot be directly observed in the course of a chemical reaction, it should be noted that molecules may be deformed by crystal forces away from their ground-state geometry towards that of a transition state. For low-energy conformational or valence-state isomerization processes, this kind of deformation in the crystal may lead to a geometry resembling that of the transition state of the isolated molecule. The transition state itself, stabilized by crystal forces, may therefore be 'observed' in a favorable crystal structure. For example, in the $Ph_3P{=}O$ stereoisomerization example mentioned above, the geometry of the transition state was inferred from the distribution of sample points and confirmed by force-field calculations, which gave the energy change along the reaction path as only a few kJ mol^{-1} [54]. Similarly, the molecular geometry observed in the crystal structure of the antiaromatic tetra-(*tert*-butyl)-*s*-indacene has been interpreted as corresponding to that of the D_{2h} transition state of the valence isomerization, stabilized by crystal forces [55]. There is a parallel here with theories of enzymatic reactions, where binding energy is 'utilized' to lower the energy of the transition state, meaning essentially that the transition state binds to the enzyme better than the ground state does. The enzyme may be considered to deform the structure of the substrate in the enzyme-substrate complex in the direction of the transition-state complex. As shown in some of these examples, small structural changes in the ground state of the reacting species can lead to dramatic changes in reaction rate.

The structure correlation approach thus provides a link between the 'statics' of crystals and the 'dynamics' of reacting chemical systems. The first steps could be made without the help of computer-assisted structure retrieval systems but today the only practicable way to find details of particular structures or classes of structures is through the various databases, of which the *Cambridge Structural Database* (*CSD*) is the most useful for organic and organometallic structures and the *Protein Data Bank* (*PDB*) for macromolecular structural data.

6. Where Are We Going?

Where are we going? Into the unknown. The influence of crystallography on 20th century chemistry has been profound – indeed, many areas of present-day chemistry are unthinkable without the contributions of X-ray analysis. It may seem ironic that this progress, almost unimaginable 50 years ago, has been accompanied by the virtual disappearance of crystallographic research and teaching in many University chemistry departments. Crystal structures are now often, perhaps even mostly, determined as part of a laboratory service, and, while this is generally done with great speed and efficiency, the service crystallographer almost always has too much to do. She has no time to think about the broader implications of her results – or even to check them for possible mistakes and misinterpretations. The world production of single-crystal X-ray analyses now runs at more than 10 000 structures a year. With the advent of area detectors and more efficient computers, the mass production of essentially unchecked crystallographic data is going to increase further. In any case, results of many (perhaps most?) current crystallographic studies remain unpublished or receive only scant mention in chemical journals – a computer drawn picture of a molecule and a brief footnote are often all the information provided. A vast amount of metrical information about molecular structure and about intermolecular interactions is being accumulated but not all of it is published and some of it will be lost. In the small-molecule area, this deficiency is due mainly to negligence but in the biomolecular area atomic coordinates and other data are sometimes withheld as a matter of policy, to prevent competing scientists from using the information, a policy which runs counter to the spirit of science and can hardly be condoned. Much of the information – the published part at least – from crystal structure analyses is collected in computer readable form in the *CSD* and the PDB. Whatever the intentions of their originators may have been, these databases have now developed into scientific instruments for studying the systematics of molecular and supramolecular structure. One could argue that the 200 000 crystal structures already stored in the CSD ought to represent a more than adequate storehouse of knowledge. The main problem, perhaps, is not to accumulate more structures but to ask the right questions. Indeed, the contents of the CSD have been aptly described as 200 000 answers waiting for questions [56]. But there is never too much knowledge, and one never knows when useless knowledge may turn out to be useful.

Forty years ago, one could read *Acta Crystallographica* with interest and even with excitement to broaden one's general education. One can still benefit from a perusal of the early issues – the 1956 volume is my personal favorite. Nowadays, information about small-molecule crystal structure analyses is confined mainly to footnotes of papers in various chemistry journals and to the brief standardized accounts in *Acta Crystallographica, Section C*, and other specialist crystallographic journals. Many of these reports are doomed never to be read by anybody, once they appear in print. More than ever, the only practicable way to find details of particular structures or classes of structures is through the *CSD* for organic and organometallic structures and through the *PDB* for biomolecular structures. It is essential that these two compilations continue to be kept running smoothly and efficiently for the foreseeable future, otherwise the only people acquainted with details of any particular crystal structure will be the people who worked on it – and then possibly only until they move on to the next problem.

REFERENCES

[1] R. G. Dickinson, A. L. Raymond, *J. Amer. Chem. Soc.* **1923**, *45*, 22.

[2] S. C. Abrahams, J. M. Robertson, J. G. White, *Acta Crystallogr.* **1949**, *2*, 233, 238; A. McL, Mathieson, J. M. Robertson, V. C. Sinclair, *Acta Crystallogr.* **1950**, *2*, 245, 241.

[3] G. Albrecht, R. B. Corey, *J. Amer. Chem. Soc.* **1939**, *61*, 1087; H. A. Levy, R. B. Corey, *J. Amer. Chem. Soc.* **1941**, *63*, 2095; J. Donohue, *J. Amer. Chem. Soc.* **1950**, *72*, 949; G. B. Carpenter, J. Donohue, *J. Amer. Chem. Soc.* **1950**, *72*, 315; D. P. Shoemaker, J. Donohue, V. Schomaker, R. B. Corey, *Acta Crystallogr.* **1953**, *45*, 22.

[4] D. Crowfoot, C. W. Bunn, B. W. Rogers-Low, A. Turner-Jones, *The Chemistry of Penicillin*, Princeton University Press, 1949, p. 310.

[5] C. H. Carlisle, D. Crowfoot, *Proc. R. Soc. London* **1945**, *A184*, 64.

[6] D. Crowfoot, J. D. Dunitz, *Nature (London)* **1948**, *162*, 608.

[7] J. H. Robertson, C. A. Beevers, *Nature (London)* **1950**, *165*, 690; J. H. Robertson, C. A. Beevers, *Acta Crystallogr.* **1951**, *4*, 270; C. Bokhoven, J. C. Schoone, J. M. Bijvoet, *Proc. Koninkl. Nederland. Akad. Wetenschap.* **1949**, *52*, 120; J. C. Schoone, J. M. Bijvoet, *Acta Crystallogr.* **1951**, *4*, 275.

[8] J. D. Dunitz, J. M. Robertson, *J. Chem. Soc.* **1947**, 142, 148, 1145.

[9] J. D. Dunitz, *Chem. Eur. J.* **1998**, *4*, 745.

[10] C. H. Carlisle, K. Dornberger, *Acta Crystallogr.* **1948**, *1*, 194.

[11] E. Grison, *Acta Crystallogr.* **1949**, *2*, 410.

[12] H. Viervoll, O. Ögrim, *Acta Crystallogr.* **1949**, *2*, 277.

[13] C. Hermann, *Acta Crystallogr.* **1949**, *2*, 139.

[14] J. D. Dunitz, *Acta Crystallogr.* **1949**, *2*, 1.

[15] C. A. Coulson, W. F. Moffitt, *J. Chem. Phys.* **1947**, *15*, 151.

[16] J. D. Dunitz, V. Schomaker, *J. Chem. Phys.* **1952**, *20*, 1703.

[17] M. Dobler, J. D. Dunitz, B. Gubler, H. P. Weber, G. Büchi, J. Padilla O., *Proc. Chem. Soc., London* **1963,** 383.

[18] D. H. R. Barton, *Chem. Britain* **1973**, *9,* 149.

[19] D. Coster, K. S. Knol, J. A. Prins, *Z. Phys.* **1930**, *63*, 345.

[20] J. M. Bijvoet, A. F. Peerdeman, A. J. van Bommel, *Nature (London)* **1951**, *168*, 271; A. F. Peerdeman, A. J. van Bommel, J. M. Bijvoet, *Proc. Koninkl. Nederland. Akad. Wetenschap.* **1951**, *B54,* 16.

[21] R. S. Cahn, C. K. Ingold, V. Prelog, *Experientia* **1956**, *12*, 81; R. S. Cahn, C. K. Ingold, V. Prelog, *Angew. Chem., Int. Ed.*, **1966**, *5*, 385.
[22] K. N. Trueblood, H.-B. Bürgi, H. Burzlaff, J. D. Dunitz, C. M. Grammacioli, H. H. Schulz, U. Shmueli, S. C. Abrahams, *Acta Crystallogr., Sect. A* **1996**, *52*, 770.
[23] D. W. J. Cruickshank, *Acta Crystallogr.* **1956**, *9*, 754.
[24] J. D. Dunitz, V. Schomaker, K. N. Trueblood, *J . Phys. Chem.* **1988**,*92*, 856; J. D. Dunitz, E. F. Maverick, K. N. Trueblood, *Angew. Chem. Int. Ed.* **1988**, *27*, 880.
[25] H. B. Bürgi, S. C. Capelli, *Acta Crystallogr., Sect. A* **2000**, *56*, 403; S. C. Capelli, M. Förtsch, H. B. Bürgi, *Acta Crystallogr., Sect. A* **2000**, *56*, 413; H. B. Bürgi, S. C. Capelli, H. Birkedal, *Acta Crystallogr., Sect. A* **2000**, *56*, 425.
[26] P. Coppens, *X-Ray Charge Densities and Chemical Bonding*, International Union of Crystallograllography, Oxford University Press 1997.
[27] J. D. Dunitz, W. B. Schweizer, P. Seiler, *Helv. Chim. Acta* **1983**, *66*, 123.
[28] F. L. Hirshfeld, *Acta Crystallogr., Sect. B* **1984**, *40*, 613.
[29] R. F. W. Bader, *Atoms in Molecules: a Quantum Theory*, Clarendon Press, Oxford. 1990.
[30] L. Pauling, R. B. Corey, *J. Am. Chem. Soc.* **1950**, *72*, 5349.
[31] L Pauling, R. B. Corey, H. R. Branson, *Proc. Natl. Acad. Sci. U.S.A.* **1951**, *37*, 205; L. Pauling, R. B. Corey, *Proc. Natl. Acad. Sci. U.S.A.* **1951**, *37*, 235, 241, 251, 256, 261, 272, 282, 729.
[32] L. Pauling, *J. Am. Chem. Soc.* **1929**, *51,* 1010.
[33] A. I. Kitaigorodskii, *Organic Chemical Crystallograllography*, Consultants Bureau, New York 1961.
[34] J. D. Dunitz, J. Bernstein, *Acc. Chem. Res.* **1995**, *28*, 193.
[35] J. P. M. Lommerse, W. D. S. Motherwell, H. L. Ammon, J. D. Dunitz, A. Gavezzotti, D. W. M. Hoffmann, F. J. J. Leusen, W. T. M. Mooij, S. L. Price, W. B. Schweizer, M. U. Schmidt, B. P. van Eijck, P. Verwer, D. E. Williams, *Acta Crystallogr., Sect. B* **2000**, *56*, 697.
[36] B. van Eijck, A. L. Spek, W. T. M. Mooij, J. Kroon , *Acta Crystallogr, Sect. B* **1998**, *54*, 291.
[37] L. Addadi, Z. Berkovitch-Yellin, N. Domb, E. Gati, M. Lahav, L. Leiserowitz, *Nature (London)* **1982**, *296,* 21.
[38] L. Addadi, Z. Berkovitch-Yellin, I. Weissbuch, M. Lahav, L. Leiserowitz, in *Topics in Stereochemistry*, Vol. 16, Eds. E. L. Eliel, S. H. Wilen, N. L. Allinger , John Wiley & Sons. New York, 1986, p. 1.
[39] R. J. Davey, N. Blagden, G. D. Potts, R. Docherty, *J. Am. Chem. Soc.* **1997**, *119,* 1767.
[40] M. D. Cohen, G. M. J. Schmidt, *J. Chem. Soc.* **1964**, 1996; M. D. Cohen, G. M. J. Schmidt, F. I. Sonntag, *J. Chem. Soc.* **1964**, 2000; G. M. J. Schmidt, *J. Chem. Soc.* **1964,** 2014.
[41] L. Pauling, *Nature of the Chemical Bond,* Cornell University Press, Ithaca, N. Y., 2nd edn. 1939; 3rd edn. 1960.
[42] H.-B. Bürgi, J. D. Dunitz, *Acta. Crystallogr., Sect. B* **1988**, *44*, 445.
[43] J. D. Dunitz, F. K. Winkler, *Acta Crystallogr., Sect. B* **1975**, *31,* 251.
[44] H. A. Bent, *Chem. Rev.* **1968**, *68*, 587.
[45] L. Pauling, *J. Am. Chem. Soc.* **1947**, *69,* 542.
[46] H. B. Bürgi, J. D. Dunitz, *J. Am. Chem. Soc.* **1987**, *108*, 2924.
[47] H. B. Bürgi, V. Shklover, in *Structure Correlation*, Eds. H. B. Bürgi, J. D. Dunitz, VCH, Weinheim, Vol. 1, 1994**,** p. 303.
[48] P. Murray-Rust, H. B. Bürgi, J. D. Dunitz, *J. Am. Chem. Soc.* **1975**, *97,* 921.
[49] H. B. Bürgi, J. D. Dunitz, E. Shefter, *J. Am. Chem. Soc.* **1973**, *95,* 5065.
[50] E. Bye, W. B. Schweizer, J. D. Dunitz, *J. Am. Chem. Soc.* **1982**, *104*, 5893.
[51] M. G. Hutchings, J. D. Andose, K. Mislow, *J. Am. Chem. Soc.* **1975,** *97,* 4553.
[52] P. G. Jones, A. J. Kirby, *J. Am. Chem. Soc.* **1984**, *106,* 6207.
[53] H. B. Bürgi, C. K. Dubler-Steudle, *J. Am. Chem. Soc.* **1988**, *110,* 7291.
[54] C. P. Brock, W. B. Schweizer, J. D. Dunitz, *J. Am. Chem. Soc.* **1985,** *107,* 7081.
[55] J. D. Dunitz, C. Krüger, H. Irngartinger, E. F. Maverick, Y. Wang, M. Nixdorf. *Angew. Chem., Int. Ed.* **1988,** *27*, 387.
[56] A. Gavezzotti, *Cryst. Rev.* 1**998**, *7*, 5.

NMR Spectroscopy as a Tool for the Determination of Structure and Dynamics of Molecules

by **Christian Griesinger**

Institut für Organische Chemie, Universität Frankfurt, Marie Curie Str. 11, D-60439 Frankfurt, and Max Planck Institute for Biophysical Chemistry, Am Fassberg 11, D-37077 Göttingen

1. Introduction

NMR Spectroscopy has developed over the last 50 years from a physical phenomenon first described by *Felix Bloch* and *Edward Purcell* to a versatile tool to elucidate the structure and dynamics of molecules in solution as well as in the solid state, to establish dynamical features of these molecules, as well as to observe and quantify interactions of molecules in solution that can be used for drug screening. This process has been possible by three parallel developments: The manufacturers have developed NMR instrumentation for the measurement of magnetic resonance into high tech devices that provide very high magnetic fields (up to 21 T) with a spatial and temporal homogeneity of better than 10^{-9}. Mainly *Bruker*, Oxford, *Magnex*, and *JMT* have pushed these developments. Radio frequency preparation is under complete computer control that allows to vary phases as well as amplitudes in any conceivable way. In addition, probes are now available that are very sensitive and will allow the structure elucidation of protein samples from less than 100 nmol. Mainly the companies *Bruker* and *Varian* push the latter developments. The fast development of computers allows acquiring and processing large amounts of data very easily and at high speed. In 1987, it took, for example, 6 h for the *Fourier* transformation of the first 3D spectra at the ETH in Zurich, the same process takes today only a few seconds. The third development essential for the progress of NMR spectroscopy of biomacromolecules lies in the progress of biotechnology to provide these molecules in an NMR adapted form, namely by labelling with NMR active nuclei like ^{13}C, ^{15}N, and ^{2}H. This labelling has allowed to push the limits of NMR feasible molecules well over 20 kD. All these developments have triggered the ingenuity of pulse-sequence developers all over the world to come up with novel arrays of radio-frequency sequences, a process that is still going on with high speed. The hallmarks of these developments for solution-state NMR have been the introduction of decoupling sequences in [3–5], pulse *Fourier* NMR in 1966 [6], two-dimensional NMR in 1976 [7], the invention of COSY to correlate nuclei that are mutually coupled *via* a scalar J coupling [7], the invention of NOESY to measure distances below 6 Å between nuclei in small and especially large molecules in 1979 [8], the invention of total correlation spectroscopy (TOCSY) in 1983 [9], of rotating frame NOESY (ROESY) in 1984 [10], three-dimensional NMR in 1987 [11], the development of multidimensional assignment experiments for labelled proteins between 1990 and 1993 [12], and labelled oligonucleotides between 1993 and 1995 [13], the use of relaxation measurements to study the dynamics of biomolecules in the 1990s [14], the use of cross correlated relaxation to obtain further dynamical and structural information [15], the use of dipolar couplings in weakly oriented media [16], the use of relaxation compensated sequences again in the late 90's [17][18], and the establishment of a host of sequences to study molecular interactions between biomolecules. NMR in the solid state will not be covered extensively in this overview. However, this technique will most probably be able to contribute to the determination of structures of membrane proteins in the future.

NMR Spectroscopy has been used by chemists and biochemists as a tool for structure elucidation. This aspect has become important due to the fact that, after the completion of the genome sequences of the most important organisms, a host of new structures will have to be determined by methods in structural biology. NMR Spectroscopy has to define its role in this postgenomic era where the structure and function of the gene-derived proteins has to be defined. It has to do so in the concert with other techniques for structure elucidation like X-ray and electron microscopy as well as with bioinformatics. The role of NMR spectroscopy in the post-genomic era lies in defining structures of proteins, DNA, RNA, and oligosaccharides, as well as of their complexes, screening small molecules, and determination of their interaction with these biomacromolecules. NMR Spectroscopy has contributed *ca.* 1/6 of *ca.* 2800 new structures submitted to the protein data bank over the last two years [19]. The role of NMR spectroscopy is shown in *Fig. 1*.

2. Interactions and Assignments

This chapter covers shortly the NMR interactions that are used for the two important steps in all NMR investigations, namely, the assignment of

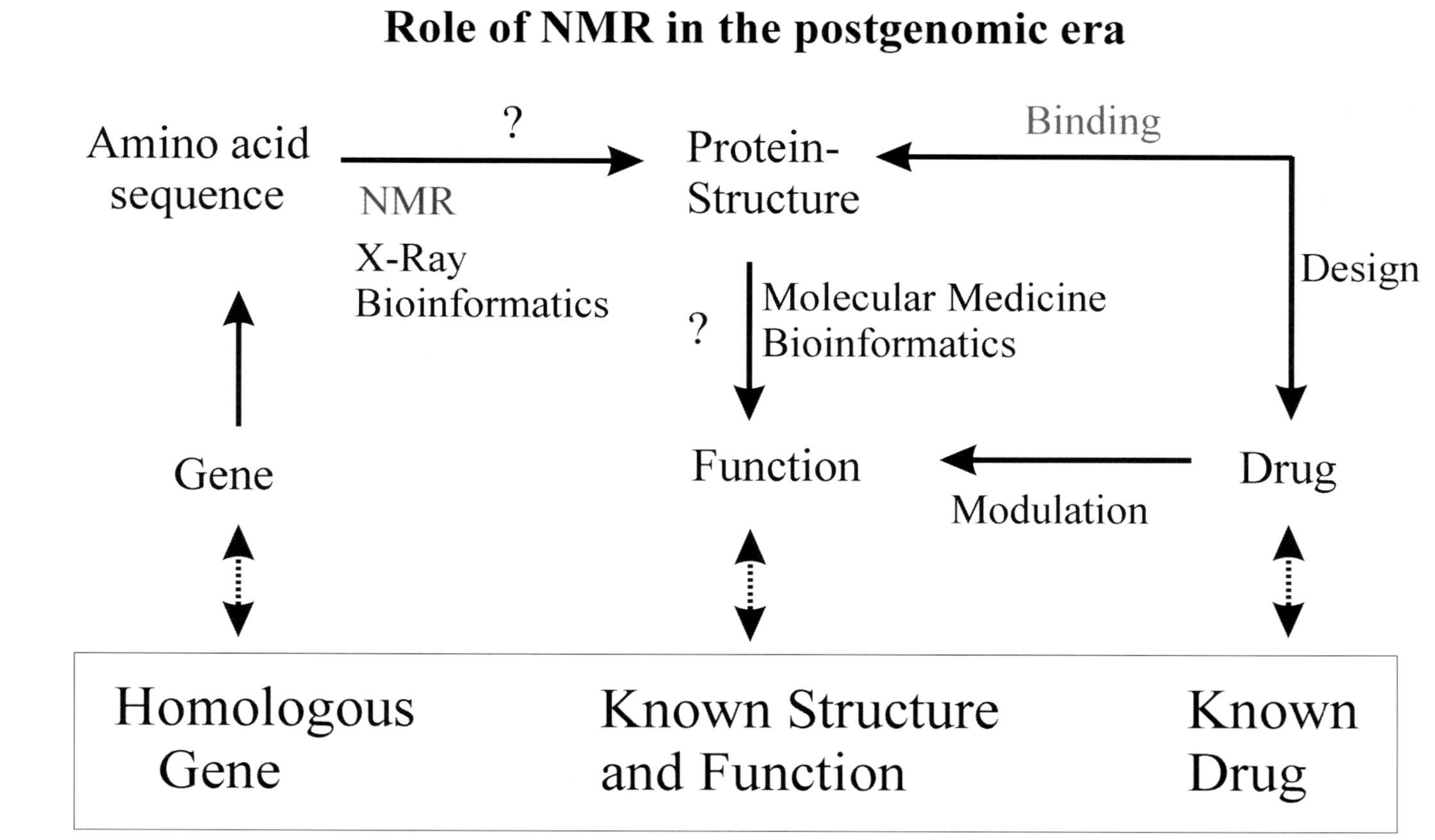

Fig. 1. *The role of NMR spectroscopy in the post-genomic era*. NMR can contribute to structure determination and to defining binding of proteins as well as drugs.

NMR resonances to the individual nuclei and the quantitative evaluation of these interactions to extract structurally and dynamically important information that is used for structure determination, for the measurement of dynamics, and for the measurement of interactions between molecules. The interactions are summarized on *Fig. 2.*

Due to the expected broad readership of this article, mathematics is almost completely excluded from this article. However, together with a pictorial representation of the interactions formulae will be given.

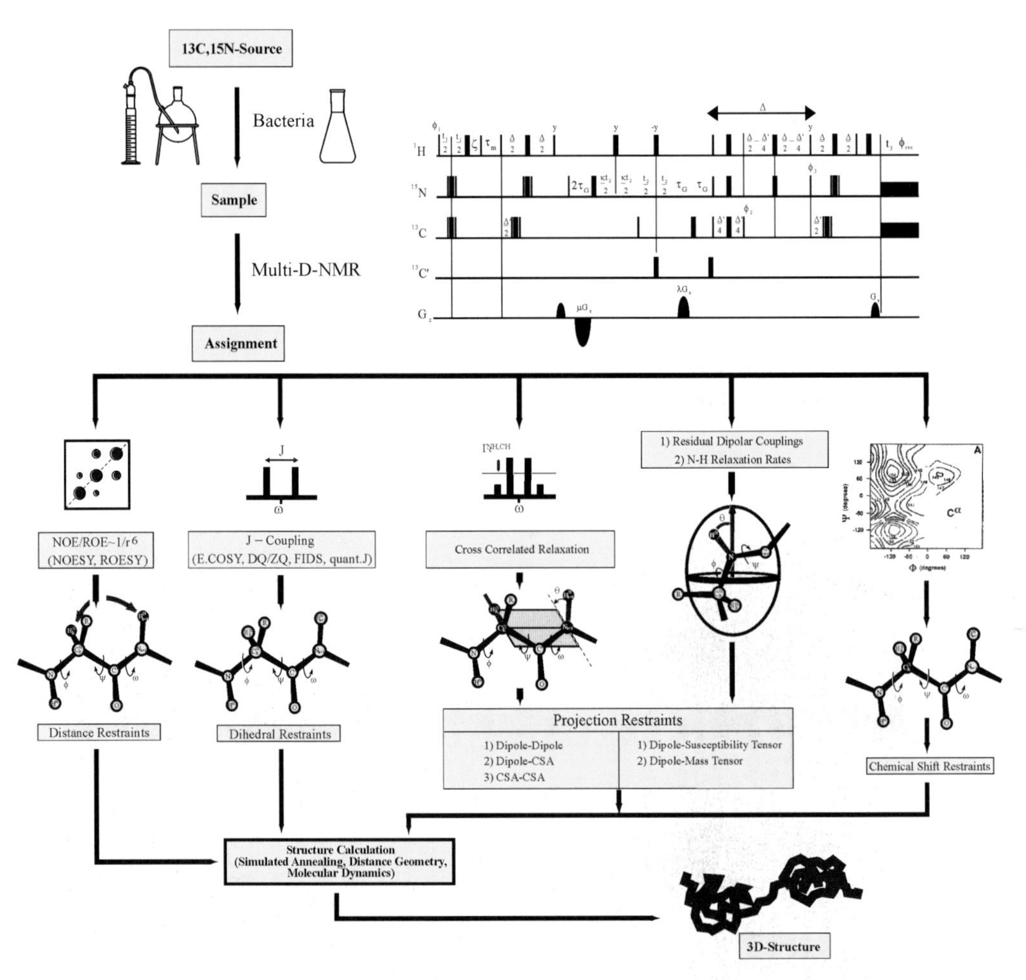

Fig. 2. *Structure determination by NMR.* Labelled samples are prepared by chemical or biochemical means. Application of multidimensional NMR pulse sequences delivers the assignment of the resonances. By measuring H,H distances from NOE or ROE, *J* couplings, cross-correlated relaxation rates, dipolar couplings, and chemical shifts, restraints are derived that can be used for structure calculation.

2.1. Chemical Shift

The chemical shift of nuclei was one of the first surprises of the physicists starting NMR, which spoiled the picture for them, and made NMR spectroscopy to be a tool for chemists and biochemists. Chemical shift is the property of nuclei to resonate in an external magnetic field at a frequency that depends on the chemical environment. Thus, physically identical nuclei such as ^{1}H in different chemical environments resonate at different frequencies. This phenomenon was first observed in 1950 on the two ^{14}N resonances of ammonium nitrate [20]. The authors did not publish the spectrum obtained. Therefore, a 30-MHz spectrum of EtOH recorded in 1951 is presented here [21] (*Fig. 3*). For comparison with the 30-MHz spectrum, a ^{1}H- spectrum of hen egg white lysozyme recorded at 900 MHz in the year 2000 is shown in *Fig. 4*. With present resolution of spectrometers of 10^{-9}, chemical shifts can in principle be resolved at the ppb level.

The chemical shift of nuclei depends on the orientation of the molecule in the magnetic field. It, therefore, has tensorial properties that can be represented by an ellipse. In isotropic solution, which is a solution in which the molecules take all orientations with respect to the external field with equal probability, only the orientation-independent (isotropic) part is observed. However, in the solid state or in anisotropic solution, the tensorial property of the chemical shift is observed, and it can be used as a structure tool.

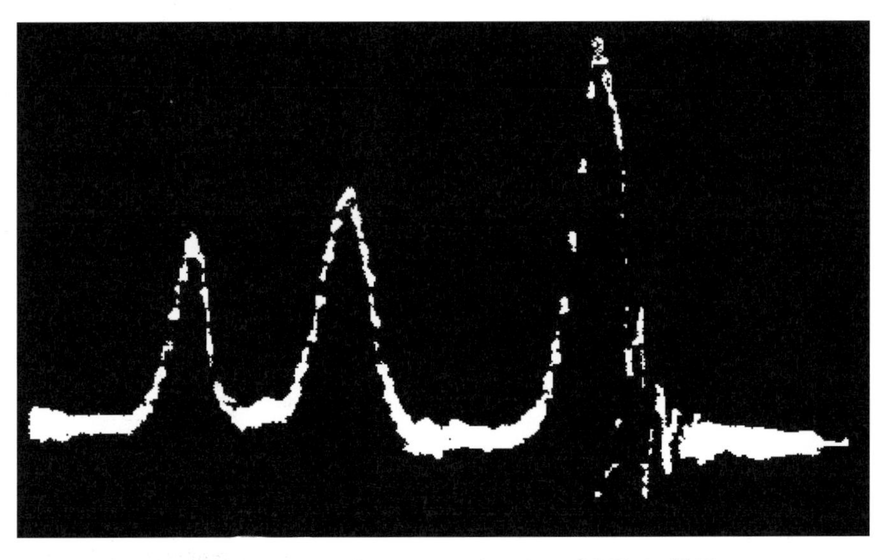

Fig. 3. *Spectrum of EtOH as Obtained at 30 MHz in 1951.*

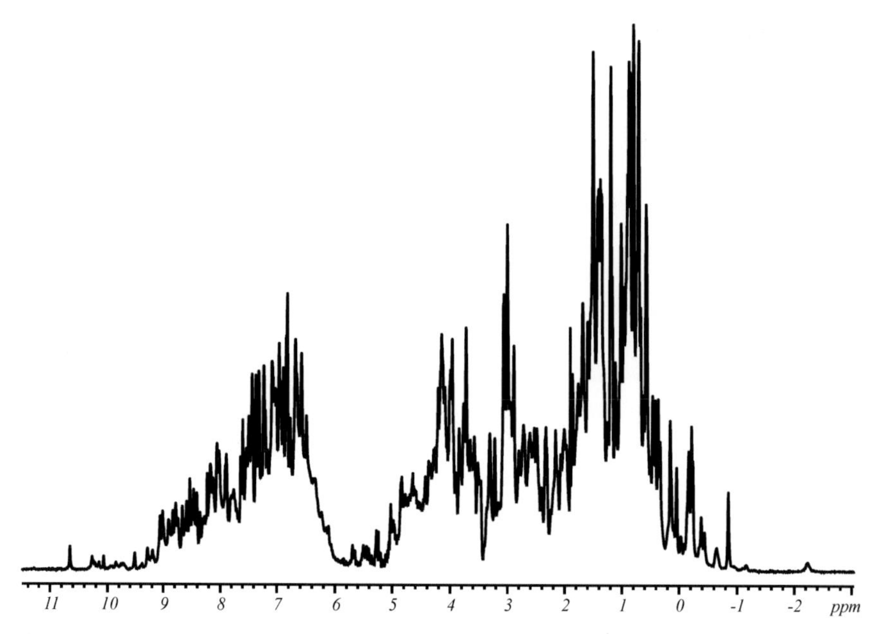

Fig. 4. *900-MHz Spectrum of lysozyme provided by* Eriks Kupce *recorded on an* Oxford Magnet *with a* Varian *console*

$$\hat{H}_k^{\text{CSA,LF}} = b_k \sum_{q=-1}^{1} F_k^{(q)}(\theta_k, \phi_k)\, \hat{A}_k^{(q)}\left(\hat{I}_k\right)$$

In the laboratory frame (LF) the CSA Hamiltonian can be separated into the time-dependent, orientational functions $F_k^{(q)}(\theta_k, \phi_k)$ and the time-independent spin operators terms $\hat{A}_k^{(q)}(\hat{I}_k)$ [22]. The expressions for the second rank tensor operators $\hat{A}_k^{(q)}(\hat{I}_k)$ and the time-dependent modified spherical harmonics $F_k^{(q)}(\theta_k, \phi_k)$ are summarized in *Table 1*.

Chemical shifts can be calculated today quite accurately. There are several tools available to do this. This includes the use of DFT [23], empirical methods such as chemical-shift calculation [24], and neuronal networks [25]. The latter approach requires a large set of molecules with known structures available, on which the network is trained. We will show the power of these calculations in the applications section.

2.2. *J* Coupling

A scalar or *J* coupling is a mutual interaction of two nuclei. It is isotropic in nature and, therefore, does not depend on the orientation of the mole-

Table 1. *Tensor Operators in the Rotating Frame and Modified Spherical Harmonics for the Dipolar and CSA Interaction*. The calibration has been chosen such that $\int F^{(q)}(\theta, \phi)\, F^{(-q)}(\theta, \phi)\, d(\cos\theta)\, d\phi$ is independent of q.

	Tensor Operators for the Dipolar Interaction $b_{kl} = -\mu_0 \frac{\gamma_k \gamma_l \hbar}{4\pi r_{kl}^3}$	Tensor Operators for the CSA Interaction $b_k = \frac{1}{3}(\sigma_{\parallel} - \sigma_{\perp})\gamma_k B_0$	Modified Spherical Harmonics	Frequency
q	$\hat{A}_{kl}^{(q)}(\hat{I}_k, I_l)$	$\hat{A}_k^{(q)}(\hat{I}_k)$	$F_k^{(q)}(\theta,\phi), F_{kl}^{(q)}(\theta,\phi)$	ω_q
−2	$\sqrt{\frac{3}{8}}\hat{I}_k^- \hat{I}_l^-$	–	$\sqrt{\frac{3}{8}}\sin^2\theta\exp(+2i\phi)$	$\omega(\hat{I}_k)+\omega(\hat{I}_l)$
−1	$\sqrt{\frac{3}{8}}\hat{I}_{k,z}^- \hat{I}_l^-$	–	$\sqrt{6}\sin\theta\cos\theta\exp(+i\phi)$	$\omega(\hat{I}_l)$
−1	$\sqrt{\frac{3}{8}}\hat{I}_k^- \hat{I}_{l,z}^-$	$\sqrt{\frac{3}{8}}\hat{I}_k^-$	$\sqrt{6}\sin\theta\cos\theta\exp(+i\phi)$	$\omega(\hat{I}_k)$
0	$\hat{I}_{k,z}\hat{I}_{l,z}$	$\hat{I}_{k,z}$	$3\cos^2\theta-1$	0
0	$\frac{1}{4}(\hat{I}_k^+\hat{I}_l^- + \hat{I}_k^-\hat{I}_l^+)$	–	$3\cos^2\theta-1$	$\omega(\hat{I}_k)-\omega(\hat{I}_l)$
+1	$\sqrt{\frac{3}{8}}\hat{I}_{k,z}\hat{I}_l^+$	$\sqrt{\frac{3}{8}}I_k^+$	$\sqrt{6}\sin\theta\cos\theta\exp(-i\phi)$	$\omega(\hat{I}_k)$
+1	$\sqrt{\frac{3}{8}}\hat{I}_{k,z}\hat{I}_l^+$	–	$\sqrt{6}\sin\theta\cos\theta\exp(-i\phi)$	$\omega(\hat{I}_l)$
+2	$\sqrt{\frac{3}{8}}\hat{I}_k^+\hat{I}_l^+$	–	$\sqrt{\frac{3}{2}}\sin^2\theta\exp(-2i\phi)$	$\omega(\hat{I}_k)+\omega(\hat{I}_l)$

cule with respect to the external field. *J* Couplings exist between nuclei that are connected by covalent bonds. They have been detected in 1950 for the first time on sodium hexafluoroantimonate [26]. It took some efforts to come up with the correct explanation of the *J* coupling as being spin-spin interactions that are mediated by the bonding electrons [27]. The *J* coupling is reflected in the spectrum as an internal splitting of a resonance line of a specific nucleus. Everybody knows the *multiplets*, *e.g.*, of EtOH the OH and Me groups of which give rise to *triplets* due to the coupling to the two enantiotopic CH_2 protons, and the CH_2 group of which appears as a *doublet* of *quadruplets* due to the coupling to the OH group and the coupling to the three homotopic Me protons. The scalar coupling is a coherent interaction and thus affects the frequency of the resonance. The size of *J* couplings depends on the involved nuclei, as well as the nature of the bonds that connect the nuclei. There is a host of empirical information about *J* couplings that has been collected over the past decades. *J* couplings have rather predictable sizes, *e.g.*, when they are observed between nuclei that are connected by one bond. These ^1J couplings, therefore, are suited in an ideal way to establish the constitution or the assignment of small and large molecules. For organic micromolecules that are available in sufficient quantities, the mapping of the connectivities by ^{1}J(C,C) in INADEQUATE-type [28] experiments and ^{1}J(H,C) correlations in HSQC-type [29] experiments is a very direct approach to obtain the constitution, since each correlation peak between the two nuclei indicates their connectivity by a bond (*Fig. 5*). The fact that the ^{1}J couplings do not vary much in size allows to adapt the NMR experiment to optimally detect these interactions. The two experiments in this context are the HSQC and the INADEQUATE, or, more recently, proton-detected ADEQUATE [30]. With modern instrumentation, HSQC spectra can be recorded on a few micrograms of compound. ADEQUATE still requires *ca.* 10 mg of compounds to obtain sufficient sensitivity.

J Coupling constants across single bonds are the most unambiguous way to infer constitution from connectivities. Of course, more long-range couplings, *i.e.*, over more than one bond can also be used. In practice, more than one bond means couplings over two or three bonds. For molecules with a lot of constitutional diversity, combinations of experiments with ^{1}H,^{1}H (COSY-type experiments) and ^{1}H,^{13}C couplings (HMBC-type experiments) [31] are used to advantage. Although these experiments provide ambiguous connectivity information due to correlations over one, two, three, and, sometimes, four bonds, the redundancy overcompensates this ambiguity. As an example, the complete analysis of the constitution of a natural product derived from correlation data is provided in *Fig. 6*.

For molecules with repetitive constitutional elements such as proteins and oligonucleotides, correlations *via* single bonds are also the established approach. This requires complete labelling of the carbon with ^{13}C and of nitrogen with ^{15}N. Then, the whole protein contains an uninterrupted chain of

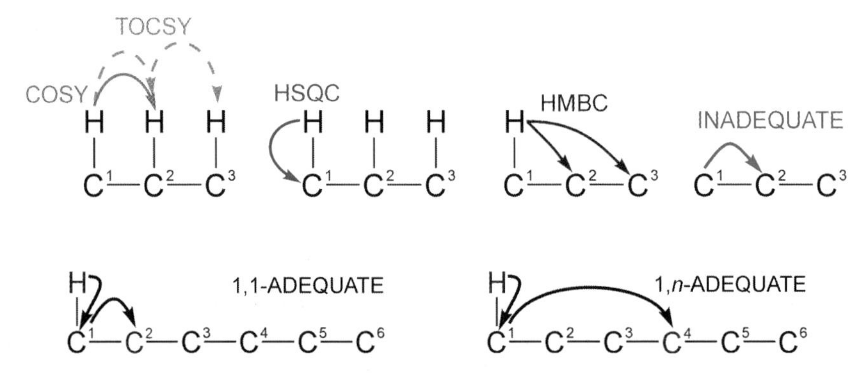

Fig. 5. *Scheme for the determination of the constitution and of the assignment of small molecules based on ^{1}H,^{1}H (COSY, TOCSY), ^{1}H,^{13}C (HSQC, HMBC), and ^{13}C,^{13}C correlations (INADEQUATE, ADEQUATE)*

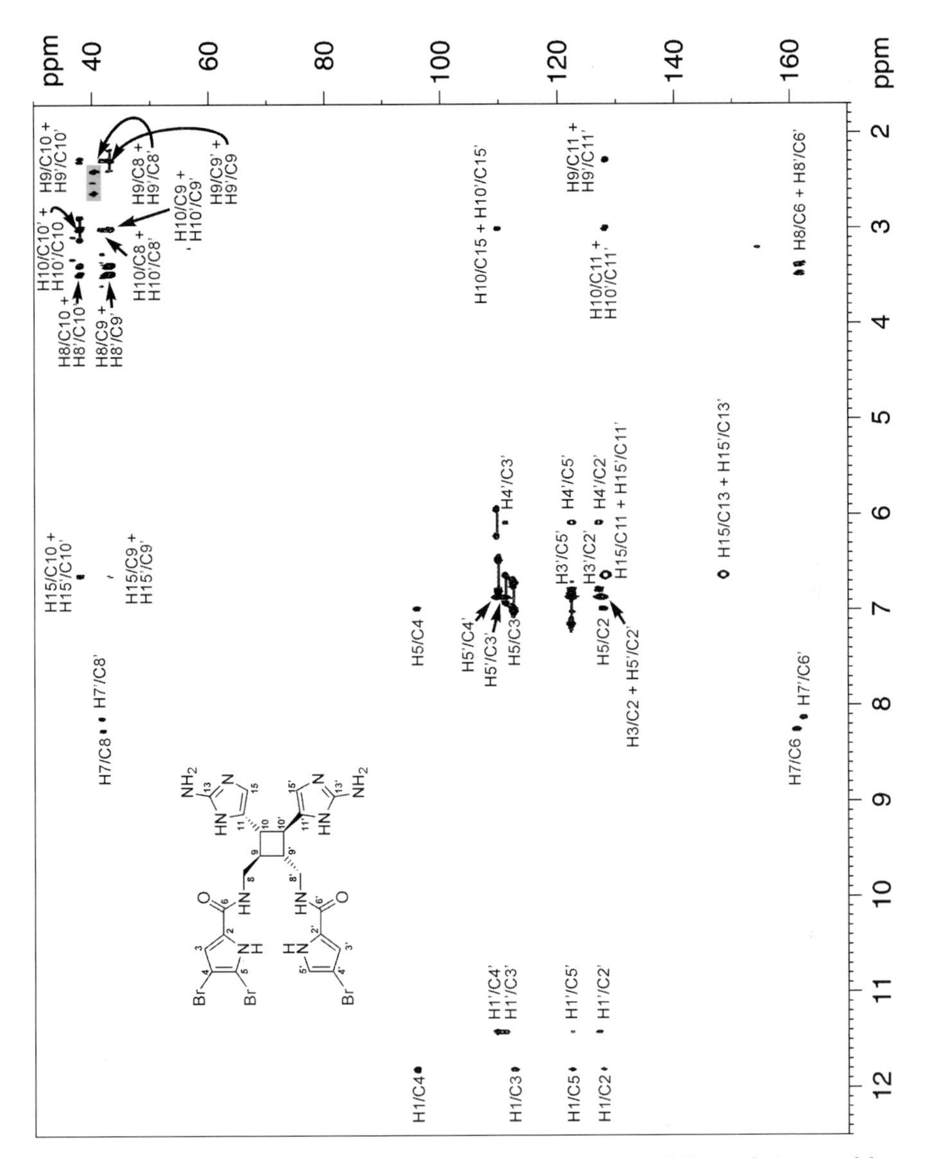

Fig. 6. *HMBC of bromosceptrin recorded at 600 MHz showing the H,C correlations used for the determination of the constitution.* Cross-peaks across a single H–C bond are indicated by red horizontal bars.

NMR-active nuclei of which each is connected by one bond at least to two others. This requires complete labelling of the proteins [32]. In this case, correlation information is inferred by essentially relating the chemical-shift information of each nucleus within an amino acid to the amide ^{15}N and ^{1}H chemical shifts. These two are taken, since they represent the least crowed region in the spectrum of a protein mainly due to the fact that the amide ^{15}N and ^{1}H chemical shifts are the least correlated ones, and most amino acids do not have more than one amide group. The coupling topology is represented for a segment of a protein chain in *Fig. 7*.

The nuclei that are related to the amide group are shown in *Fig. 8*. Representative slices through two three-dimensional experiments a CBCA(CO)NH [33] and a CBCANH [34] are shown in *Fig. 9* for a complex

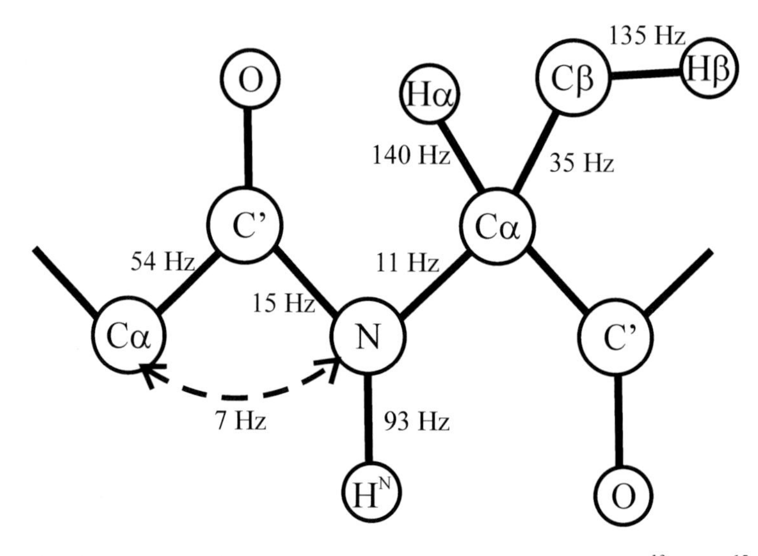

Fig. 7. *Coupling topology of amino acids in proteins after labelling with ^{13}C and ^{15}N.* The one-bond couplings are given with the bonds, the two-bond coupling is indicated with broken double arrow.

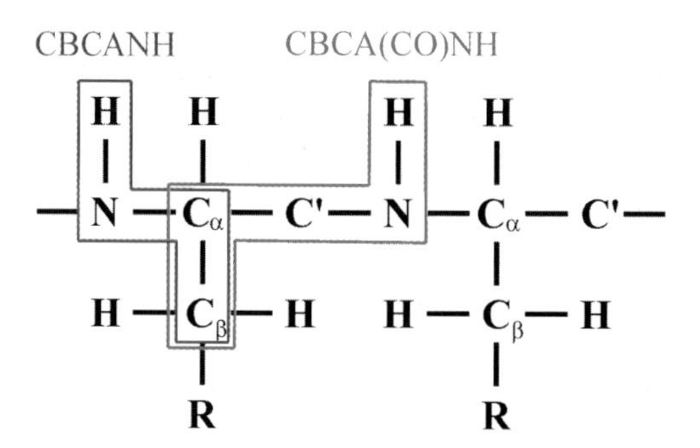

Fig. 8. *The most frequently used assignment experiments, CBCANH and the CB(CACO)NH, which correlate the C_β and the C_α with the NH of the same and of the preceding amino acid, respectively*. By searching for identical C_β and C_α chemical shifts in the two experiments, a sequential assignment is derived.

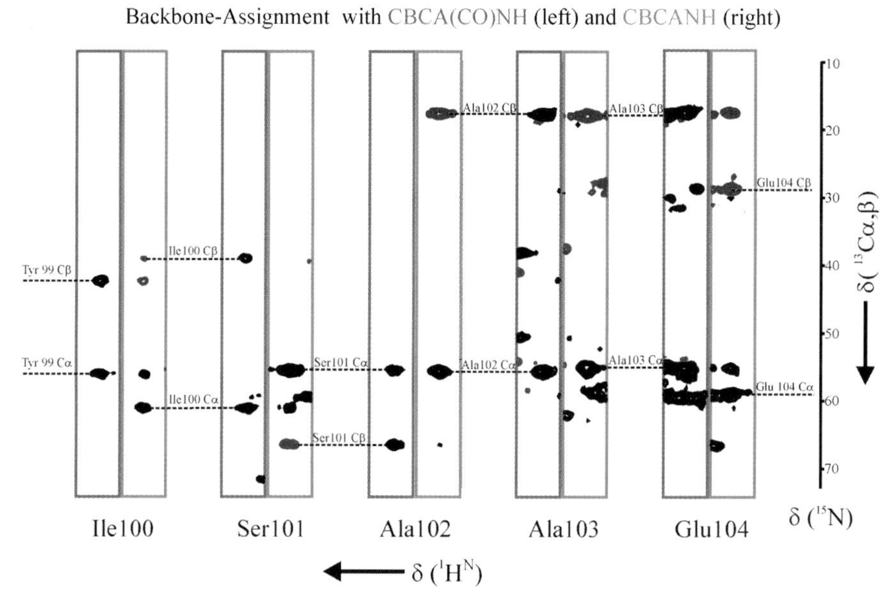

Fig. 9 *Strips from CBCANH and the CB(CACO)NH of calmodulin in the complex with the peptide C20W*. The two strips are taken at the same ^{15}N and ^{1}H chemical shifts, thus connecting two sequential NH resonances. By detecting identical C_β and C_α chemical shifts, the sequential assignment can be completed.

between calmodulin and a cognate peptide [35]. The lines connecting the cross-peaks are used to establish the sequential assignment of resonances for a known primary structure of the protein.

More special experiments are also possible, *e.g.*, the establishment of assignments for Me groups in methionines. These groups often entertain *van der Waals* contacts to other hydrophobic groups and are, therefore, important to establish a structure using ^{1}H,^{1}H NOEs. *Fig. 10* shows a correlation spectrum in which the assignment of the methionine Me groups in the CaM/C20W complex is obtained by relating their chemical shifts to the C_γ and C_ε resonances of the same methionine [36].

Assignment of biomolecules in the solid state by correlations *via* single bonds using magic-angle spinning are available very recently as well. *Fig. 11* shows the aliphatic region of a 2D heteronuclear correlation experiment of a solid 62 residue (*ul*-^{13}C,^{15}N)-labelled protein containing the α-spectrin SH3 domain. The spectrum was recorded at 750 MHz with a MAS frequency of 8 kHz and at a temperature of 278 K [37]. After a ^{15}N evolution period, the so-called band-selective zero-quantum polarization transfer [38] is used to transfer magnetization to C_α resonances (blue peaks). Subsequently, so called band-selective double-quantum polarization transfer [39] is used to observe

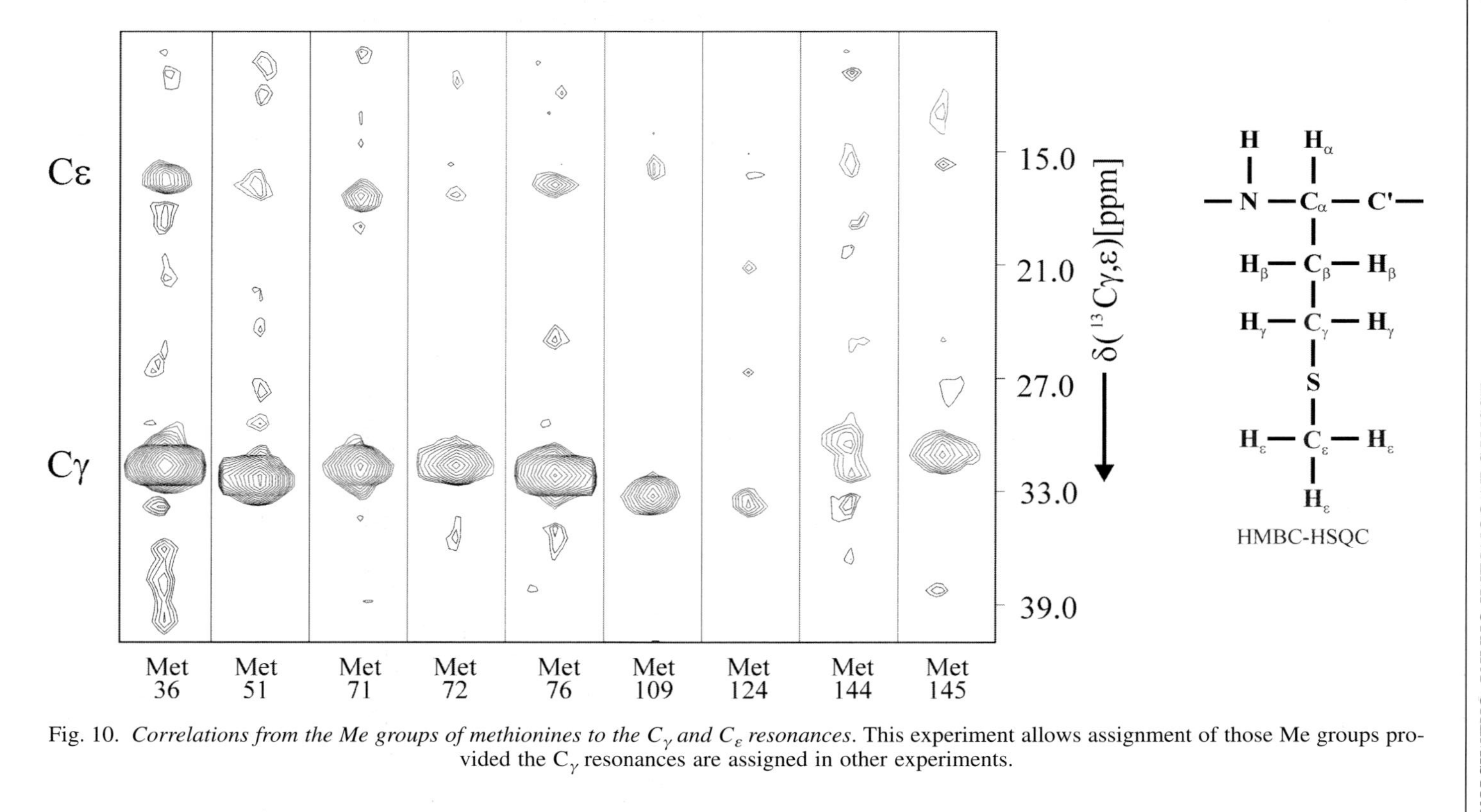

Fig. 10. *Correlations from the Me groups of methionines to the C_γ and C_ε resonances*. This experiment allows assignment of those Me groups provided the C_γ resonances are assigned in other experiments.

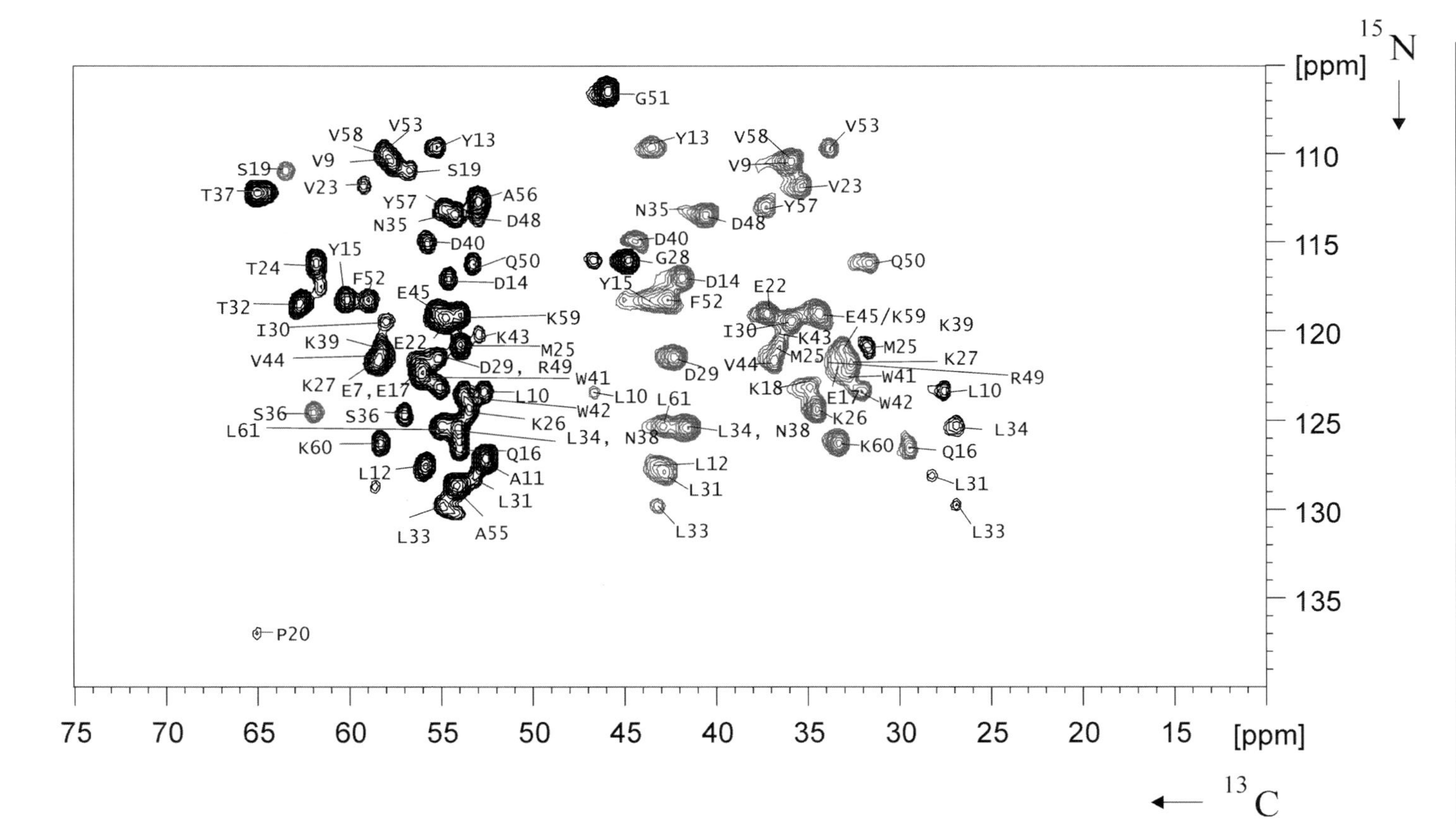

Fig. 11. *Aliphatic region of a 2D heteronuclear correlation experiment of a solid 62 residue* (ul-^{13}C, ^{15}N)*-labelled protein containing the α-spectrin SH3 domain*. The spectrum was recorded at 750 MHz with a MAS frequency of 8 kHz and at a temperature of 278 K [37]. A detailed discussion of the indicated assignments is given in [37].

C_β correlations with negative intensities [40]. Thus, it can be expected that, in a few years, structural information will be available also for proteins that cannot be dissolved in isotropic solvents.

Assignment of oligonucleotides using labelled compounds is performed by a similar strategy based on one bond correlations. The only obstacles are the O-atoms in the phosphodiester bond [13].

2.3. *J* Couplings as a Structural Tool

J Coupling constants, mostly between nuclei separated by three bonds, the so-called ^3J coupling constants can be used to obtain directly information on torsional angles. This can be done through a *Karplus* relation [41] that allows to predict the value of a coupling constant provided the *Karplus* curve could be calibrated by a more or less large number of test compounds whose structure is known, *e.g.*, *via* an X-ray analysis. Such calibrated *Karplus* curves exist for ^{1}H, ^{1}H couplings as well as for a lot of coupling constants in proteins and RNA. The best calibrated coupling constant is the vicinal ^{1}H,^{1}H coupling ^3J(H,H) for which multiparameter calibration curves exist [42]. As an example, we show *Karplus* curves for the measurement of the ϕ torsional angle in proteins (*Fig. 12*) together with experimental values measured on a protein to establish unambiguously this torsional angle from a set of couplings constants measured [43].

There is almost no angle in biomolecules that cannot be defined by *J* coupling constants quite accurately.

It has been possible recently to perform DFT calculations [44] for *J* couplings as well and obtain rather reliable values. This has opened the possibility to infer from couplings constants structural information without the need for calibration. A rather new and important example of this breakthrough is the measurement and structural interpretation of scalar couplings across H-bridges in biomolecules. It turns out that ^{15}N,^{15}N couplings across H-bridges in a N–H···N moiety is rather large and assumes a value of *ca.* 7 Hz [45]. As an example, we show correlations in AT base pairs of RNA as well as of two histidines in a apomyoglobin [46] (*Figs. 13* and *14*).

It turns out that measured coupling constants depend on the geometry of the H-bridge. Thus, it has been possible to obtain a new and very well measurable parameter to define the H-bridge geometry. This is difficult to obtain with other methods except for neutron diffraction studies.

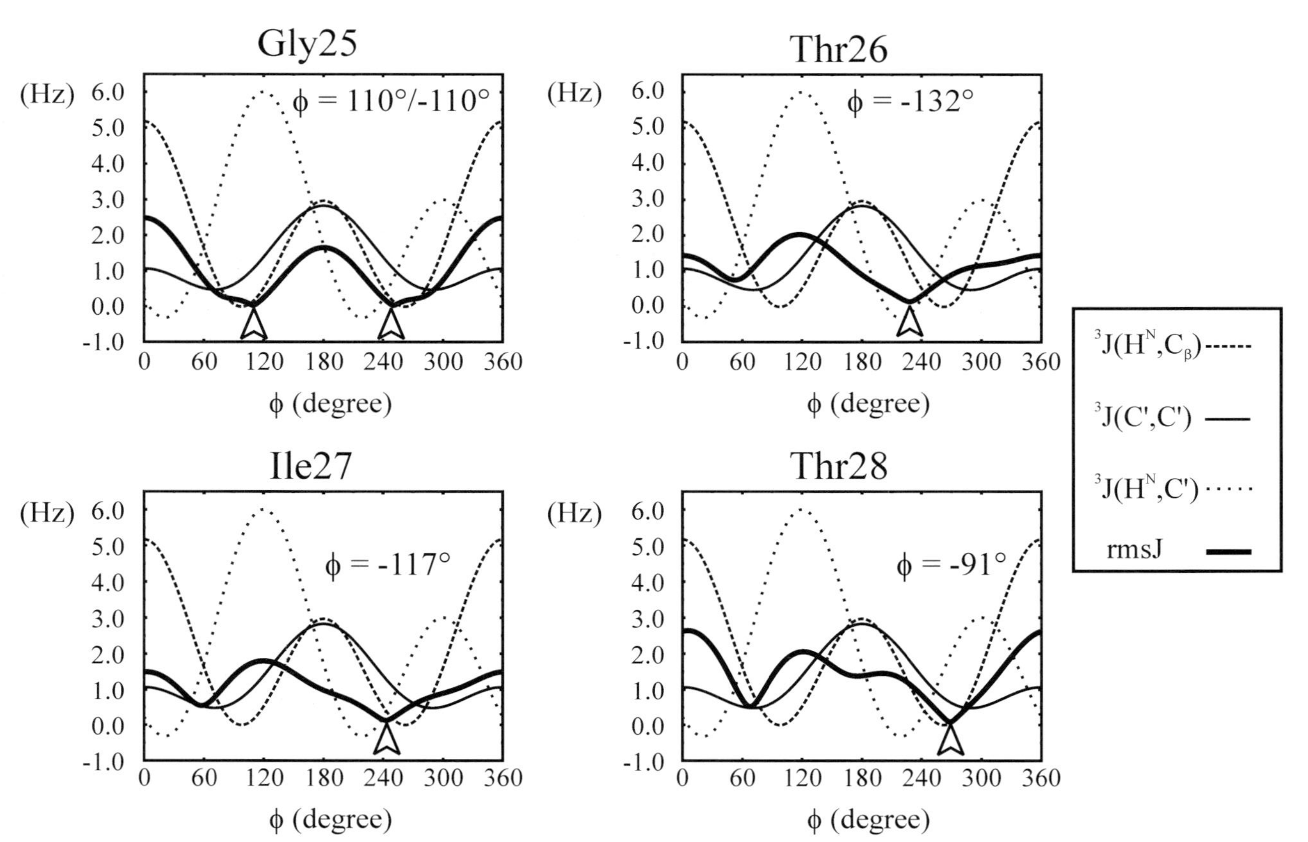

Fig. 12. *Dependence of three* J *coupling constants on the torsional angle of the* N–C_α *bond.* Measuring several couplings allows us to unambiguously define the torsional angle ϕ as shown here for four amino acids in calmodulin.

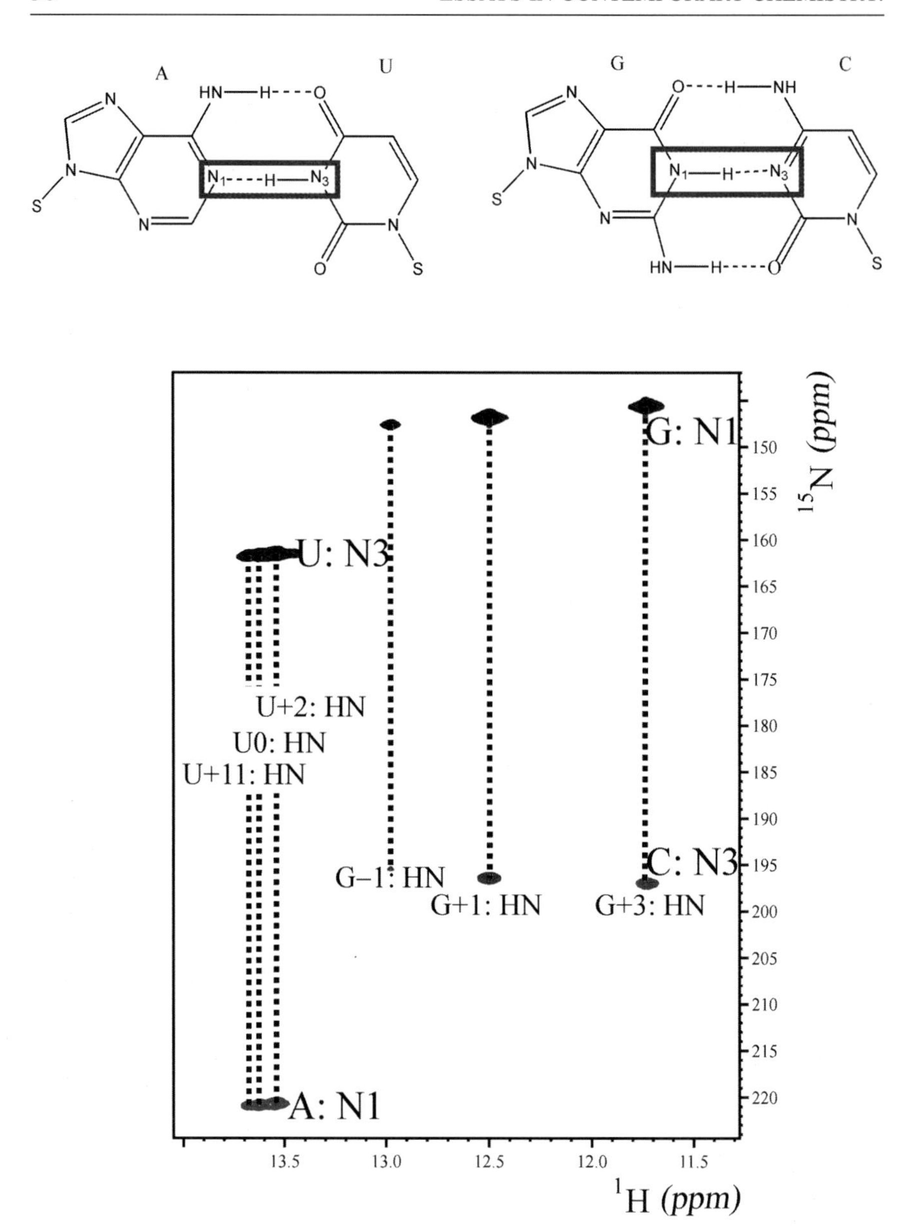

Fig. 13. *Correlation of N_1 and N_3 of A and U, or G and C, respectively,* via *the H-bond-mediated scalar couplings between these two nuclei.* The observation of the scalar couplings proves the covalent character of the H-bond and is very useful for assignment of oligonucleotides.

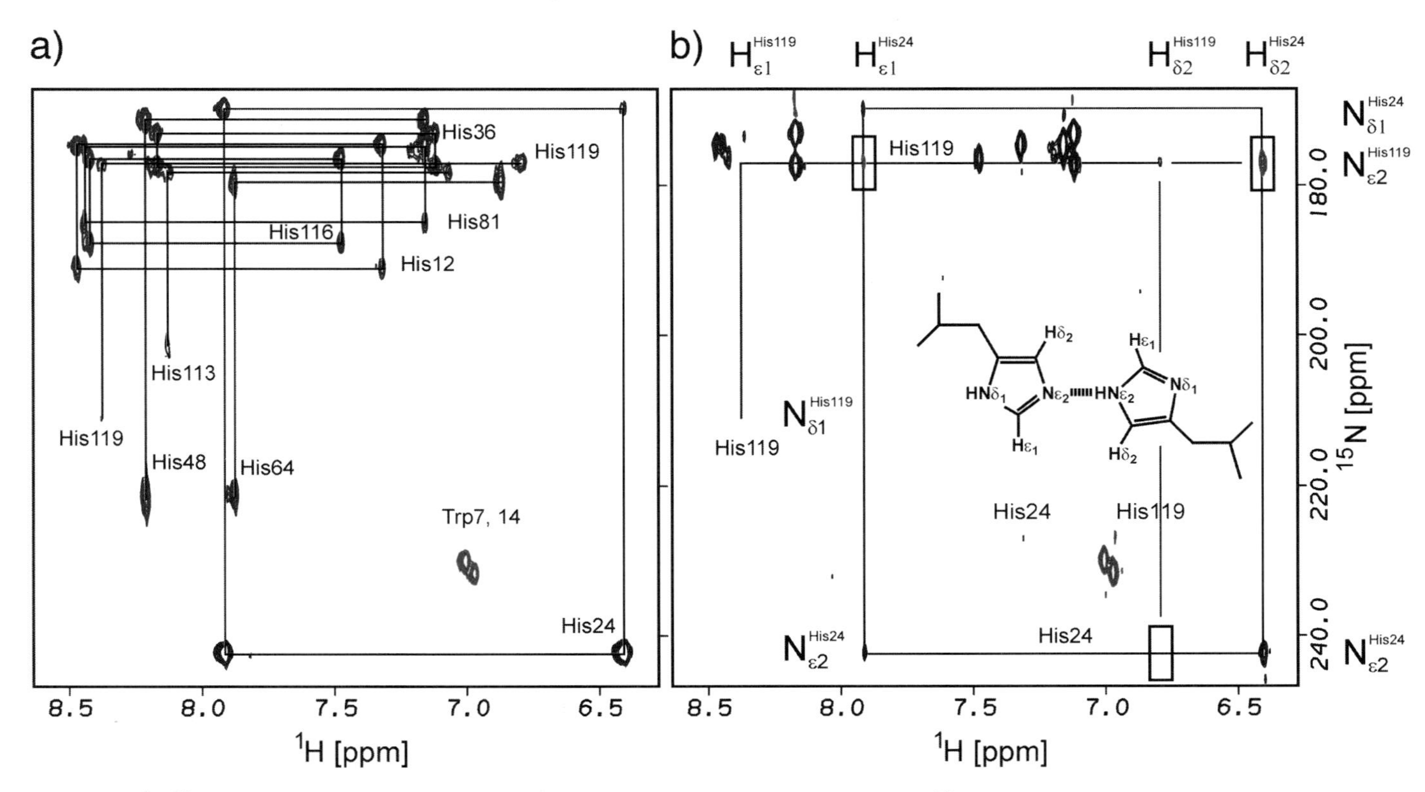

Fig. 14. a) *$^1H,^{15}N$ Correlation spectrum correlating the aromatic protons of histidines with the ^{15}N resonances.* b) *HNN-COSY Spectrum correlating two histidines* via *the N,N coupling constant between the two $N_{\varepsilon 2}$-atoms of His 24 and His 119.* This proves this H-bridge in apomyoglobin.

2.4. Dipolar Couplings

Dipolar couplings exist in addition to the scalar couplings between nuclei [22]. As opposed to the scalar coupling that is transmitted between nuclei *via* the bonding electrons, the dipolar coupling is a pure internuclear interaction. Therefore, it can be calculated directly and is found to depend on the distance between the nuclei, as well as the angle of the internuclear vector with respect to the external magnetic field. The dipolar coupling in Hz is found to be:

$$D_{\mathrm{H,C}} = -\frac{\gamma_{\mathrm{H}}\,\gamma_{\mathrm{C}}\,\mu_0}{4\pi^2\, r_{\mathrm{H,C}}^3}\,(3\cos^2\theta - 1)$$

θ is the angle between the external magnetic field B_0 and the bond vector between C and H, $\boldsymbol{r}_{\mathrm{C,H}}$. Dipolar couplings cannot be observed in isotropic solution, since the tumbling of molecules occurs on a pico- to nanosecond time scale, whereas the dipolar couplings are in the order of up to some 100 kHz, and the average over the angular term vanishes after averaging over the sphere. However, anisotropic solutions can be prepared in various ways, *e.g.*, by putting mixtures of lipids into the H_2O solution which form so-called bicelles [16] (*Fig. 15*). Also many other different ways are established to align biomacromolecules in solution like phages [47], surfactant lipids [48] or even crystallite cellulose [49].

In such solutions, the dipolar couplings are reintroduced and can thus be observed on top of the scalar couplings. A spectrum of ubiquitin measured in anisotropic solution is shown in *Fig. 16*.

Anisotropic solutions impose a distribution of orientations of the solute that is no longer isotropic and can be described by the so-called order parameters. It turns out that the orientational anisotropy relevant for measuring dipolar couplings has the functionality like a $d_{zz} = (3\cos^2\theta - 1)/2$ or $d_{xx-yy} = \sin^2\theta \cos 2\phi$ orbital. The angles θ and ϕ are measured now in the coordinate system of the molecule and no longer with respect to the external magnetic field. Thus, orientations of internuclear vectors in a molecule can be measured based on the dipolar couplings. This is shown schematically in *Fig. 17*.

2.5. Proton, Proton Cross Relaxation: the Nuclear *Overhauser* Effect (NOE)

The NOE is an additional NMR parameter for structure elucidation [8]. It describes the flow of z-magnetization of one nucleus to a neighboring one with the so-called cross-correlated relaxation rate σ. This rate depends on the distance between two protons *via* the inverse 6th power, as well as the diffu-

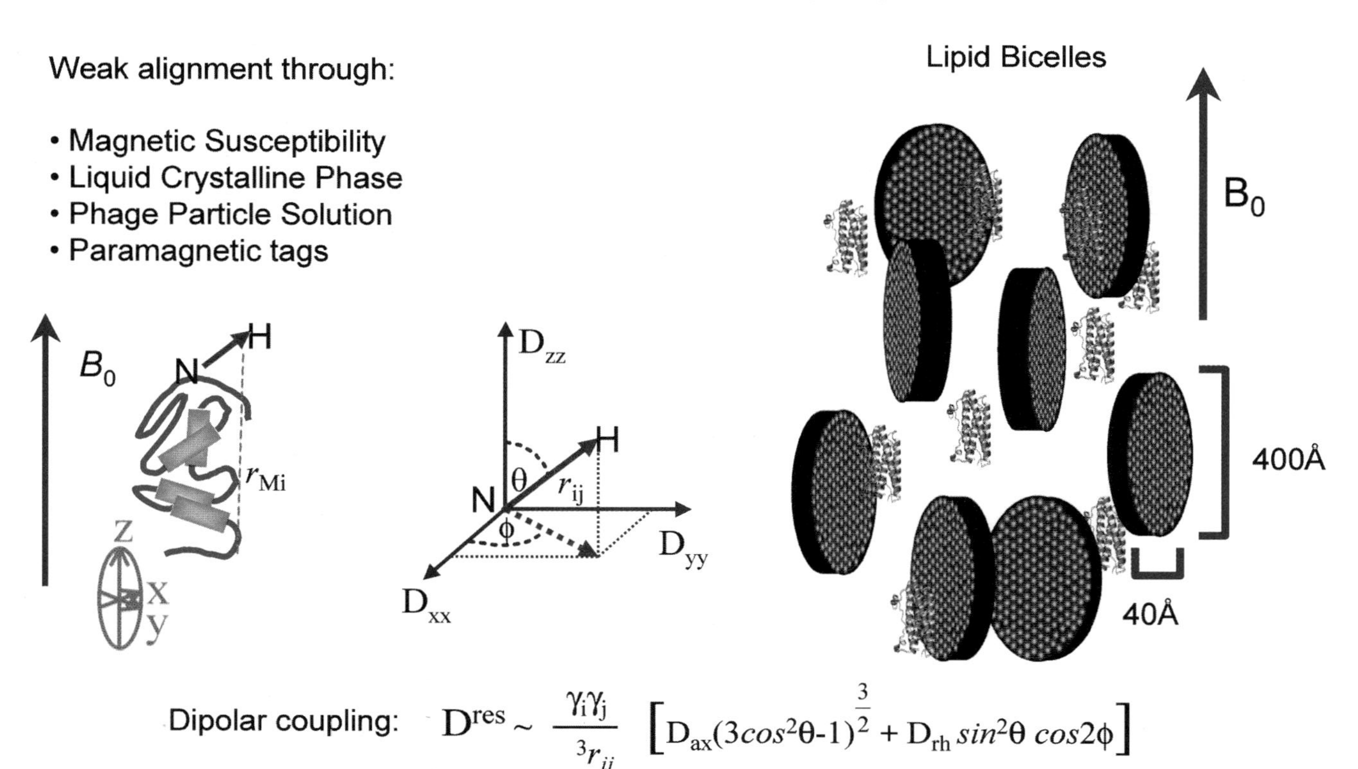
Residual Dipolar Coupling
Weak alignment through:
• Magnetic Susceptibility
• Liquid Crystalline Phase
• Phage Particle Solution
• Paramagnetic tags
Lipid Bicelles
B0
N
H
rMi
Z
X
y
Dzz
Dxx
Dyy
θ
φ
rij
400Å
40Å
Dipolar coupling: $D^{res} \sim \frac{\gamma_i \gamma_j}{{}^3 r_{ij}} \left[D_{ax}(3cos^2\theta - 1)^{\frac{3}{2}} + D_{rh}\, sin^2\theta\, cos 2\phi \right]$

Fig. 15. *Weak alignment through addition of lipid bicelles or magnetic anisotropy.* The preferential orientation of the protein is depicted. The dipolar coupling D^{res} is observable in this case according to the formula given in the *Fig.*

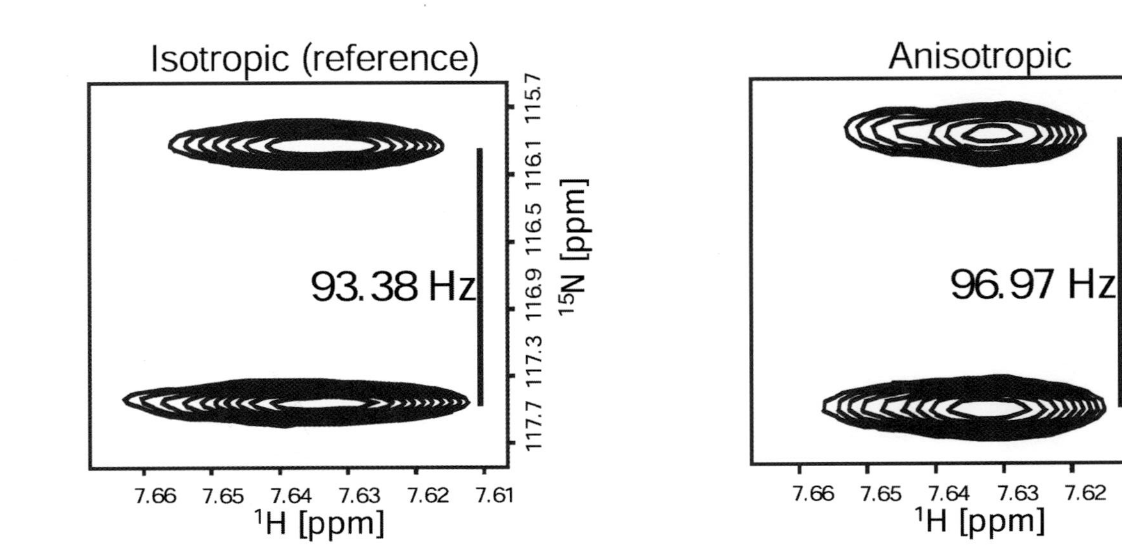

Fig. 16. *Extraction of dipolar couplings that add to the normal scalar couplings.* By forming the difference between the observed splitting in anisotropic and isotropic medium, the coupling can be derived. It reflects orientation information of the corresponding NH vector.

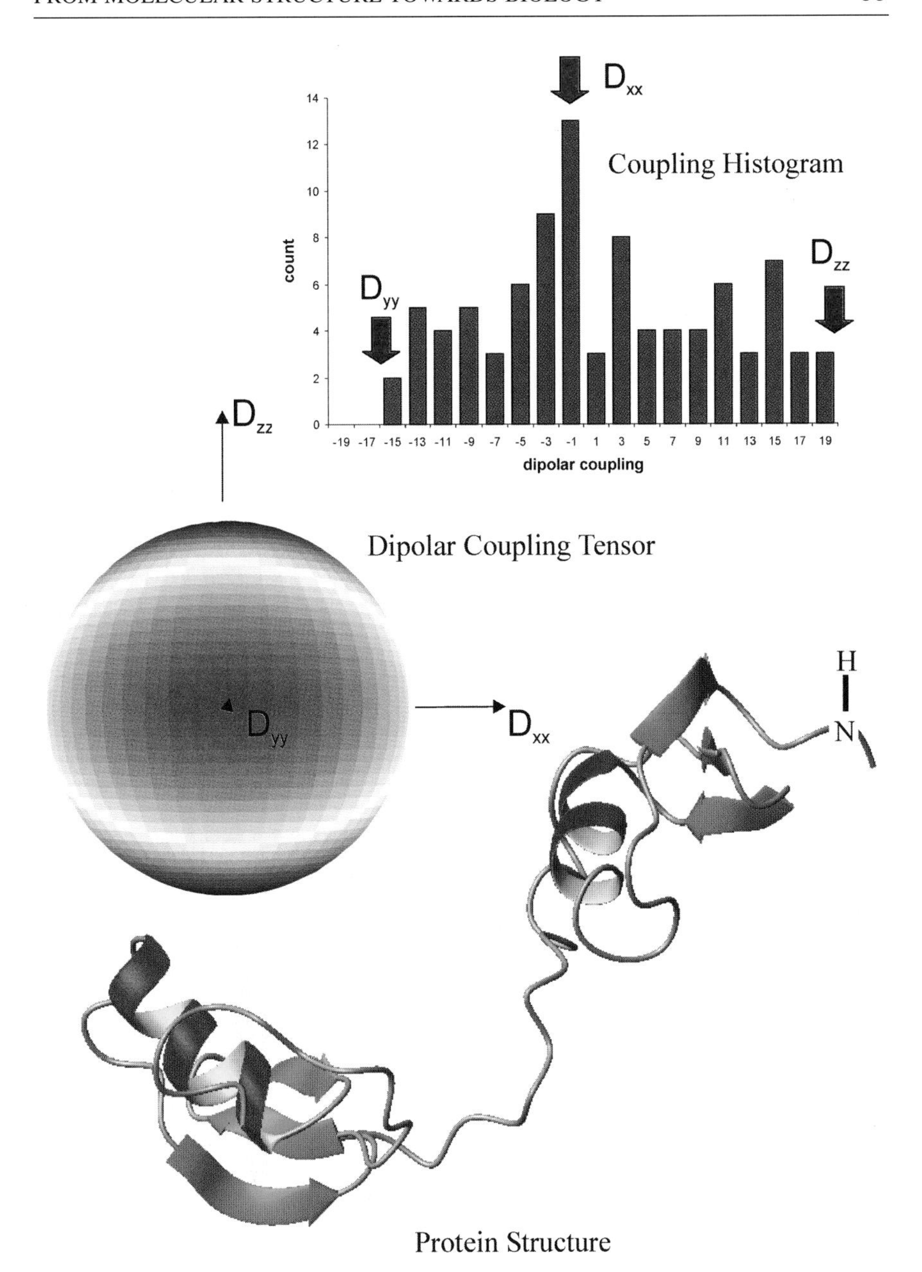

Fig. 17. *Dipolar coupling tensor depicted with a color coding on a sphere.* Equal colors mean equal size of the dipolar coupling. Blue is negative, red is positive dipolar coupling. For uniformly distributed NH vectors in a molecule, the tensor produces a dipolar coupling histogram from which the principal components of the tensor can be read directly. The structure of rhodniin, from which the data have been generated, is shown at the bottom.

sional rotational correlation time of the molecule. The formula for the cross relaxation rate σ is given by:

$$\sigma_{\mathrm{HH}} = \frac{\gamma_{\mathrm{H}}^4}{10 r_{\mathrm{H,H}}^6} \left(\frac{\hbar \mu_0}{4\pi} \right)^2 \left(-\tau_c + 6 \frac{\tau_c}{1 + 4\omega_{\mathrm{H}}^2 \tau_c^2} \right)$$

It is possible to derive the absolute value of the cross relaxation rate from NOESY spectra provided one diagonal peak is resolved. Most of the time, this is not the case or not done for the following reason: when one internuclear distance is known in a molecule, *e.g.*, between two geminal protons or two vicinal protons on an aromatic ring, the cross-relaxation rates of all proton proton pairs in this molecule can be calibrated with this distance and thus be translated directly into the internuclear distance.

NOESY Spectra of proteins tend to overlap due to the large number of protons. Three- and four-dimensional experiments are, therefore, mostly necessary to disentangle and assign the cross peaks [50]. As an example, NOESY spectra of the rhodniin complex are shown in two and three dimensions in *Fig. 18*. The enhancement of the resolution is obvious. The assignment of the peaks is indicated in the spectrum.

Through filter experiments, it is possible to detect exclusively cross-peaks between protons that are attached, *e.g.*, to a ^{15}N or ^{13}C, and protons that are not attached to ^{15}N or ^{13}C [51]. Now, if one investigates a complex between two molecules, of which one is labelled with ^{15}N and ^{13}C, and the other one not, this allows to detect intermolecular H,H proximity exclusively and thus allows mapping of the contact surfaces in macromolecular assemblages. The mapping yields the amino acids in both molecules that interact with each other in the molecular-recognition process. As an example, the filtered NOESY spectrum between CaM and C20W (*Fig. 19*) is shown, as well as the contacts between the two moieties in a schematic representation derived from the NOESY cross peaks (*Fig. 20*).

2.6. Cross-Correlated Relaxation

The nuclear *Overhauser* effect is used to measure distances between nuclei. It is a relaxation effect that, like a chemical reaction, is incoherent in nature as opposed to the scalar coupling that is coherent and is detected as a frequency splitting of resonances in the spectrum. Relaxation can be best viewed at as the flow of population between different states as it would occur in a chemical reaction where one molecule is transformed, *e.g.*, by a reaction of zero-order into a second molecule. The same is true for the NOE where magnetization of spin A by a zero-order reaction is transferred to mag-

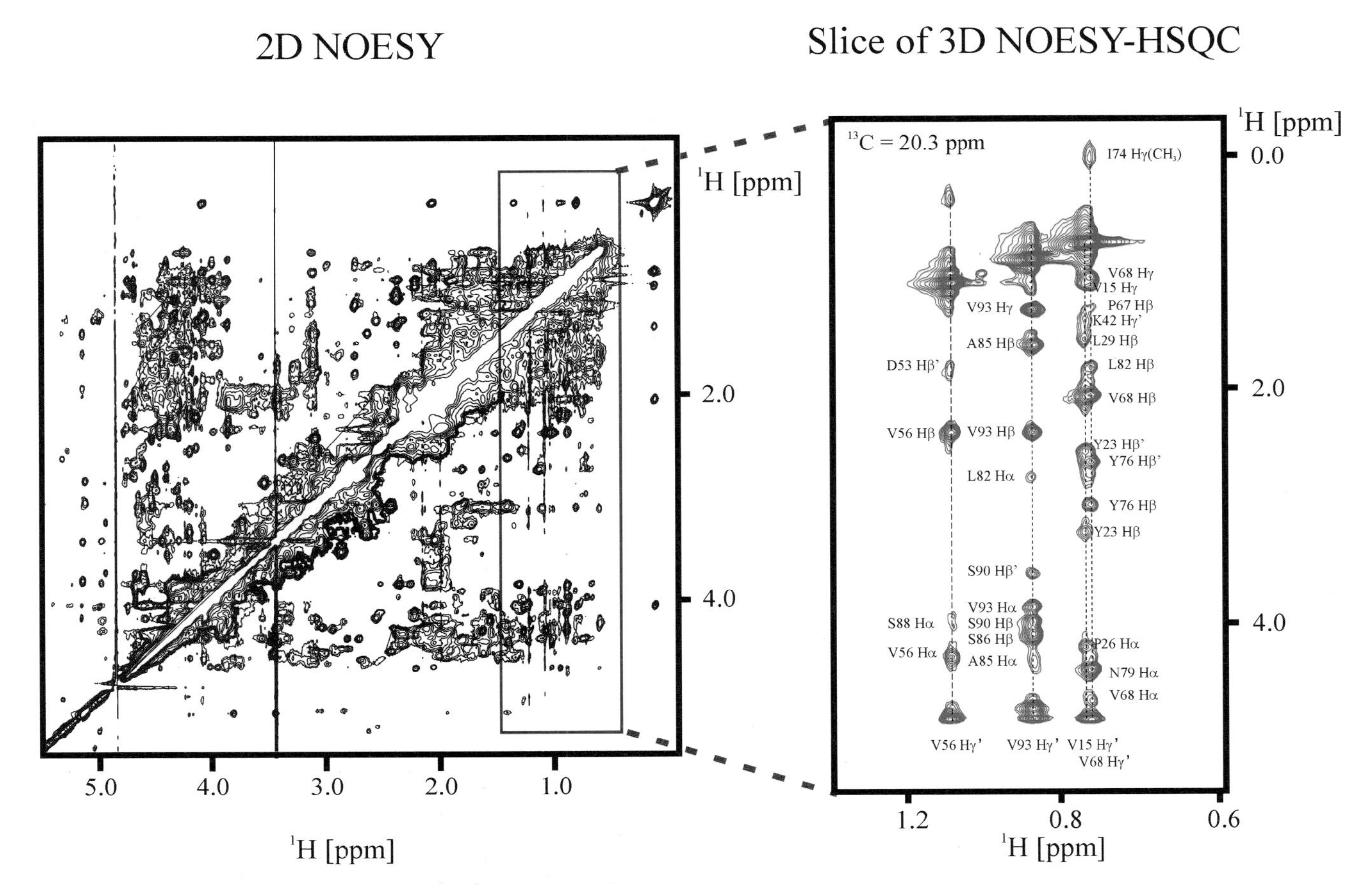

Fig. 18. *Comparison of the 2D NOESY of rhodniin and a slice through the 3D NOESY.* Especially in the aliphatic region, the overlap is unresolvable in the 2D whereas it can be nicely resolved in 3D.

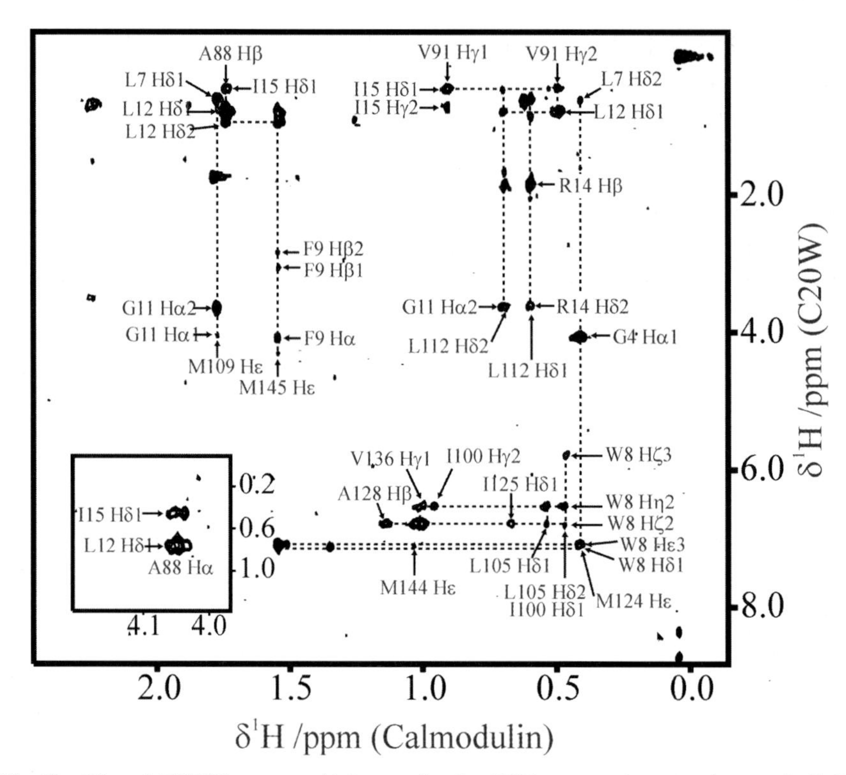

Fig. 19. *Filtered NOESY spectra with intermolecular NOE contacts between the protein CaM and the bound peptide C20W.* The intermolecular NOEs can be detected exclusively due to the different labelling of the peptide and the protein.

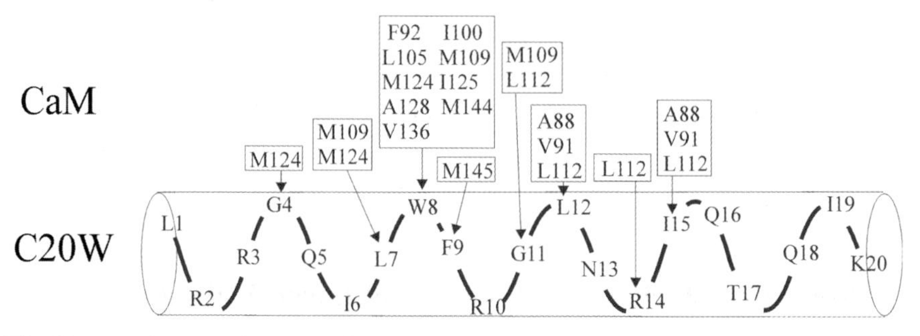

Fig. 20. *Translation of the NOEs of* Fig. 19 *into amino acid contacts between CaM and C20W*

netization of spin B. The kinetic equation of the NOE and such a zero order reaction are identical.

There are other types of relaxation, *e.g.*, cross-correlated relaxation in which resonance lines in a *multiplet* are differentially broadened due to the cross correlation of two dipolar couplings between the nuclei [15][52]. Provided we have two pairs of nuclei, *e.g.*, CH and NH, of which each pair defines one internuclear vector, namely the CH and the NH vector, we can excite a so-called double-quantum coherence between two out of the four nuclei, *e.g.*, between C and N. Such double-quantum coherence behaves as if there was a combined spin including all the couplings C and N entertain with other nuclei. Thus, in our case, the spectrum of this double-quantum coherence will be a *doublet* of *doublets* due to the CH and the NH coupling. Then, the differential relaxation observed on the *multiplet* of this double-quantum coherence can be used to measure the angle between the two internuclear vectors, CH and NH. The formula for the differential relaxation $\Gamma^{c}_{\mathrm{NH,CH}}$ is given by:

$$\Gamma^{c}_{\mathrm{NH,CH}} = \frac{\gamma_{\mathrm{H}}\gamma_{\mathrm{N}}}{(r_{\mathrm{NH}})^3}\frac{\gamma_{\mathrm{H}}\gamma_{\mathrm{C}}}{(r_{\mathrm{CH}})^3}\left(\frac{\mu_0}{4\pi}\hbar\right)^2 \frac{1}{5}(3\cos^2\theta_{\mathrm{NH,CH}}-1)\,\tau_c$$

It is obvious that the cross-correlated relaxation rate depends on natural constants, material constants, distances $\boldsymbol{r}_{\mathrm{NH}}$ and $\boldsymbol{r}_{\mathrm{CH}}$ that need to be known, as well as the desired intervector angle $\theta_{\mathrm{NH,CH}}$ and the correlation time that can be determined experimentally. *Fig. 21* shows a schematic representation of C,N double-quantum-coherence *multiplets* depending on the intervectorial angle $\theta_{\mathrm{NH,CH}}$. If the angle is 0 or 180°, the outer *multiplet* lines are broadened whereas the inner lines are narrowed. For 90°, the situation is reversed. For the magic angle $\theta^{\mathrm{magic}}_{\mathrm{NH,CH}} = 54°$ no effect is observed since $(3\cos^2\theta^{\mathrm{magic}}_{\mathrm{NH,CH}} - 1) = 0$.

Due to the fact that the cross-correlated relaxation like the NOE depends linearly on the correlation time of the molecule, it grows with the correlation time and thus the size of the molecule. This is the first and so far only NMR parameter with this property that delivers angular information and thus behaves like the NOE that delivers distance information and also scales with the correlation time. This is an interesting property and helps to study molecules that are weakly bound to larger molecules.

3. Structure Elucidation of Small and Large Molecules

Structure comes in three layers of complexity, constitution, configuration; and conformation, and structures are dynamic referring to the fact that the constitution can change (*e.g.*, tautomerism), that the configuration can change (*e.g.*, mutarotation), or the conformation can change. All these changes can be detected by NMR spectroscopy, since the NMR parameters are sensitive to the conformation of the molecule and much more to its configuration and even more to its constitution. All parameters, not only the chemical

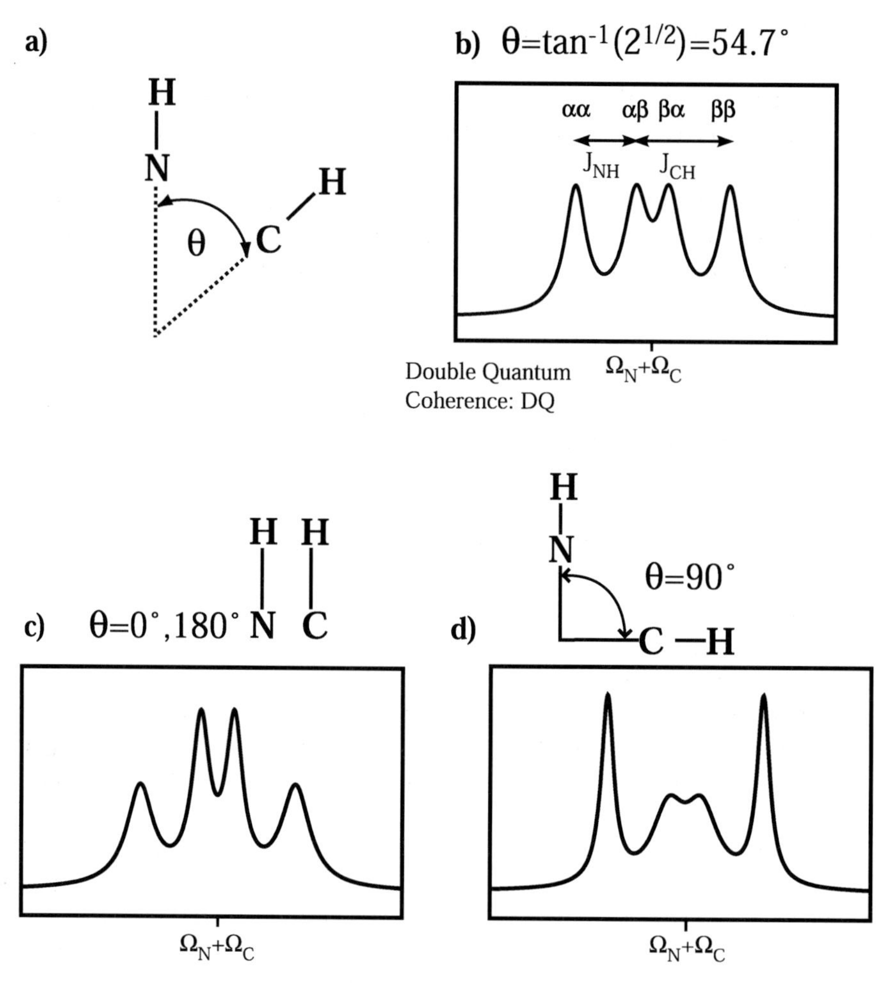

Fig. 21. *Measurement of cross-correlated relaxation from double-quantum coherences between a nitrogen and a carbon.* Depending on the intervector angle θ between the NH and the CH vector, the *multiplet* has different linewidths from which θ can be derived without any calibration.

shifts, are affected. What matters is the time scale with which the dynamics takes place. If the rates involved with the dynamics are smaller than the difference of the NMR parameters in Hz (slow exchange) between the different structures, one will obtain spectra like from a mixture of molecules. If the rates of the structural change are faster than the NMR parameters (fast exchange), they will be averaged, and the detection of the dynamics is more difficult than in the slow exchange regime.

3.1. Determination of Constitution

The constitution of repetitive biomolecules like proteins or oligonucleotides can be detected by more sensitive methods than NMR. Thus, NMR is not the way to determine the sequence of a protein or RNA. This is simply because NMR requires some hundred nanomoles of substance of repetitive biomolecules, and other methods such as *Edman* degradation, mass spectroscopy, or oligonucleotide sequencing are much more sensitive.

The constitution is often desired, however, for all non-repetitive biomolecules or synthetic compounds although, for the latter, the chemist mostly has some preknowledge about the compound in question. HSQC [29], COSY [7], TOCSY [9], and HMBC [31] experiments are the experiments of choice for such molecules. They reflect the connectivities of atoms in the molecule from which the constitution can be derived. With correlation spectra as discussed in *Sect. 2.2*, connectivity information is obtained. With an intelligent structure builder like *Cocon* [53], an especially powerful program, that takes the connectivity information and the rules of bonding between atoms into account, all constitutions that are in agreement with the provided correlation data are proposed. These are frequently more than one. Chemical-shift information can be used in addition to single out the most probable constitution. To use chemical shift information, data bases, *ab initio* calculations, or neuronal networks are available. As an example, the ascididemin constitution has been derived from connectivity information as well as with chemical shifts derived from neuronal networks (*Fig. 22*) [25].

Of course, this is just one way to derive constitution from NMR data, and there are several others in use. However, the scheme remains that connectivity information needs to be translated into constitution, and that chemical shifts can further narrow down the possible constitutions. With presently available computers, the molecular weight for this approach is not much limited, and molecules with more than 100 heavy atoms could be successfully treated.

3.2. Determination of Configuration

NMR Spectroscopy is not able to provide information about the relative configuration of stereogenic centers in a molecule unless one knows the conformation of the molecule. Thus, I will discuss approaches for the determination of configuration together with those for the determination of conformation.

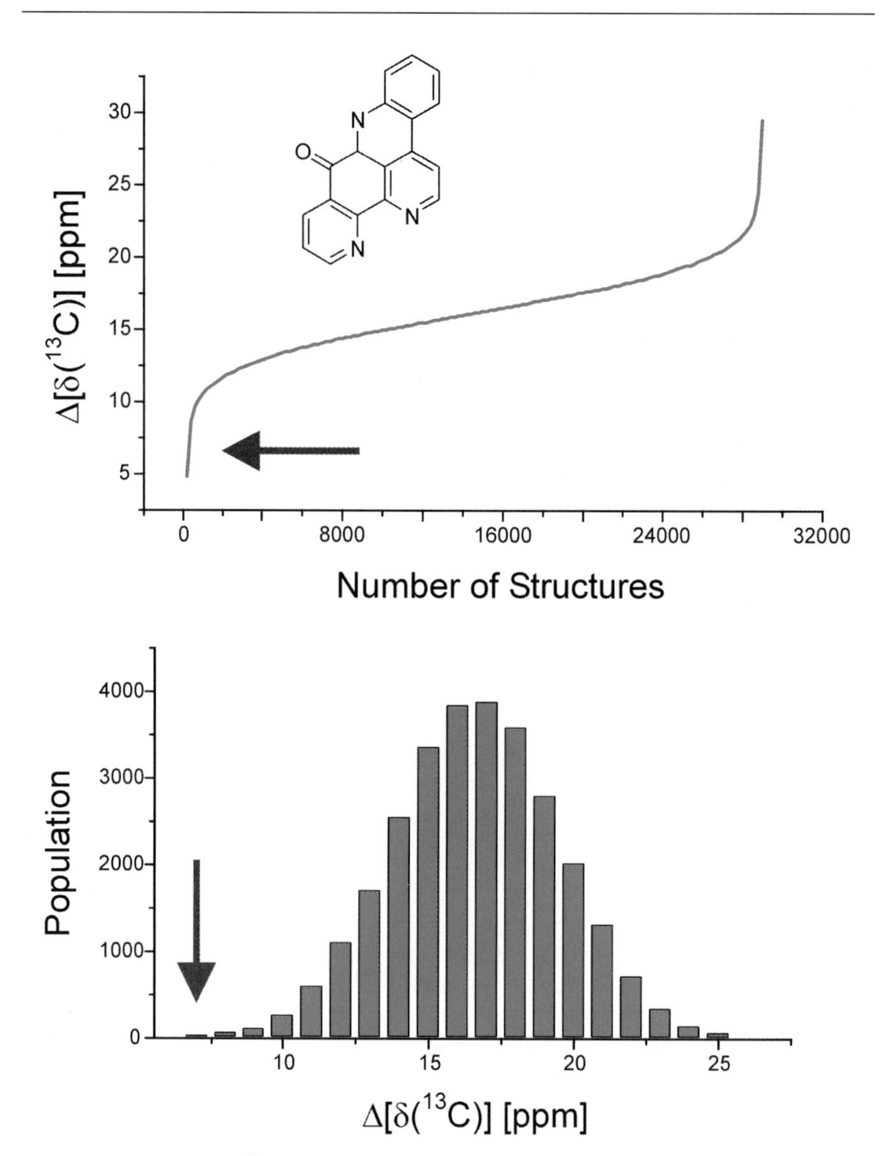

Fig. 22. *Results of the $\delta(^{13}C)$ calculation for all 28672 structure proposals of ascididemin generated by* Cocon *using all possible $^{3}J_{HH}$ and $^{2/3}J_{CH}$ correlations.* The correct constitution is ranked 25th (red arrow). The calculation time for the $\delta(^{13}C)$ was 30 min on a PC.

3.3. Determination of Conformation

The NMR parameters for the determination of conformation have all been introduced in *Sect. 2*. To translate them into conformation of biomolecules, two protocols have survived and are applied in general. One is the distance geometry protocol [54] that translates interatomic distances into conformation. Thus, it is predestined for using NOEs that provide distance information and was indeed implemented when this was the main parameter to use for structure elucidation of proteins. However, also J couplings can be translated into distance restraints between the distal atoms. Other parameters like dipolar couplings or cross-correlated relaxation are more difficult to use directly in the distance geometry protocol. They can be used in molecular-dynamics protocols much easier.

Molecular dynamics (MD) protocols [55] use *Newton*'s equations for all atoms of the molecule and look for energy minima when all forces are zero. The energy potentials describing the bond distances and bond angles are derived from crystal structures and known vibration and rotational spectra [56]. They are truncated to be parabolic. Charge and *van der Waals* energies are calculated from the formulae known in physical chemistry.

$$E_{\text{empirical}}(R) = \sum_{\text{bonds}} K_b(b-b_0)^2 + \sum_{\text{angles}} K_\Theta(\Theta-\Theta_0)^2 + \sum_{\substack{\text{dihedral and}\\ \text{improper angles}}} K_\phi\left[1+\cos(n\Phi-d)\right] + \sum_{\text{pairs}(i,j)}\left(\frac{\text{A}}{r_{ij}^{12}}-\frac{\text{B}}{r_{ij}^{6}}+\frac{q_i q_j}{D\cdot r_{ij}}\right)$$

The charges are approximated for each atom and, as an assumption, do not depend on the conformation. Most of the time a dielectric constant that is distance dependent: $\varepsilon = r/\text{Å}$ is used to simulate the electric shielding of charges that are more distant in space either by the solvent or by the induceable electric dipoles of the molecule. The restraints due to experimental NMR parameters are introduced as parabolic potentials as well. NOEs are frequently translated into distance restraints [57], whereas, for other parameters like J couplings, dipolar couplings, and cross-correlated relaxation rates, a fit against the parameter is performed. For example, using cross correlated relaxation between the four nuclei A^1, A^2, B^1, B^2 as described in *Sect. 2.6* the potential energy onto each of the four atoms derived from this parameter is expressed as [52]:

$$V^{\text{c}}_{\text{A}^1\text{A}^2,\text{B}^1\text{B}^2} = k\left[\frac{\gamma_{\text{A}^1}\gamma_{\text{A}^2}}{(r_{\text{A}^1,\text{A}^2})^3}\frac{\gamma_{\text{B}^1}\gamma_{\text{B}^2}}{(r_{\text{B}^1,\text{B}^2})^3}\left(\frac{\hbar\mu_0}{4\pi}\right)^2 P_2(\cos\theta_{\text{A}^1\text{A}^2,\text{B}^1\text{B}^2})J(0) - \Gamma^{\text{c}}_{\text{A}^1\text{A}^2,\text{B}^1\text{B}^2}\right]^2$$

$$= V^{\text{c}}_{\text{A}^1\text{A}^2,\text{B}^1\text{B}^2} = k\left[\frac{\kappa}{r^2_{\text{A}^1,\text{A}^2}r^2_{\text{B}^1,\text{B}^2}}\left\{3\left[(\vec{\text{A}}^1-\vec{\text{A}}^2)^2(\vec{\text{B}}^1-\vec{\text{B}}^2)\right]^2 - (\vec{\text{A}}^1-\vec{\text{A}}^2)^2(\vec{\text{B}}^1-\vec{\text{B}}^2)^2\right\} - \Gamma^{\text{c}}_{\text{A}^1\text{A}^2,\text{B}^1\text{B}^2}\right]^2$$

with $\kappa = \frac{\gamma_{A^1}\gamma_{A^2}}{(r_{A^1,A^2})^3}\frac{\gamma_{B^1}\gamma_{B^2}}{(r_{B^1,B^2})^3}\left(\frac{\hbar\mu_0}{4\pi}\right)^2 J(0)$

The forces are calculated from:

$$\boldsymbol{F}_{A^1} = -\vec{\nabla}_{A^1}\boldsymbol{V}^{c}_{A^1A^2,B^1B^2},\ \boldsymbol{F}_{A^2} = -\vec{\nabla}_{A^2}\boldsymbol{V}^{c}_{A^1A^2,B^1B^2},$$
$$\boldsymbol{F}_{B^1} = -\vec{\nabla}_{B^2}\boldsymbol{V}^{c}_{A^1A^2,B^1B^2},\ \boldsymbol{F}_{B^2} = -\vec{\nabla}_{B^2}\boldsymbol{V}^{c}_{A^1A^2,B^1B^2}$$

The structure optimizer in most NMR laboratories makes use of the so-called simulated annealing protocol [58]. This is done by heating the molecule to an elevated temperature that is accomplished by defining velocities for the atoms that obey the *Maxwell* distribution for the given temperature. The molecule is then able to cross high barriers between different conformations. Cooling the molecule in successive steps allows the molecule less and less to travel over high energy potentials and forces it to settle into energy minima. The procedure is repeated several times from different non-structured conformations and form the family of structures. Normally, a subset of structures with the lowest energies and the least violations of experimental restraints is selected. From this subset, the root mean standard deviation (rmsd) or the atomic positions is calculated as one quality factor of the family of structures. There are other protocols to check, *e.g.*, the distribution of ϕ and ψ angles that should fall into the allowed regions of the *Ramachandran* space for most amino acids, chirality checks *etc.* [29].

A special notice is given to the chiral and prochiral groups in molecules. Mathematically, chirality can be endowed with a potential by defining the so called chiral volume $\boldsymbol{V}^{c}_{ABCD}$ that is the spade product defined as:

$$\boldsymbol{V}^{c}_{ABCD} = \vec{\boldsymbol{a}}(\vec{\boldsymbol{b}}\times\vec{\boldsymbol{c}}) \quad \text{where} \quad \vec{\boldsymbol{a}} = \vec{\boldsymbol{A}} - \vec{\boldsymbol{D}},\ \vec{\boldsymbol{b}} = \vec{\boldsymbol{B}} - \vec{\boldsymbol{D}} \quad \text{and} \quad \vec{\boldsymbol{c}} = \vec{\boldsymbol{C}} - \vec{\boldsymbol{D}}$$

and the vectors $\vec{\boldsymbol{A}}$, $\vec{\boldsymbol{B}}$, $\vec{\boldsymbol{C}}$, and $\vec{\boldsymbol{D}}$ reflect the coordinates of the atoms A, B, C, and D required to define the chiral center. For trigonal pyramidal coordination, A, B, and C are the ligands, and D is the center atom; for tetrahedral coordination, A, B, C, and D are the ligands. The spade product is a pseudoscalar that changes sign for a pairwise exchange of two atoms out of A, B, and C, while a cyclic permutation of A, B, and C does not change the result. Thus, the chiral volume is a continuous parameter for chirality, and an energy potential can be derived from it. Now, defining for the chiral volume a double well potential leaves the chirality of a stereogenic center open while forcing it to assume either of the two possible chiralities. This can be done for chiral centers whose chirality is not known, as well as for prochiral centers where the two diasterotopic groups produce different resonances but can still not be stereochemically assigned. Both protocols are identical and are called floating chirality [60]. They imply that, for a prochiral center, the two diasterotopic groups can be interchanged together with all their NMR parameters. For chiral centers, interchange of non-identical groups is done to accomplish the same. Again, when the family of conformations is analyzed, it is checked that the prochiral groups are assigned in the same way, and, if the relative chirality of centers is not known, whether this is consistent in all structures constituting the family.

3.4.1. *Determination of Conformation and Configuration of an Organic Molecule*

As an example for the use of floating chirality, we analyze a synthetic compound whose structure is determined by distance geometry [54] using NOEs only and then a so-called distance-driven dynamics that uses the structures fulfilling the NOEs and optimizes them further allowing for floating chirality of the undetermined six chiral centers [61] (*Fig. 23*).

As can be nicely seen in *Fig. 24*, the configuration of two enantiomers is found to be unique in all structure calculations using the distance geometry, floating chirality protocol, whereas a quasi-homogenous distribution of chiralities is found without experimental restraints. Thus, it is possible to determine relative configurations in rather complicated molecules by this approach.

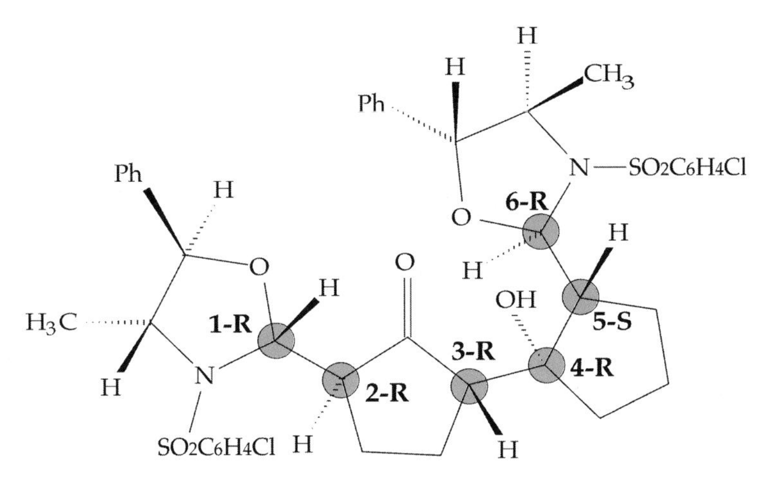

Fig. 23. *Structure of a synthetic compound the configuration of which was unknown.* The six centers were determined by NOEs.

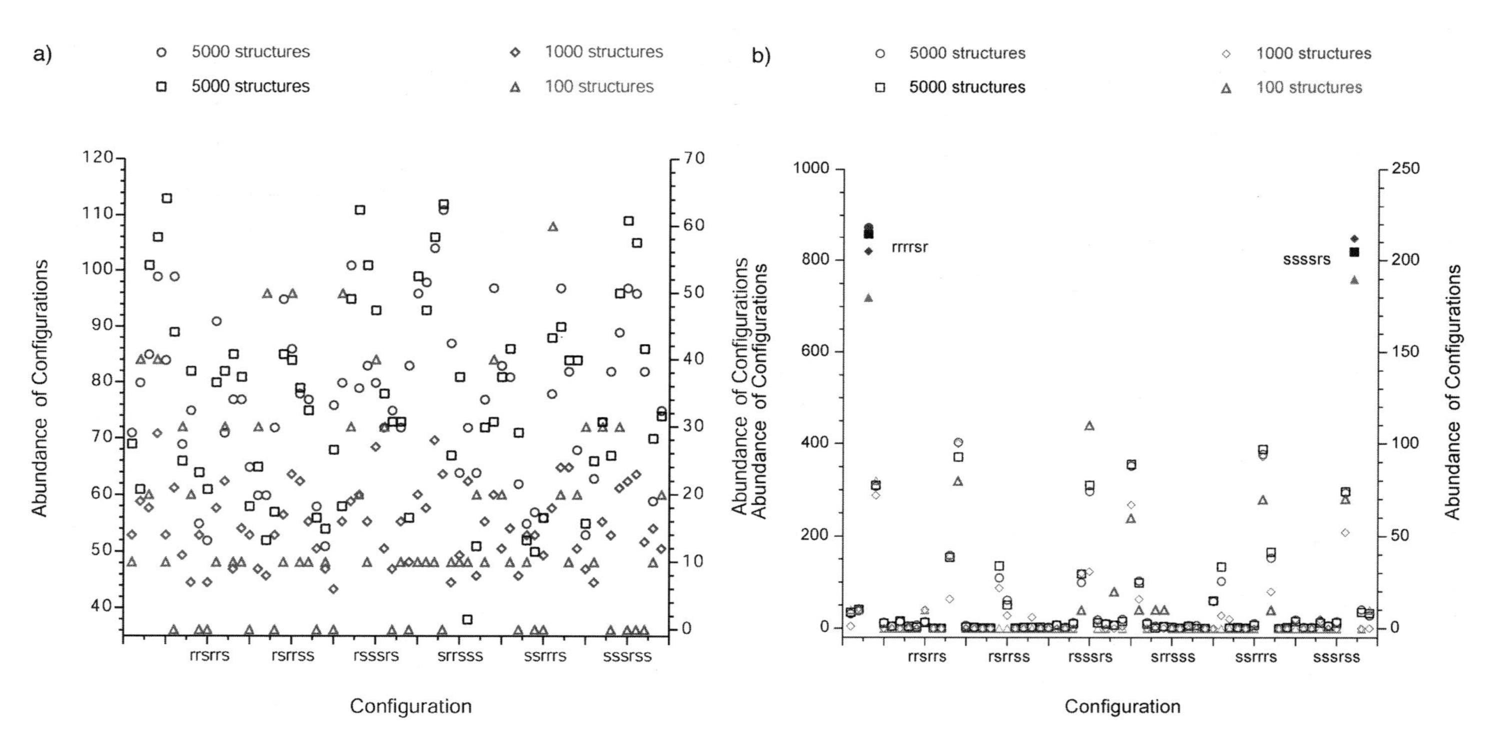

Fig. 24. a) *Abundance of configurations of the six stereogenic centers without using NOEs.* The 64 possible configurations have approximately the same probability showing that there is little bias towards certain configurations. b) *Abundance of configurations based on NOEs as restraints to determine the relative configurations.* As expected from the impossibility to determine absolute configurations from NMR, the two enantiomeric configurations *R,R,R,R,S,R* and *S,S,S,S,R,S* are equally abundant provided the number of structures calculated is high enough.

3.4.2. *Determination of an Organometallic Intermediate Using Cross-Correlated Relaxation*

The next example describes the structure determination of two complexes that was difficult to obtain due to lack of conventional restraints as well as good force fields for the metal coordination. Therefore, all available NMR parameters had to be used to define the structures of the two complexes well. Complex **2** is formed by nucleophilic attack of sodium dimethyl malonate onto the *exo*-diastereoisomer of the allyl complex **1** [62] (*Fig. 25*).

The reaction occurs nearly exclusively from the *re*-side, *trans* to the P-atom, of the allylic moiety giving rise to the uniformly configurated substitution product **2**, which can be regarded as the primary Pd^0 complex. It turned out to be possible to prepare a sample containing *ca.* 75% of the phosphorus in the form of this Pd^0-olefin complex **2**. The assignment of the protons and the interesting C-atoms was possible by conventional 2D-NMR methods for both complexes [62]. However, the structures calculated from the conventional NMR data were not very precise for the interesting region, namely the olefin with respect to the rest of the molecule. This is due to the fact that the distance between the olefin part and the catalytic auxiliary group is rather large, and the NOEs are not sufficient to define the conformation of the olefin (*Fig. 26*).

As additional restraint, cross-correlated relaxation could be measured between the two moieties and for the olefinic bond. Indeed, for complex **1** and **2** the rates in *Table 2* were measured.

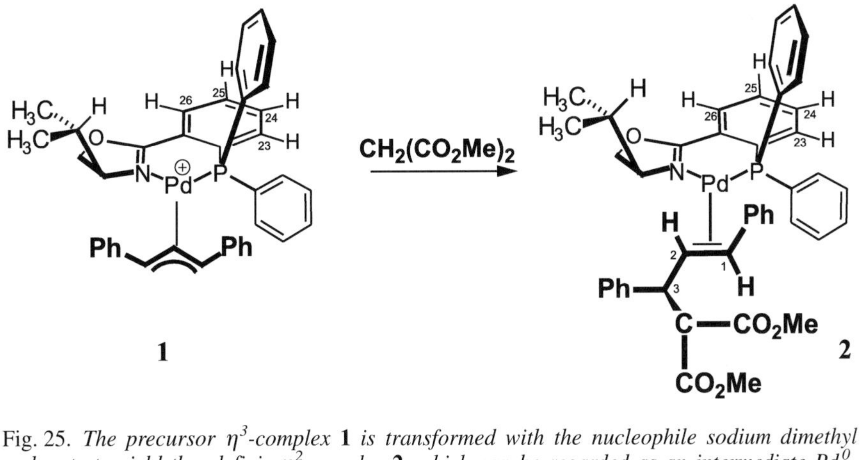

$CH_2(CO_2Me)_2$

Fig. 25. *The precursor η^3-complex* **1** *is transformed with the nucleophile sodium dimethyl malonate to yield the olefinic η^2-complex* **2**, *which can be regarded as an intermediate Pd^0-complex*

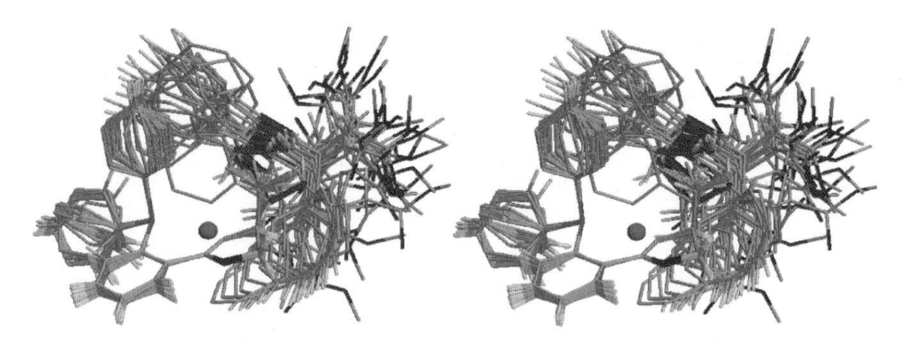

Fig. 26. *Structure of the olefinic η^2-complex* **2** *determined from NOEs and coupling constants.* The precision of the structure, especially concerning the olefinic bond, is low.

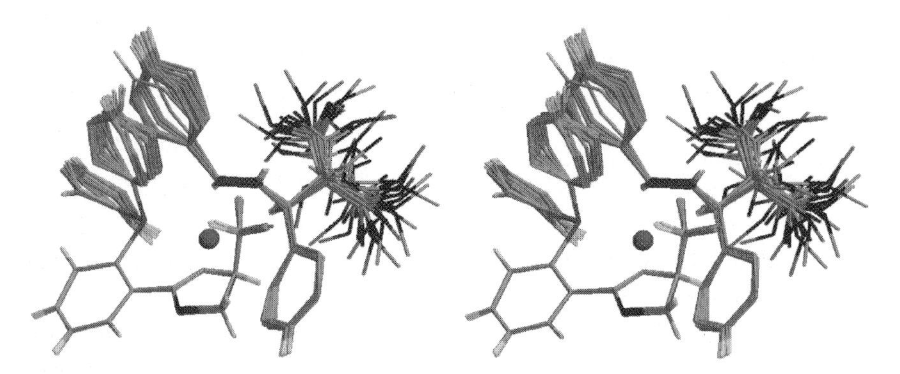

Fig. 27. *Same structure as in* Fig. 26, *however, including the cross-correlated relaxation rates.* The precision of the olefinic bond has increased dramatically.

Table 2. *Nuclear Assemblies for the Double- and Zero-Quantum Coherences Measured in Complex* **1** *and* **2**.

Complex	A^2	A^2	B^1	$B^{2\wedge}$	$\Gamma^c_{A^1A^2,B^1B^2}$ [a]	$\theta^{NMR}_{A^1A^2,B^1B^2}$	$\theta^{X\text{-}Ray}_{A^1A^2,B^1B^2}$
1	H^{26}	H^{25}	C^3	H^3	(+0.33 ± 0.30) Hz	52°, 128°, ± 3° [b]	54° [c]
1	H^{23}	H^{24}	C^3	H^3	(+1.60 ± 0.10) Hz	41°, 139°, ± 1° [b]	140° [c]
2	H^{26}	H^{25}	C^2	H^2		48° / 132°, ± 3° [b]	
2	H^{23}	H^{24}	C^2	H^2		61° / 119°, ± 1° [b]	
2	H^2	C^2	C^1	H^1	3.5 ± 1 Hz	35° ± 3° [b]	

[a]) The determined rates.

[b]) The angles were determined from the measured relaxation rates according to:

$$\Gamma^c_{A^1A^2,B^1B^2} = \frac{\gamma_{A^1}\gamma_{A^2}}{(\boldsymbol{r}_{A^1,A^2})}\frac{\gamma_{B^1}\gamma_{B^2}}{(\boldsymbol{r}_{B^1,B^2})^3}\left(\frac{\hbar\mu_0}{4\pi}\right)^2 \boldsymbol{P}_2(\cos\theta_{A^1A^2,B^1B^2})\left[\boldsymbol{J}(0)+\tfrac{3}{4}\boldsymbol{J}(\omega_H)\right].$$

[c]) The projection angles were taken from the crystal structure of **1**.

The internuclear angles measured for complex **1** are precise to within less than 3° and accurate within 1°, if one assumes that the X-ray structure of the complex is the same as in solution. For complex **2**, the inclusion of the cross correlated relaxation rates improves the precision of the structures quite dramatically [63] as shown in *Fig. 27*.

3.4.3. *Structure Determination of an Oligosaccharide*

The structure determination of oligosaccharides has always been difficult due to the lack of NMR parameters across the glycosidic bonds. There are only few distances to be measured between sugars and few couplings, mainly heteronuclear coupling constants. Therefore, structures of sugars in solution have been much more difficult than those of proteins with a lot of NOEs for each proton. The ability to orient sugars weakly in solution and thus make dipolar couplings available has changed this situation [64]. As an example, the determination of the conformation of the trisaccharide raffinose (*O*-α-D-galactopyranosyl-1-6-α-D-glucopyranosyl-β-1-2-D-fructofuranoside) in H_2O was accomplished in our laboratory (*Fig. 28*) [65].

Dipolar couplings can be measured between ^{13}C and ^{1}H as well as between protons. The results of these measurements are shown in *Fig. 28*, in which each dipolar couplings is identified by a bar connecting the nuclei that entertain this dipolar coupling.

Now, first, from each of the dipolar couplings of the monosaccharides, the amount of orientation or, more quantitatively, the size of the alignment tensor can be back-calculated. If the molecule is completely rigid, all the monosaccharides would have the same alignment tensor. If there is local mobility of one sugar with respect to the other sugars, its tensor should be scaled down. The sizes of the back-calculated tensors for the three monosaccharides in raffinose are given also in *Fig. 28*. As can be seen, the galactose moiety has a tensor size that is scaled to *ca.* 60% of the tensor size of the glucose and saccharose. Indeed, looking at carbon relaxation data, there is a fast motion of the galactose around its C1–O1 bond leading to this tensor-size reduction. A structure calculation with these dipolar couplings as an additional restraint can be undertaken. The result is shown in *Fig. 29*. The structure is much more precise than without dipolar couplings. The rmsd of the heavy atoms is reduced from 1.08 Å to 0.28 Å. Some dipolar couplings that could not be used for the structure refinement were used for cross validation of the structure, *i.e.*, it was checked whether they were reproduced from the structure without forcing the structure to fulfil them. This is summarized in *Table 3*. The raffinose structure in solution turns out to be different from that in the crystal. *Table 4* shows the values of the angles around the two glycosidic linkages for the NMR structure and the X-ray structure.

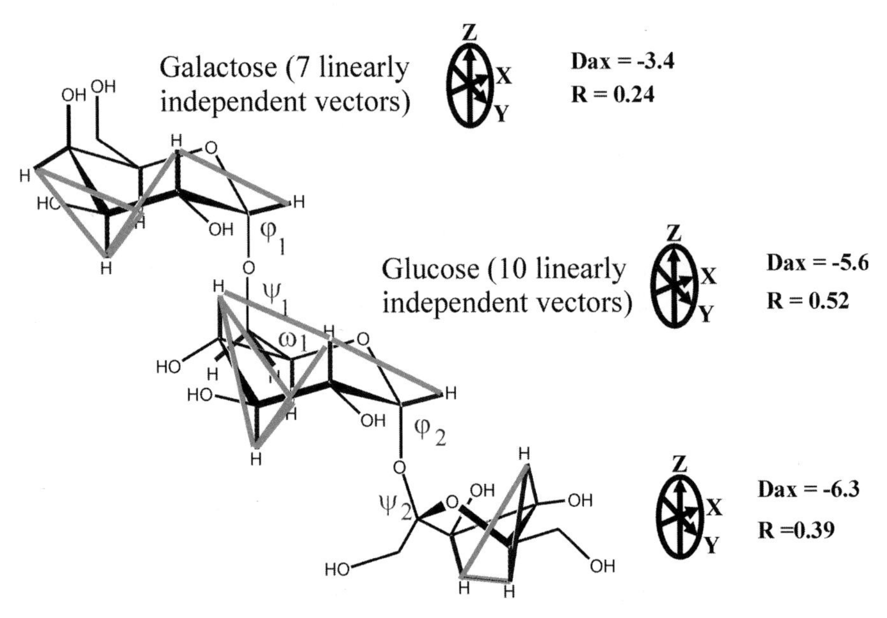

Fig. 28. *Configuration of raffinose together with the glycosidic dihedral angles.* The measured dipolar couplings are shown in blue and in purple. The alignment tensors determined for each of the sugars are shown to the right.

Table 3. *Cross Validation of Dipolar Couplings that Were Not Used in the Structure Calculation but Reproduced from the Derived Structure.* The fact that the couplings are fulfilled indicates that the derived structure is valid.

	Exp. [Hz]	Theory [Hz]
Gal-C6/H6;H6′	−2.34	−2.32 ± 0.17
Fru-C1/H1;H1′	4.55	4.72 ± 0.15
Fru-C6/H6	1.8	2.42 ± 0.09
Fru-C6/H6′	−2.0	−2.27 ± 0.05
Fru-H5/H6′	−1.7	−1.78 ± 0.0
Fru-H5/H6′	−1.24	−1.78 ± 0.02
Fru-H6/H6′	1.7	0.83 ± 0.06
Gal-H1/Glc-H6′	±1.82	−1.62 ± 0.06
Gal-H5/Glc-H4	±1.56	−1.66 ± 0.06
Gal-H1/Glc-H6	±1.45	0.81 ± 0.02

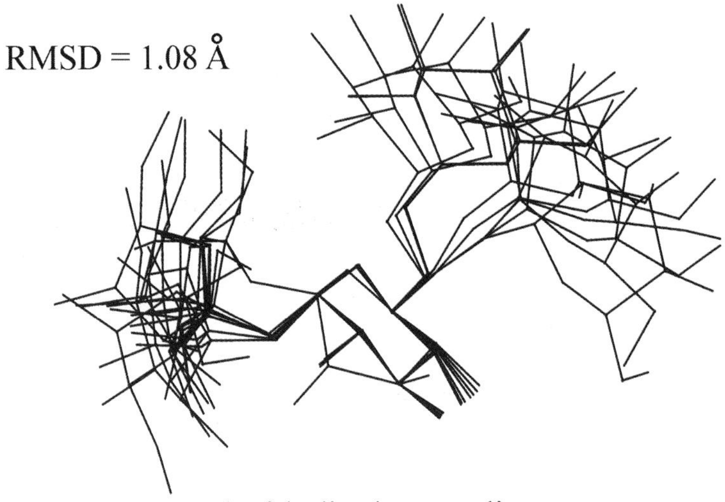

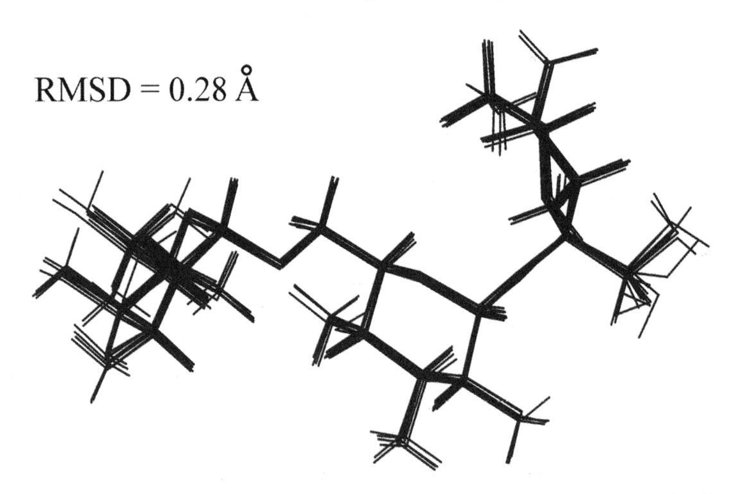

Fig. 29. *Structures of raffinose determined with NOEs alone and using the dipolar couplings as well.* The rmsd of the structures decreases from 1.08 Å for the heavy atoms to 0.28 Å.

Table 4. *Comparison of the NMR-Derived Raffinose Angles across the Glycosidic Linkages with Those Derived from the Crystal Structure.* There are clear deviations for both linkages, the Glc-Fru being more pronounced.

	ω_1	φ_1	ψ_1	φ_2	ψ_2
X-Ray	–64.8°	71.9°	–169.6°	81.6°	–105.5°
NMR-Derived	–62.8 ± 0.5°	57.3 ± 1.5°	–178.3 ± 0.7°	98.4 ± 1.6°	–151.9 ± 1.1°

3.4.4. *Structure Determination of a Protein/Peptide Complex*

Structure determination of repetitive biomolecules has become the main thrust of NMR in solution. Based on assignment experiments discussed in *Sect. 2*, the several hundred to few thousand resonances of a protein can be assigned. This approach will be presented here on the example of a complex between calmodulin (CaM) and C20W, a cognate peptide out of the plasma membrane Ca^{2+}-ATPase, which is responsible for pumping calcium ions out of the cell. The complex is formed from the completely ^{15}N- and ^{13}C-labelled CaM and unlabelled C20W [35].

After assignment of the resonances of the protein, secondary-structure information can be retrieved from chemical shifts quite directly. For H_α, C_α, C_β, and carbonyl resonances, one observes conformation induced changes of the chemical shifts that can be referenced to chemical shifts measured from the same amino acid in a peptide. As a rule, these structure-induced chemical shifts are low field for H_α and C_β, and high field for C_α and carbonyl in a β-strand. For α-helices, the structure-induced chemical shift is opposite [66][67]. Thus, the secondary-structure elucidation is possible based on the assignment of the resonances and the determination of the chemical shifts. The result from this analysis for the CaM/C20W complex is shown in *Fig. 30*

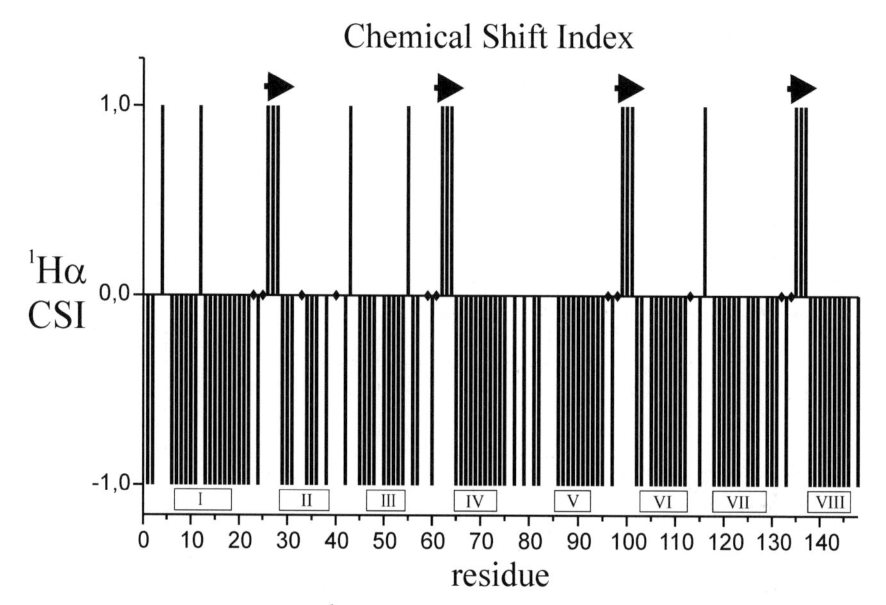

Fig. 30. *Chemical-shift index of $^1H_\alpha$ resonances of CaM complexed to the peptide C20W.* Low-field shift are indicative for α-helices and high field shifts for β-strand. The chemical-shift index is correlating with the secondary structure found in the NMR-derived structure (arrows for β-strands, boxes for α-helices).

for the structure-induced shifts of the H_α resonances. A clear correlation with the secondary-structure elements is found.

However, tertiary-structure information cannot be derived from this information yet. NOESY Spectra need to be recorded and analyzed in order to obtain a sufficient amount of proton pairs that are close in space to build a structure. Assignment of NOESY spectra is still a time-consuming job, since there is a lot of overlap even in 3D NOESY spectra. Tools for a handy interpretation of the spectra exist that assist in assigning the NOESY cross-peaks and in bookkeeping of them. As an example, a sequence of strips through the 3D NOESY of the CaM/C20W spectrum is shown in *Fig. 31*.

There are additional tools available, for example, the program Aria [68], to use NOESY cross-peaks with still ambiguous assignments that become assigned only during the structure calculation.

It is found that the number of NOEs per residue varies frequently along the sequence (*Fig. 32*). Normally, one finds that the more NOEs are observed the better-structured a region in the spectrum is.

The summary of the NOE restraints differentiating the long-range and the sequential restraints is shown in the contact diagram in *Fig. 33*. Helices show up as 'thick' diagonals, antiparallel β-sheets as antidiagonals.

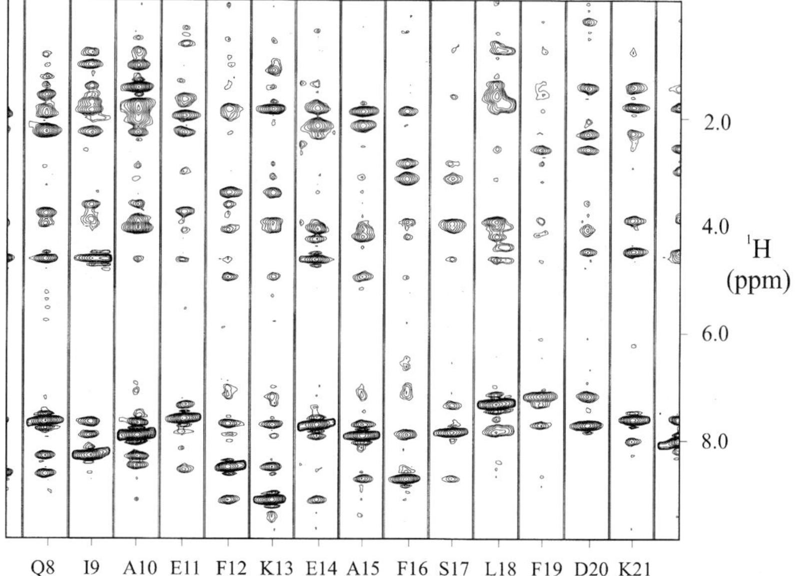

Fig. 31. *Strip plot of the three-dimensional NOESY-^{15}N,^{1}H-HSQC spectra for residues E7 to S17 of the CaM/C20W complex.* Dashed lines indicate sequential NOEs between residues.

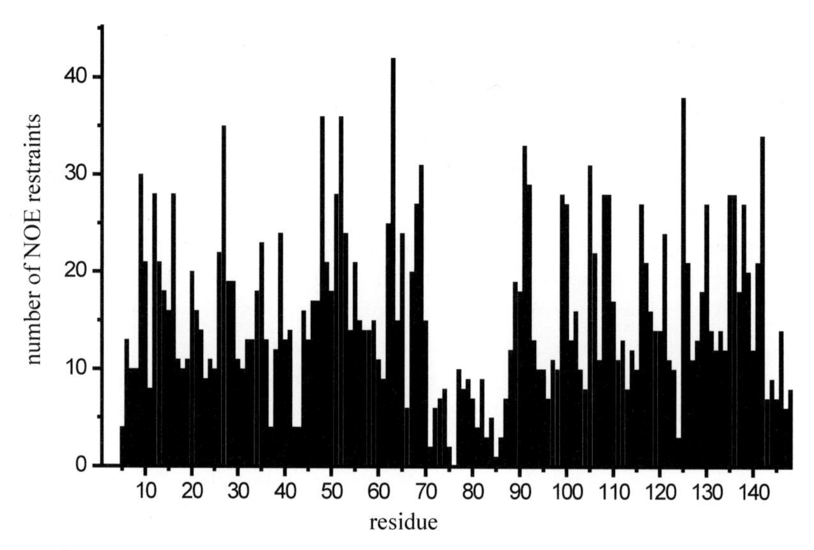

Fig. 32. *Numbers of NOEs per residue of CaM complexed to the peptide C20W*

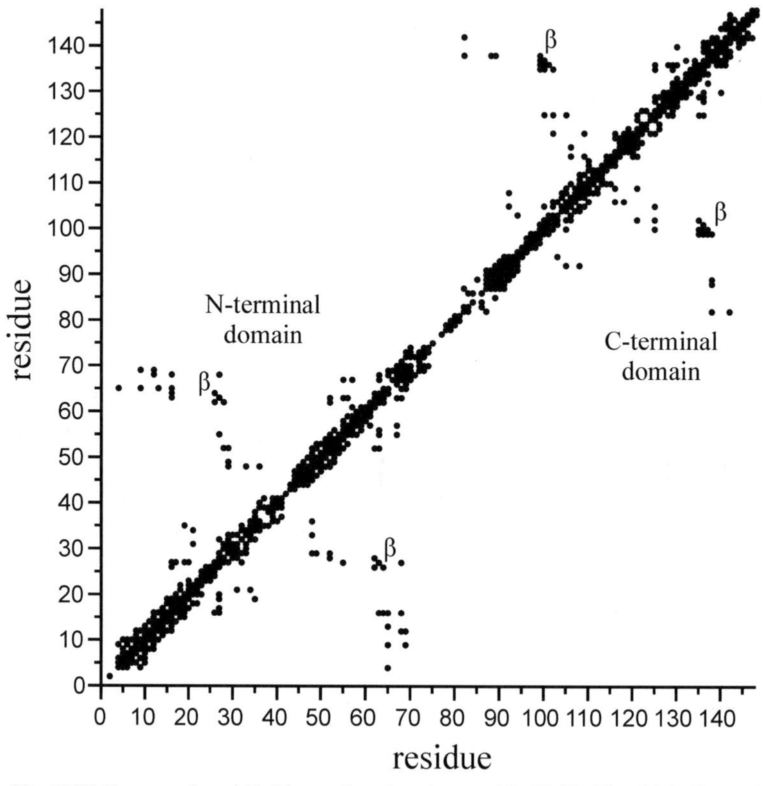

Fig. 33. *NOE-Contact plot of CaM complexed to the peptide C20W.* The thick diagonals indicate α-helices, the antidiagonal β-sheets.

Further restraints can be obtained by measurement of coupling constants of the protein. The relevant dihedral angles are shown in *Fig. 34*. As an example traces from an HNHA experiment [69] [70] are shown that restrict the ϕ angle according to the *Karplus* curve as shown before.

As a further example for the usage of coupling constants, ${}^3J_{NC\gamma}$ couplings have been measured to define the angle χ_1. Cross-peaks in the respective experiment occur if the C_γ is *trans* to the amide N-atom. *Fig. 35* represents the

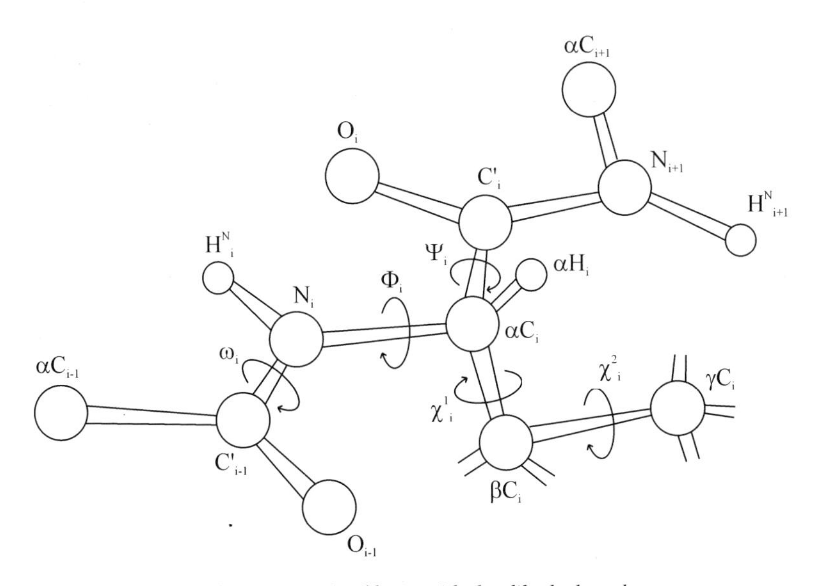

Fig. 34. *Protein backbone with the dihedral angles*

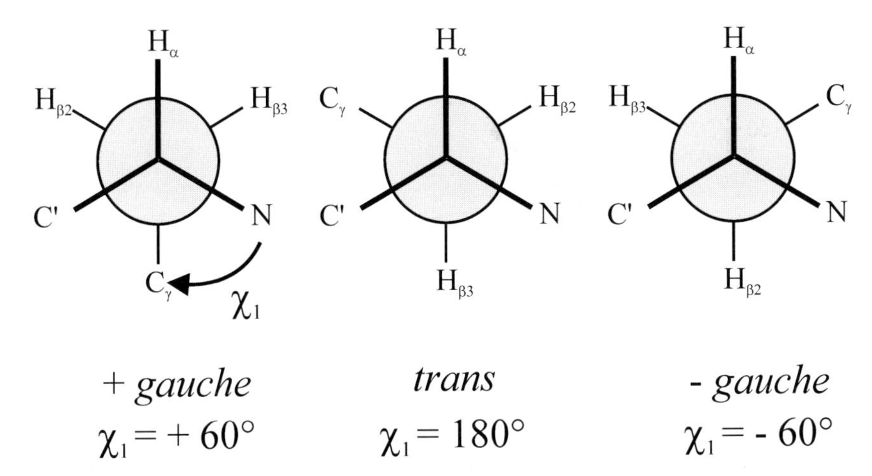

Fig. 35. *Three staggered conformations of an amino acid side chain around the C_α–C_β bond*

staggered conformations about the C_α–C_β bond. Thus, simple inspection of the spectrum in *Fig. 36* indicates those amino acids that are in the *trans*-conformation.

To determine the conformation of our complex CaM/C20W, the C20W peptide needs to be assigned as well. The techniques that were used for the protein cannot be applied, since the peptide is not labelled with ^{15}N and ^{13}C. Thus, the traditional assignment techniques based on COSY, NOESY, and TOCSY are applied for the assignment [71]. The COSY and TOCSY spectra reveal the peak patterns of the amino acids, whereas the NOESY provides the sequential assignment of the peptide. The pulse sequences need to filter all

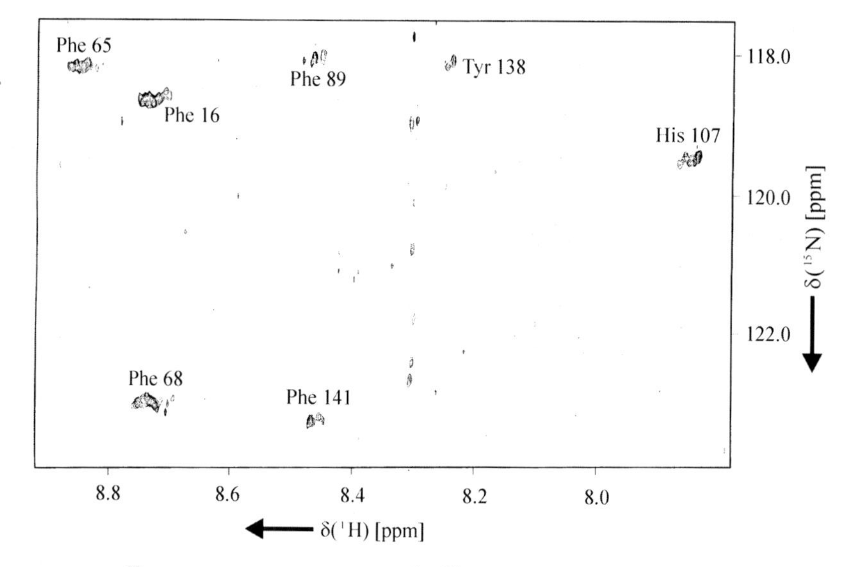

Fig. 36. *2D-{$^{13}C_\gamma^{arom}$}-Spin-echo-difference $^1H,^{15}N$-CT-HSQC spectra.* Only residues with large $^3J_{NC\gamma}$ couplings give signals in the spectrum indicating an *antiperiplanar* conformation about this bond.

Table 5. $^3J_{NC\gamma}^{arom}$ *Coupling Constants that Allow to Define the χ_1 Dihedral Angle*. The couplings indicate a *trans*-arrangement.

Amino acid residue	$^3J_{NC\gamma}^{arom}$ [Hz]
Phe 16	2.7
Phe 65	2.3
Phe 68	2.4
Phe 89	2.2
His 107	2.3
Tyr 138	2.4
Phe 141	2.5

the signals stemming from the protein to obtain signals exclusively from the peptide [72]. As an illustration of the efficiency of these filter techniques, the 1D spectrum with suppression of all resonances belonging to protons that are bound to ^{15}N or ^{13}C (blue) is shown in comparison with the conventional 1D spectrum (red) in *Fig. 37*. The sharp lines stem from a fraction of peptide that is not bound to CaM. In contrast, the resonances of the peptide that is bound to the protein are rather broad, since they adopt the correlation time of the protein. The complex has a dissociation constant in the nM range, and, therefore, the exchange between the bound and free conformation of the peptide is slow.

As an example of a NOESY spectrum of the peptide, applying the same filtering technique, the amide proton/proton region is shown in the spectrum. It shows strong sequential amide/amide cross-peaks that are indicative of a helical conformation of the peptide. The number of NOEs per amino acid of the peptide is reflected in *Fig. 38* showing up to 20 NOEs per amino acid.

The structure calculation of the whole complex is performed according to the protocols described in *Sect. 3.3*. *Fig. 39* shows the protocol used for the structure calculations in which the course of the energy potentials as well as the temperature are depicted.

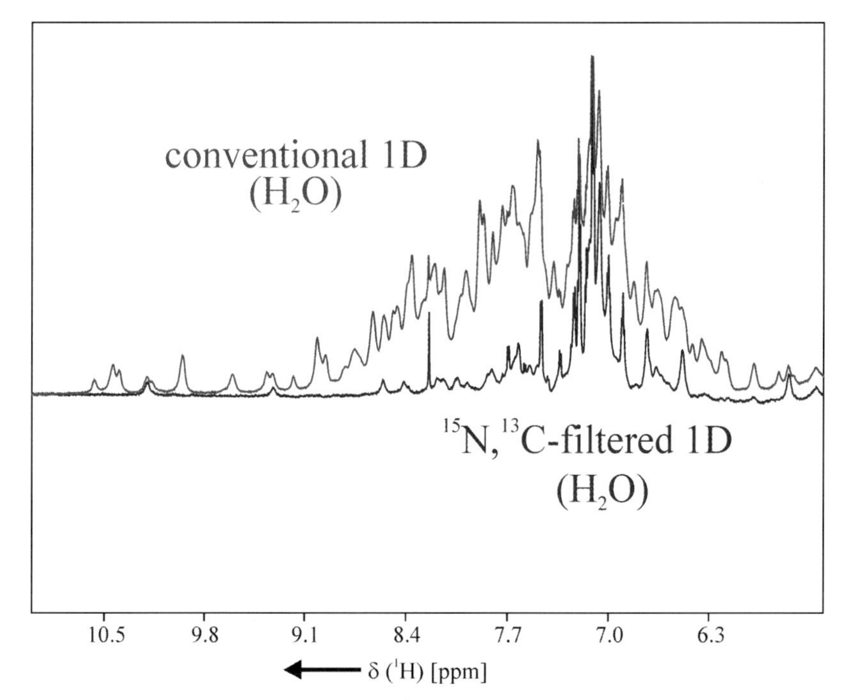

Fig. 37. *Comparison of a conventional 1D-^1H spectra of the CaM/C20W complex* (red) *with ^{15}N,^{13}C-filtered 1D-^1H spectra showing only resonances of the unlabeled peptide C20W*

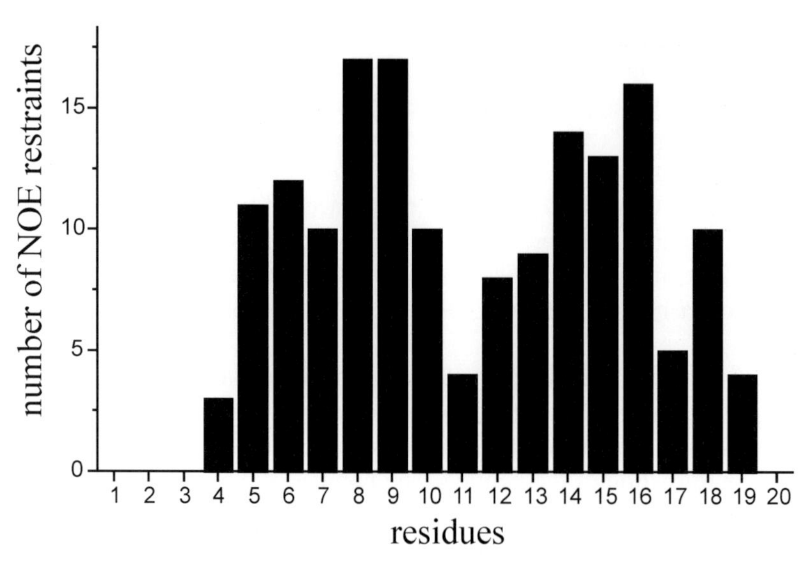

Fig. 38. *Number of NOEs per residue of the peptide C20W in complex with CaM*

The resulting structure family is shown in *Fig. 40* for the C-terminal domain together with the peptide. It is customary to represent averaged structures calculated from the family of structures. Such a structure is shown in *Fig. 41*.

What does this structure tell about molecular recognition: The first thing is that the peptide in solution is random coil with a propensity to form an α-helix. Thus, it adopts the helical structure only when it is in contact with the CaM. The structural changes of CaM, free and when in complex with C20W, can be discussed by comparison with a high resolution X-ray structure of the free CaM [73]. It is mainly in the side chains that structural changes occur. Out of 128 amino acids in well-structured regions of the protein, *ca.* 30% change their local conformation when the peptide binds.

The peptide is recognized mainly by hydrophobic interactions between Phe, Leu, Ala, Val, Ile, and Met residues. The binding mode is such that only half of the helical residues of C20W are involved in the binding, whereas the other half does not entertain contacts. This is in strong contrast to the structure, *e.g.*, of CaM/M13. M13 is the cognate peptide from the myosin light chain kinase [74]. There, the protein wraps around the helix of the peptide making an uninterrupted row of contacts to both sides of the helical wheel formed by the peptide.

CaM recognizes a series of peptides with different sequences, which is called fine tuning of the protein. Such a fine tuning can be analyzed by comparison of the CaM/C20W structure with CaM alone and other CaM/peptide complexes. The most dramatic local conformational changes are observed for

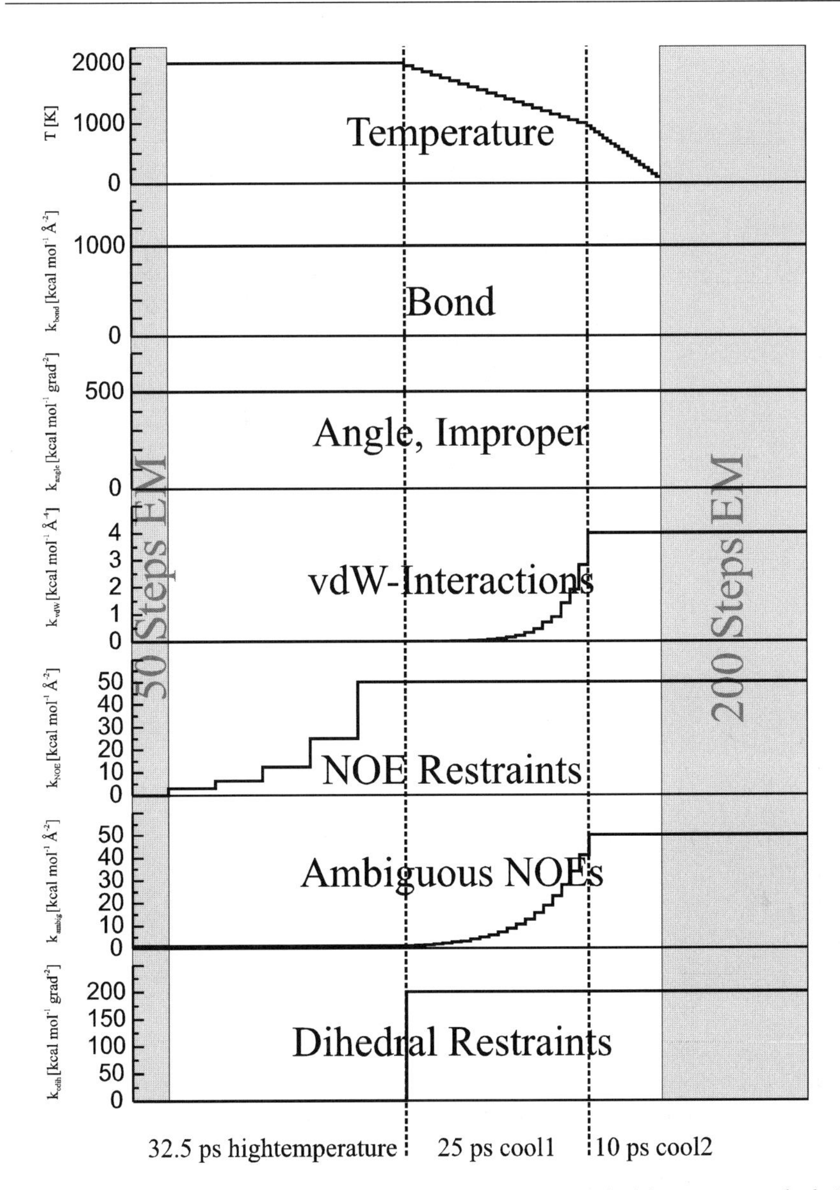

Fig. 39. *Different parameters of the simulated annealing protocol of the structure calculation of CaM/C20W*

the conformation of the Trp 8 of the peptide as compared to the same Trp of the structure of CaM/M13. As can be seen in *Fig. 42*, the tryptophan aromatic rings are rotated by almost 90° between the two conformations. This ensues changes of the conformations of the hydrophobic residues about the tryptophan.

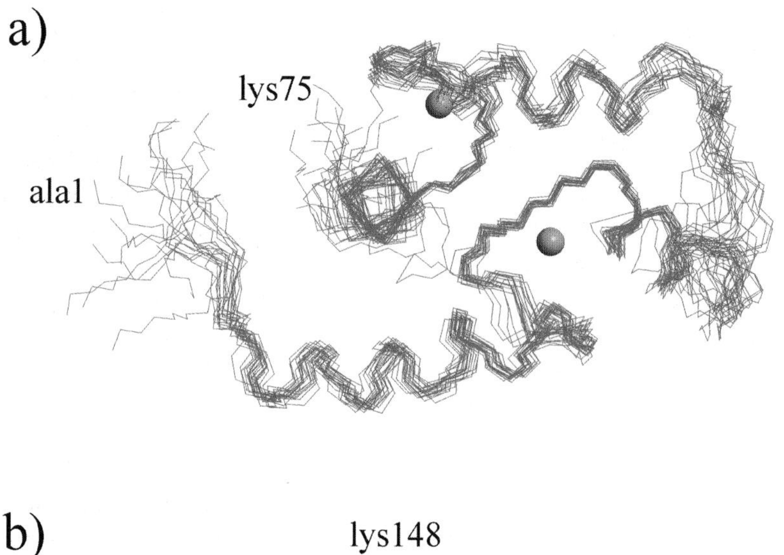

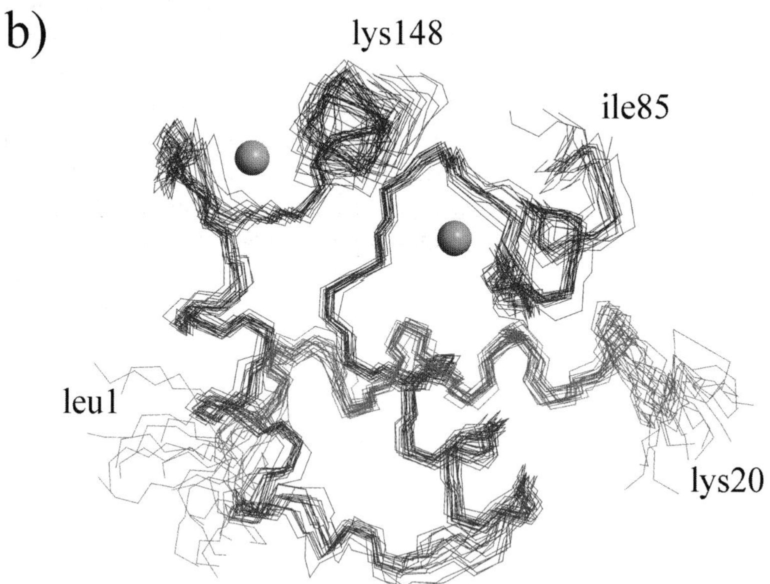

Fig. 40. *Best-fit superposition of the backbone atoms of the final 26 NMR-derived structures representing the CaM/C20W complex. a*) N-Terminal domain of CaM (orange); *b*) C-Terminal domain of CaM (red) and the bound peptide C20W (blue). Ca^{2+} Ions are depicted as blue balls.

The global conformation of the CaM/C20W complex as compared to CaM alone shows a compaction of the structure that is put into evidence by SAXS data (small angle X-ray scattering, *Fig. 43*) [75]. CaM alone shows two maxima and a radius of gyration of more than 60 Å, whereas the structure of CaM/C20W shows a less pronounced double-maximum curve, and the

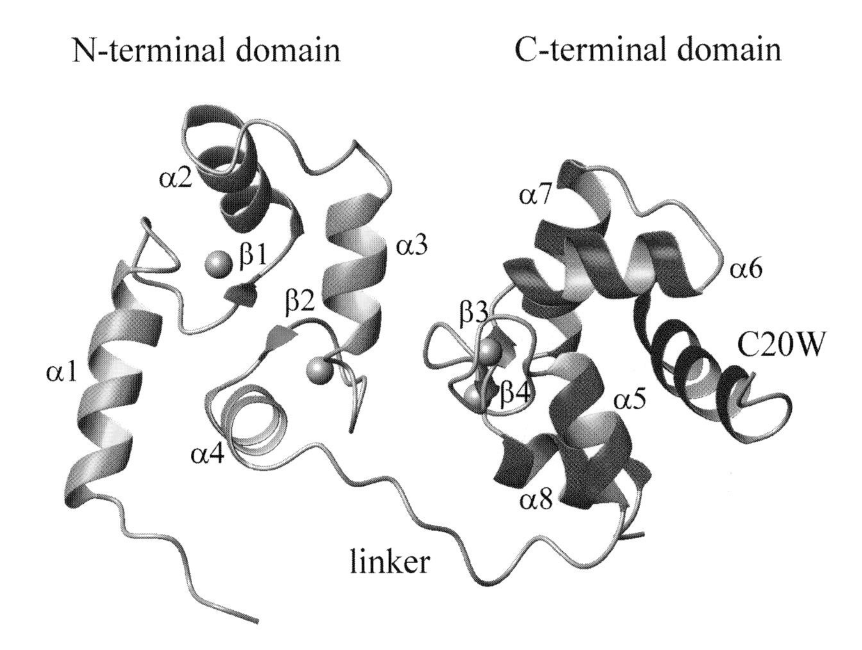

Fig. 41. *Ribbon representation of the CaM/C20W complex.* The eight α-helices are depicted as ribbons, the four β-sheets as arrows. The N-terminal domain of CaM (*orange*) is connected over a flexible linker to the C-terminal domain of CaM (*red*) which is complexed to the helical peptid C20W (*blue*).

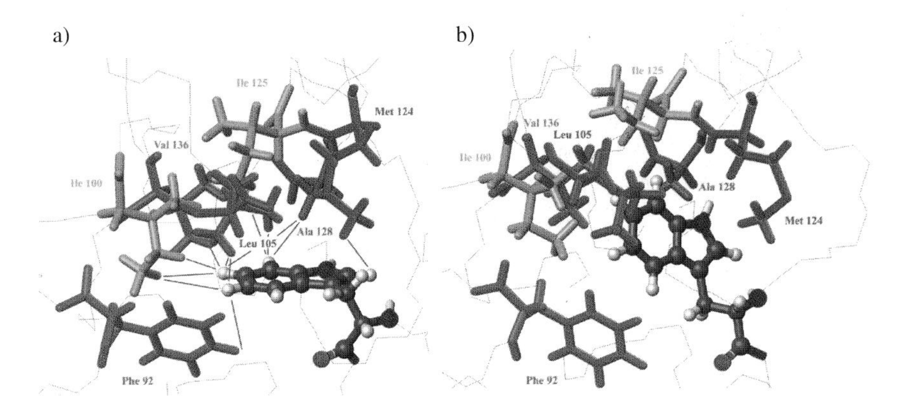

Fig. 42. *Comparison of tryptophan environments in the CaM/C20W* (a) *and CaM/M13* (b) *complexes.* The plane of the indole ring of Trp 8 in CaM/C20W is rotated by almost 90° with respect to the one in CaM/M13. Experimental NOEs observed between Trp 8 and the residues of CaM are shown as green lines for CaM/C20W.

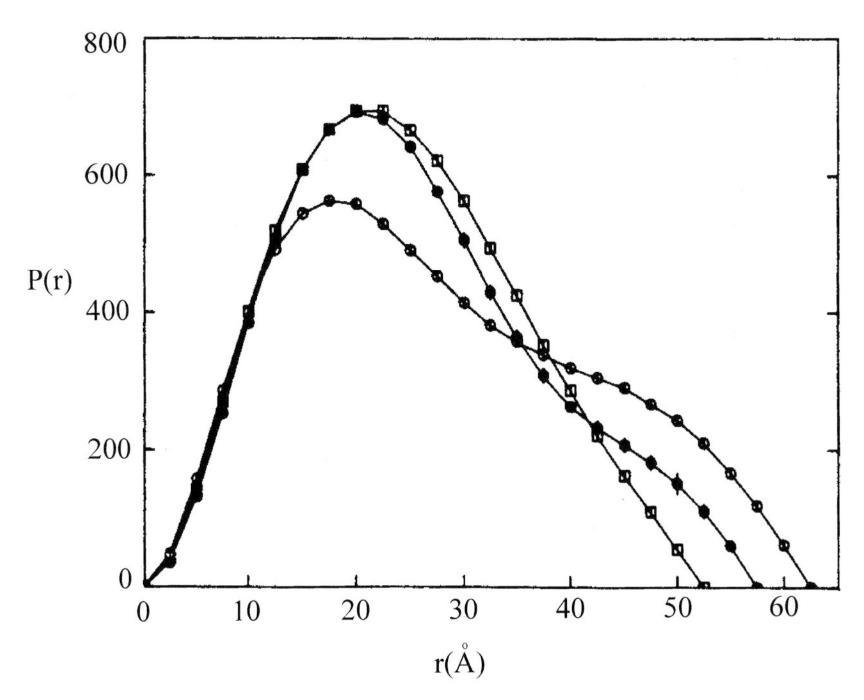

Fig. 43. *Interatomic-length-distribution function P*(r) *of free CaM* (open circles), *of CaM/C20W complex* (black circles) *and of CaM/C24W complex* (open squares) *derived from SAXS experiments*

radius of gyration is well under 60 Å. The NOE-derived structure of CaM/C20W does not reproduce this result. Comparison of its radius of gyration with that of CaM alone shows no alteration. However, looking at relaxation data, there is a marked decrease of the local mobility of the linker when C20W binds CaM (*Fig. 44*). Both the heteronuclear $^{1}H,^{15}N$-NOE and the ^{15}N-T_1 time show less change in the complex than in free CaM [76].

Coming back to the CaM/C20W complex whose structure was determined quite accurately and with big effort, the question which residues of CaM are involved in the binding can be answered with much less effort. When the peptide binds to CaM, the chemical shifts of those residues that are involved in the binding change more dramatically than those that are distant from the binding epitope. This can be seen when comparing the chemical shifts of the CaM free [77], in the complex with C20W, and in the complex with M13 that binds both domains of CaM (*Fig. 45*).

For CaM/C20W *vs.* CaM, only the residues of the N-terminal domain are appreciably affected, whereas, for CaM/M13 *vs.* CaM, the $^{1}H^{N}$-chemical shifts of both domains change quite severely. Thus, the global binding mode of the ligand to the protein is obvious as soon as one has the assignment of

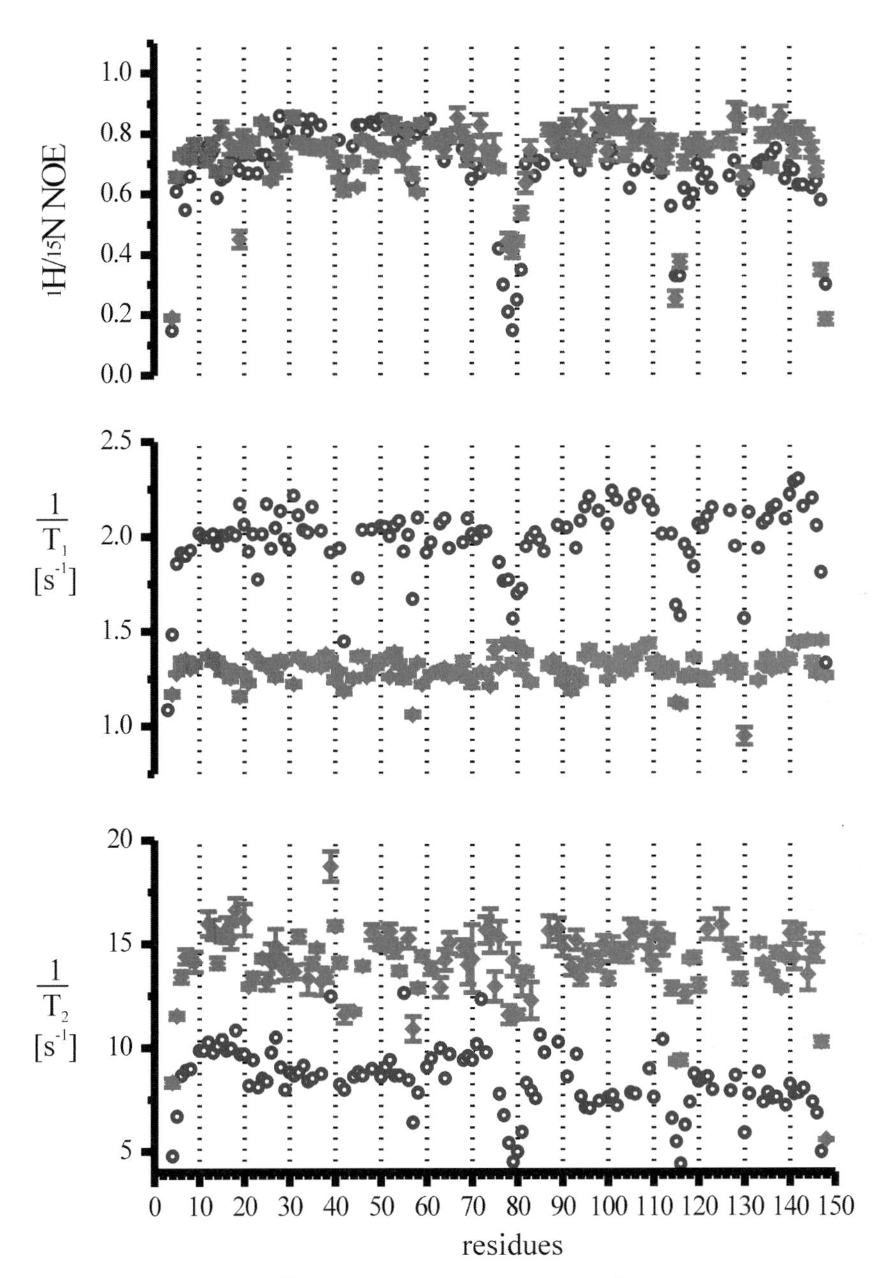

Fig. 44. *Heteronuclear* $\{^1H\}^{15}$*N-NOE,* T_1 *and* T_2 *data of free* $(Ca^{2+})_4$*-CaM* (blue circles) *and of CaM/C20W complex* (red rhombuses)

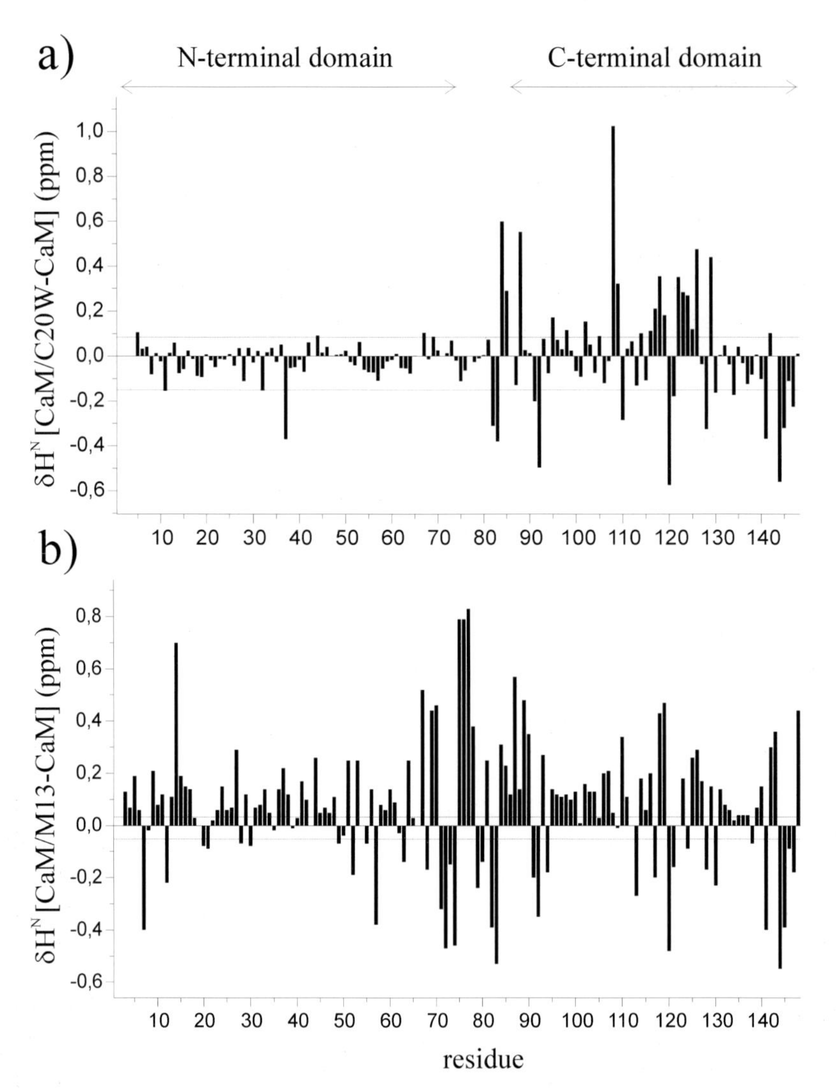

Fig. 45. *Chemical-shift differences of the amide protons of* a) *CaM/C20W and* b) *CaM/M13 complex with respect to free* $(Ca^{2+})_4$*-CaM.* For the CaM/C20W complex, major changes occur only in the C-terminal domain of CaM, indicating that the peptide C20W binds only to this domain. This is in contrast to the CaM/M13 complex where both domains are involved in binding the peptide M13, and hence chemical-shift changes can be observed over the complete amino acid sequence of CaM.

resonances, thus at an early stage of the structure-elucidation process. This simple analysis of chemical shifts should be borne in mind for the next section of tracing and searching for molecular recognition between molecules by NMR.

4. Molecular Interactions and Their Importance for Drug-Screening

Interactions between molecules can be studied by NMR spectroscopy in a large variety of binding constants. Although NMR was used for a long time just as a tool to determine the structure, the unique ability of NMR to detect molecular interactions have recently established its importance in the field of drug screening and optimization [78]. For example, the complex between CaM and C20W was a tight complex with a binding constant in the nM range. However, also much weaker binding in the mM range can easily be detected by NMR.

4.1. Tight Binding

Tight binding of molecules to other molecules requires structural analysis very much along the lines as described for the CaM/C20W complex. In this case, the dissociation constant of the complex is so small that the rate of dissociation is normally found to be in the order of at least seconds or even longer. Thus, the lifetime of NMR observable (magnetizations, coherences) is shorter than the lifetime of the molecule in the complex bound form. In the context of drug/target interactions, the drug can be a small, light-weight molecule (*e.g.*, below 1 kD) and the target a large, heavy molecule (*e.g.*, above 10 kD). However, the target would still determine the spectroscopic properties of the drug. The fact that the drug interacts with the target can be detected easily due to chemical-shift changes in the target spectrum when the drug is added. This can be done with concentrations of the protein/drug complex in the 100 μM range, since only very sensitive experiments need to be recorded. If the drug does not bind to the target, there will be only minor shift changes; if the drug binds one expects major shift changes. The use of $^{15}N,^{1}H$ correlation spectroscopy for the detection of binding has been introduced by the laboratory of *Steven W. Fesik* under the name of SAR by NMR [79]. The chemical-shift differences of CaM and CaM/C20W can serve as an example. Another example would be the binding of nicotinic acid and phenylimidazole to FKBP, of which the first one does not interact with FKBP, while the second does (*Fig. 46*).

With an assignment of the protein, the surface residues of FKBP that interacts with the drug can be mapped qualitatively according to the assumption that their resonances are affected by the binding of the drug. Coloring of these residues leads to *Fig. 47*.

However, this very fast and simple kind of measurements does not yet allow knowing the conformation of the drug when bound to the protein. This information still requires a full-blown structure analysis as described for CaM/C20W.

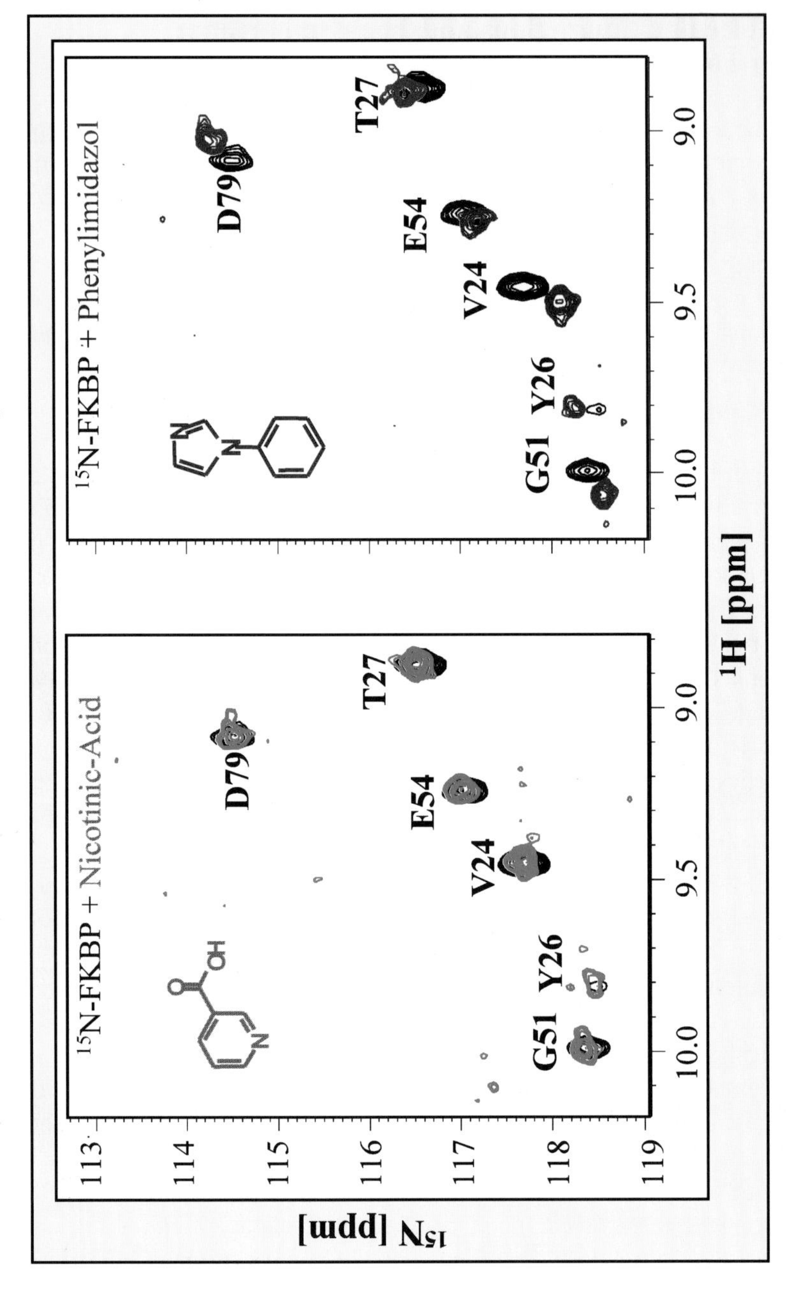

Fig. 46. *Screening of small organic compounds with ^{15}N-labelled FKBP.* Phenylimidazole induces a shift of the N,H resonances of FKBP upon binding, whereas nicotinic acid does not.

Mapping the FKBP Binding Site

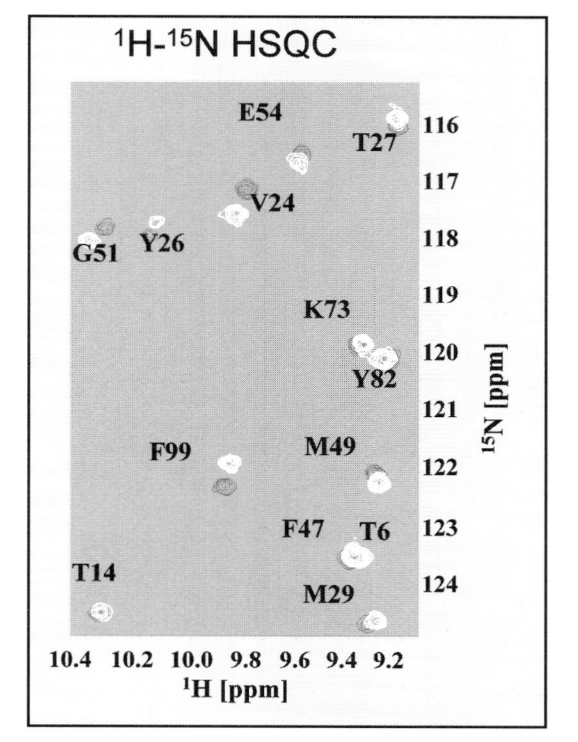

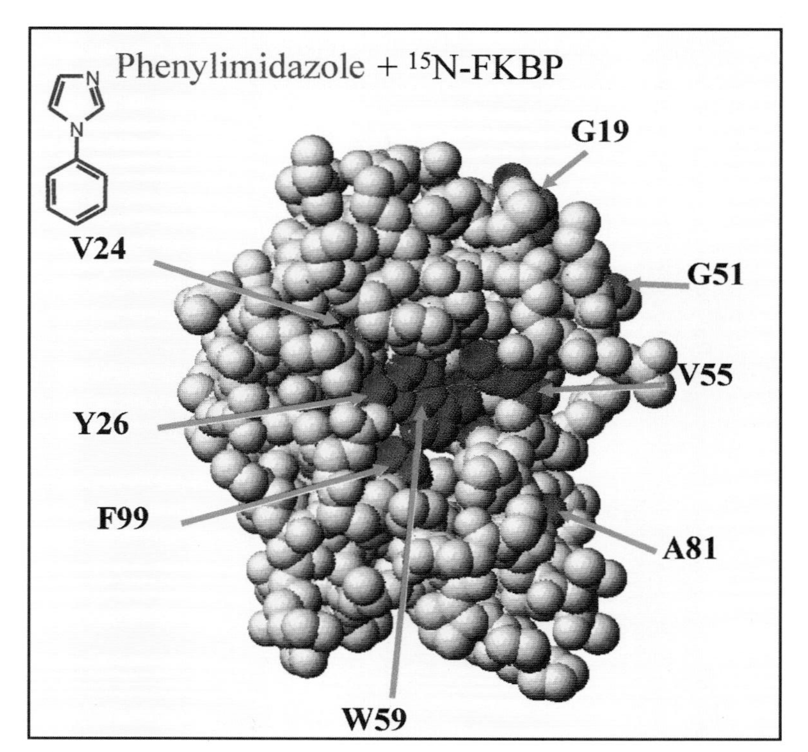

Fig. 47. *Mapping of the binding site of phenylimidazole based on the chemical shifts observed in* Fig. 46. The amino acids that show shifts are colored in red.

If, in the screening process, two ligands are found that bind to neighboring or overlappling surface areas of the protein, and if the structures of the ligand complexes can be determined, the two ligands can be connected by a short linker to increase the binding affinity dramatically as has been shown on FKBP (*Fig. 48*) [79].

4.2. Weak Binding

If the binding constant of the drug to the target is of the order of μM to mM techniques can be used for the detection of the binding as well as for the determination of the structure that rely on the fact that the drug in its target-bound form as compared to the free form has largely different translational as well as rotational diffusion times (correlation time), which will affect the relaxation times of the drug. The different relaxation of the bound *vs.* the free form is reflected in broader line widths and cross-correlated relaxation rates as well as an increase in the cross relaxation rates (NOE). The fast exchange between bound and free form leads to an averaging of the properties of the bound and the free form. For weak binding, the experiments are performed in such a way that the drug is present in excess (normally mM concentration) compared to the target protein (normally μM concentration). Observing the drug resonances then yields a strong signal due to the rather high concentration of the drug. Frequently, the bound form is present to less than 10% in solution. However, still the bound form dominates the relaxation properties provided the molecular mass of the drug is much less than that of the target protein. A quick calculation provides insight: if the target weighs 100 times more than the drug, and taking into account that the mentioned relaxation mechanisms are proportional to the correlation time, which is in turn proportional to the molecular mass relaxation, and if we assume 10% abundance of the bound form of the drug, the relaxation contribution of the bound form will be $10\% \cdot 101 = 10.1$, which is eleven times larger than the contribution of the free form: $90\% \cdot 1 = 0.9$. Thus, the bound form can be detected, although its population is only minor in solution. This fact can be used for a variety of techniques of which we will discuss transfer NOEs [80], transfer cross-correlated relaxation [81] [82], and saturation transfer NMR (STD) [83].

4.2.1. *Transfer NOE*

Transfer NOE measures the NOESY of a drug in the presence and absence of the target molecule. The cross-relaxation rates are averages due to the conformation of the free and the bound form. Comparison of the NOEs

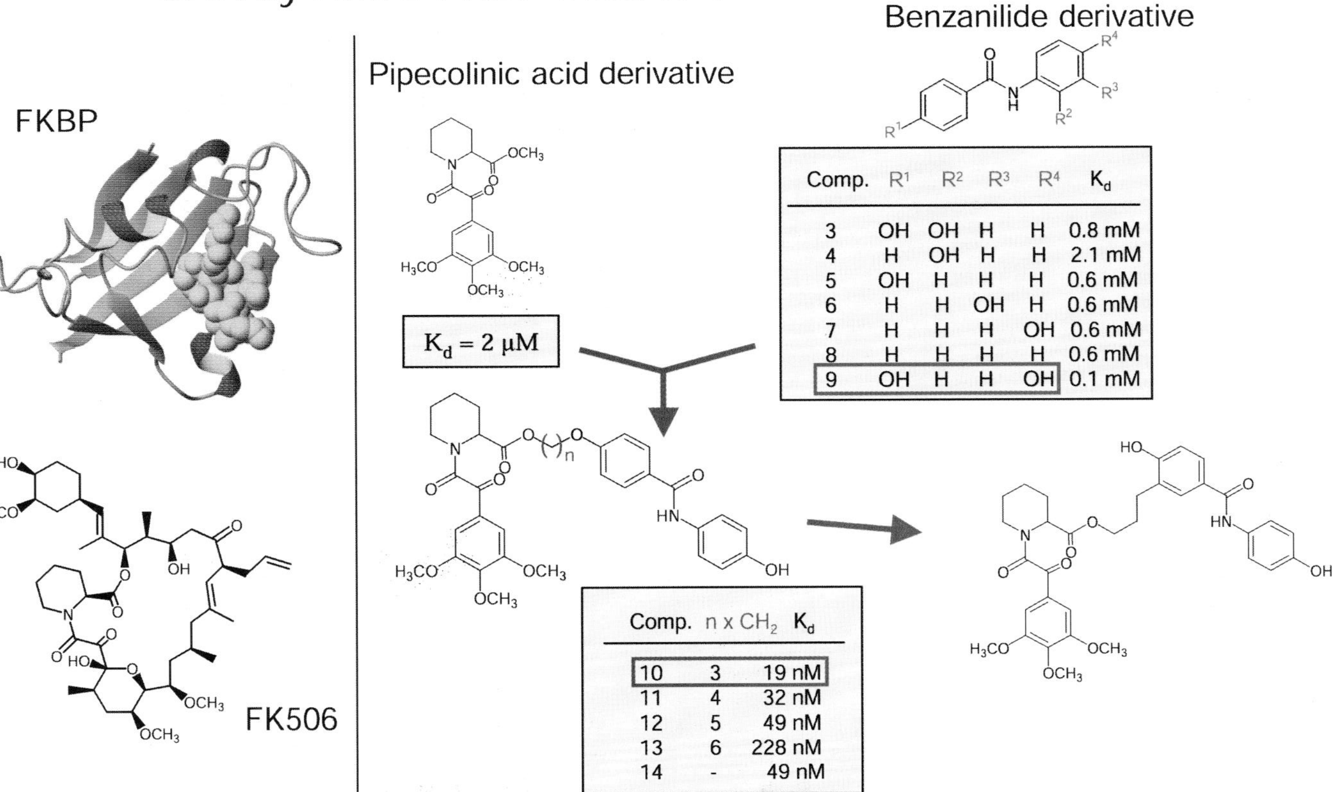

Fig. 48. *SAR by NMR as described by* Fesik *and co-workers* [79]. By recording HSQC spectra, two ligands binding at different sites of FKBP are identified. After structure determination of the two ligand/protein complexes, a compound composed from the two ligands with an optimized linker is made that binds in the nM range.

in the two cases allows one to 'isolate' the information on the bound form and then perform a normal structure calculation. Potential artefacts can stem from the fact that target protein protons may interfere with the flow of magnetization in the drug and thus introduce systematic errors. Also, the drug may bind not only to the site of interest but also, less specifically, to other sites on the target molecule. All this cannot be distinguished in the transfer NOE spectra.

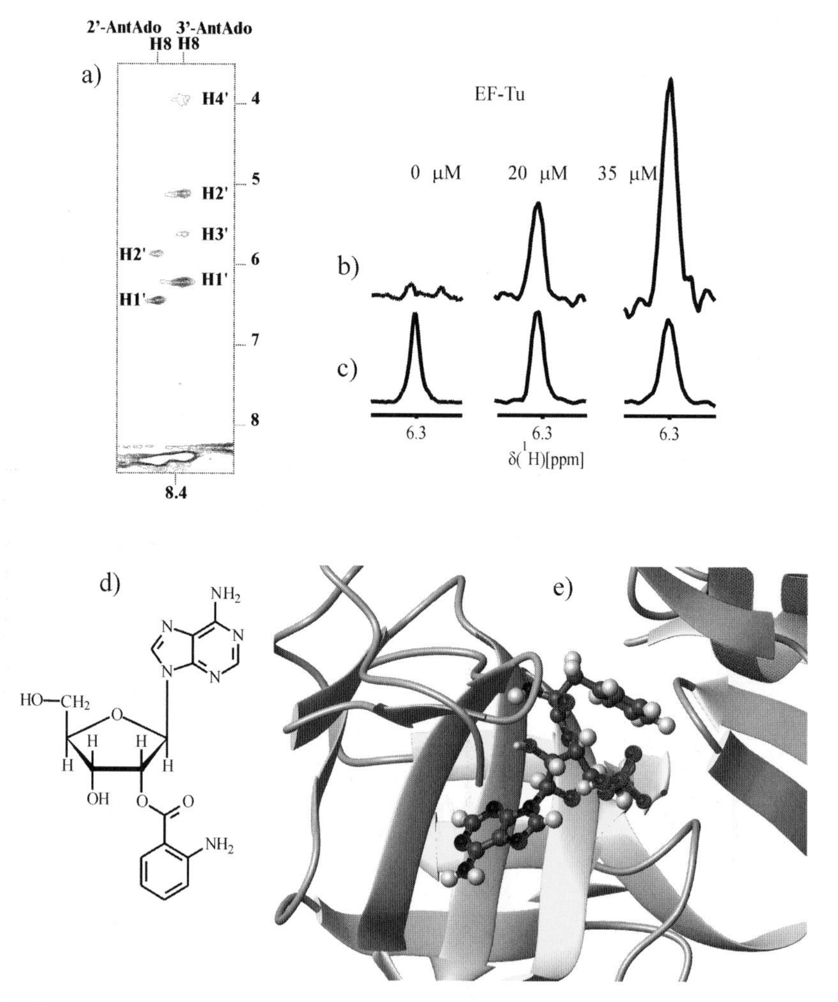

Fig. 49. *Determination of the structure of Ant-Ado* (d) *in the complex with EF-Tu. a*) The transfer NOESY experiment is shown with the cross-peaks between H8 of adenine and the sugar protons. *b*) shows the titration of the cross-peaks originating from cross-correlated relaxation between C3,H3 and C4,H4 vectors with increasing concentration of the protein. *c*) The reference peak. *e*) The conformation of the molecule in the protein.

As an example, the transfer NOE spectrum of Ant-Ado (anthranilic acid adenine) a mimetic of an amino-acid-loaded tRNA, bound to the protein Elongation Factor Thermo-unstable (EF-Tu) is shown in *Fig. 49,a*. The NOEs can be used to determine the conformation of the bound form.

4.2.2. *Transfer Cross-Correlated Relaxation Rates*

Not only the NOE but also cross correlated relaxation rates can be used for the structure elucidation of weakly bound small molecules to target proteins. This is due to the fact that the cross-correlated relaxation scales linearly with the correlation time and, therefore, is much bigger for the bound form than for the free form of a small drug molecule. As an example, the same Ant-Ado will be used in the complex with EF-Tu. The sugar conformation can thus be determined by the comparison of the cross-correlated relaxation rates between the two CH,CH pairs: C1-H1,C2-H2 and C3-H3,C4-H4 [81]. The first reports on the torsion angle ν_1 in the sugar, the second on ν_3. The two cross correlated relaxation rates are very different for the two canonical conformations of ribose, C2′*endo* and C3′*endo*, thus allowing the differentiation by measuring the indicated cross-correlated relaxation rates (*Fig. 50*).

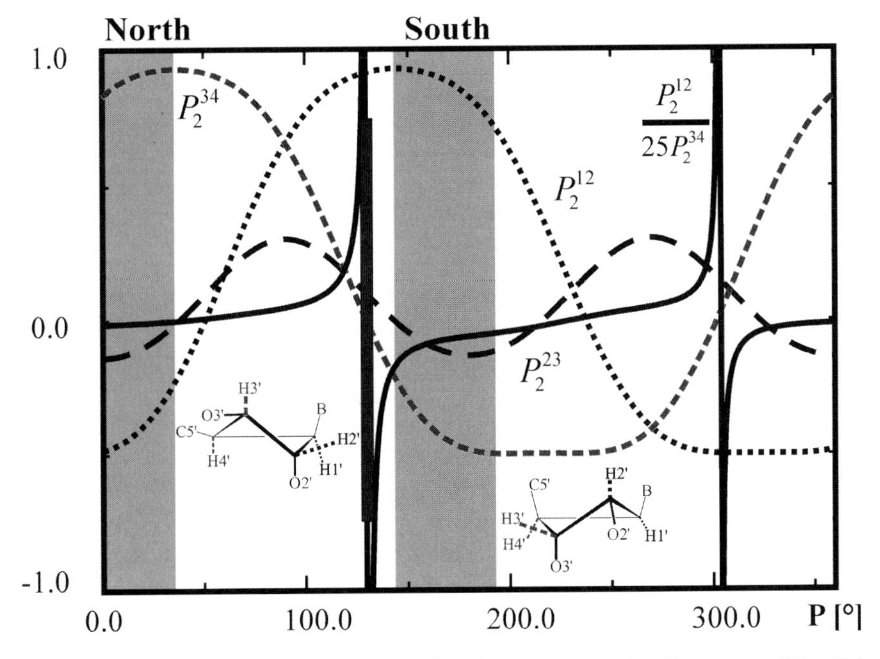

Fig. 50. *Dependence of the cross-correlated relaxation rate measured in the spectra of* Fig. 49,b *and* c, *on the sugar pucker*

Measuring those rates with and without the EF-Tu clearly indicates that the effect is only visible when the Ant-Ado binds to the EF-Tu (*Fig. 49,b* and *c*). The result of this measurement is that the sugar is in the C2' conformation when bound to EF-Tu (*Fig. 49,e*).

Another example of this technique involves a peptide weakly binding to STAT6 (Signal Transducer and Activator of Transcription). In this case, measurement of the backbone angle ψ was performed in the presence and absence of the protein, thus allowing to define the conformation of the bound form better (*Fig. 51*) [82].

4.2.3. *Saturation **T**ransfer **D**ifference Measurements (STD)*

For fast exchange between free and bound form of a ligand, one can also observe NOEs between the target protein and the drug molecule (*Fig. 52*) [83]. This comes from the fact that, during the contact time, magnetization can be transferred from the protein to the drug molecule. For screening purposes, the experiment is performed with non-selective saturation of a region in the protein spectrum that is void of any drug molecule resonances. Intermolecular interactions can then be observed in the spectrum of the drug provided it binds to the target molecule. This is a quite efficient screening tool of ligand mixtures, since the binding ligands can be identified from the STD spectrum, provided their spectrum is known. As an example, the binding of a mixture of small molecules to FKBP is shown in *Fig. 53*.

5. Future Directions

There are three challenges for structure determination of biomolecules in solution: Increasing the speed of structure elucidation, increasing the molecular weight of the molecules or complexes to be investigated, and learning more about dynamics of the molecules. I will describe directions along these three lines.

Structure elucidation requires still the very time consuming analysis of distance information from NOESY spectra. Therefore, it would be highly desirable to obtain information from different sources. This is possible by the measurement of dipolar couplings in partially oriented samples. The use of orientational restraints can enhance the speed of structure determination quite dramatically. In *Fig. 54*, the amount of time for the different steps in structure elucidation are summarized.

It is shown that the backbone assignment is rather quick and requires *ca.* 10% of the total amount of time needed to determine the structure. At least

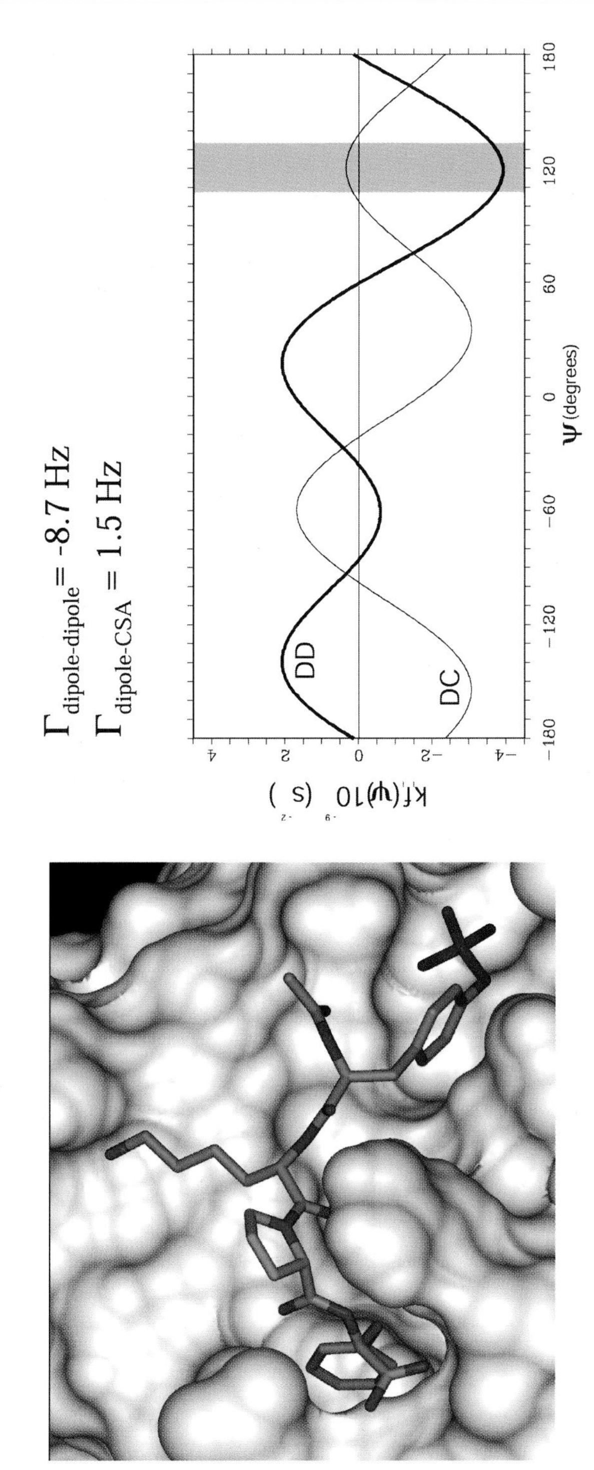

Fig. 51. *Determination of the conformation of a peptide binding to Stat6.* The backbone angle ψ has been determined from the cross-correlated relaxation rates between C_α,H_α and NH (DD) of the following amino acid, as well as the cross-correlated relaxation between the C_α,H_α vector and the carbonyl CSA (DC).

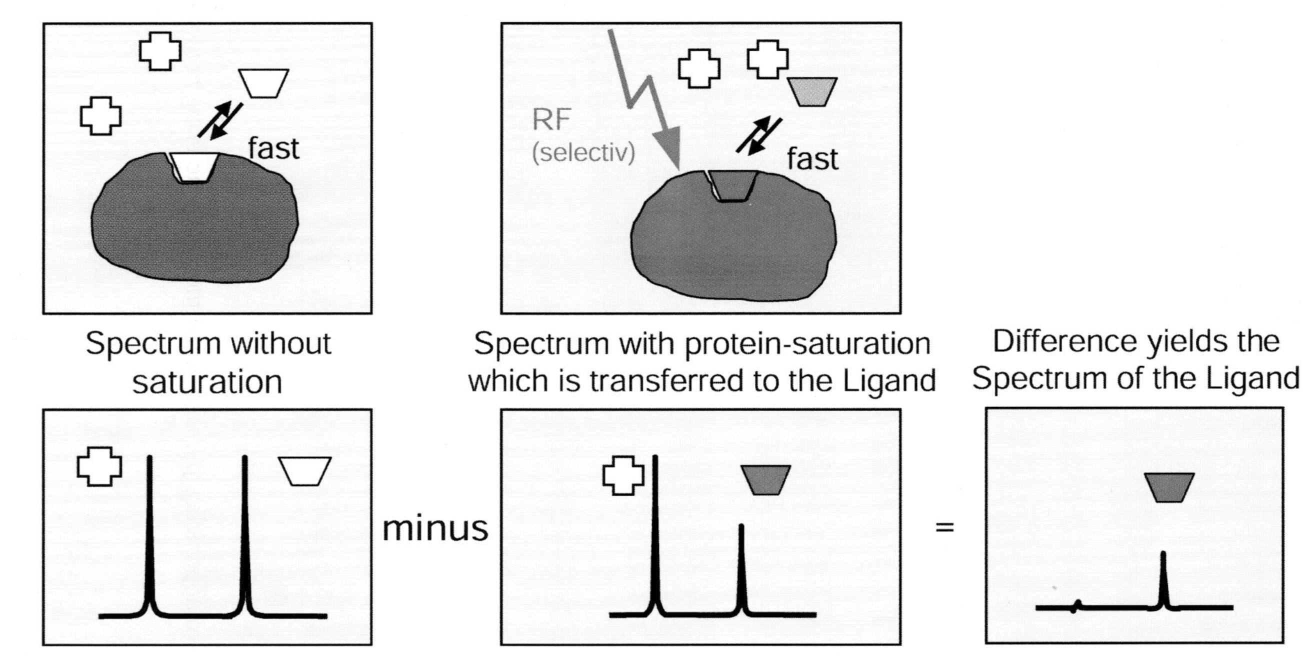

Fig. 52. *Saturation-Transfer-Difference Spectroscopy.* By saturation of the protein resonances, those ligands that bind weakly to the protein are saturated as well *via* saturation transfer. Thus, ligands that bind the protein will show up in the saturation difference experiment, while those that do not are invisible in this experiment.

FKBP:
Saturation Transfer and Line-Broadening

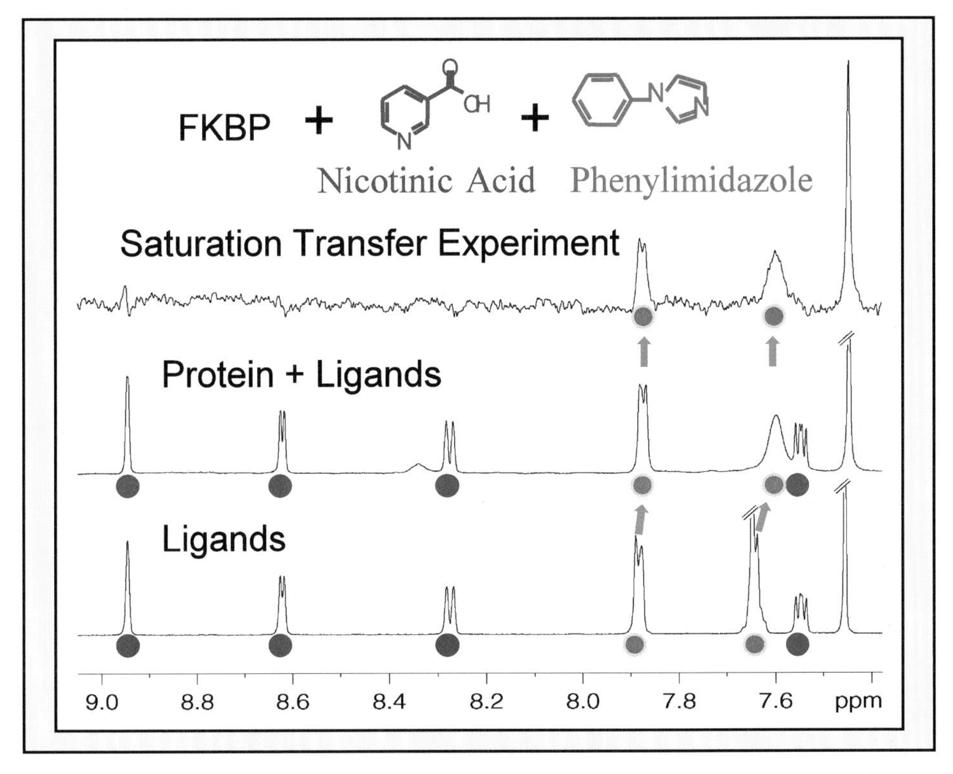

Fig. 53. *Saturation-transfer-difference spectra of the same compounds as used in the* $^{1}H,^{15}N$-*HSQC spectra of* Fig. 46. Only the resonances of phenylimidazole show up, whereas those of nicotinic acid are invisible in the STD spectrum. The spectrum of the ligands and with the protein are given as reference.

the backbone structure can be obtained by using dipolar couplings [84]. There are two approaches pursued at the moment, one using homology data from known X-ray or NMR structures of the proteins [85] [86], the other by using different orientations and measuring a host of dipolar coupling data for the determination of the structure from scratch [87].

We demonstrate the latter approach on a small protein domain, which is homologous to proteins that are present in the protein data bank [86]. As can be nicely seen in *Fig. 55*, rhodniin, which is homologous to other *Kazal*-type inhibitors, is found in the protein data bank. Here, the sequence homology is quite high. However, also structures with almost no sequence homologies are detected, as is the case for dihydrofolate reductase and D-xylose isomerase.

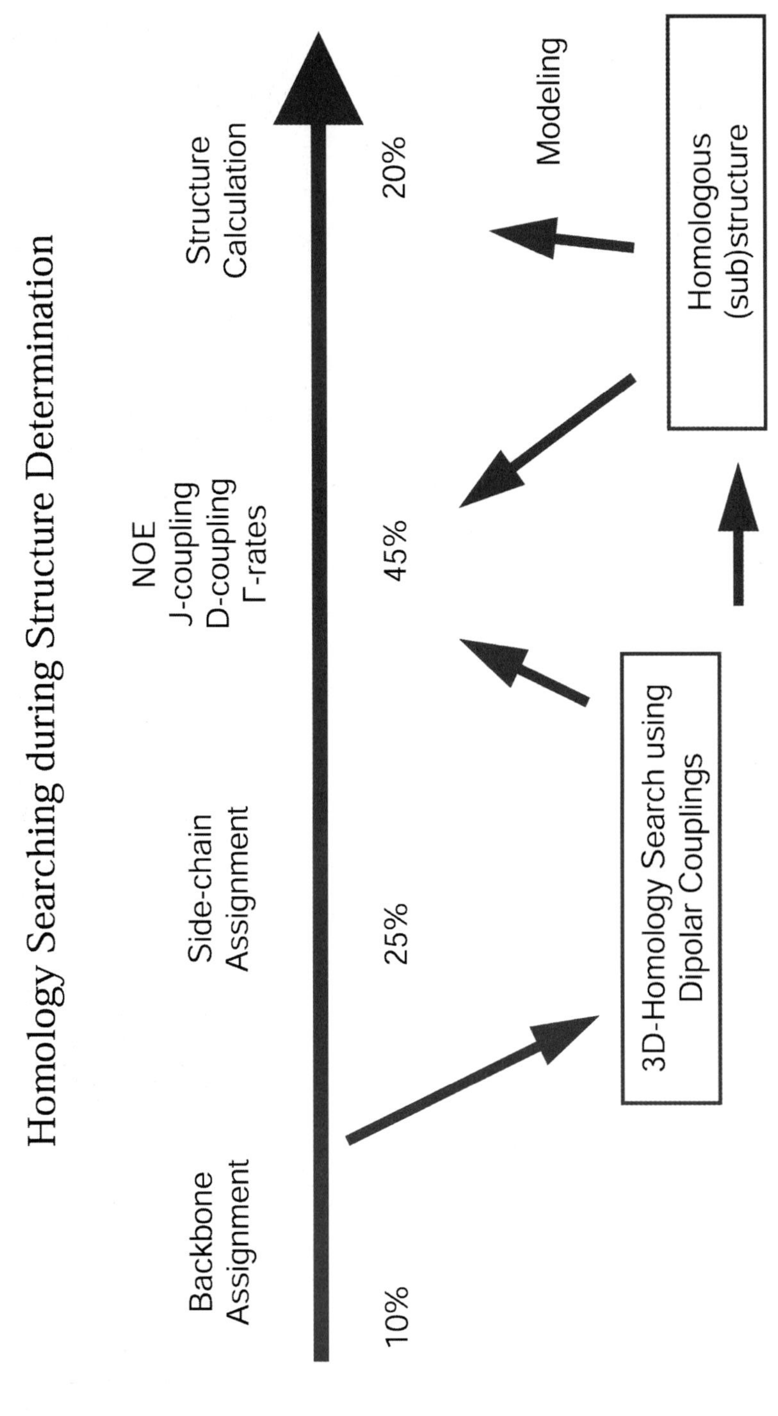

Fig. 54. *Speed of structure determination with NMR.* The backbone assignment takes the minor part of the total work. Homology searching using dipolar couplings would speed up the structure-determination process quite dramatically.

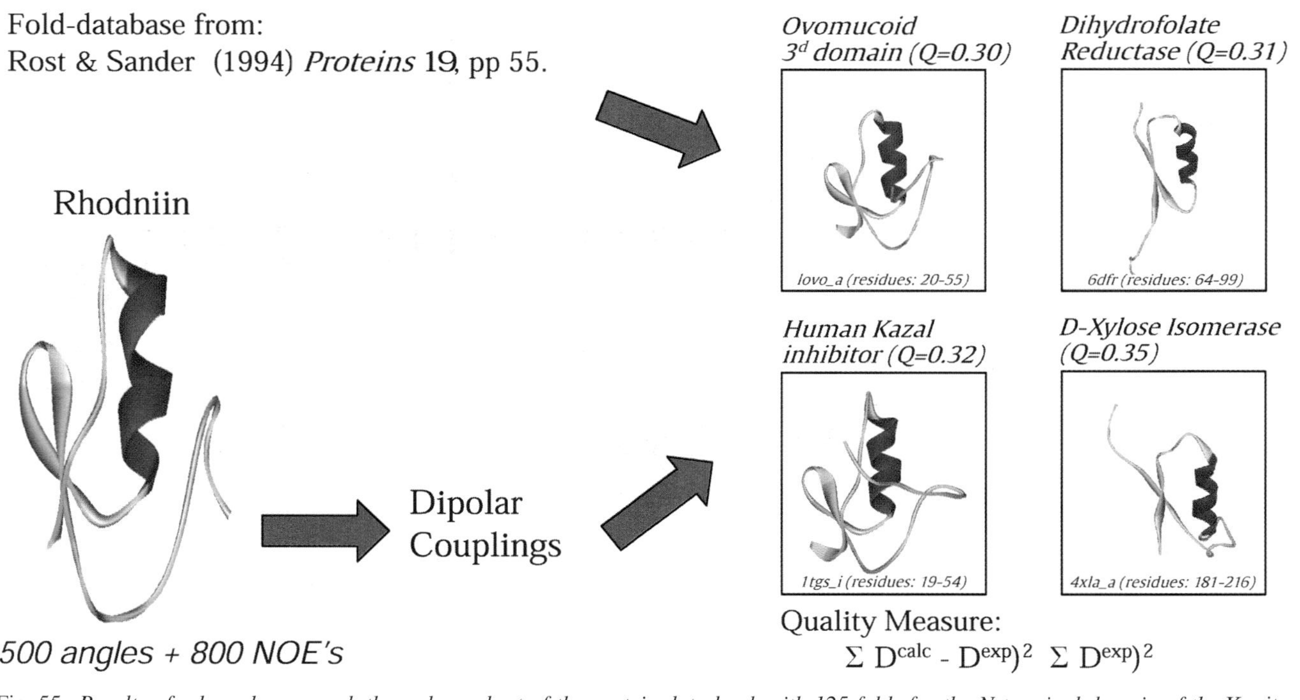

Fig. 55. *Results of a homology search through a subset of the protein data bank with 125 folds for the N-terminal domain of the* Kunitz *protease inhibitor rhodniin*

The protocol uses alignment of secondary-structure elements and looks for identical three-dimensional orientation as depicted in *Fig. 56*. The best scoring proteins are selected in this approach. It is no risk to assume that this type of approach will be used for structure elucidation in the post genomic era where minimal effort will be spent on proteins with homologous structures.

The obstacle for larger systems lies in the following facts: the larger the systems become the more NMR resonances will be congested in the same ppm area. In addition, the size of the molecule correlates in a linear way with the correlation time for rotational diffusion that, in turn, is linearly related with the line width of the resonances. Thus, not only does the number of lines per ppm increase with increasing molecular weight, but also these lines become broader and broader, which makes overlap even more likely and at the same time reduces the signal-to-noise ratio. Therefore, strategies have been developed to reduce the number of resonances as well as to try to reduce the line width of the resonances of interest.

Reducing the number of active resonances in a molecule of large size can only be achieved in the process of the synthesis of the molecule. The most promising strategy is block labelling that introduces ^{13}C and ^{15}N labels in only part of the molecules while leaving the rest of the molecule unlabelled. Application of pulse sequences that observe only ^{13}C- and ^{15}N-bound protons will thus reduce the number of resonances in the spectrum. The most promising approach towards block labelling of proteins is the use of inteins [88] and chemical ligation [89]. Inteins at the C-terminus of a protein allow to produce a C-terminal thioester that can be chemically ligated with another protein fragment that contains an N-terminal cystein (*Fig. 57*).

The reduction of the line width of resonances can be achieved by reducing the cause of relaxation. Most of the relaxation that broadens the lines of the nuclear resonances comes from dipolar interactions. In fact, other protons in the vicinity mainly govern the proton line width. Substitution of most protons by deuterium that has a gyromagnetic ratio of 1:6.5 compared with protons will reduce relaxation due to the fact that relaxation depends quadratically on the gyromagnetic ratios of the relaxing nuclei. Thus, partial or total deuteration [90] allows to go to very high molecular weights, however, at the expense of loosing information due to the lack of observable protons. Techniques that combine NOEs of the still available protons with orientational restraints look very promising that they provide structures even with this lack of information [83].

An additional option is the use of relaxation-compensating pulse sequences. This is the domain of the pulse-sequence developers. At the moment, there are two approaches available. One uses heteronuclear multiple-quantum coherences, in which the dipolar relaxation is removed [17]. The achieved gain in intensity is shown on the example of an RNA.

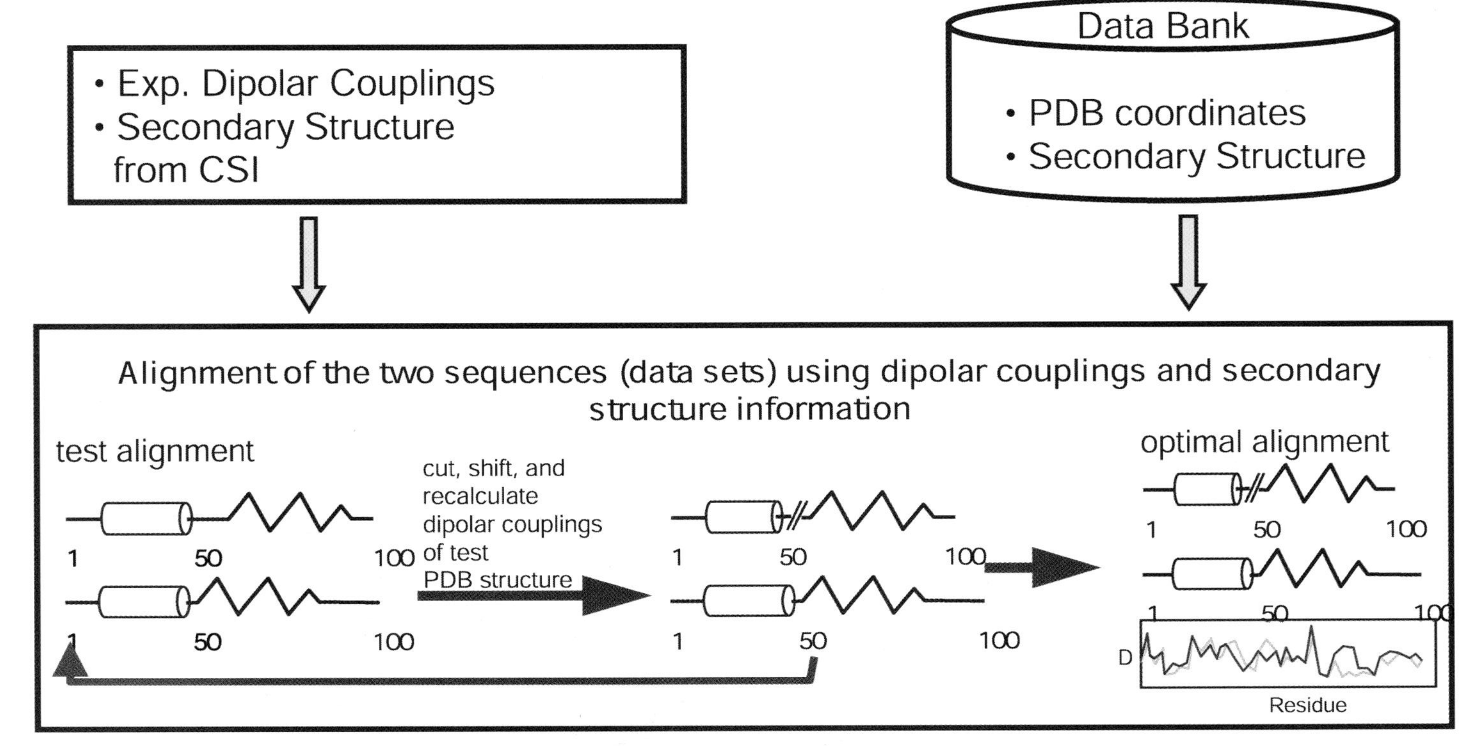

Fig. 56. *Ingredients of the 3D homology-modelling program.* The experimental dipolar couplings of the amides as well as the information about the secondary structure as derived from the chemical-shift analysis is used as experimental input. The pdb data bank is used in addition. Alignment of secondary-structure elements while optimizing the fit of the dipolar couplings yields the scoring function for the homology fitting.

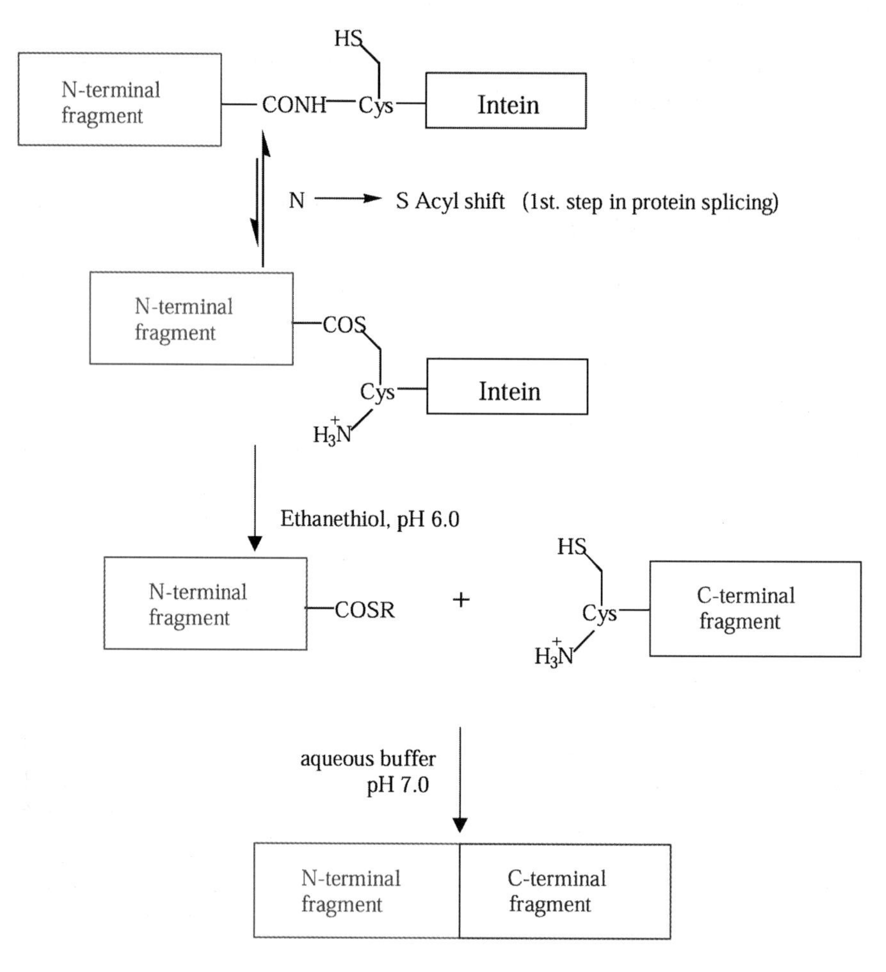

Fig. 57. *Chemical ligation for the reduction of the number of resonances in a protein.* Synthesis of, *e.g.*, the N-terminal part of the protein using ^{13}C- and^{15}N-labelling, and of the C-terminal part without labelling will produce spectra that originate only from the N-terminal but not from the C-terminal domain of the protein.

The other approach relies on self compensation of relaxation contributions. This approach has been named TROSY for transverse relaxation optimized spectroscopy [18]. It works very well for amide NH's, aromatic CH's, however, not for aliphatic CH pairs where the multiple-quantum approach works best [17].

Comparison of 1D Traces through the MQ-HCN and SQ-HCN experiment applied to the 36mer RNA Hairpin

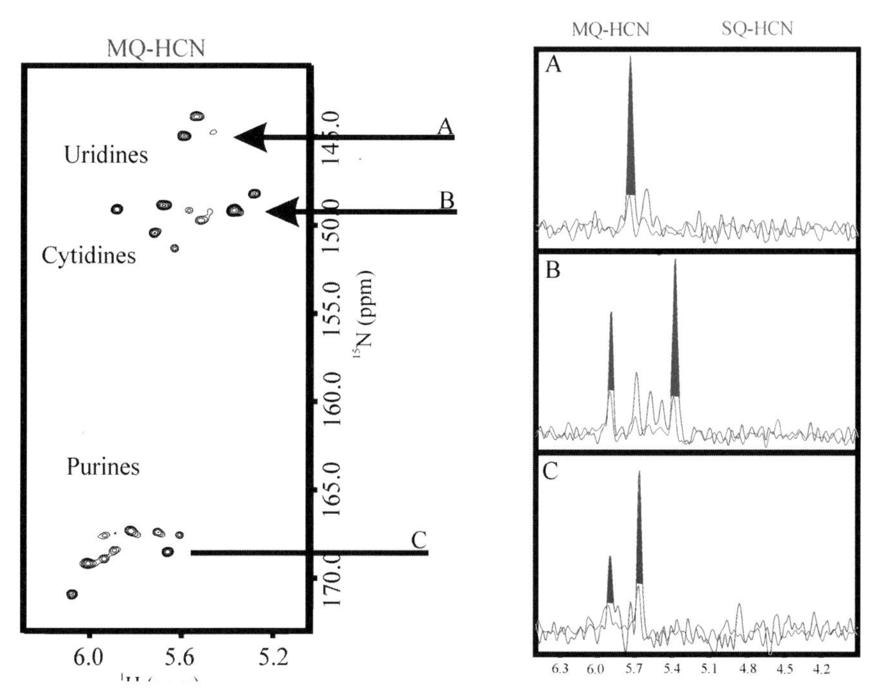

Fig. 58. *Multiple-quantum spectra of a 36mer RNA with correlations from the sugar protons into the bases.* The enhancement of signal to noise using the multiple quantum (*blue*) approach as opposed to the conventional single quantum approach (*red*) is clearly put into evidence in the spectra.

6. Conclusion

The overview given in this article reports on NMR spectroscopy that is still explosively developing. While I was writing this article, papers appeared in the literature that allow to determine for special molecules their conformations without or with only very few NOEs. The world record in structure determination that has been published is 4 days [83]. These developments will be essential for NMR to play an important role in the effort to solve protein folds in the post genomic era. At the same time, NMR techniques are being developed that allow to measure unfolded states of proteins [91], including the observation of site specific kinetics [92]. Methods are developed also to obtain structures from membrane proteins [93], which are difficult to crystallize for X-ray structure determinations. Therefore, it is no risk to expect that

NMR will continue to play an important role at the interface of chemistry, biochemistry, medicinal chemistry, and physical chemistry as the prime structural-biology tool for biomolecules in solution.

This article is based mainly on work that my group in Frankfurt has provided over the last years. The prime contributors to this development have been my former co-workers and colleagues Prof. *Harald Schwalbe*, now MIT, Prof. *Steffen Glaser*, now University of Munich, Prof. *Michael Reggelin*, now University of Darmstadt, Dr. *Matthias Köck*, now Marine Research Institute, Bremerhaven, Dr. *John Marino*, now CARB, USA, and Dr. *Michael Sattler*, now EMBL, Heidelberg. Further, I would like to thank Dr. *Klaus Fiebig* and Dr. *Martin Vogtherr*, Aventis cooperation, Frankfurt, Dr. *Marcus Maurer*, ASTA, Frankfurt, Dr. *Teresa Carlomagno* and Dr. *Mirko Hennig*, both at Scripps Research Institute, and Dr. *Bernhard Geierstanger*, Novartis Institute at Scripps, and Dr. *Christian Richter*, Bruker, Fällanden. The long-term collaborations with Dr. *Wolfgang Bermel*, Bruker, Karlsruhe, as well as with Prof. *E. Carafoli* and Dr. *J. Krebs*, ETH-Zürich, are gratefully acknowledged, as well as collaboration with Dr. *Marcel Blommers*, *Novartis*, Basel. The solid-state spectrum for protein assignment has been contributed by my colleague Dr. *Marc Baldus*, Göttingen. I would like to thank, in addition, *Wolfgang Peti*, *Stefan Bartoschek*, *Jens Meiler*, *Heike Neubauer*, *Bettina Elshorst*, Dr. *Jochen Junker*, *Bernd Junker*, Dr. *Tanja Parac*, and Dr. *Hennig Steinhagen* for their contributions to this article. I would like to thank Dr. *Eriks Kupce*, Varian, Oxford, UK, for kindly providing the 900-MHz 1D spectrum. Funding by the *Deutsche Forschungsgemeinschaft*, the *Max Planck Gesellschaft*, the *Fonds der Chemischen Industrie*, the *Humboldt Foundation*, the *VW Foundation*, *Aventis Research and Technology*, the *BMBF*, and the *European Union* is gratefully acknowledged.

REFERENCES

[1] E. M. Purcell, H. C. Torrey, R. V. Pound, *Phys. Rev.* **1946**, *69*, 37.
[2] F. Bloch, W. W. Hansen, M. Packard, *Phys. Rev.* **1946**, *69*, 127.
[3] J. S. Waugh, J. *Magn. Reson.* **1982**, *49*, 517.
[4] J. S. Waugh, J. *Magn. Reson.* **1982**, *50*, 30.
[5] A. J. Shaka, J. Keller, R. Freeman, *J. Magn. Reson.* **1983**, *53*, 313.
[6] R. R. Ernst, W. A. Anderson, *Rev. Sci. Instrum.* **1966**, *37*, 93.
[7] W. P. Aue, E. Bartholdi, R. R. Ernst, *J. Chem. Phys.* **1976**, *64,* 2229.
[8] J. Jeener, B. H. Meier, P. Bachmann, R. R. Ernst, *J. Am. Chem. Soc.* **1979**, *101,* 6441.
[9] L. Braunschweiler, R. R. Ernst, *J. Magn. Reson.* **1983**, *53,* 521.
[10] A. A. Bothner-By, R. L. Stephens, J. Lee, *J. Am. Chem. Soc.* **1983**, *106*, 811.
[11] C. Griesinger, O. W. Sørensen, R. R. Ernst, *J. Magn. Reson.* **1987**, *73*, 574; G. W. Vuister, R. Boelens, R. Kaptein, *J. Magn. Reson.* **1988**, *80*, 176.
[12] M. Ikura, L. E. Kay, A. Bax, *Biochemistry* **1990**, *29*, 4659.
[13] A. Pardi, *Methods Enzymol.* **1995**, *261*, 350.
[14] G. M. Clore, A. Szabo, A. Bax, L. E. Kay, P. C. Driscoll, A. M. Gronenborn, *J. Am. Chem. Soc.* **1990**, *112*, 4989; N. Tjandra, P. Wingfield, S. Stahl, A. Bax, *J. Biomol. NMR* **1996**, *8*, 273.
[15] B. Reif, M. Hennig, C. Griesinger, *Science* **1997**, 276, 1230.
[16] J. R. Tolman, J. M. Flanagan, M. Am. Kennedy, J. H. Prestegard, *Proc. Natl. Acad. Sci.* USA **1995**, *92*, 9279; N. Tjandra, A. Bax, *Science* **1997**, *278*, 1111.
[17] J. P. Marino, J. Diener, P. B. Moore, C. Griesinger, *J. Am. Chem. Soc.* **1997**, *119*, 7361.
[18] K. Pervushin, R. Riek, G. Wider, K. Wüthrich, *Proc. Natl. Acad. Sci.* USA **1997**, *94*, 12366.
[19] http://www.rcsb.org/pdb/
[20] W. G. Proctor, F. C. Yu, *Phys. Rev.* **1950**, *77*, 717.

[21] J. T. Arnold, S.S. Dharmatti, M. E. Packard, *J. Chem. Phys.* **1951**, *19*, 507.
[22] R. R. Ernst, G. Bodenhausen, A. Wokaun, 'Principles of Nuclear Magnetic Resonance in One and Two Dimensions', Clarendon Press, Oxford, 1989; J. Cavangh, W. J. Fairbrother, A. G. Palmer III, N. J. Skelton, 'Protein NMR Spectroscopy', Academic Press, San Diego, 1996; M. Goldman, 'Quantum Description of High-Resolution NMR in Liquids', Clarendon Press; Oxford, 1988.
[23] V. G. Malkin, O. L. Malkina, L. A. Eriksson, D. R. Salahub, in 'Theoretical and Computational Chemistry', Eds. J. M. Seminario and P. Politzer, *2*, Elsevier Science, 1995, p. 273.
[24] D. A. Case, *J. Biomol. NMR* **1995**, *6*, 341.
[25] J. Meiler, R. Meusinger, M. Will, *J. Chem. Inf. Comput. Sci.* **2000**, *40*, 1169.
[26] W. G. Proctor, F. C. Yu, *Phys. Rev.* **1950**, *77*, 717.
[27] H. S. Gutowsky, D. W. McCall, *Phys. Rev.* **1951**, *82*, 748; E. L. Hahn, D. E. Maxwell, *Phys. Rev.* **1951**, *84*, 1246.
[28] A. Bax, R. Freeman, S. P. Kempsell, *J. Am. Chem. Soc.* **1980**, *102*, 4849.
[29] G. Bodenhausen, D. J. Ruben, *Chem. Phys. Lett.* **1980**, *69*, 185.
[30] B. Reif, M. Köck, R. Kerssebaum, H. Kang, William H. Fenical, C. Griesinger, *J. Magn. Reson. Ser. A* **1996**, *118*, 282.
[31] A. Bax, M. F. Summers, *J. Am. Chem. Soc.* **1986**, *108*, 2093.
[32] B. H. Oh, W. M. Westler, P. Derba, J. L. Markley, *Science* **1988**, *240*, 908.
[33] S. Grzesiek, A. Bax, *J. Am. Chem. Soc.* **1992**, *114*, 6291.
[34] S. Grzesiek, A. Bax, *J. Magn. Reson.* **1992**, *99*, 201.
[35] B. Elshorst, M. Hennig, H. Försterling, A. Diener, M. Maurer, P. Schulte, H. Schwalbe, C. Griesinger, J. Krebs, H. Schmid, T. Vorherr, E. Carafoli, *Biochemistry* **1999**, *38*, 12330.
[36] A. Bax, F. Delaglio, S. Grzesiek, G. W. Vuister, *J. Biomol. NMR* **1994**, *4*, 787.
[37] J. Pauli, M. Baldus, B. van Rossum, H. de Groot, H. Oschkinat, *Chem.BioChem.* **2000**, in press.
[38] M. Baldus, A. T. Petkova, J. Herzfeld, R. G. Griffin, *Mol. Phys.* **1998**, *95*, 1197.
[39] R. Verel, M. Baldus, M. Ernst, B. H. Meier, *Chem. Phys. Lett.* **1998**, *287*, 421.
[40] M. Baldus, M. Tomaselli, B. H. Meier, R. R. Ernst, *Chem. Phys. Lett.* **1994**, *230*, 329.
[41] M. Karplus, *J. Chem. Phys.* **1959**, 30, 11.
[42] C. A. G. Haasnoot, F. A. A. M. de Leeuw, C. Altona, *Tetrahedron.* **1980**, *36*, 2783.
[43] C. Griesinger, M. Hennig, J. P. Marino, B. Reif, H. Schwalbe, in 'Biomolecular NMR', Eds. L. Berliner and R. Krishna, 'Biological Magnetic Resonance', *16*, Kluwer Academic – Plenum Press, 1999, p. 259-367.
[44] A. J. Dingley, E. Masse, R. D. Peterson, M. Barfield, J. Feigon, S. Grzesiek, *J. Am. Chem. Soc.* **1999**, *121*, 6019.
[45] A. J. Dingley, S. Grzesiek, *J. Am. Chem. Soc.* **1998**, *120*, 8293.
[46] M. Hennig, B. H. Geierstanger, *J. Am. Chem. Soc.* **1999**, *121*, 5123.
[47] M. R. Hansen, L. Mueller, A. Pardi, *Nat. Struct. Biol.* **1998**, *5*, 1065.
[48] S. R. Prosser, J. A. Losonczi, I. V. Shiyanovskaya, *J. Am. Chem. Soc.* **1998**, *120*, 11010.
[49] K.Flemming, D. Gray, S. Prasannan, S. Matthews, *J. Am. Chem. Soc.* **2000**, *122*, 5224.
[50] S. W. Fesik, E. R. P. Zuiderweg, *J. Am. Chem. Soc.* **1988**, *785*, 588.
[51] G. Otting, K. Wüthrich, *Q. Rev. Biophys.* **1990**, *23*, 39; C. Griesinger, H. Schwalbe, J. Schleucher, M. Sattler, in 'Two Dimensional NMR Spectroscopy: Applications for Chemists and Biochemists', Eds. Dr. William R. Croasmun and Dr. Robert Carlson, VCH Publishers Inc., New York, Weinheim, Cambridge, 1994, p. 457–580.
[52] B. Reif, A. Diener, M. Hennig, M. Maurer, C. Griesinger, *J. Magn. Reson.* **2000**, *143*, 45.
[53] T. Lindel, J. Junker, M. Köck, *J. Mol. Model.* **1997**, *3*, 364.
[54] G. M. Crippen, 'Distance Geometry and Conformational Calculations', Wiley, Chichester, England, 1981; G. M. Crippen, T. F. Havel, 'Distance Geometry and Molecular Dynamics', John Wiley & Sons, New York, Chichester, Brisbane, Toronto, Singapore, 1988.
[55] W. F. van Gunsteren, H. J. C. Berendsen, *Angew. Chem.* **1990**, *102*, 1020; J. M. Haile, 'Molecular Dynmamics Simulation. Elementary Methods', John Wiley & Sons, New York, Chichester, Brisbane, Toronto, Singapore 1992.

[56] A. T. Brünger, 'XPLOR Version 3.1, A System for X-Ray Crystallography and NMR', Yale University Press, New Haven, London, 1992.

[57] W. F. van Gunsteren, H. J. C. Berendsen, *Angew. Chem.* **1990**, *102*, 1020; J. M. Haile; Molecular Dynmamics Simulation. Elementary Methods, John Wiley & Sons, New York, Chichester, Brisbane, Toronto, Singapore 1992.

[58] M. Nilges, *J. Mol. Biol.* **1994**, *245*, 645; M. Nilges, S. I. O'Donoghue, *Prog. Nucl. Magn. Reson. Spectrosc.* **1998**, *32*, 107.

[59] R. A. Laskowski, J. A. C. Rullmann, M. W. Mac Arthur, R. Kaptein, J. M. Thornton, *J. Biomol. NMR* **1996**, *8*, 477.

[60] P. L. Weber, R. Morrison, D. Hare, *J. Mol. Biol.* **1988**, *204*, 483.

[61] M. Reggelin, H. Hoffmann, M. Köck, D. F. Mierke, *J. Am. Chem. Soc.* **1992**, *114*, 3272.

[62] H. Steinhagen, M. Reggelin, G. Helmchen, *Angew. Chem.* **1997**, *109*, 2199; *Angew. Chem., Int. Ed.* **1997**, *36*, 2108.

[63] J. Junker, B. Reif, H. Steinhagen, B. Junker, I. C. Felli, M. Reggelin, C. Griesinger, *Chem. Eur. J.* **2000**, *6*, 3281.

[64] G. R. Kiddle, S. W. Hommans, *FEBS Letters* **1998**, *436*, 128; T. Rundlöf, C. Landersjö, K. Lycknert, A. Maliniak, G. Widmalm, *Magn. Reson. Chem.* **1998**, *36*, 773.

[65] H. Neubauer, M. Meiler, W. Peti, C. Griesinger, in preparation.

[66] D. S. Wishart, B. D. Sykes, F. M. Richards, *J. Mol. Biol.* **1991**, *222*, 311.

[67] D. S. Wishart, B. D. Sykes, F. M. Richards, *Biochemistry* **1992**, *31*, 1647.

[68] M. Nilges, M. J. Macias, S. I. O'Donoghue, H. Oschkinat, *J. Mol. Biol.* **1997**, *269*, 408.

[69] A. Rexroth, P. Schmidt, S. Szalma, T. Geppert, H. Schwalbe, C. Griesinger, *J. Am. Chem. Soc.* **1995**, *17*, 10389.

[70] G. W. Vuister, A. Bax, *J. Am. Chem. Soc.* **1993**, *115*, 7772.

[71] K. Wüthrich, NMR of Proteins and Nuclei Acids, John Wiley, New York, 1986.

[72] A. L.Breeze, *Prog. Nucl. Magn. Reson. Spectrosc.* **2000**, *36*, 323.

[73] M. A. Wilson, A. T. Brunger, *J. Mol. Biol.* **2000**, *301*, 1237.

[74] M. Ikura, G. M. Clore, A. M. Gronenborn, G. Zhu, C. B. Klee, A. Bax, *Science* **1992**, *256*, 632.

[75] M. Kataoka, J. F. Head, T. Vorherr, J. Krebs, E. Carafoli, *Biochemistry* **1991**, *30*, 6247.

[76] G. Barbato, M. Ikura, L. E. Kay, R. W. Pastor, A. Bax, *Biochemistry* **1992**, *31*, 5269.

[77] M. Ikura, L. E. Kay, A. Bax, *Biochemistry* **1990**, *29*, 4659.

[78] G. C. K. Roberts, *Curr. Opin. Biotech.* **1999**, *10*, 42; J. M. Moore, *Curr. Opin. Biotech.* **1999**, *10*, 54; G. C. K. Roberts, *Drug Discovery Today* **2000**, *5*, 230.

[79] S. B. Shuker, P. J. Hajduk, R. P. Meadows, S. W. Fesik, *Science* **1996**, *274*, 1531; P. J. Hajduk, G. Sheppard, D. G. Nettesheim, *J. Am. Chem. Soc.* **1997**, *119*, 5818.

[80] F. Ni, *Prog. Nucl. Magn. Reson. Spectrosc.* **1994**, *26*, 517; L. Y. Lian, I. L. Barsukov, M. J. Sutcliffe, K. H. Sze, G. C. K. Roberts, *Methods Enzymol.* **1994**, *239*, 657. G. M. Clore, A. M. Gronenborn, *J. Magn. Reson.* **1982**, *48*, 402; G. M. Clore, A. M. Gronenborn, *J. Magn. Reson.* **1983**, *53*, 423.

[81] T. Carlomagno, I. C. Felli, M. Czech, R. Fischer, M. Sprinzl, C. Griesinger, *J. Am. Chem. Soc.* **1999**, *121*, 1945.

[82] M. J. J. Blommers, W. Stark, C. E. Jones, D. Head, C. E. Owen, W. Jahnke, *J. Am. Chem. Soc.* **1999**, *121*, 1949.

[83] M. Mayer, B. Meyer, *Angew. Chem. Int. Ed.* **1999**, *35*, 1784.

[84] F. Delaglio, G. Kontaxis, A. Bax, *J. Am. Chem. Soc.* **2000**, *122*, 2142.

[85] H. Aito, A. Annila, S. Heikkinen, E. Thulin, T. Drakenberg, I. Kilpeläinen, *Protein Sci.* **1999**, *8*, 2580; A. Annila, H. Aito, E. Thulin, T. Drakenberg, *J. Biomol. NMR* **1999**, *14*, 223; F. Delaglio, G. Kontaxis, A. Bax, *J. Am. Chem. Soc.* **2000**, *122*, 2142.

[86] J. Meiler, W. Peti, C. Griesinger, *J. Biomol. NMR* **2000**, *17*, 283.

[87] A. Medek, E. T. Olejniczak, R. P. Meadows, S. W. Fesik, *J. Biomol. NMR* **2000**, *18*, 229.

[88] R. Xu, B. Ayers, D. Cowburn, T. W. Muir, *Proc. Natl. Acad. Sci.* USA **1999**, 96, 338.

[89] P. E. Dawson, T. W. Muir, I. Clark-Lewis, S. B. H. Kent, *Science* **1994**, *266*, 776.

[90] K. H. Gardner, M. K. Rosen, L. E. Kay, *Biochemistry* **1997**, *36*, 1389; N. K. Goto, L. E. Kay, *Curr. Opin. Struct. Biol.* **2000**, *10*, 585.

[91] H. Schwalbe, K. M. Fiebig, M. Buck, J. A. Jones, S. B. Grimshaw, A. Spencer, S. J. Glaser, L. S. Smith, C. M. Dobson, *Biochemistry* **1997**, *36*, 8977.
[92] C. Frieden, S. D. Hoeltzli, I. J. Ropson, *Protein Sci.* **1993**, *2*, 2007; J. Balbach, V. Forge, N. A. van Nuland, S. L. Winder, P. J. Hore, C. M. Dobson, *Nature Struct. Biol.* **1995**, *2*, 865; S. D. Hoeltzli, C. Frieden, *Proc. Natl. Acad. Sci.* U.S.A. **1995**, *92*, 9318; S. Balbach, V. Forge, W. S. Lau, N. A. van Nuland, K. Brew, C. M. Dobson, *Science* **1996**, *274*, 1161; T. Kühn, H. Schwalbe, *J. Am. Chem. Soc.* **2000**, *122*, 6169.
[93] T. A. Egorova-Zachernyuk, J. Hollander, N. Fraser, P. Gast, A. J. Hoff, R. Cogdell, H. J. M. de Groot, M. Baldus, in preparation.

New Methods in Electron Paramagnetic Resonance Spectroscopy for Structure and Function Determination in Biological Systems

by **Thomas F. Prisner**

Institute of Physical and Theoretical Chemistry, Johann Wolfgang Goethe-Universität,
Marie Curie Strasse 11, D-60439 Frankfurt am Main
(Phone: 49-69-798 29406, Fax: 49-69-798 29404; e-mail: Prisner@Chemie.Uni-Frankfurt.de)

1. Introduction

With the development of protein crystallization and X-ray crystal-structure determination, the understanding of proteins has started on a molecular level. Nowadays almost 10000 X-ray structures of proteins are listed in the *Brookhaven Protein Data Bank*, accompanied by more than 1600 NMR structures. These structures are not only the starting point for a detailed understanding of the function of proteins on a molecular level, but they also provide very helpful information for biochemical mutagenesis work, molecular-dynamics studies, and for other spectroscopical studies. Despite all the success of these two methods in obtaining global structures of proteins, there are still limitations for both of them. For X-ray crystallography (see *Chapt. 1*), the crystallization of large proteins or protein complexes with sufficiently high quality is still the work of an artist and very time-consuming. More importantly, the protein is not necessarily in its physiological phase in the crystal, and packing effects as well as additional interactions may distort the tertiary structure of the protein. Additionally, the structure is static and, in almost all cases, in the most stable ground state (inactive state or resting state) of the protein. The NMR method determines the protein structure in its natural liquid phase. In addition, the NMR technique can also give access to dynamic aspects of the structure, observable *via* relaxation effects. Unfortunately, even with the spectral resolution enhancement obtained by multi-dimensional FT-NMR spectroscopy, this method is still limited to proteins with a molecular weight below 50 KD for structural applications. For larger macromolecules, a selective spin-isotope-labelling technique is required to 'highlight' specific parts of the protein.

This is also the case for some other spectroscopical techniques, which allow the examination of structural and dynamic aspects of specifically localized sites within a protein. One of these methods is Electron Paramagnetic Resonance (EPR) spectroscopy, which highlights the local structure around paramagnetic molecules within the protein. Such paramagnetic molecules can be natural stable cofactors (such as heme molecules, iron-sulfur clusters, or metal ions), transiently generated radicals within a reaction cycle (cofactors or amino acid radicals of the polypeptide chain), or artificial spin labels attached to the protein. In contrast to NMR, EPR is not restricted by the size of the protein, because only the paramagnetic centres and their interaction with the protein are spectroscopically visible. In cases where only one localized paramagnetic centre is located within the protein, a simple one-dimensional continuous-wave (cw) EPR experiment (where the sample is irradiated continuously with a constant microwave frequency – typically 9 GHz – and the magnetic field is swept slowly through the resonance) can give detailed information about the paramagnetic species, such as its oxidation state, symmetry, and concentration. In more realistic cases in enzymes, the situation is far more complex: often, more than one paramagnetic species is involved (some of them may be transiently formed within the catalytic cycle of the enzyme), and they will experience internal dynamics that complicates and broadens their EPR spectra. In these cases, the simple cw-EPR spectra will experience the same difficulties as NMR for large molecules: spectral lines overlap and broaden. Therefore, it becomes very difficult to quantitatively analyze the spectra and to obtain a unique solution; the problem is ill-defined. As in NMR, multidimensional methods allow the resolution of overlapping spectral components by distributing them in a more than one-dimensional frequency space and by additionally diluting the spectra by specific suppression of unwanted spectral components with the aid of appropriate pulse sequences. Whilst the methodological principles for the manipulation of the spin system (electron spin S in EPR, nuclear spin I in NMR) are very similar in NMR and EPR spectroscopy, the technical requirements and, therefore, the practical realizations are quite different in both fields. Due to the much larger magnetic moment of the electron spin S (nearly a factor of 1000 larger as compared to a proton nuclear spin), the technical requirements (resonance frequency, relaxation times, pulse lengths) for EPR are much more demanding as compared with NMR spectroscopy. Nevertheless, most of the technical restrictions have been overcome by the development of specific methods and techniques. These advanced EPR techniques have dramatically increased the potential of EPR spectroscopy, especially in the field of biochemical applications. The potential of these methods will be demonstrated by some selected applications within this essay.

2. Physical Origin of Spectral Contributions

2.1. Scaling of Interactions

The EPR spectral line shape is influenced by a number of different interactions such as the hyperfine coupling to the surrounding nucleus, the dipolar coupling to further paramagnetic centres in the macromolecule, the local internal fields for high-spin systems, and by relaxation effects. Unfortunately, their effect in most cases is to broaden spectral lines; in many cases, to an extent that the extractable information content is very limited or even totally lost. Nevertheless, all of them bear important information for a detailed characterization of the paramagnetic site in question. For example, relaxation contributions to the line shape may contain important information on local dynamics between conformational states, motion of the paramagnetic molecule, and the local protein environment or of the whole macromolecule. But as long as these contributions are convoluted with other broadening mechanisms, such as, for example, heterogeneity in site geometry, different nuclear spin configurations or isotope statistics, it will be probably impossible to extract such interesting information from the spectra. The success of modern pulse-, multi-frequency-, and multi-dimensional-EPR spectroscopy methods is to unravel these different contributions to the EPR line shape in a unambiguous way. This can be done, because, very often, these interactions are quite different in their strength and in their dynamic behavior. Interactions can initially be divided into static and dynamic interactions, as compared with the time scale of a pulsed EPR experiment (0.1–100 μs). For pulsed experiments, such static interactions can be refocused in many cases, and, therefore, selectively eliminated from the spectra. With pulse experiments, dynamic interactions can be further distinguished to be effective on a time scale of the electron spin coherence (T_2 transversal relaxation time, typically 100 ps – 1 μs) or on the time scale of the inverse microwave excitation frequency (typically 10 GHz, called X-frequency band), affecting the T_1 longitudinal relaxation time (ranging from sub-ns up to 10 ms and more). With specific pulse sequences, even rather slow dynamics on the timescale of the T_1 relaxation time can be monitored.

Furthermore, interactions can be distinguished by the radiation frequency that affects these interactions. Whereas interactions with nuclei are toggled by a radiofrequency radiation, microwave frequencies are needed for the manipulation of the interaction between closeby paramagnetic centres. The reason is the large difference in the magnetic moment of the nuclear spin (I) and the electron spin (S) of a factor of *ca.* 1000. Because of these large differences in magnetic moments, the distance at which other spins can be observed is much larger for another electron spin S (up to 5 nm) than for a nuclear spin I (up to 0.6 nm). Therefore, for structural investigations on proteins, the interactions to which the EPR line will be sensitive can also be distinguished by their radii around the paramagnetic centre. As explained in more detail below, the Zero-Field Splitting (ZFS) tensor and the g-tensor will be the narrowest sensors – mostly influenced by the paramagnetic molecule itself and only sensitive to its surrounding up to a distance of less than 0.2 nm. They will serve mainly to characterize the paramagnetic molecule itself. Hyperfine interactions can be detected up to distances of 0.6 nm from the paramagnetic molecule, as mentioned above. The strong hyperfine coupling of nuclei of the paramagnetic molecule itself can be typically seen directly in the EPR spectra and will help to further characterize the paramagnetic molecule. Hyperfine couplings to more distant nuclei in the first shell of molecules around the paramagnetic molecule can be probed by ENDOR or ESEEM spectroscopy, two more advanced EPR methods explained later. Finally, the coupling of the paramagnetic molecule to other paramagnetic centres in the protein can be probed up to distances as much as 5 nm! Again the cw-EPR spectra will show such interactions directly for distances up to 2 nm, and only pulse methods can extend this range up to 5 nm. These three different spheres, which are accessible by EPR spectroscopy, are shown schematically in *Fig. 1* for a semiquinone anion radical $Q_A^{-\cdot}$ of a bacterial reaction centre. It is obvious from this figure that important structural aspects of active sites in proteins can be obtained by a detailed analysis of all these interactions *via* modern EPR methods. Before an introduction to these advanced EPR methods is given, the most important line-shape contributions for paramagnetic centres in proteins will be explained in the following *Sect.* [1].

2.2. Anisotropy of Interactions

In most cases, the biological macromolecules will be disordered in the sample tube, which means that the molecules in the spectrometer will have random orientation with respect to the external magnetic field B_0 applied to the sample. With typical sample volumes for EPR experiments of between 100 µl and 10 nl, and with protein concentrations in liquid solutions of 100 µmol/l, the sample consists of 10^{16}–10^{13} molecules with a statistical equal distribution of all possible orientations with respect to the magnetic field axis B_0. If these macromolecules are static in their orientation, as is the case for frozen samples, a sample with this random orientation distribution is called a powder sample. EPR Line shapes in such powder samples will typically be enormously broadened as compared with crystalline samples, where all molecules have the same orientation with respect to the magnetic-field axis, or as compared with liquid samples, where the molecules are reorient-

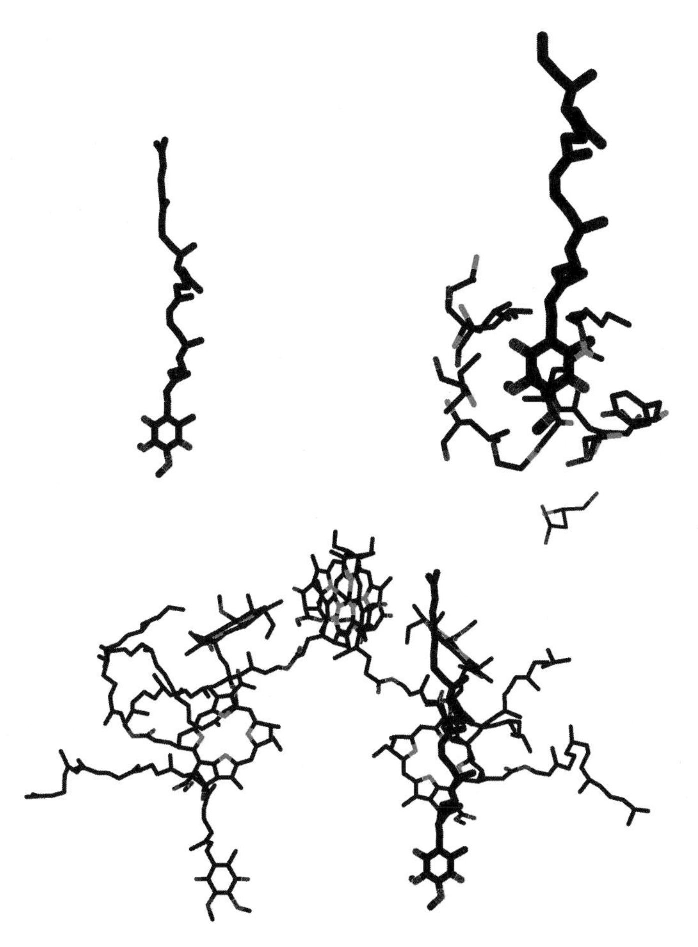

Fig. 1. *The three different interaction ranges of EPR spectroscopy are shown on a transient semiquinone anion radical $Q_A^{-\cdot}$ of a bacterial reaction centre of* Rhodobacter sphaeroides (R26). The structural data for this illustration are taken from the X-ray structure of *Ermler et al.* [2]. *Upper left:* The local interactions such as intermolecular hyperfine couplings and *g*-tensor of the ubisemiquinone-10 anion radical characterized by multifrequency-EPR. *Upper right:* The local protein surrounding (within a distance of 0.6 nm to the semiquinone radical) as observable by ENDOR and ESEEM spectroscopies. *Lower picture:* The interaction of the semiquinone anion radical with the other transient paramagnetic centres in the protein complex as observable with pulse-EPR methods.

ed on a very fast timescale. The reason for this broadening is the anisotropy of the relevant interactions, which arises from the fact that the paramagnetic molecule and its protein surrounding is not spherically symmetric. Therefore, the EPR line position for these molecules will change for different orientations of the molecule with respect to the B_0 axis. Because the experimentally observed spectrum is a linear superposition of all molecules, the spectral lines will become broad and the line-width is given by the maximal shift of the line positions with orientation. Analysis of such spectra should take into account the contributions of all possible orientations. It should be obvious that an evaluation of structural information is drastically hindered by such an orientational distribution within the sample.

The analysis and information content of an EPR spectrum from single-crystal samples is much more obvious. For proteins, this possibility is limited by the difficulties involved in growing of large-enough crystals ($>10^{-3}$ mm^3 for EPR applications). The analysis of these spectra may even be complicated by many non-equivalent sites in the unit cell of such a crystal. Additionally, one has to consider that crystallization effects may potentially change the structure of a protein or hinder internal conformational dynamics. Samples may also be partially aligned, for example, by external fields, on surfaces, membranes, or on thin films. In these cases, the ordering of the molecules is not perfect, and additional information on order parameters is necessary to simulate the observed spectra correctly.

Another possibility to simplify the EPR spectra is to make the molecules tumble fast enough. If the rotational correlation time of a molecule is much shorter than the inverse spectral anisotropy (in frequency units) the anisotropy will be averaged out (fast motional limit) and only isotropic interactions will contribute to its line position. This is the case for many organic radicals and small molecules in solution at room temperature, leading to highly resolved EPR spectra with very narrow lines. Unfortunately this is no longer true for large macromolecules such as proteins. In many protein samples the EPR lineshape will be a powder spectrum even for a liquid sample at room temperature, because the tumbling of the protein is too slow for efficient averaging of the anisotropic interactions.

2.3. Interactions with Local Nuclei – Hyperfine Coupling

One of the most important interactions of the electron spin of the paramagnetic molecule, which influences its line position, is that with the nearby nuclear spins. These are the nuclei of the molecule itself and nuclei in the local surrounding upto a distance of *ca.* 0.6 nm. There are two different contributions to this interaction: an isotropic part (*Fermi*-contact interaction), which arises from electron spin density of the unpaired electron at the nucleus, and an anisotropic part, which arises from through-space magnetic dipole-dipole interaction between the electron and nuclear magnetic moments. Whereas for intramolecular nuclei, the isotropic part can be very large, interactions with intermolecular nuclei further apart are mainly anisotropic. This offers the possibility to measure directly the distance to the nucleus by the strong distance dependence of the dipole-dipole interaction Δ:

$$\Delta = \frac{\mu_0}{4\pi} (3\cos^2\theta - 1) \frac{\mu_S \mu_I}{r^3}$$

where μ_S and μ_I are the electron and nucleus magnetic moments, respectively and θ is the angle between the distance vector and the external magnetic field axis. Additionally, the orientation of these nuclei with respect to the paramagnetic molecule g-tensor axis system can be determined in specific cases, where the angle θ is not averaged over all possible orientations for the randomly oriented molecules, as described later.

2.4. Electron Orbital – *g* Tensor

The interaction of the spin magnetic moment with the orbital magnetic moment of the unpaired electron leads to an orientationally dependent shift of the resonance frequency. This effect is normally described by an effective spin operator S and an anisotropic g matrix. The quantum-mechanical *Hamilton* operator for the interaction of the electron spin with the external magnetic field (*Zeeman* interaction) can, therefore, be described by

$$H_Z = \vec{S} \cdot \hat{g} \cdot \vec{B}_0$$

where g is a 3×3 matrix, with eigenvalues g_{xx}, g_{yy}, and g_{zz}. The anisotropy Δg (deviation from mean g value) for typical paramagnetic centres in biological systems ranges from almost 100% for metal centres (heme centres, metal ions, iron-sulfur complexes) to less than 1‰ for aromatic organic radicals (amino acid radicals, organic co-factors). Whereas, for the former group, this anisotropy, in most cases, dominates the spectra and is easily observable, the anisotropy of the g-tensor of the second group of paramagnetic molecules is totally obscured by other larger interactions at typical external magnetic field strengths (< 1 T). It can, however, be easily observed by modern high-frequency/high-field EPR spectroscopy (magnetic field of 3–25 T) because this interaction, in contrast to other interactions, scales with the applied magnetic field. It is not easy to calculate, quantitatively, the g-tensor of paramagnetic molecules with *ab initio* quantum-theoretical methods accurately. The reason is that the *Eigen* energies of excited electronic orbitals have to be known accurately for open-shell systems, which is difficult even with advanced MO methods such as density functional theory. Nevertheless, resolving the anisotropy of the g-tensor is useful for several reasons. First, the main g values are a fingerprint for specific paramagnetic molecules, and furthermore their values are sensitive to the specific surroundings, such as for example H-bonds of the molecule. It is, therefore, a very helpful parameter for the unambiguous identification and characterization of the paramagnetic molecule. Additionally, the symmetry of the tensor (spherical, axial, rhombic) provides some information on the symmetry of the paramagnetic site. Another important point is that the anisotropy of the resonance conditions is related to the g-tensor, which can be used as a tool to selectively excite, with the microwave pulses, only macromolecules with a specific orientation with respect to the external magnetic field axis. This can be used very elegantly to obtain crystal like, orientation-dependent information even in disordered samples, which is extremely helpful in extracting structural information from a paramagnetic site.

2.5. Interactions to Other Paramagnetic Centres – Dipolar Coupling

The interaction of two electron spins S_1 and S_2 with each other is very similar to the interaction of an electron spin S with a nuclear spin I, except that the strength is much larger for a given distance, as described above. This interaction can, therefore, be observed over extremely large distances (<2 nm with cw methods and up to 5 nm with pulse methods), which makes this a very powerful tool for the determination of long-range distances in proteins. For short distances (<1 nm), where both electronic orbitals start to overlap, there is also an isotropic part to this interaction (exchange interaction), which can be described by a parameter J. If the g-tensor anisotropy is resolved for both paramagnetic molecules, it is possible, with orientation-selective excitation of the two paramagnetic molecules, to measure not only the distance, but also the relative orientation of the two molecules with respect to each other, as described above. This is especially helpful for a detailed understanding of electron-transfer reactions in proteins that occur along chains of paramagnetic molecules, where distance, relative orientation, and overlap of electronic orbitals play a crucial role for the reaction rates.

2.6. Orbital Symmetry of High-Spin Systems – Zero-Field Splitting (ZFS)

A number of paramagnetic metals, for example V, Mn, Fe, Co, and Ni, have in some oxidation states, more than one unpaired electron, which might lead to high-spin states ($S > 1/2$). This may be seen as an extreme case of the interaction between unpaired electrons on different molecules, where the interaction now becomes so strong that coupled spin states are generated. In this case, the different m_S sublevels of the electronic ground state of the molecule may become split by spin-orbit coupling and crystal-field effects even in the absence of an external magnetic field. This leads to an anisotropic line shape related to the symmetry of the metal complex. Whereas this can provide valuable information on the coordination symmetry of the metal ion, the anisotropy is often so large that it will become difficult or even impossible to observe the EPR transition. Fortunately, for systems with an odd number of unpaired electrons, at least a twofold degeneracy of levels will remain in the absence of an external magnetic field (*Kramer*'s theorem). For such systems, the central $m_S = -1/2$ to $m_S = +1/2$ transition is normally easily observable and only distorted in second order by the ZFS. Thus, the ZFS contribution to the line-width becomes negligible when the external magnetic field exceeds the ZFS. Again, this can be achieved by performing the experiment at magnetic fields as high as possible.

2.7. Dynamic Effects of the Site – Relaxation Effects

If the molecule has dynamic motions on the timescale of the EPR experiment, this motion will lead to relaxation effects on the EPR line. Depending on the timescale and size of these motions, these effects may be observable directly in the cw-EPR spectrum or indirectly by pulsed EPR measurements of the relaxation times. In many cases, different dynamics may simultaneously contribute to the relaxation behavior of the electron spin system, as, for example, vibrational and rotational motion, conformational dynamics, phonon coupling to the frozen solvent, and nuclear spin dynamics. In these cases, it will be difficult to obtain specific information from these relaxation measurements. On the other hand, it is possible to highlight a specific time-scale window by the selection of pulse sequences and microwave frequencies that can lead, in favourable cases, to a direct relation between measured relaxation times and interesting molecular dynamics at the paramagnetic site. In these cases, very interesting molecular dynamical aspects of electron-transfer, catalytic, or photo-reactions, unobservable by other structural methods, can be studied directly by pulse-EPR techniques.

3. Spectral Resolution Enhancement Methods

After this brief characterization of the main interactions contributing to the EPR spectrum, it should be clear that a simple continuous-wave (cw) EPR spectrum at a single microwave frequency will not have enough resolution to allow the unraveling of all these different contributions in a unique way. Advanced techniques are necessary to separate and distinguish these contributions and to obtain an unambiguous assignment of the spin Hamiltonian parameters. These parameters will then serve as constrains to obtain a molecular structure model of the centre. In recent years, three advanced methods (and combinations of them) have mainly proven to be of great importance for unravelling complex spectra in biological samples. They will be briefly introduced in the following chapter.

3.1. Electron Nuclear Double Resonance Spectroscopy (ENDOR)

As described above, the hyperfine interaction between the unpaired electron spin and closeby nuclear spins provides detailed information on the nuclei of the paramagnetic molecule and its close surrounding. Unfortunately, this interaction is normally obscured in macromolecules because of the line-broadening mechanisms as described above. Additionally, in proteins there are a large number of nuclei interacting with the electron spin, which again leads to unresolved spectra caused by a strong overlap of the individual hyperfine lines. ENDOR is a very powerful tool, however, for extracting this information by the reduction of the number of hyperfine lines, suppression of other linewidth contributions, and by spreading the spectra into a second spectroscopic dimension (in this case radiofrequency). This is very similar to the principles and goals of more-dimensional NMR spectroscopy, which also is used for macromolecular systems.

ENDOR Experiments can be performed with simultaneous cw microwave and radiofrequency excitation, or in a pulsed fashion, where microwave and radiofrequency pulses are applied in a specific time sequence [3]. The pulse timing table for a *Davies*-ENDOR sequence is shown in *Fig. 2*. The pulse spacing is kept constant, while the radiofrequency is varied slowly. Only if the radiofrequency hits an NMR transition of one of the nuclei interacting with the paramagnetic species, a change in the EPR echo signal can be observed. In this sense, the ENDOR experiment is an NMR experiment, where only nuclei in the surrounding of the paramagnetic centre are selected. Because of this selection, the ENDOR experiment can be performed on macromolecules of arbitrary size, in contrast to NMR spectroscopy. In this case, the NMR signal is observed as a change of the EPR echo signal and not directly as an NMR radiofrequency signal. Therefore, the much higher sensitivity of EPR, as compared with NMR spectroscopy, also holds for this experiment, despite the fact that each additional dimension in a magnetic resonance experiment (here the radiofrequency) increases spectral resolution at the cost of sensitivity. To perform pulsed ENDOR experiments, the microwave resonant cavity has to be modified to additionally support radiofrequency excitation. Several such resonant structures have been developed for microwave excitation frequencies between 9 and 140 GHz [4].

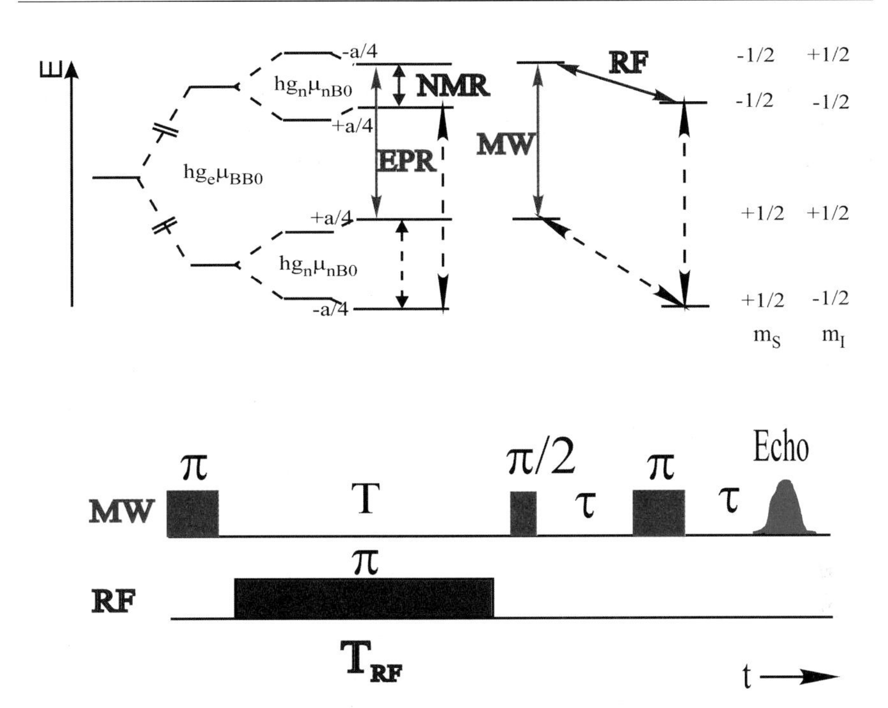

Fig. 2. *Energy diagram for an electron spin* S = *1/2 coupled to a nuclear spin* I = *1/2* (top) *and* Davies-*ENDOR pulse sequence* (bottom). The microwave (mw) pulses are shown on the upper line and the radiofrequency (rf) pulses on the lower line. Typical pulse lengths for the mw pulses are on the order of 100 ns and for the rf pulse on the order of 10 μs.

The ENDOR experiment allows us to distinguish different nuclei with a nuclear spin $I \neq 0$, and, therefore, their structural assignment with respect to the paramagnetic centre can be evaluated. While different nuclei can be easily distinguished in NMR spectroscopy by their different magnetic moments and, therefore, different resonance frequencies, this is more complicated for nuclei coupled to a paramagnetic centre. In this case, the internal magnetic field from the magnetic moment of the unpaired electron may exceed the applied external magnetic field for typical EPR spectrometers (9 GHz microwave frequency (X-band) corresponding to an external field of *ca.* 0.3 T). In these cases, it may become difficult or impossible to distinguish different nuclei. ENDOR Experiments at different external magnetic fields (and therefore different microwave frequencies) will help greatly to unravel these ambiguities: the nuclear *Zeeman*-splitting scales linearly with the external magnetic field, whereas the hyperfine coupling to the unpaired electron is independent of the external magnetic field. At very high magnetic-field values, where the external magnetic-field value exceeds the internal hyperfine fields, the situation becomes much clearer, and different nuclei can be separated by their NMR resonance frequency. This situation is depicted in *Fig. 3*, where typical ranges of ENDOR frequencies for some important nuclei in proteins are shown for an external field of 0.3 T (X-band ENDOR) and for 6.4 T (G-band ENDOR). Clearly, at high magnetic fields different nuclei are much easier to distinguish by their nuclear *Zeeman* frequencies.

Another advantage can be envisaged at high magnetic fields for organic aromatic radicals in disordered protein samples. As described above, the anisotropy of the *g*-tensor will be the major source of line broadening in contrast to lower magnetic field values. In this case, it is possible to excite with the microwave pulses only those paramagnetic molecules that have a specific orientation with respect to the external magnetic field. This additional selectivity for protein bound paramagnetic molecules will permit us to obtain even better resolved and more distinct information for the individual hyperfine tensors [5].

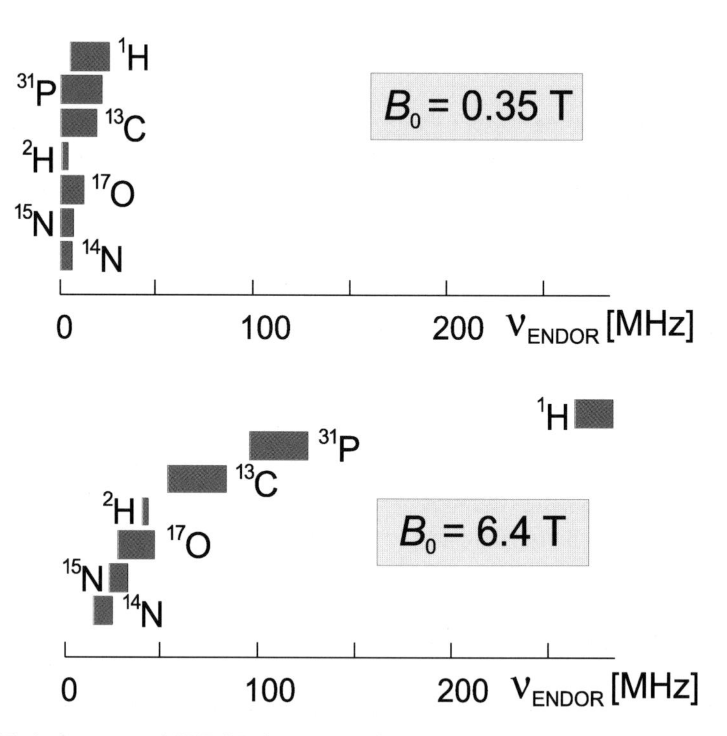

Fig. 3. *Typical ranges of ENDOR frequencies for some important nuclei in proteins.* The length of the bar represents a typical distribution of hyperfine couplings for the nuclei. The upper diagram shows the strong overlap for a typical X-band EPR magnetic-field strength of 0.3 T. The lower diagram shows the much better separation of different nuclei at G-band EPR magnetic-field strengths of 6.4 T.

3.2. Pulse EPR Spectroscopy

In enzymes, very often more than one paramagnetic species is present under specific conditions. This can lead to difficulties in the interpretation if their spectra overlap. Pulse experiments [6] allow the separatation of such different paramagnetic species, for example, by their relaxation behavior or by their spin state *S* [7]. Additionally, the examination of these parameters will give more insight into the nature of the paramagnetic species and possibly its interaction with the surrounding. For nuclei coupled to the electron spin *via* an anisotropic hyperfine interaction, one- and two-dimensional pulsed Electron Spin Echo Envelope Modulation (ESEEM) experiments have proven to be very powerful tools to examine these interactions, especially for weak couplings and for nuclei with a small *Zeeman* splitting. In contrast to the ENDOR experiment, no radiofrequency irradiation is needed; the nuclear spins are only indirectly affected by microwave toggling of the coupled electron spin. With similar concepts, it is also possible to observe small magnetic dipolar couplings between very distant paramagnetic centres. With such Double Electron-Electron Resonance experiments (DEER), interactions between radicals have been measured for distances up to 6 nm [8]. In this case, two different microwave frequencies have to be applied to the sample *via* the microwave resonator. Because of the much shorter relaxation times of electron spins compared to nuclear spin systems – on the microsecond timescale for EPR in contrast to a millisecond to second timescale for NMR! – technical requirements for the microwave pulses and for the time resolution of the spectrometer are much more demanding for pulse-EPR. Pulse lengths have to be shorter than 100 ns for useful work, and much pulsed EPR work is done at very low temperatures (2–100 K) in order to increase the relaxation times of the investigated centre. Because all the technical limitations for pulse-EPR have been overcome in the last decade, pulsed experiments are now starting to outrun the traditional cw experiments for many applications, in a fashion similar to the field of NMR. This is also true for the combination of pulsed methods with ENDOR and multifrequency EPR.

3.3. Multifrequency EPR Spectroscopy

Multifrequency EPR and especially high-frequency EPR has proven to be an important additional tool for the interpretation of complex EPR spectra. If the experiment is performed with different microwave excitation frequencies (ranging from *S*-band, 3 GHz, up to sub-mm wavelengths, 360 GHz) the magnetic field for the resonance condition with the *Zeeman*-split electron spin levels will also differ over a large range (0.1 to 14 T). The different interactions can thus be distinguished by multi-frequency EPR spectroscopy by their specific magnetic field dependence (ranging from B_0^2 to B_0^{-4}). Zero-field splitting, for example, can be best observed at very low magnetic field values, whereas the anisotropic g-tensor contribution will be best observed at high magnetic fields. Usually, EPR experiments on proteins have to be performed with a resonant microwave cavity for sensitivity reasons; therefore, for each microwave frequency a new experimental setup is required. For the 2–50-GHz microwave-frequency range, electromagnets and classical microwave semiconductor technology can be used. Above this frequency, superconducting magnets have to be used, and, above 100 GHz, also the microwave technology has to be partially replaced by quasioptical far-infrared technology. For this very high frequency range above 100 GHz, microwave technology is still very demanding, and instruments are not commercially available, but still home built by EPR research groups [7].

4. Application to Active Sites in Enzymes

The literature on applications of advanced EPR methods for the characterization of paramagnetic sites in proteins is already quite large and growing fast. Instead of giving a review of all of these applications, the potential of these methods will be demonstrated on some of our own research examples. Examples will be shown on protein complexes of *a*) photosynthetic reaction centres of purple bacteria, *b*) soluble G-protein-nucleotide complexes with p_{21}^{ras}, and *c*) cytochrome *c* oxidase, a membrane-bound protein of the mitochondrial respiration chain.

In all three proteins, several kind of paramagnetic centres are involved, transient organic radicals created within the reaction cycle and metal ion centres.

4.1. Characterization of the Paramagnetic Centre

For organic aromatic radicals, the EPR spectra at X-band frequencies (9 GHz) are difficult to distinguish, all of them are in resonance approximately at the free electron g value (2.0023), which indicates that the angular momentum is strongly quenched for these molecules, and a typical line-width of 1–2 mT. They can be characterized by ENDOR spectroscopy but also very efficiently distinguished by their g-tensor anisotropy observable by means of high-field EPR spectroscopy. This is demonstrated (*Fig. 4*) for two radicals created transiently within the photocycle of photosynthetic purple bacteria and higher plants [10]. After photoexcitation, within less than a ns, an electron is transferred in the protein complex over 2 nm, and a chlorophyll dimer cation radical and an semiquinone anion radical are transiently created. With pulsed EPR methods, it is possible to directly observe these intermediate radicals, their mobility, and their interaction with the protein surrounding within the reaction. The *g*-tensor can be used, in this case, to identify and distinguish the paramagnetic molecules, to select specifically orientated molecules from the disordered protein sample, and to observe the mobility of the paramagnetic molecule within its protein surrounding, as will be described later.

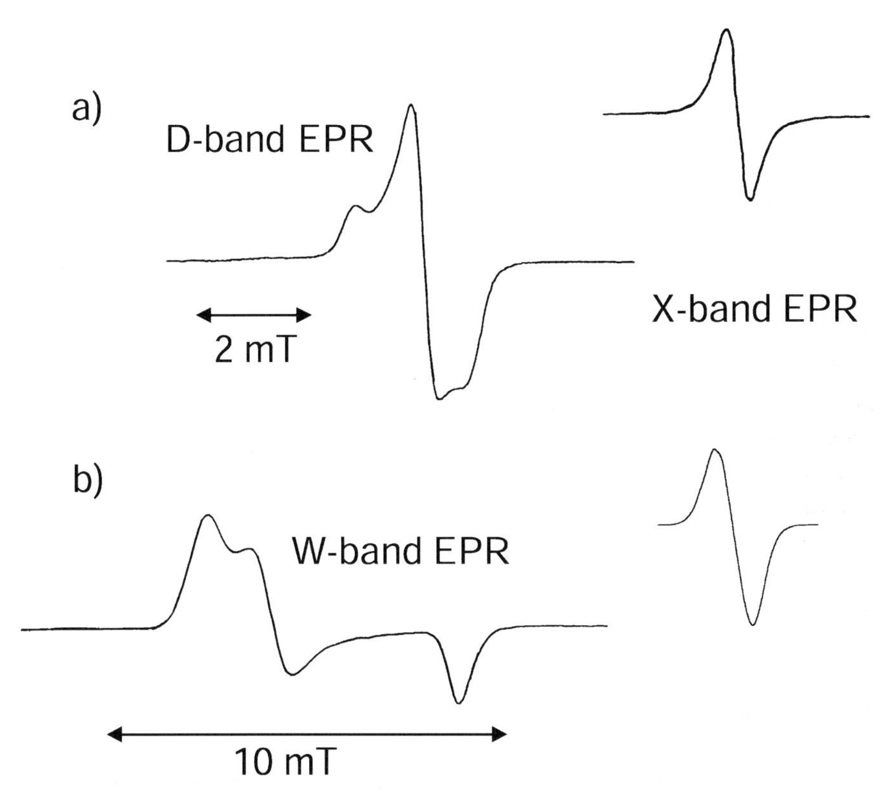

Fig. 4. *High-field* (left side) *and X-band* (right side) *EPR spectra on two transient radicals created within the photocycle in photosynthetic proteins. a)* $P_{700}^{+\cdot}$ chlorophyll dimer cation radical of deuterated cyanobacteria, measured at 140 and at 9 GHz. *b)* $Q_A^{-\cdot}$ semiquinone radical of bacterial reaction centres of *Rhodobacter sphaeroides R26* measured at 95 and at 9 GHz, respectively.

4.2. Characterization of the Local Environment

Even more interesting than a characterization of the paramagnetic centre itself is often information on the local structure and the dynamics of the protein surrounding the paramagnetic centre. This includes, for example, information on metal-ion ligands and on H-bonds between the paramagnetic centre and the protein, and, therefore, on the specific function of the paramagnetic centre in the enzymic reaction.

4.2.1. *Ligands*

Ligands to paramagnetic metal centres play an important role on the parameters describing the EPR spectra. So, for instance, they will influence the *g*-tensor as well as the metal hyperfine splitting. These changes are related to the covalency of the metal–ligand bond, charge-transfer, and spin-orbit contributions. For transition-metal ions with spin states *S* larger than $^1/_2$, the ZFS tensor can, similar to the *g*-tensor, give information on the paramagnetic centre and its ligand sphere. While the symmetry of the *g* and ZFS tensors can easily be related qualitatively to the symmetry of the paramagnetic centre; quantum-theoretical calculations for the quantitative relation between these parameters and the molecular structure of the centre are still difficult. Nevertheless, empirical correlations exist to relate these tensors to symmetry and type of ligands. If the ligand has a nuclear spin ($I \neq 0$), it may be possible to identify the nucleus directly by its hyperfine coupling. Where the number of hyperfine lines will unravel the multiplicity of the nuclear spin *I*, multifrequency ENDOR or ESEEM can unravel the gyromagnetic ratio (γ_I) of the nuclear spin. Taken together this information will help to identify uniquely the ligand. For a number of important nuclei, isotopes exist with different nuclear spin numbers (*e.g.*, $^1H/^2H$, $^{12}C/^{13}C$, $^{14}N/^{15}N$, $^{16}O/^{17}O$). In these cases, the ligand can be identified by isotope exchange. This is shown in *Fig. 5* for a Mn metal centre in the G-protein complex of $p_{21}^{ras} \cdot Mn^{+II} \cdot$ GDP. At high magnetic fields, the ZFS contribution to the central EPR transition ($m_S = +1/2$ to $m_S = -1/2$) is strongly reduced, therefore, the hyperfine coupling of a ^{17}O ligand can easily be observed in the EPR spectra. Where an ^{17}O-isotope labelling at the α-O-position of the GDP nucleotide did not show an additional hyperfine splitting on the line, isotope exchange at the β-O-position of the GDP clearly shows an oxygen hyperfine coupling and, therefore, identifies the β-O of the nucleotide as a direct ligand to the metal ion [11] This assignment has been confirmed by X-ray crystal structures [12].

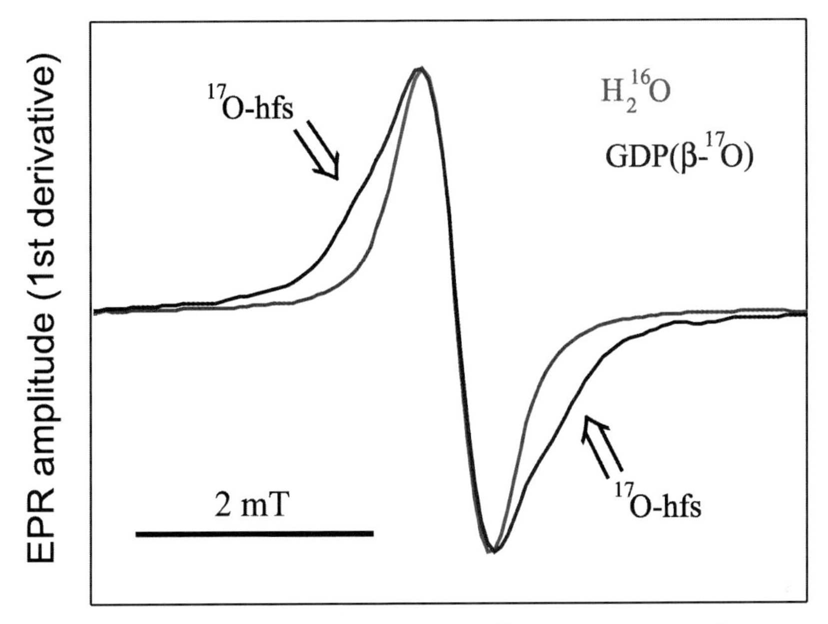

Fig. 5. *High-field* (95 GHz) *EPR spectra of* $p_{21}^{ras} \cdot Mn^{+II} \cdot$ *GDP complex with* ^{17}O *labeling of the GDP at specific positions.* Shown is only one Mn hyperfine line ($m_I = 5/2$) of the central EPR transition ($m_s = +1/2$ to $m_s = -1/2$). The experiment is performed at room temperature, and only 10 pmol of protein is used for this measurement.

4.2.2. *Protein Surrounding*

With modern methods such as ENDOR or ESEEM, it is not only possible to identify direct ligands to the paramagnetic centre, but also to detect nuclei with a nuclear spin I up to distances of *ca.* 0.6 nm. Because, for these more distant nuclei, the interaction with the electron spin is mostly dipolar, the distance and, in specific cases, also the orientation with respect to the paramagnetic molecule can be determined. This has been done, for example, in great detail for the transient semiquinone radical $Q_A^{-\cdot}$ of bacterial reaction centres [5][12].

4.2.3. *Hydrogen Bonding*

An especially important case is H-bonding of a paramagnetic molecule to the protein. In membrane-located proteins, quinone molecules play an important role in electron-transfer reactions. These quinone molecules can be bound to the protein by H-bonds or may diffuse through the membrane. By resolving the dipolar hyperfine coupling of protons to the quinone molecule by means of ENDOR or ESEEM methods it can be clarified whether the observed semiquinone radical is H-bound to the protein or freely diffusing in the membrane. Again this has been observed, for example, on the quinone cofactors of bacterial reaction centres [5][12].

4.2.4. *Distance to Other Paramagnetic Sites*

For chemical redox reactions at an active site of membrane-bound proteins, electrons and protons have to be channelled to this site. For the electron transfer, often a chain of paramagnetic centres with distances in the order of 1–3 nm leads from the membrane surface to the active site. The determination of these molecules, their relative distances and orientations with EPR spectroscopy, is an important tool to locate the active site within the protein and to understand the electron-transfer reaction in detail. Whereas, for short distances (<2 nm), the dipolar coupling of two adjacent paramagnetic molecules may be directly observable on the cw-EPR spectra, for longer distances, pulsed methods, as described above, have to be used to measure distances up to 6 nm. An example of such a dipolar coupling between a Cu centre and a Mn^{+II} ion is shown in *Fig. 6*. The binuclear Cu centre in cytochrome *c* oxidase consists of two strongly coupled Cu-atoms, called Cu_A, which is involved in the electron-transfer reaction within the protein. In the oxidized state, the electron spin state of this centre is $S = 1/2$, whereas, for the reduced state, the centre is diamagnetic ($S = 0$). By multifrequency EPR spectroscopy, the coupling of the Cu_A centre to the Mn ion could be determined [13]. A detailed analysis of the spectra allowed a determination of the distance and the relative orientation of the two metal centres, in very good agreement to the X-ray structure [14]. The analysis showed that the exchange interaction J between these two centres has to be rather small (<3 G), which excludes the possibility that the Mn ion participates directly in the electron-transfer reaction from Cu_A. This again is in good agreement with biochemical assumptions about the electron-transfer mechanism, which predict a heme *a* molecule to be the electron acceptor from the Cu_A centre. Distance and relative-orientation measurements between the cofactors in photosynthetic proteins have also been performed for distances up to 2.5 nm [12] and have led to a very detailed understanding of their arrangement in the protein, with a superior resolution to that seen in X-ray structure models.

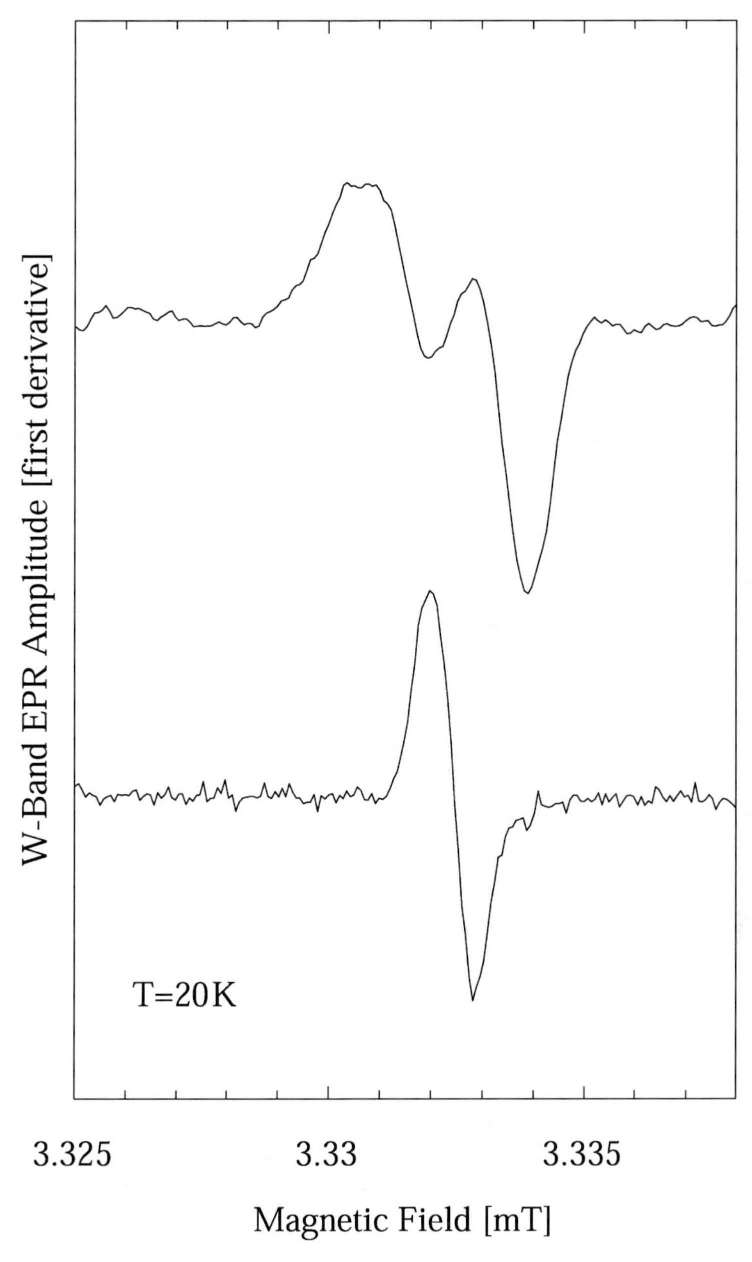

Fig. 6. *High-field* (95 GHz) *EPR spectra of the* Mn^{+II} *metal ion in cytochrome* c *oxidase of* Paracoccus denitrificans. Shown is only one Mn hyperfine line ($m_I = 5/2$) of the central EPR-transition ($m_s = +1/2$ to $m_s = -1/2$).The splitting from the dipolar coupling to the Cu_A centre can be observed in the oxidized enzyme below 60 K (upper spectrum). Lower spectrum is of the reduced enzyme where the Cu_A centre is diamagnetic.

4.2.5. *Dynamics of Local Structure*

EPR Spectroscopy can describe not only the structure of paramagnetic molecules and their local surrounding, but also dynamic aspects of the paramagnetic molecule, as well as its surrounding. As has been known from time-resolved optical experiments, these dynamics play an important role in the kinetics of electron-transfer reactions in proteins. Dynamics on different timescales can be probed by EPR experiments. Whereas dynamics on the sub-nanosecond timescale influences directly the EPR-spectral line shapes *via* relaxation broadening, dynamics on the nanosecond to microsecond timescale can be probed by two-pulse echo experiments. Dynamics on the microsecond to millisecond timescale can be probed by stimulated echo, inversion recovery, and spinlock experiments. Therefore, dynamics ranging from fast reordering of the paramagnetic molecule, for example, after light excitation, to relaxation processes of the local protein surrounding and finally to slow tumbling of the whole protein or site can be observed by EPR spectroscopy. As an example, the librational motion of the semiquinone radical $Q_A^{-\bullet}$ of bacterial reaction centres is shown in *Fig. 7* [15]. The quinone molecule in its protein pocket is still mobile even at temperatures as low as 100 K. But because this quinone molecule is H-bound to the protein surrounding, its librational motion is restricted to a uniaxial motion around the quinone C–O axis (x-axis). The librational motion on the timescale of a few nanoseconds leads to a T_2 relaxation, defined by:

$$\frac{1}{T_2} = \langle \Delta\omega^2 \rangle \, \tau_c$$

where τ_c is the rotational correlation time of the librational motion and $\langle \Delta\omega^2 \rangle$ is the average quadratic deviation of the resonant *Larmor* frequency by the motional process. At high magnetic fields, the modulation of the resonance frequency by the motional process is directly related to the anisotropy of the g-tensor, as this becomes the dominant orientation-dependent interaction. Additionally, because the EPR spectrum resolves the g-tensor at high magnetic fields, as described above, the relaxation time T_2 can now be measured selectively for different field values B_0 (and, therefore, ordered sub-ensembles of the quinone molecules with respect to the magnetic field axis). Because the slope of the anisotropy of the g-tensor is different for all directions, the modulation in *Larmor* frequency $\langle \Delta\omega^2 \rangle$ and, therefore, the relaxation time T_2 is dependent on the resonance position within the EPR line:

$$\langle \Delta\omega(B_0)^2 \rangle = \langle (\omega_{\text{static}}(B_0) - \omega(B_0, t))^2 \rangle \propto (g_{\text{static}}(B_0) - g(B_0, t))^2 \rangle \cdot B_0^2$$

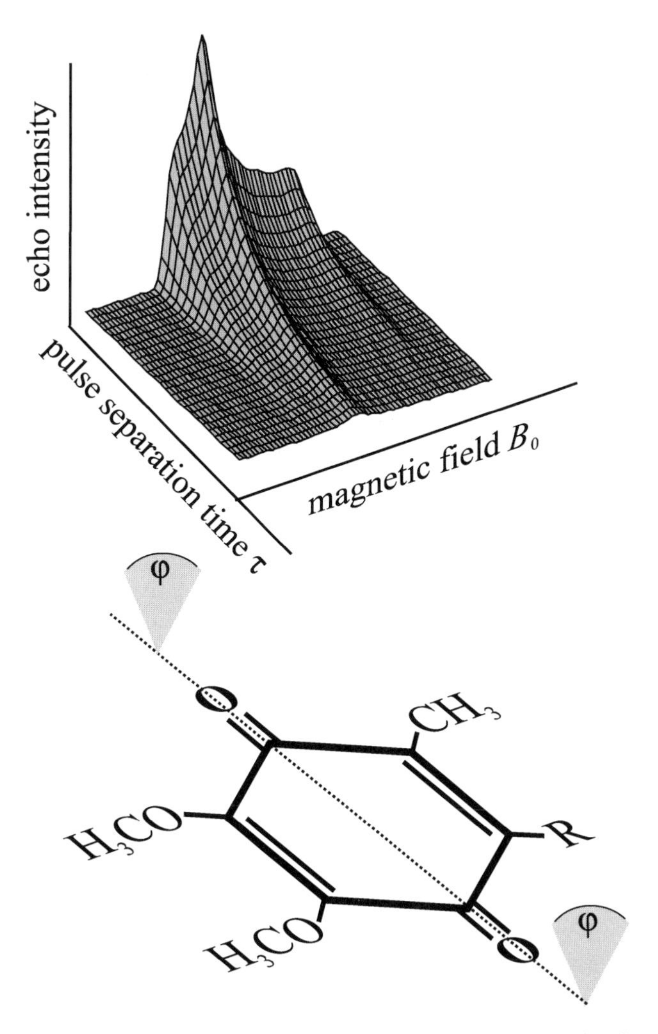

Fig. 7. *High-frequency* (95 GHz) *two-pulse echo decay of the* $Q_A^{-\cdot}$ *radical in bacterial reaction centres of* Rhodobacter sphaeroides. The echo decay is measured at 110 K for different resonant field positions within the spectra as explained in the text. The anisotropy of the relaxation time T_2 can be explained by a uniaxial librational motion of the semiquinone molecule along its C–O bond direction (x-axis) as shown at the bottom of the figure.

where $g_{\text{static}}(B_0)$ is the g value for the specific resonance field position of a static sample. The librational motion changes the orientation of the molecule with respect to the external magnetic field on the timescale of a two-pulse echo experiment, leading to an anisotropic relaxation time $T_2(B_0)$ over the EPR spectra. This relaxation pattern can now be compared with different possible modes of librational motion. Different motions of the quinone molecule can, therefore, be distinguished by these measurements. For example, a uniaxial motion along the *x*-axis of the quinone molecule can be easily distinguished from an isotropic Brownian tumbling motion. These measurements can provide a detailed insight on the timescale and amount of librational motion, related to the H-bonding strength and steric restrictions imposed by the protein pocket. They can also be simply used as a fingerprint of a protein-bound quinone molecule, which cannot easily be deduced by other techniques.

5. Summary and Outlook

EPR Spectroscopy has played a role as an spectroscopical technique in the understanding of biochemical processes in enzymes on a molecular level ever since its development. Nevertheless, the complexity of biological systems together with the low spectral resolution of cw-EPR spectra at a single microwave frequency (typically 9 GHz/X-band) have hindered, in many cases, a unique description of the structure of the paramagnetic site by EPR spectroscopy alone. Similar to NMR spectroscopy, specific advanced techniques have recently been developed in the field of EPR, which allow the determination of much more detailed information on complex paramagnetic spin systems such as those in enzymes. The consequent application of these techniques, pulse-EPR and ENDOR, multifrequency and especially high-frequency EPR, allows the confirmation of the assignments obtained, disentangles the different contributions to the spectra in a unique way, and also detects interactions with the surrounding of the paramagnetic molecule in a much more sensitive and direct fashion. I hope that the selected examples of our own work on structural and dynamic aspects of paramagnetic centres in proteins has helped to give an impression of the enormous possibilities these techniques offer to the detailed investigation on paramagnetic centres in biological systems. It should be clear that these examples nevertheless only cover a small fraction of the growing field of modern EPR applications to biology. Whereas the focus of this article was on the applications of advanced EPR techniques, the improved modern techniques of biochemical manipulation of proteins, such as site-directed mutagenesis, selective isotope labelling, and incorporation of spin labels, have also raised numerous new applications of EPR spectroscopy in this field. Therefore, it is my strong belief that EPR spectroscopy will gain increased importance in the characterization of the active sites, redox centres, and electron-transfer reactions in large protein complexes in the future.

The work on photosynthetic proteins were performed together with the groups of *K. Möbius* and *D. Stehlik* (FU Berlin), and the group of *W. Lubitz* (TU Berlin), and was supported by the *DFG* (Sfb337). The high-field EPR experiments on G-proteins were performed by

M. Rohrer in cooperation with the group of *H. R. Kalbitzer* (University of Regensburg) and the group of *A. Wittinghofer* (MPI for Molecular Physiology, Dortmund), and were supported by the 'DFG-Schwerpunktprogramm' *'High-field EPR in Biology, Chemistry and Physics'*. The work on cytochrome *c* oxidase was performed by *F. MacMillan* in cooperation with the group of *B. Ludwig* (Department of Biochemistry) and the group of *H. Michel* (MPI for Biophysics, Frankfurt), and was supported by the *DFG* (Sfb472). The group of *K. P. Dinse* (University of Dortmund) is acknowledged for W-band measurements on cytochrome *c* oxidase and on the G-proteins.

REFERENCES

[1] C. P. Poole, H. A. Farach, *The Theory of Magnetic Resonance,* John Wiley & Sons, New York, 1972; N. M. Atherton, *Principles of Electron Spin Resonance*, E. Horwood and PTR Prentice Hall, 1993.

[2] U. Ermler, G. Fritzsch, S. K. Buchanan, H. Michel, *Structure*, **1994**, *2*, 925–936.

[3] H. Kurreck, B. Kirste, W. Lubitz, *ENDOR Spectroscopy of Radicals in Solution,* VCH, 1988; E. R. Davies, *Phys. Lett. A* **1974**, *47*, 1–2; Arthur Grupp and Michael Mehring, in *'Modern Pulsed and Continuous Wave ESR'*, 1979, 195–229.

[4] J. Forrer, S. Pfenninger, J. Eisenegger, A. Schweiger, *Rev. Sci. Instrum.* **1990**, *61*, 3360–3367; M. Rohrer, M. Plato, F. MacMillan, Y. Grishin, W. Lubitz, K. Möbius, *J. Magn. Reson., Ser. A* **1995**, *116*, 59–66.

[5] M. Rohrer, F. MacMillan, T. F. Prisner, A. T. Gardiner, K. Möbius, W. Lubitz, *J. Phys. Chem.* **1998**, *102,* 4648–4657.

[6] A. Schweiger, *Angew. Chem.* **1991**, *103*, 223–250.

[7] S. Stoll, G. Jeschke, M. Willer, A. Schweiger, *J. Magn. Reson.* **1998**, *130*, 86–96.

[8] R. G. Larsen, D. J. Singel, *J. Chem. Phys.* **1993**, *98*, 5134–5146.

[9] Klaus Möbius, *Biol. Magn. Reson.* **1993**, *13*, 253–274; T. F. Prisner, *Adv. Magn. Opt. Reson.* **1997**, *20*, 245–300.

[10] O. Burghaus, M. Plato, M. Rohrer, K. Möbius, F. MacMillan, W. Lubitz, *J. Phys. Chem.* **1993**, *97*, 7639–7647; T. F. Prisner, A. E. McDermott, S. Un, J. R. Norries, M. C. Thurnauer, R. G. Griffin, *Proc. Natl. Acad. Sci. U.S.A.* **1993**, *90*, 9485–9488.

[11] J. Feuerstein, H. Kalbitzer, J. John, R. S. Goody, A. Wittinghofer, *Eur. J. Biochem.* **1987**, *162*, 49–55; T. F. Prisner, *Adv. Magn. Opt. Reson.* **1997**, *20*, 245–300.

[12] D. Stehlik, K. Möbius, *Annu. Rev. Phys. Chem.* **1997**, *48*, 739–778.

[13] H. Käss, F. MacMillan, B. Ludwig, T. F. Prisner, *J. Phys. Chemistry* **2000**, *104*, 5362–5371.

[14] C. Ostermeier, A. Harrenga, U. Ermler, H. Michel, *Proc. Natl. Acad. Sci. U.S.A.* **1997**, *94,* 10547.

[15] M. Rohrer, P. Gast, K. Möbius, T. F. Prisner, *Chem. Phys. Lett.* **1996**, *259*, 523–530.

Reactivity Concepts for Oxidation Catalysis: Spin and Stoichiometry Problems in Dioxygen Activation

by **Detlef Schröder** and **Helmut Schwarz**

Institut für Organische Chemie der Technischen Universität Berlin,
Straße des 17. Juni 135, D-10623 Berlin

Es ist der Valenzwechsel einer Eisenverbindung, auf dem die katalytische Oxydation in der lebendigen Substanz beruht.

Otto Warburg, Les Prix Nobel 1931

1. Introduction

Life on earth is by and large based on metastability; organic matter would spontaneously combust in our O_2-rich atmosphere unless significant kinetic restrictions were operative. The origin of this metastability is fairly simple. It is associated with the electronic ground state of molecular oxygen O_2 ($^3\Sigma_g^-$); according to *Wigner*'s rule of spin conservation, the triplet configuration imposes a spin-inversion bottleneck to the oxidation of organic matter [1]. In contrast, singlet oxygen lacks such spin constraints and is well-known to react rapidly with oxidizable substrates [2].

Consider the hypothetical perhydroxylation of methane as an example[1]). Although this leads to a peroxide, the activation of $^3O_2$ in *Reaction 1a* is highly exothermic due to the formation of strong O–H and C–O bonds [4]. However, the process is spin-forbidden, as methane and methyl hydroperoxide are singlets while dioxygen is a triplet. In contrast, the spin-conserving *Reaction 1b*, though, is considerably endothermic [5], and consequently both reactions are hampered. Stepwise, radical-type processes, such as the combination of *Reactions 2a* and *2b*, circumvent the spin bottleneck, but here the first step is quite endothermic. Hence, oxidation of CH_4 by dioxygen – and also the onset of combustion of hydrocarbons – is likely to coincide with the thermal activation of *Reaction 2a*.

---

[1]) Throughout this paper, a simplified notation of spin states by superscripts preceding the formula is used, *e.g.*, ground state O_2 ($^3\Sigma_g^-$) is referred to as $^3O_2$.

$$^{1}CH_4 + {}^{3}O_2 \rightarrow {}^{1}CH_3OOH \qquad \Delta_rH = -14 \text{ kcal/mol} \qquad (1a)$$

$$\rightarrow {}^{3}CH_3OOH \qquad \Delta_rH = +28 \text{ kcal/mol} \qquad (1b)$$

$$^{1}CH_4 + {}^{3}O_2 \rightarrow {}^{2}CH_3^{\bullet} + {}^{2}HOO^{\bullet} \qquad \Delta_rH = +55 \text{ kcal/mol} \qquad (2a)$$

$$^{2}CH_3^{\bullet} + {}^{2}HOO^{\bullet} \rightarrow {}^{1}CH_3OOH \qquad \Delta_rH = -69 \text{ kcal/mol} \qquad (2b)$$

Nevertheless, direct use of dioxygen as a terminal oxidant would be very attractive for both economic and ecologic reasons. For example, easy and safe procedures for the partial oxidation of methane according to *Reactions 3* and *4* would offer tremendous possibilities for the on-site conversion of CH_4, stemming from biological or geological resources, into more valuable feedstocks.

$$CH_4 + \tfrac{1}{2}O_2 \rightarrow CH_3OH \qquad (3)$$

$$CH_4 + O_2 \rightarrow CH_2O + H_2O \qquad (4)$$

In the course of evolution, Nature has developed ways to mediate these oxidation processes precisely when and where they are required. For this purpose, $^{3}O_2$ needs to be converted into formal singlet O-atom equivalents $^{1}\langle O \rangle$: *e.g.*, μ^2-peroxo ligands, free or complexed peroxides, high-valent metal-oxo species *etc.*

Another concern in an atom-economic design of oxidations involves the use of dioxygen for the selective monoxidation of a substrate ***S*** according to *Reaction 5*.

$$2\,S + O_2 \rightarrow 2\,SO \qquad (5)$$

This reaction appears straightforward, and appropriate processing may permit the prevention of over-oxidation – that is, subsequent degradation of the product ***S***O – as long as the rate for the oxidation of ***S*** is not much slower than that for ***S***O. Furthermore, the use of dioxygen provides a formidable thermochemical driving force such that *Reaction 5* is exothermic for almost every substrate.

A third topic, not specific to the use of dioxygen as oxidizing agent, concerns selectivity and, in particular, the closely related aspect of over-oxidation. This problem is most severe in the partial oxidation of saturated hydrocarbons, because activation of alkanes is usually more difficult than that of any of the oxygenated products (*e.g.*, alcohols and aldehydes), which are easily oxidized further. The reason is simply that C–H bonds of functionalized alkanes are generally weaker than those of the parent hydrocarbons. Moreover, as far as metal catalysis is concerned, the polar oxidation products can coordinate to the active sites better than the hydrocarbons. For example, the bond-dissociation energies of CH_3OH and CH_2O to bare Fe^+ cation in the gas phase exceed that of CH_4 more than twofold, *i.e.*, $D(Fe^+–CH_3OH)$ = 34.4 kcal/mol [6] and $D(Fe^+–CH_2O)$ = 33.0 kcal/mol [7] *vs.* $D(Fe^+–CH_4)$ = 13.6 kcal/mol [3b]. While this effect is particularly pronounced for charged species because of ion/dipole interactions, similar trends can be expected for formally neutral transition-metal centers in the active sites of real catalysts, giving rise to increased residence times for the oxygenated products and so further increasing the risk of over-oxidation.

Last but not least, the metastability of substrate/oxygen mixtures is associated with non-trivial problems in processing and handling. Specifically, since the presence of oxidation catalysts may significantly lower ignition temperatures, any formation of hazardous mixtures needs to be rigorously excluded under all operating conditions. Thus, the seemingly 'ideal' atom-economic oxidation of substrates with dioxygen imposes considerable challenges in chemical engineering.

From a mechanistic point of view, the direct use of O_2 as oxidant has to meet two major problems. These are outlined below, followed by strategies for possible solutions. To this end, two representative case studies are discussed, illustrating the scope and limitations of gas-phase experiments. We omit experimental details and restrict ourselves to the fundamental requirements in the activation of O_2 as deduced from model studies conducted with isolated species in the highly diluted gas phase, thus excluding complicating effects due to the presence of solvents, counter ions, aggregation, bulk phenomena *etc.* (for reviews, see [7]).

2. The Problems

There are two fundamental problems in the direct use of dioxygen as terminal oxidant in partial oxidations: *i*) The spin dilemma, *i.e.*, how to invert spin in the transformation $^3O_2 \rightarrow 2\ ^1\langle O \rangle$; *ii*) The stoichiometry aspect, *i.e.*, how to use both $^1\langle O \rangle$ equivalents in partial oxidations. We need to develop systems for achieving selectivity and for preventing, or at least limiting, over-oxidation. In this respect, it is essential to avoid reactive radical intermediates, especially oxygen-centered radicals because these tend to promote atom-abstraction reactions. Radical pathways often result in low selectivities and loss of stereochemical integrity. They may even provoke the undesired ignition of substrate/oxygen mixtures.

The Spin Dilemma

To employ triplet dioxygen directly for selective oxidations, spin inversion is mandatory. This can proceed either through single-electron transfer (SET) or through intersystem crossing by spin-orbit coupling (SOC). In both cases, transition metals come into play, because they can easily undergo redox reactions, thereby affecting SET, and also because they exhibit significant SOC to mediate formally spin-forbidden reactions. Furthermore, transition metals may coordinate and possibly activate the substrates, thereby increasing the number of tunable modules for the achievement of selectivity by ligand effects.

First, we consider the activation of dioxygen *via* a sequence of SET and protonation steps to produce (singlet) hydrogen peroxide, from which a $^1\langle O\rangle$-atom equivalent can be delivered without involving spin-forbidden reactions in the elementary steps (*Reactions 6a–6f*).

$$^3O_2 + e^- \rightarrow {}^2O_2^{-\bullet} \tag{6a}$$
$$^2O_2^{-\bullet} + H^+ \rightarrow {}^2HOO^\bullet \tag{6b}$$
$$^2HOO^\bullet + e^- \rightarrow {}^1HOO^- \tag{6c}$$
$$^1HOO^- + H^+ \rightarrow {}^1HOOH \tag{6d}$$
$$^1HOOH \rightarrow {}^1H_2O + {}^1\langle O\rangle \tag{6e}$$
$$^1\mathbf{S} + {}^1\langle O\rangle \rightarrow {}^1\mathbf{S}O \tag{6f}$$

Here, transition metals – as well as other reductants – can serve as electron donors in the activation of dioxygen (*Reaction 6a*). Instead of the formal H-atom equivalents, *i.e.*, $H^+ + e^-$, used as co-reductants in *Reactions 6a/6b* and *6c/6d*, the substrate itself may induce the reduction of O_2 to $^1\langle O\rangle$ equivalents (see below). Transition metals can serve to fix the $^1\langle O\rangle$ equivalent formed in *Reaction 6e* as a metal peroxide or high-valent metal-oxo species. They may also help to mediate subsequent oxygenation of the substrate in *Reaction 6f.*

From a reactivity/selectivity point of view, however, the generation of intermediate superoxide $^2O_2^{-\bullet}$ and hydroperoxy $^2HOO^\bullet$ radicals poses a dilemma, as these radicals may also act as H-atom abstractors. This is because the corresponding O–H bond strengths, $D(H{-}OO^-) = 80$ kcal/mol and $D(H{-}OOH) = 90$ kcal/mol [8], are close to the strengths of activated C–H bonds in organic compounds; *e.g.*, $D(C_2H_3CH_2{-}H) = 88.2$ kcal/mol, $D(C_6H_5CH_2{-}H) = 88.5$ kcal/mol, $D(CH_3C(O){-}H) = 89.4$ kcal/mol [9]. Furthermore, side reactions may give rise to the production of even more reactive species such as the hydroxy radical ($D(HO{-}H) = 118.1$ kcal/mol), due to homolytic O–O bond cleavage of hydrogen peroxide. Accordingly, reduced selectivities are expected unless the sequence of *Reactions 6a–6d* is fast compared with the reactions of the substrate with the intermediate peroxy radicals. Nature often solves this particular problem by compartmentalization; *i.e.*, physical, spatial separation of SET steps, and oxidation of the substrate. For oxidation catalysis in chemical technology, however, this option is less favorable as it would require elaborate processing procedures associated with considerable fixed costs. While the SET route and related radical-type processes are certainly viable options for oxidations of specific substrates using dioxygen – for example, the Ag-mediated epoxidation of ethene, the *Wacker* oxidation of olefins, or the production of phenol *via* the *Hock* process – they cannot provide general strategies because selectivity depends crucially on the relative rates of *Reactions 6a–6f* compared with side reactions. Of course, there do exist some examples of selective oxidations using O_2 in 'ideally' atom-economic procedures, *e.g.*, as the epoxidation of ethene on silver contacts. However, this very reaction is limited to the C_2H_4/O_2 couple; other olefins give poor yields due to competing processes such as allylic oxidation, epoxide isomerization *etc.* A prominent example is the large-scale epoxidation of propene using O_2, a process for which no satisfactory solution has so far been achieved (for a recent example, see [10]). Hence, we require alternative routes that allow for the activation of O_2 in a formally concerted manner without involving radical intermediates.

Circumvention of the spin problem by intersystem crossing (ISC) between surfaces of different multiplicities can be mediated by compounds with sizable spin-orbit coupling constants. While heavy elements exhibit significant SOC, transition metals are particularly attractive in the context of catalytic oxidation, because, in addition to the mere mediation of spin changes, they can accept the resulting redox equivalents by appropriate changes in their valence states. Among the d-block series, the 3d metals appear most suitable, not only because of their abundance, but also because their matrix elements relevant for SOC are not too large. In contrast, spin inversion approaches unit efficiency for 4d and 5d metals; in fact, spin is no longer a good quantum number for 5d elements. For example, ever since *Döbereiner*'s invention of the lighter elemental platinum has been known to activate O_2. Indeed, platinum is too efficient in this respect and likely to bring about complete combustion of the organic substrate [11]. Hence, as the spin-inversion barriers are unlikely to play any role for these heavy metals, thermochemical equilibrium is established once dioxygen activation has been achieved. However, the intermediate SOC of the 3d metals offers the opportunity to modulate spin inversion in dioxygen activation by means of ligand effects.

Among the 3d series, we may further restrict the choice of metals for circumvention of the spin problem to those that do so by means of SOC rather than SET[2]). Specifically, early transition metals, because of their large oxophilicity, tend to exist in their highest oxidation states in the presence of oxidizing agents. These formal $3d^0$ compounds are, in turn, singlet coupled – *e.g.*, Sc^{III} in Sc_2O_3, Ti^{IV} in TiO_2, V^V in V_2O_5, and Cr^{VI} in CrO_3 – so that spin-orbit coupling is small in their ground states. Hence, more attractive candidates for selective oxidation *via* the SOC route are the late 3d transition metals, which are well-known to undergo facile transitions between low- and high-spin states. This reasoning finds phenomenological support in the metals Nature has chosen as components of metallo-enzymes involved in the generation, transportation and activation of O_2: *e.g.*, manganese, iron, and copper [1c][14]. However, it can also be argued that these particular metals were selected in the course of evolution simply because of their high abundance in the Earth's crust and their mobility in aquifers and soils.

Another manifestation of precisely the same spin problem is shown by the catalase activity of several synthetically useful catalysts, where peroxides are employed as terminal oxidants. In analogy to the corresponding enzyme function, catalase activity refers to the decomposition of hydrogen peroxide according to $H_2O_2 \rightarrow H_2O + \frac{1}{2}O_2$ – another example of a highly exothermic process ($\Delta_rH = -26$ kcal/mol) associated with a spin-inversion bottleneck if dioxygen is to be formed in its triplet ground state. In these synthetic procedures, decomposition of the peroxide into dioxygen constitutes an undesired side-reaction that lowers efficiency and may reduce both selectivity and turnover number through radical pathways, such as *Fenton*-style formation of hydroxyl radicals. Catalase activity should thus be prevented for these catalysts. If we exclude SET routes, release of $^3O_2$ from a singlet peroxide requires a spin crossover, *e.g.*, the coupling of two $^1\langle O\rangle$ equivalents or the interconversion of a singlet coupled μ^2-peroxo ligand to a triplet dioxygen complex (*Scheme 1*). Both intersystem crossings can be mediated by metals possessing significant spin-orbit coupling and variability in accessible oxidation states; typical examples are the decompositions of H_2O_2 catalyzed by manganese and iron salts. In contrast, it is not surprising that several d^0-metal compounds lack catalase activity and can thus be used as efficient catalysts in oxidations employing peroxidic reagents, such as alkene epoxidations and dihydroxylations employing H_2O_2 or organic peroxides as oxidants in combination with Ti^{IV}, V^V, Mo^{VI}, Re^{VII}, or Os^{VIII} compounds. In summary, the ability of the catalysts to mediate spin inversion is desired in the use of dioxygen as terminal oxidant, whereas SOC of the metal should be as small as possible in metal-mediated oxidations with peroxides.

---

[2]) The famous photo-oxidation of organic compounds on rutile (TiO_2) involves SET as a consequence of lattice defects [12]; a gas-phase variant of this behavior is provided by the radical-type chemistry of the TiO_2^+ cation, which exhibits the characteristic reactivity features of an oxygen-centered radical [13].

Scheme 1

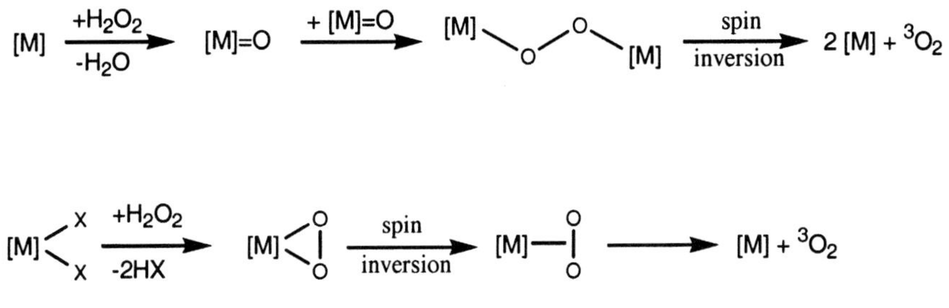

$$[M] \xrightarrow[-H_2O]{+H_2O_2} [M]{=}O \xrightarrow{+[M]=O} [M]{-}O{-}O{-}[M] \xrightarrow[\text{inversion}]{\text{spin}} 2\,[M] + {}^3O_2$$

$$[M]X_2 \xrightarrow[-2HX]{+H_2O_2} [M](\eta^2\text{-}O_2) \xrightarrow[\text{inversion}]{\text{spin}} [M]{-}O{-}O \longrightarrow [M] + {}^3O_2$$

(X=Cl, OH, *etc.*)

The Stoichiometry Aspect

Although SET can circumvent the spin-inversion bottleneck, selective oxidation of organic compounds is difficult to achieve by this means (see above). In addition, the SET pathway for substrate oxidation according to *Reactions 6a–6f* reveals the existence of yet another obstacle to the use of O_2 as terminal oxidant. The overall reaction involves the action of two formal H-atom equivalents as co-reductants *en route* to the production of the reactive ${}^1\langle O\rangle$ equivalent, which eventually brings about oxidation of the substrate; note also that Nature applies a formal co-reduction using two H-atom equivalents in the activation of the resting state of cytochrome P-450 [15]. The desired, atom-economic usage of both oxidation equivalents of dioxygen according to *Reactions 5* would instead involve O–O bond cleavage and subsequent transfer of the O-atoms to *two* substrate molecules.

As far as two-electron oxidations such as hydroxylations, epoxidations, or sulfoxidations are concerned, stoichiometry considerations give rise to a mechanistic complication, because two different elementary steps are operative in the oxidation of the substrate. To clarify this argument, let us imagine a metal catalyst which can mediate the transformation ${}^3O_2 \rightarrow 2\,{}^1\langle O\rangle$. For example, if a mononuclear metal complex [M] is involved, activation of O_2 leads to a metal dioxide (or a metal peroxide) which may transfer one O-atom to the substrate in a first step, thus producing the corresponding metal monoxide. A subsequent O-atom transfer from [M]O to the substrate regenerates [M], thus closing the catalytic cycle (*Scheme 2,a*). The key issue in what we term the stoichiometry problem is that two different elementary steps are involved in the oxidation of the substrate by O_2, *i.e.*, a first O-atom transfer from a dioxo (or peroxo) species and then a second from a monoxo species. Activation parameters are unlikely to be comparable in both these steps. Hence, optimization of selectivity for a given substrate is difficult, if not impossible. In a *'Gedankenexperiment'*, this scenario can be extended to apply to metal clusters or surfaces. For example, the reactivities of the dinuclear clusters $[M_2]O_2$ and $[M_2]O$ towards a substrate ***S*** are likely to differ because of proximity effects (*Scheme 2,b*). This reactivity difference persists on a surface when the $^1\langle O\rangle$ equivalents remain in close contact (*Scheme 2,c*). Although diminishing in importance, this situation also pertains to larger cluster sizes. It vanishes only when complete dissipation of the $^1\langle O\rangle$ equivalents is reached; *e.g.*, *via* O-atom migration on the surface of a heterogeneous catalyst (*Scheme 2,d*), or *via* dissociation of an intermediate $[M_2]O_2$ species into two identical [M]O entities which then serve as oxidizing agents under homogeneous conditions (*Scheme 2,e*). There are other, termolecular approaches towards solutions of the stoichiometry problem, such as the separation of oxidation and reduction steps by means of two different redox couples (*e.g.*, Cu^I/Cu^{II} and Pd^0/Pd^{II} in the *Wacker* oxidation[3]). At a molecular level, however, no general approach to the stoichiometry aspect is known at this time as far as single O-atom transfer is concerned.

For a mononuclear metal complex, direct conversion of $^3O_2$ into two $^1\langle O\rangle$ equivalents would require a change of the formal valence of the metal by +IV. While such reactions have been reported to occur in solution (*e.g.*, $Cr^{II} + O_2 \rightarrow Cr^{VI}O_2$ [16]), they are unlikely to play a role in catalysis, unless strong reducing agents provide access to the required low-valent precursors (*e.g.*, Cr^{II}). This dilemma leads to the idea of co-reduction as an approach to solving the stoichiometry problem (see below).

In summary, it seems obvious that a generally applicable, simultaneous solution of the spin and stoichiometry problems is rather difficult to achieve. Indeed, as far as single O-atom transfer is concerned, the two problems are inherently linked. Thus, a metal catalyst capable of both activating $^3O_2$ and mono-oxygenating a substrate has to release one O-atom equivalent. For example, in the oxygenation of an alkane (R–H) by dioxygen, the residual O-atom may be bound as a metal-oxo species or as a peroxide (*Reaction 7*).

$$[M] + R{-}H + {}^3O_2 \rightarrow [M]{=}O + R{-}OH \qquad (7a)$$

$$\rightarrow [M] + R{-}OOH \qquad (7b)$$

$$[M]{=}O \rightarrow [M] + \tfrac{1}{2}\,{}^3O_2 \qquad (8a)$$

$$[M] + R{-}OOH \rightarrow [M] + R{-}OH + \tfrac{1}{2}\,{}^3O_2 \qquad (8b)$$

[3]) Some bimetallic Pd/Cu and Pd/Fe catalysts were recently reported to be capable of activating methane under ambient conditions using carbon monoxide as a coreductant (see M. Lin, T. Hogan, A. Sen, *J. Am. Chem. Soc.* **1997**, *119*, 6048). Also of interest is the recently described bimetallic Pd/Cu system for the catalytic hydroxylations of remote C–H bonds using O_2 (C. Shen, E. A. Garcia-Zayas, A. Sen, *J. Am. Chem. Soc.* **2000**, *122*, 4029).

Scheme 2

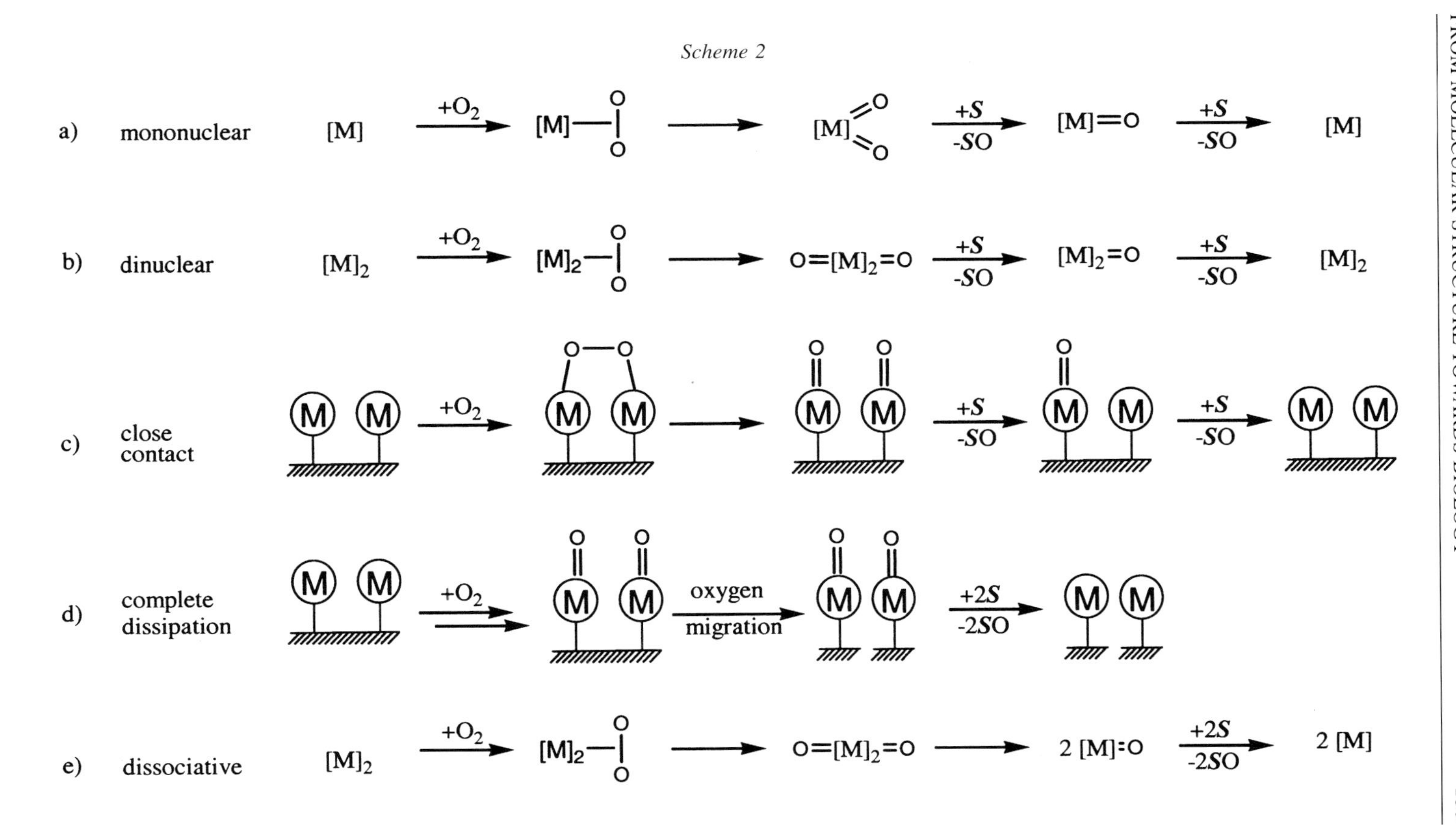
a) mononuclear
[M]
+O2
[M]=O
+S
-SO
b) dinuclear
[M]2
O=[M]2=O
[M]2=O
c) close contact
d) complete dissipation
oxygen migration
+2S
-2SO
e) dissociative
2 [M]:O
2 [M]

In turn, the very same catalyst is also likely to exhibit catalase activity, *i.e.*, the reversal of $^3O_2$ activation, thereby depleting the free $^1\langle O \rangle$ equivalents necessarily formed as intermediates (*Reaction 8*). Moreover, metal-catalyzed decomposition of peroxides according to *Reaction 8b* often involves radical pathways (*e.g.*, *Fenton* chemistry), thereby increasing the risk of the occurrence of non-selective pathways. Accordingly, systems for selective mediation of reactions with $^3O_2$ need to display a delicate balance of rate constants if, for example, *Reactions 7a* and *8a* have to be fast compared with all possible competing channels.

3. Can Gas-Phase Experiments Clarify the Problems?

We now ask whether and how gas-phase studies can offer conceptual models to illustrate, and so possibly help to suggest solutions to the mechan-

Scheme 3

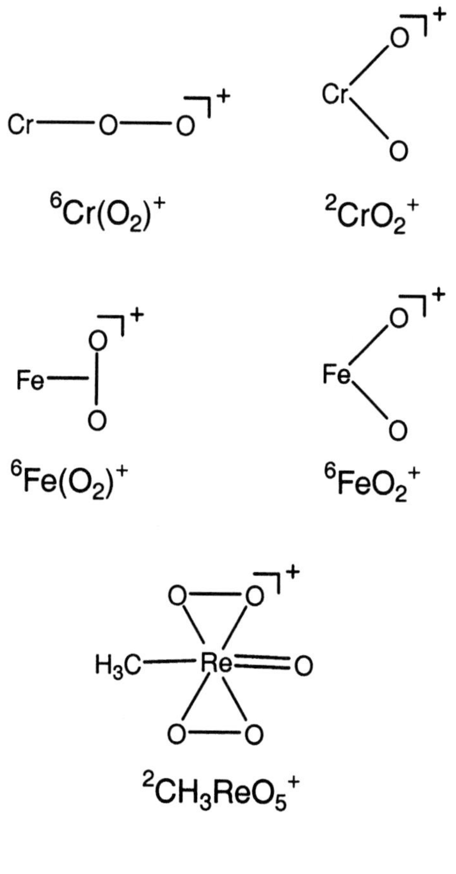

$^6Cr(O_2)^+$ $^2CrO_2^+$

$^6Fe(O_2)^+$ $^6FeO_2^+$

$^2CH_3ReO_5^+$

istic problems described above. Can understanding of the obstacles be enhanced by molecular models, or are extended bulk ensemble properties essential? Some gas-phase experiments concerning the reactivity of dioxygen towards organometallic ions are now described phenomenologically; the mechanistic background is postponed to a later section.

Structural Polychotomy and Reactivity of Transition-Metal Oxides

The first issue is the structural multitude of polyatomic $[M,O_n]$ species ($n \geq 2$). Even for the simplest case ($n = 2$), there are several conceivable bonding schemes: *i*) side-on or end-on high-spin dioxygen complexes, *ii*) the corresponding low-spin variants, *i.e.*, superoxo and peroxo species, and *iii*) metal-dioxide structures. In fact, we have demonstrated the existence of all these types of bonding in a comparative study of $[Cr,O_2]^+$, $[Fe,O_2]^+$, and $CH_3Re(O_2)_2O^+$ cations (*Scheme 3* [17]). This structural polychotomy needs to be kept in mind in the appraisal of experimental studies dealing with $[M, O_n]^{+/0/-}$ species, because more than one isomer and/or state may be sampled, depending largely on the mode of generation. For example, a plethora of oxo and peroxo species, dioxygen complexes, *etc.* has been found for $[V, O_n]^{+/0/-}$ species, of which a selection is displayed in *Scheme 4* [18].

Scheme 4. *Selected Structures of Mononuclear Vanadium Oxides in Different Charge States According to Experimental and Theoretical Studies* (see [18])

(Matrix isolation [18a–c])

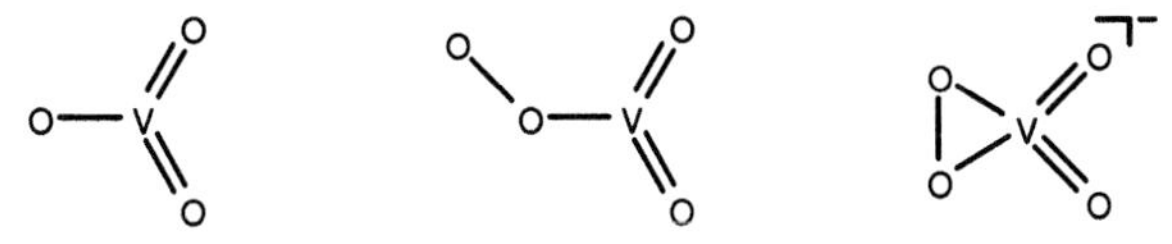

(Photodetachment [18d])

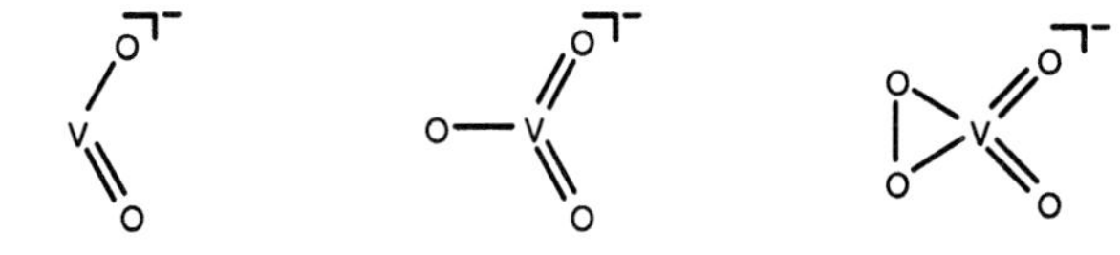

(Mass spectrometry, theory [18e])

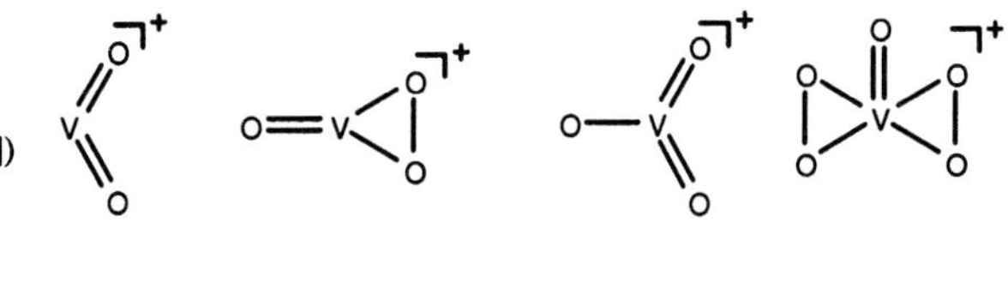

This structural variation notwithstanding, only a few cationic transition-metal ions react efficiently with molecular oxygen under gas-phase conditions (see below). In contrast, many anionic metal complexes and clusters are readily oxidized by O_2 to afford various metal-oxide anions [19]. From a conceptual point of view, however, anionic species appear to be inadequate reagents for the activation of hydrocarbons, because they generally require electrophilic attack. At present, only a few oxidations by transition-metal oxide anions have been reported to occur in the gas phase, and they are mostly limited to relatively polar substrates, such as the $CH_3OH \rightarrow CH_2O$ conversion [20]. Because of the lower reactivity of hydrocarbons, their C–H bond activation by metal-oxide anions is likely to be limited to radical pathways driven purely thermodynamically, *i.e.*, when D(O–H) exceeds D(C–H) of the substrate [21]. As radical-type pathways are prone to create selectivity problems, and over-oxidation is particularly difficult to control, the anionic route appears less attractive as far as partial oxidation of alkanes is concerned.

Examination of neutral transition-metal oxides would circumvent the particular restrictions associated with the net *Coulomb* charges of the species. In the gas phase, however, the reactivity of neutral transition-metal species is quite difficult to study, and, even more importantly, characterization of reaction products is often impossible. While matrix-isolation studies can fill this gap to some extent, these experiments are less general with respect to the variability in the range of metals, substrates, and possible ligands than are salient mass spectrometric investigations. In the context of catalysis, the low temperatures of matrix studies, in conjunction with the thermal coupling to the dense bulk material of the matrices, limit the scope of thermally driven reactions that can be probed using this technique.

Spin Inversion

An example of the spin dilemma is provided by the gas-phase chemistry of the transition-metal alkyl ions MCH_3^+ (M = Mn, Fe, Co). These MCH_3^+ species are capable of activating a broad variety of organic substrates including alkanes [22], but they fail to react with O_2 at appreciable rates [23]. Even the (cyclopentadienyl)magnesium cation $Mg(C_5H_5)^+$ – a prototype organometallic species – remains unoxidized by O_2 in the gas phase [24]. Lack of reactivity towards O_2 in the diluted gas phase is in marked contrast to the well-known, often vigorous decomposition of metal alkyls in solutions exposed to air. In view of the favorable thermochemistry of M–C bond oxidation by O_2 [7b], significant kinetic barriers must be operative in ion/molecule reactions in the gas phase. Since MCH_3^+ species can by no means be termed unreactive [22], failure to circumvent the spin-inversion bottleneck in dioxygen activation would appear the most obvious reason for their inertness towards O_2. Unlike the formal Fe^{II} compound $FeCH_3^+$, the formal Fe^I species $Fe(L)^+$ (with L = olefins, arenes, ketones [25][26]) and the formal Fe^{III} compound $FeCH_2^+$ [27] are both efficiently oxidized by molecular oxygen in the gas phase. Hence, the ability to mediate dioxygen activation is not simply a matter of choice of metal; rather, ligand effects need to be taken into account (see below).

Stoichiometry Problem

The reactions of $Fe(L)^+$ complexes with molecular oxygen [25–27] serve to illustrate the stoichiometry problem. For example, $Fe(C_2H_4)^+$ is capable of O–O bond activation, *i.e.*, the spin dilemma is surmounted. However, eight different products are formed, by pathways including O-atom transfer (*Reaction 9a*), C–H bond activation (*Reaction 9b*), and C–C bond cleavage (*Reaction 9c*); even the iron-carbene cation $FeCH_2^+$ is formed to some extent (*Reaction 9d*). Moreover, the huge exothermicity of olefin oxidation in *Reaction 9* renders the experimental assignment of the neutral product structures difficult; for example, the 'CH_2O_2' neutral(s) formed in *Reaction 9d* could equally well correspond to HCOOH, CO + H_2O, or CO_2 + H_2.

$$Fe(C_2H_4)^+ + O_2 \rightarrow FeO^+ + C_2H_4O \quad (9a)$$
$$\rightarrow FeOH^+ + CH_3CO \quad (9b)$$
$$\rightarrow Fe(CH_2O)^+ + CH_2O \quad (9c)$$
$$\rightarrow FeCH_2^+ + \text{`}CH_2O_2\text{'} \quad (9d)$$

The situation becomes even worse for higher alkenes; more than 12 product channels are involved in the oxidations of $Fe(C_3H_6)^+$ and $Fe(C_4H_8)^+$ by O_2, for example [27]. Furthermore, some of the initially formed reaction products continue to react with O_2, thereby increasing the complexities of the product distributions. In other words, while Fe can obviously overcome the problem of spin-inversion in these compounds, it fails insofar as control of selectivity is concerned. The system cannot accommodate the two O-atoms in a selective manner, thus demonstrating the stoichiometry problem. In fact, most gas-phase oxidations of alkanes with better selectivities apply single O-atom donors such as dinitrogen oxide [7] as oxidants, thereby circumventing the stoichiometry problem associated with the use of O_2 as terminal oxidant.

$$Pt^+ + CH_4 \rightarrow PtCH_2^+ + H_2 \quad (10)$$

$$PtCH_2^+ + O_2 \rightarrow PtO^+ + CH_2O \quad (11a)$$

$$\rightarrow Pt^+ + HCOOH \quad (11b)$$

$$PtO^+ + CH_4 \rightarrow Pt^+ + CH_3OH \quad (12a)$$

$$\rightarrow PtCH_2^+ + H_2O \quad (12b)$$

Selectivities better than those obtained with iron have been achieved in the Pt^+-catalyzed oxidation of methane by O_2 [28]. Here, methane is activated by bare Pt^+ to afford the $PtCH_2^+$ cation in the first step (*Reaction 10*). The subsequent oxidation *Reactions 11a* and *11b* occur in a 3:7 ratio. Because PtO^+ continues to activate methane (*Reaction 12*), Pt^+ and $PtCH_2^+$ cations are effectively regenerated, thus giving rise to one of the rare examples of genuine catalytic cycles in the gas phase; nevertheless, side reactions limit the turnover number to *ca.* 6. Instructive in this context is the observation [29] that the putative neutral products of *Reaction 11*, *i.e.,* formaldehyde and formic acid, efficiently react with Pt^+ as well as PtO^+, eventually leading to CO_2 and H_2O. Thus, combustion rather than partial oxidation of methane takes place if no particular precautions are applied.

4. Towards Possible Solutions

In this section, we discuss selected gas-phase studies that may provide possible problem solutions. Here, the term 'solution' does not refer to any specific applications but rather to options that may help to circumvent the conceptual mechanistic dilemmas outlined above.

Spin Inversion

In the gas phase, the spin-inversion bottleneck in the activation of O_2 can be overcome by certain transition-metal complexes [25][27–31]. For example, experimental and theoretical studies of the $[Cr, O_2]^+$ system indicate that the formal four-electron oxidation $^6Cr(O_2)^+ \rightarrow {}^2CrO_2^+$ (*i.e.*, the conversion of a high-spin-coupled, low-valent dioxygen complex to a low-spin-coupled, high-valent metal dioxide) is indeed feasible [31]. However, the metal must not necessarily undergo a four-electron oxidation from a dioxygen complex to a dioxide. Instead, activation of dioxygen can occur in the presence of the oxidizable substrate, thereby combining spin inversion, O–O bond activation, and oxidation of the substrate in a single step. Thus, various cationic Fe^I

complexes can be oxidized using O_2 [25–27] (*Reaction 9*, for example). Although the reactions of Cr and Fe cations with dioxygen have some shortcomings (see above), these two metals are obviously capable of activating $^3O_2$ and have, therefore, been chosen in the case studies described further below.

Formally, one might attribute the ability to mediate O_2 activation either to a particular metal or to a specific electronic structure. For example, of all the $M(C_2H_4)^+$ complexes of the 3d metals, only those of iron (M = Fe) permit efficient oxidation of the coordinated ethene molecule by O_2. However, oxidation by O_2 of the ligand L in $Fe(L)^+$ complexes is not restricted to olefins; it also occurs with other ligands such as dienes, arenes, ketones, alkenones, and nitriles [25–27]. While these oxidizable substrates range from two-electron ones, such as ethene and acetone, to multidentate ones, such as butadiene and benzene, the corresponding bis-ligated Fe^I ions – *e.g.*, $Fe(L)_2^+$ with L = ethene, acetone, and benzene – no longer react with O_2. Similarly, the formal Fe^{II} ions – *e.g.*, $XFe(C_2H_4)^+$ with X = OH, Cl, and Br – are inert towards dioxygen. The fact that even $FeCH_3^+$ does not react with O_2 further refutes steric congestion as the decisive parameter. Clearly, circumvention of the spin problem and activation of dioxygen are not only matters of the metal itself, but depend crucially on its formal valence, the architecture of the coordination spheres *etc.* As a general solution cannot be provided at present, this topic deserves much more attention in future.

Co-reduction

One method of bypassing the stoichiometry problem involves the deliberate addition of a sacrificial reagents, which accept one of the O-atom equivalents. Examples are metal-catalyzed oxidations of alkenes with O_2 in the presence of alcohols, aldehydes *etc.* as co-reductants. While this approach is acceptable on a small-scale [32], it is impracticable for the production of bulk chemicals. More practicable co-reductants are molecular hydrogen and carbon monoxide, provided that *Reactions 13* and *14* are appropriately mediated by a catalyst.

$$H_2 + O_2 \rightarrow H_2O + O \qquad \Delta_r H = +2 \text{ kcal/mol} \qquad (13)$$

$$CO + O_2 \rightarrow CO_2 + O \qquad \Delta_r H = -8 \text{ kcal/mol} \qquad (14)$$

Gas-phase reductions of metal-dioxide cations with dihydrogen under thermal conditions have been achieved for CrO_2^+ [31], OsO_2^+ [33], and PtO_2^+ [11b][34]. The cationic metal monoxides thus formed, *i.e.*, CrO^+, OsO^+, and PtO^+, are, in turn, capable of activating saturated and unsaturated hydrocarbons [28][33][35]. Once the transformation $M^+ + O_2 \rightarrow MO_2^+$ is feasible, the formation of MO^+ in the reactions of MO_2^+ with dihydrogen provides a gas-phase variant of *Reaction 13*. While the monoxide cations of Os and Pt undergo reduction to the bare metal cations in the presence of dihydrogen, a spin-inversion barrier [36] hinders the reaction of thermalized CrO^+ with H_2 [35b], although formation of Cr^+ and H_2O would be highly exothermic [3b]. Accordingly, the $Cr^+/O_2/H_2$ system fulfills an important requirement for co-reduction, in that H_2 promotes formation of the metal-oxo species CrO^+, which does not itself react with the co-reductant, but is still capable of oxidizing hydrocarbons. A gas-phase model of *Reaction 14* is provided by the generation of FeO^+ from the $FeCO^+/O_2$ couple [37]. However, the FeO^+ cation thus formed also oxidizes CO at an appreciable rate [38], thereby limiting prospects with regard to catalytic co-reduction in the gas phase. Future studies will attempt to unravel the mechanistic details and electronic requirements of those systems in which the co-reductants react only with the dioxo and not with the monoxo species.

Four-Electron Oxidation

One reaction worth considering, even though it is not very selective, is the direct oxidation of methylene to keto groups according to *Reaction 15* by means of *Gif* reagents [39].

$$R_2CH_2 + O_2 \rightarrow R_2CO + H_2O \quad (15)$$

In *Gif* systems – these are variable mixtures containing *inter alia* iron compounds, carbonic acids, aromatic amines, dioxygen, and some reducing agents – the substrates undergo formal four-electron oxidations, thus essentially avoiding the stoichiometry problem. Of course, this variant cannot be used for hydroxylations; nevertheless, the direct conversion of methylene into keto groups is quite attractive for the production of bulk chemicals, such as cyclohexane → cyclohexanone. Interestingly, the *Gif* oxidation of cyclohexane to cyclohexanone sparks another, remarkable feature as far as selectivity is concerned. When (labeled) cyclohexanol – which had been viewed as a perfectly conceivable intermediate in the oxidation of cyclohexane to cyclohexanone – is added to a *Gif* mixture, its oxidation does not compete with that of the parent hydrocarbon [40], even though the alcohol is more polar and so has weaker C–H bonds than the alkane. While the *Gif* reagents seem to involve radical intermediates [41][4]), this particular observation points to-

[4]) The current view is that the *Gif* systems proceed *via* initial formation of an alkyl radical, which is then trapped by O_2 to yield a hydroperoxide, followed by a rearrangement $R_2CHOOH \rightarrow R_2CO + H_2O$.

wards mechanistic strategies for limiting over-oxidation of hydrocarbon substrates by modulating the approach of the substrate to the reactive center, rather than by tuning activation parameters. We have recently proposed a working hypothesis in which the presence of an adjacent leaving group plays a decisive role in this remarkable reaction [7b]; nevertheless, no closely related gas-phase mimic of this intriguing iron-based oxidant has yet been achieved.

5. Case Studies

In this chapter, two examples of gas-phase oxidations using O_2 are discussed in order to illustrate the performance as well as the limitations of gas-phase studies. While some of the experimental data have been reported previously, the key experiments in this work were conducted by using *Fourier*-transform ion-cyclotron resonance (FTICR) mass spectrometry [42][5]).

Chromium-Mediated Spin Inversion

As mentioned above, fundamental studies of the $[Cr, O_2]^+$ system indicate that low-valent Cr may be able to mediate the spin inversion of dioxygen in the transformation $^6Cr(O_2)^+ \rightarrow {}^2CrO_2^+$. The high-valent chromium-dioxide cation is in turn a powerful oxidant in the gas phase, capable even of activating methane [31]. However, previous evidence for the crucial spin inversion in the $[Cr, O_2]^+$ system remained circumstantial. We, therefore, wished to

[5]) The experiments were performed with a *Spectrospin CMS 47X* FTICR mass spectrometer, which has been described in [43]. Briefly, Cr^+ and Fe^+ ions were generated *via* laser desorption/laser ionization by focusing the beam of a Nd:YAG laser (*Spectron Systems*, λ = 1064 nm) onto a target made of the corresponding metal. The ions were extracted from the source and transferred into the analyzer cell by a system of electrostatic potentials and lenses. After deceleration, the ions were trapped in the field of a superconducting magnet (maximum field strength 7.05 T). Prior to ion/molecule reactions, the most abundant isotopes, $^{52}Cr^+$ and $^{56}Fe^+$, were mass-selected by using the FERETS technique [44]. The ligated ions were subsequently generated by pulsing-in appropriate neutral reagents, *i.e.*, $Cr(C_6H_5C_2H_5)^+$ from Cr^+ and ethylbenzene [45], FeO^+ from Fe^+ and N_2O [38], and $Fe(CH_3OH)^+$ from Fe^+, and a *ca.* 10:1 mixture of propane and methanol [5]. The ions of interest were then isolated using FERETS and reacted with continuously leaked-in oxygen at $p = 5 - 20 \cdot 10^{-9}$ mbar. Data were accumulated and processed by means of an *ASPECT 3000* minicomputer. Analysis of the pseudo-first-order kinetics of the ion/molecule reactions provides branching ratios and effective bimolecular rate constants k, which are reported within experimental errors of ±10 and ±50%, respectively. Due to unfavorable pumping characteristics, pulsed-in ethylbenzene and methanol cannot be removed completely during the duty cycle, giving rise to competitive reactions with background methanol, which were corrected for by employing previously published procedures [46].

design a more direct, chemical probe for this transformation. To this end, it appears attractive to assist spin inversion by an appropriate ligand attached to the chromium and then to use the oxidation of the ligand by the chromium-dioxo species as a monitor. Ethylbenzene was chosen as a model ligand for the following reasons:

i) Monoligated $Cr(C_6H_5C_2H_5)^+$ can be prepared easily in the low-pressure regime, merely by associating Cr^+ with ethylbenzene.

ii) The activated C–H bonds of the benzylic positions offer easily oxidizable sites, so that substrate oxidation is likely, once dioxygen activation is achieved.

iii) The ionization energy of ethylbenzene ($IE = 8.77 \pm 0.01$ eV [3a]) is significantly lower than that of CrO_2 ($IE = 9.7 \pm 0.5$ eV [31]), so that dioxygen activation, resulting in the formation of neutral CrO_2, is indicated by the generation of ionized ethylbenzene.

Note that partial oxidation of ethylbenzene is also of technological relevance.

While $Cr(C_6H_5C_2H_5)^+$ has so far not been studied theoretically, the bare metal cation as well as the benzene complex $Cr(C_6H_6)^+$ possess sextet ground states [46]. Given that $D(Cr^+{-}C_6H_5C_2H_5)$ is only by 3 kcal/mol [48] greater than $D(Cr^+{-}C_6H_6) = 41$ kcal/mol [47b][48], similar bonding schemes for both arene ligands are implied. Therefore, it appears safe to assume that $Cr(C_6H_5C_2H_5)^+$ also has a sextet ground state. As CrO_2^+ is low-spin-coupled in its ground state [31], activation of dioxygen by $Cr(C_6H_5C_2H_5)^+$ is formally spin-forbidden. Accordingly, evidence for the generation of CrO_2 (monitored by the formation of $C_8H_{10}^+$) can be regarded as a chemical probe whether or not chromium is able to circumvent the spin-inversion problem in this particular system.

$$\begin{array}{llllr}
Cr(C_6H_5C_2H_5)^+ + O_2 & \rightarrow & Cr^+ + \text{`}C_8H_{10}O_2\text{'} & 30\% & (16a) \\
 & \rightarrow & C_7H_7^+ + [Cr, C, H_3, O_2] & 5\% & (16b) \\
 & \rightarrow & C_8H_9^+ + OCrOH & 50\% & (16c) \\
 & \rightarrow & C_8H_{10}^+ + CrO_2 & 5\% & (16d) \\
 & \rightarrow & [Cr, C_6, H_4, O]^+ + \text{`}C_2H_6O\text{'} & 10\% & (16e)
\end{array}$$

On reacting $Cr(C_6H_5C_2H_5)^+$ with dioxygen in the gas phase, *Reaction 16* is observed with an overall rate constant of $k_{16} = 1 \cdot 10^{-10}$ cm^3 molecules^{-1} s^{-1}. The cationic products can safely be assigned as bare chromium cation Cr^+, the benzylium ions $C_6H_5CH_2^+$ and $C_6H_5CHCH_3^+$ (and/or their tropylium isomers) formed *via* formal hydride and methanide transfer, respectively, and ionized ethylbenzene *via* electron transfer (ET). By analogy to the reaction of CrO_2^+ with benzene [31], the $[Cr,C_6,H_4,O]^+$ ion formed in *Reaction 16e* is assumed to be a benzochromaoxetene. As far as the neutral products are concerned, the species formed in *Reactions 16a*, *16b*, and *16e* remain unknown, while the formation of the organic cations in *Reactions 16c* and *16d* can occur only if the indicated neutrals are generated, *i.e.*, chromium oxide hydroxide OCrOH and chromium dioxide CrO_2, respectively. For example, *Reaction 16d* is exothermic by *ca.* 21 kcal/mol if ionized ethylbenzene and neutral CrO_2 are formed [3a][31], whereas the generation of $Cr + O_2$ or $CrO + O$ as neutral products is predicted to be highly endothermic (86 and 122 kcal/mol, resp. [3a]). According to the criteria outlined above, the $Cr(C_6H_5C_2H_5)^+$ complex can undergo intersystem crossing (ISC), leading to spin inversion of molecular oxygen and oxygenation of the ligand (*Scheme 5*). While selectivity is rather small in *Reaction 16*, a preference for benzylic C–H bond activation is suggested by the branching ratios of *Reactions 16a–16e*.

Scheme 5

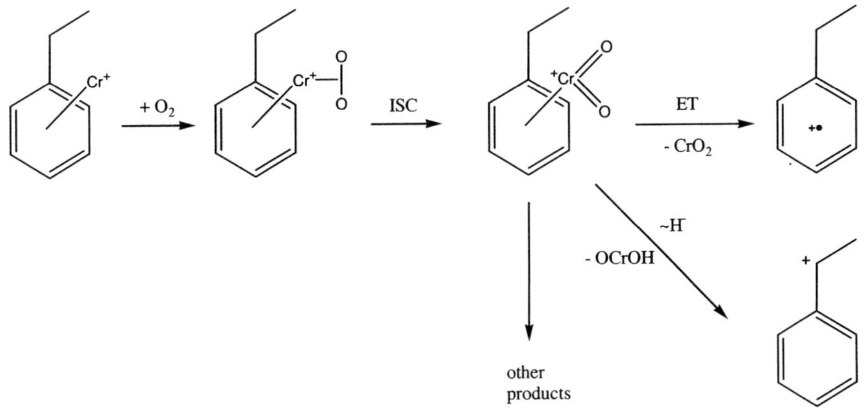

We are currently extending these studies to other arene complexes of Cr^+ in order to address some of the more peculiar aspects such as the following questions:

i) Are the putative high-spin encounter complexes $(arene)Cr(O_2)^+$ intrinsically short-lived intermediates due to their rapid intersystem crossing, affording the low-spin dioxides $(arene)CrO_2^+$, or is intersystem crossing and/or O–O bond cleavage associated with a notable barrier?

ii) If the latter holds true, which of these steps is rate-determining, and what is the precise role of substituents on the arene?

To this end, extensive labeling studies and experiments aimed at further characterization of the product ions are definitely required.

Co-reduction with Methanol:
A Case of Autocatalysis in the Partial Oxidation of Methane?

An appealing variant of co-reduction is the use of CH_3OH as co-reductant in the partial oxidation of CH_4 to CH_2O (*Reaction 17*). Thus, CH_3OH may act as a co-catalyst for the generation of reactive metal-oxo species capable of activating CH_4, thereby suggesting a route for the conversion $CH_4 \rightarrow CH_2O$ according to *Reaction 4*.

$$CH_4 + {}^1\langle O\rangle \rightarrow CH_3OH \quad (17a)$$

$$CH_3OH + O_2 \rightarrow CH_2O + H_2O + {}^1\langle O\rangle \quad (17b)$$

$$CH_4 + O_2 \rightarrow CH_2O + H_2O \quad (4)$$

Earlier gas-phase studies had already demonstrated that FeO^+ cation hydroxylates methane (*Reaction 18*) and thus acts as a reagent for O-atom transfer according to *Reaction 17a* [7][50]. However, for thermochemical reasons, FeO^+ cation cannot be generated directly from O_2 (*Reaction 19*), because $D(O-O) = 118$ kcal/mol [3a] fairly well exceeds $D(Fe^+-O) = 80$ kcal/mol [3b]. The key experiment is therefore a gas-phase mimic of *Reaction 17b*, *i.e.*, a methanol-mediated activation of dioxygen in which the alcohol serves as a co-reductant. For this purpose, $Fe(CH_3OH)^+$ was reacted with dioxygen; formation of FeO^+ according to *Reaction 20* is indeed the only product channel. Since $D(Fe^+-CH_3OH) = 34$ kcal/mol [5] and $D(Fe^+-O) = 80$ kcal/mol [3b] and using auxiliary thermochemical data [3a], *Reaction 20* is predicted to be highly exothermic. Nevertheless, the rate constant of *Reaction 20* is only about one tenth of the collision frequency. This is presumably due to a spin barrier, since CH_3OH is otherwise known to react rapidly with cationic iron oxides and hydroxides [51–53].

$$FeO^+ + CH_4 \rightarrow Fe^+ + CH_3OH \quad (18)$$
$$\Delta_rH = -7 \text{ kcal/mol} \quad k = 0.4 \cdot 10^{-10} \text{ cm}^3 \text{ molecules}^{-1} \text{ s}^{-1}$$

$$Fe^+ + O_2 \rightarrow FeO^+ + O \quad (19)$$
$$\Delta_rH = 39 \text{ kcal/mol} \quad k \ll 10^{-13} \text{ cm}^3 \text{ molecules}^{-1} \text{ s}^{-1}$$

$$Fe(CH_3OH)^+ + O_2 \rightarrow FeO^+ + CH_2O + H_2O \quad (20)$$
$$\Delta_rH = -24 \text{ kcal/mol} \quad k = 0.6 \cdot 10^{-10} \text{ cm}^3 \text{ molecules}^{-1} \text{ s}^{-1}$$

$$Fe^+ + CH_3OH \rightarrow Fe(CH_3OH)^+ \quad (21)$$
$$\Delta_rH = -34 \text{ kcal/mol} \quad k < 1 \cdot 10^{-13} \text{ cm}^3 \text{ molecules}^{-1} \text{ s}^{-1}$$

Combination of these steps leads to a viable sequence for the Fe^+-mediated oxidation of CH_4 to CH_2O according to *Reaction 4*. Here, the CH_3OH molecule, coordinated to the metal, plays a central role both as a precursor for CH_2O as the final oxidation product as well as a crucial intermediate for the activation of O_2 (*Fig.*). Further, the formation of FeO^+ in *Reaction 20* is remarkable in that over-oxidation (*i.e.*, formation of Fe^+ and HCOOH) does not occur in this experiment, even though bare FeO^+ cation rapidly reacts with formaldehyde [23].

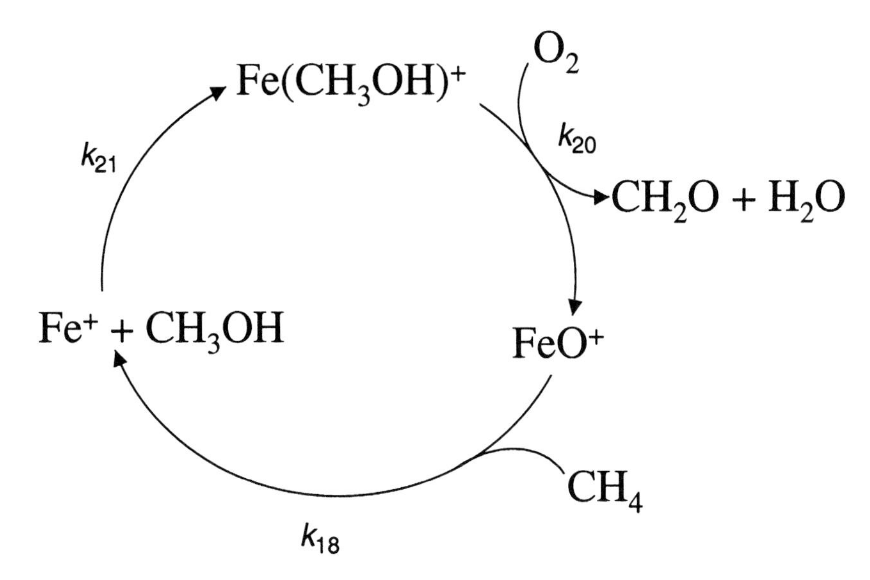

Figure. *Catalytic sequence for the Fe^+-mediated oxidation of methane by molecular oxygen with methanol as a catalytic co-reductant.* The side reaction of the FeO^+/CH_4 couple, which leads to $FeOH^+$ cation [50], is omitted for the sake of clarity.

It seems ironic, however, that the seemingly most simple step of the sequence turns out to be impossible in the highly diluted gas phase (typical pressures below 10^{-7} mbar). Specifically, the mere complexation of methanol by the iron cation according to *Reaction 21* does not occur with measurable rate constant in this pressure regime. The reason is simply that ligand association often proceeds only *via* termolecular collisions, which have negligible probabilities under these conditions [53]. Further, selectivity is limited by the fact that FeO^+ also acts as a H-atom abstractor with CH_4 to afford $FeOH^+$ together with a methyl radical [50]. There is also the possibility of over-oxidation, *i.e.*, FeO^+ also reacts with CH_3OH to yield CH_2O with a rate constant of $14 \cdot 10^{-10}$ cm^3 $molecules^{-1}$ s^{-1} [51]. Although the latter rate exceeds that of *Reaction 18* by more than an order of magnitude, the ratio of the two rate constants is not such as to prevent the achievement of reasonable selectivities in partial oxidation of methane at low conversions by optimizing mixing ratios, residence times *etc.*

As far as the mechanism of *Reaction 20* is concerned, we are by and large left to extrapolate from the related behaviors of the Fe^+/O_2 and FeO^+/CH_3OH couples. For example, O–O bond insertion in the reaction $Fe^+ + O_2 \rightarrow OFeO^+$ to afford a high-spin-coupled dioxide occurs almost barrierless [54]. Further, attack on CH_3OH by the metal-oxo species FeO^+ is assumed to involve proton transfer in the first step to yield the insertion species $HO–Fe^+–OCH_3$ as a key intermediate *en route* to product formation [51]. Absence of over-oxidation in the $Fe^+/CH_3OH/O_2$ system can be explained by invoking an intermediate lacking a reactive metal-oxo unit, such as an iron dihydroxide rather than a ligated oxo species [55]. Combining these results, we propose by analogy the mechanism for *Reaction 20* as shown in *Scheme 6*; a more detailed treatment, including electronic structure calculations, is clearly needed[6]).

Scheme 6

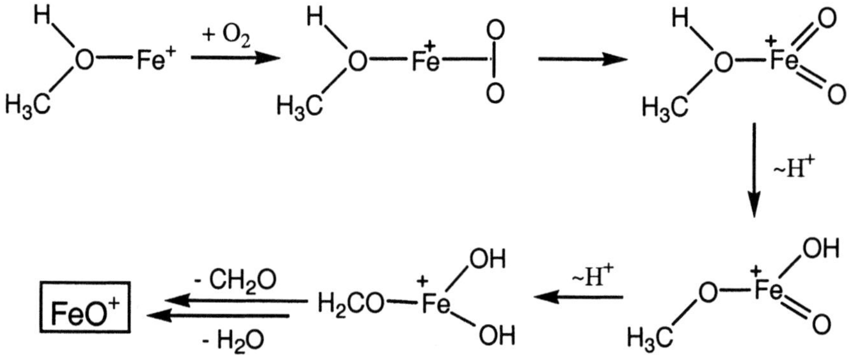

One mechanistic implication is that the sequence of *Reactions 18*, *20*, and *21* involves an autocatalytic step in the CH_4/O_2 oxidation. While neither bare Fe^+ nor $Fe(CH_4)^+$ is oxidized by O_2 at appreciable rates [54][56], a catalytic sequence for the activation of O_2 must evolve, once the primary oxidation product, CH_3OH, is formed. Hence, the system exhibits autocatalysis as far as formation of CH_3OH is concerned. If similar principles are operative in technical catalysts for the partial oxidation of CH_4, re-feeding the reactant

[6]) For enhancement effects of added methanol on selective CH_4 oxidations, see: Y. Teng, Y. Yamaguchi, T. Takemoto, K. Tabata, E. Suzuki, *Phys. Chem. Chem. Phys.* **2000**, *2*, 3429.

stream with traces of CH_3OH, might increase catalytic activity, thereby possibly allowing less drastic operating conditions.

Finally, *Reaction 22*, observed by *Burnier et al.* [57], is of interest as it represents the reversal of a Fe^+-mediated coupling of methane and formaldehyde to dimethyl ether, according to *Reaction 23*.

$$Fe^+ + H_3COCH_3 \rightarrow Fe(CH_2O)^+ + CH_4 \quad \Delta_rH = -34 \text{ kcal/mol} \quad (22)$$

$$CH_2O + CH_4 \rightarrow H_3COCH_3 \quad \Delta_rH = +1 \text{ kcal/mol} \quad (23)$$

If this coupling could be realized, combination with *Reaction 4* leads to a possible route for the generation of dimethyl ether as a product of partial oxidation of methane (*Reaction 24*).

$$2\,CH_4 + O_2 \rightarrow H_3COCH_3 + H_2O \quad \Delta_rH = -65 \text{ kcal/mol} \quad (24)$$

Although gas-phase activation of methane by formaldehyde complexes has not yet been reported, transfer hydrogenations from several substrates to ketones and other coupling reactions of ligated formaldehyde have been observed [58].

6. Conclusions

Gas-phase studies provide mechanistic insight into the elementary steps of oxidation reactions at a molecular level [59]. With respect to the use of dioxygen as terminal oxidant, spin inversion and reaction stoichiometry are identified as key obstacles to the achievement of selectivity. Both problems can be tackled by means of transition-metal catalysis, which offers flexible means for optimization of a given substrate/oxygen mixture. Identification of the problems has been achieved by a mechanistic analysis based on gas-phase experiments and complementary theoretical studies, but general paradigms for circumventing both problems simultaneously are still lacking [60].

Of course, the examination and identification of elementary steps in idealized gas-phase experiments do not apply directly to the questions posed in applied catalysis. Nevertheless, some of the concepts arising from this approach may be useful in research aimed at rational design of oxidation catalysts. We hope that the present contribution poses challenges to search for unifying, interdisciplinary approaches that could satisfy economic as well as environmental demands.

We thank the *Deutsche Forschungsgemeinschaft*, the *Volkswagen-Stiftung*, the *Fonds der Chemischen Industrie*, and the *Gesellschaft der Freunde der Technischen Universität Berlin* for sponsoring our work. Further, we appreciate Dr. *I. Kretzschmar* for assistance in the MCH_3^+/O_2 studies, thank Dr. *H. Mestdagh* for providing us with unpublished results [53], are grateful to Dr. *P. Gentz-Werner* for drawing our attention to *Otto Warburg*'s 1931 lecture in Stockholm, and acknowledge Dr. *G. Fengler*, *BAYER AG*, for helpful comments.

REFERENCES

[1] a) E. F. Elstner, 'Der Sauerstoff: Biochemie, Biologie, Medizin', Wissenschaftsverlag, Mannheim, 1990; b) C. de Duve, 'Ursprung des Lebens', Spektrum Akademischer Verlag, Berlin, 1994; c) S. S. Stahl, S. J. Lippard in 'Dioxygen and Alkane Activation by Iron-Containing Enzymes', Eds. G. C. Ferreira, J. J. G. Moura, R. Franco, Wiley-VCH, Weinheim, Germany, 1009, pp. 303.

[2] P. R. Ogilby, *Acc. Chem. Res.* **1999**, *32*, 512, and ref. cit. therein.

[3] a) S. G. Lias, J. E. Bartmess, J. F. Liebman, J. L. Holmes, R. D. Levin, W. G. Mallard, *J. Phys. Chem. Ref. Data, Suppl. 1* **1988,** *17*; b) B. S. Freiser, Ed., 'Organometallic Ion Chemistry', Kluwer, Dordrecht, 1996, p. 283.

[4] a) C. A. Schalley, 'Gas-Phase Ion Chemistry of Peroxides', Shaker, Aachen, 1997; b) C. A. Schalley, J. N. Harvey, D. Schröder, H. Schwarz, *J. Phys. Chem. A* **1998**, *102*, 1021.

[5] R. Wesendrup, C. A. Schalley, D. Schröder, H. Schwarz, *Organometallics* **1996**, *15*, 1435.

[6] a) D. Schröder, H. Schwarz, *J. Organomet. Chem.* **1995**, *504*, 123; b) B. L. Tjelta, P. B. Armentrout, *J. Phys. Chem. A* **1997**, *101*, 2064.

[7] a) D. Schröder, H. Schwarz, *Angew. Chem.* **1995**, *107*, 2126; *Angew. Chem., Int. Ed.* **1995**, *34*, 1973; b) D. Schröder, S. Shaik, H. Schwarz, in 'Structure and Bonding', Eds. B. Meunier, Springer, Berlin, 2000, pp. 91.

[8] D. T. Sawyer, *J. Phys. Chem.* **1989**, *93*, 7977.

[9] J. Berkowitz, G. B. Ellison, D. Gutman, *J. Chem. Phys.* **1994**, *98*, 2744.

[10] H. Yoshida, C. Murata, T. Hattori, *Chem. Commun.* **1999**, 1551.

[11] a) J. M. Thomas, *Angew. Chem.* **1994**, *106*, 963; *Angew. Chem., Int. Ed.* **1994**, *26*, 913; b) M. Brönstrup, D. Schröder, I. Kretzschmar, H. Schwarz, J. N. Harvey, *J. Am. Chem. Soc.* **2001**, *123*, 142.

[12] H. Kisch, L. Zang, C. Lange, W. F. Maier, C. Antonius, D. Meissner, *Angew. Chem.* **1998**, *110*, 3201; *Angew. Chem., Int. Ed.* **1998**, *37*, 3034.

[13] J. N. Harvey, M. Diefenbach, D. Schröder, H. Schwarz, *Int. J. Mass. Spectrom.* **1999**, *182/183*, 85.

[14] W. Kaim, B. Schwederski, 'Bioanorganische Chemie', Teubner, Stuttgart, 1991.

[15] a) 'Cytochrome P-450: Structure, Mechanism and Biochemistry', Ed. P. R. Ortiz de Montellano, 2nd edn., Plenum, New York, 1995; b) I. Schlichting, J. Berendzen, K. Chu, A. M. Stock, S. A. Maves, D. E. Benson, R. M. Sweet, D. Ringe, G. A. Petsko, S. G. Sligar, *Science* **2000**, *287*, 1615.

[16] a) A. Bakac, J. H. Espenson, *Acc. Chem. Res.* **1993**, *26*, 519; b) A. Bakac, S. L. Scott, J. H. Espenson, K. R. Rodgers, *J. Am. Chem. Soc.* **1995**, *117*, 6483, and refs. cit. therein.

[17] D. Schröder, A. Fiedler, W. A. Herrmann, H. Schwarz, *Angew. Chem.* **1995**, *107*, 2714; *Angew. Chem., Int. Ed.* **1995**, *34*, 2517.

[18] a) M. J. Almond, R. W Atkins, *J. Chem. Soc., Dalton Trans.* **1994**, 835; b) L. B. Knight, Jr., R. Babb, M. Ray, T. J. Banisaukas III, L. Russon, R. S. Dailey, E. R. Davidson, *J. Phys. Chem.* **1996**, *105*, 10237; c) G. V. Chertihin, W. D. Bare, L. Andrews, *J. Phys. Chem. A* **1997**, *101*, 5090; d) H. Wu, L.-S. Wang, *J. Chem. Phys.* **1998**, *108*, 5310; e) I. Kretzschmar, D. Schröder, H. Schwarz, D. K. Bohme, unpublished results.

[19] a) R. R. Squires, *Chem. Rev.* **1987**, *87*, 623; b) C. E. C. A. Hop, T. B. McMahon, *J. Am. Chem. Soc.* **1992**, *114*, 1237.

[20] M. C. Oliveira, J. Marcalo M. C. Vieira, M. A. Almoster Ferreira, *Int. J. Mass Spectrom.* **1999**, *185–187*, 825.
[21] J. M. Mayer, *Acc. Chem. Res.* **1998**, *31*, 441.
[22] D. B. Jacobsen, J. R. Gord, B. S. Freiser, *Organometallics* **1989**, *8*, 2957, and refs. cit. therein.
[23] I. Kretzschmar, D. Schröder, H. Schwarz, unpublished results.
[24] R. K. Milburn, V. Baranov, A. C. Hopkinson, D. K. Bohme, *J. Phys. Chem. A* **1999**, *103*, 6373.
[25] D. Schröder, H. Schwarz, *Angew. Chem.* **1993**, *105*, 1493; *Angew. Chem., Int. Ed.* **1993**, *32*, 1420.
[26] P. Boissel, P. Marty, A. Klotz, P. de Parseval, B. Chaudret, G. Serra, *Chem. Phys. Lett.* **1995**, *242*, 157.
[27] D. Schröder, Dissertation, TU Berlin D83, 1993.
[28] R. Wesendrup, D. Schröder, H. Schwarz, *Angew. Chem.* **1994**, *106*, 1232; *Angew. Chem., Int. Ed.* **1994**, *33*, 1174.
[29] M. Pavlov, M. R. A. Blomberg, P. E. M. Siegbahn, R. Wesendrup, C. Heinemann, H. Schwarz, *J. Phys. Chem. A* **1997**, *101*, 1567.
[30] R. Wesendrup, Diplomarbeit, TU Berlin, 1994.
[31] A. Fiedler, I. Kretzschmar, D. Schröder, H. Schwarz, *J. Am. Chem. Soc.* **1996**, *118*, 9941.
[32] T. Mukaiyama, in 'The Activation of Dioxygen and Homogeneous Catalytic Oxidation', Eds. D. H. R. Barton, A. E. Martell, D. T. Sawyer, Plenum, New York, 1993, p. 133.
[33] K. K. Irikura, J. L. Beauchamp, *J. Am. Chem. Soc.* **1989**, *111*, 75.
[34] M. Brönstrup, Dissertation, TU Berlin, D83, 1999.
[35] a) H. Kang, J. L. Beauchamp, *J. Am. Chem. Soc.* **1986**, *108*, 5663; b) H. Kang, J. L. Beauchamp, *J. Am. Chem. Soc.* **1986**, *108*, 7502.
[36] A. Irigoas, J. E. Fowler, J. M. Ugalde, *J. Am. Chem. Soc.* **1999**, *121*, 8549.
[37] M. Brönstrup, I. Kretzschmar, D. Schröder, H. Schwarz, *Helv. Chim. Acta* **1998**, *81*, 2348.
[38] M. M. Kappes, R. H. Staley, *J. Am. Chem. Soc.* **1981**, *103*, 1286.
[39] D. H. R. Barton, *Tetrahedron* **1998**, *54*, 5805.
[40] D. H. R. Barton, J. Boivin, M. Gastiger, J. Morzycki, R. S. Hay-Motherwell, W. B. Motherwell, N. Obzalik, K. M. Schwarzentruber, *J. Chem. Soc. Perkin Trans. 1* **1986**, 947.
[41] a) F. Minisci, F. Fontana, S. Araneo, F. Recupero, L. Zhao, *Synlett* **1996**, 119; b) P. A. MacFaul, K. U. Ingold, D. D. M. Wayner, L. Que, *J. Am. Chem. Soc.* **1997**, *119*, 10594.
[42] A. G. Marshall, C. L. Hendrickson, G. S. Jackson, *Mass Spectrom. Rev.* **1998**, *17*, 1.
[43] a) K. Eller, W. Zummack, H. Schwarz, *J. Am. Chem. Soc.* **1990**, *112*, 621; b) K. Eller, H. Schwarz, *Int. J. Mass Spectrom. Ion Processes* **1989**, *93*, 243.
[44] R. A. Forbes, F. H. Laukien, J. Wronka, *Int. J. Mass Spectrom. Ion Processes* **1988**, *83*, 23.
[45] C.-Y. Lin, R. C. Dunbar, *Organometallic* **1997**, *16*, 2691.
[46] a) C. Heinemann, H. H. Cornehl, D. Schröder, M. Dolg, H. Schwarz, *Inorg. Chem.* **1996**, *35*, 2463; b) I. Kretzschmar, A. Fiedler, J. N. Harvey, D. Schröder, H. Schwarz, *J. Phys. Chem. A* **1997**, *101*, 6252.
[47] a) C. W. Bauschlicher, Jr., H. Partridge, S. R. Langhoff, *J. Phys. Chem.* **1992**, *96*, 3273; b) C. N. Yang, S. J. Klippenstein, *J. Phys. Chem. A* **1999**, *103*, 1094.
[48] K. Schroeter, R. Wesendrup, H. Schwarz, *Eur. J. Org. Chem.* **1998**, 565.
[49] F. Meyer, F. A. Khan, P. B. Armentrout, *J. Am. Chem. Soc.* **1995**, *117*, 9740.
[50] D. Schröder, H. Schwarz, D. E. Clemmer, Y.-M. Chen, P. B. Armentrout, V. I. Baranov, D. K. Bohme, *Int. J. Mass Spectrom. Ion Processes* **1997**, *161*, 177, and refs. cit. therein.
[51] D. Schröder, R. Wesendrup, C. A. Schalley, W. Zummack, H. Schwarz, *Helv. Chim. Acta* **1996**, *79*, 123.
[52] A. Fiedler, D. Schröder, H. Schwarz, B. L. Tjelta, P. B. Armentrout, *J. Am. Chem. Soc.* **1996**, *118*, 5047.
[53] M. Heninger, P. Pernot, H. Mestdagh, P. Boissel, J. Lemaire, R. Marx, G. Mauclaire, Université Paris-Sud, Orsay, unpublished results.

[54] D. Schröder, A. Fiedler, J. Schwarz, H. Schwarz, *Inorg. Chem.* **1994**, *33*, 5094.
[55] S. Bärsch, D. Schröder, H. Schwarz, *Chem. Eur. J.*, **2000**, *6*, 1789.
[56] V. I. Baranov, G. Javahery, A. C. Hopkinson, D. K. Bohme, *J. Am. Chem. Soc.* **1995**, *117*, 12801.
[57] R. C. Burnier, G. D. Byrd, B. S. Freiser, *J. Am. Chem. Soc.* **1981**, *103*, 4360.
[58] a) S. Karraß, D. Schröder, H. Schwarz, *Chem. Ber.* **1992**, *125*, 751; b) J. Schwarz, R. Wesendrup, D. Schröder, H. Schwarz, *Chem. Ber.* **1996**, *129*, 1463.
[59] a) D. Schröder, S. Shaik, H. Schwarz, *Acc. Chem. Res.* **2000**, *33*, 139; b) F. Ogliaro, N. Harris, S. Cohen, M. Filatov, S. P. de Visser, S. Shaik, *J. Am. Chem. Soc.* **2000**, *122*, 8977.
[60] A. Sen, in 'Activation of Unreactive Bonds and Organic Synthesis', Ed. S. Murai, Springer, Berlin, 1999, pp. 81, and refs. therein.

Femtosecond Activation of Reactions: The Concepts of Nonergodic Behavior and Reduced-Space Dynamics

by **Klaus B. Møller** and **Ahmed H. Zewail**

Arthur Amos Noyes Laboratory of Chemical Physics, California Institute of Technology, Pasadena, CA 91125, USA

1. Introduction

Our basic understanding of chemical reactions relies on concepts such as reaction mechanisms and time scales. In a macroscopic kinetic picture of the *Arrhenius* type, the reaction rate (constant) is determined by energetics, that is, the internal energy content of the reactant species as well as the relation between the available energy for reaction and the activation energy for the process. This phenomenological description is not concerned with the microscopic (nuclear) motion as chemical transformations take place. As a first step towards a microscopic treatment, one may define a nuclear configuration that separates reactants from products – the transition state – and the reaction rate is determined as the flux (speed times density) of molecules that passes through the transition state (see *Fig. 1*).

In the framework known as transition state theory (TST) it is assumed that crossing of the transition state is an irreversible process; this is termed the 'TST assumption'. Under the additional assumption that the energy of the reactant molecules is distributed uniformly among the different degrees of freedom, computation of the reaction rate (constant) becomes a 'single-point' calculation involving only characteristics of the transition-state nuclear configuration. In other words, no dynamics is involved in determining the rate constant. The celebrated *Rice-Ramsberger-Kassel-Marcus* (*RRKM*) theory [1] provides one such scheme for computation of microcanonical rate constants. In fact, this scheme is so commonly used that the term '*RRKM* behavior' now covers, in general, the dynamics of reactions for which these two assumptions are valid.

However, a true microscopic (dynamical) treatment of chemical transformations is desirable in order to obtain 'exact' reaction rates and to give in-

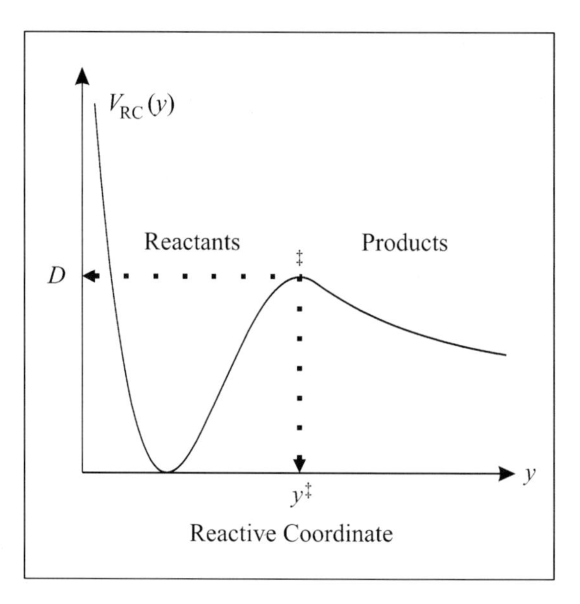

Fig. 1. *A typical potential energy profile along the internuclear distance* (y) *of a reactive coordinate* (RC). The figure illustrates the dissociation barrier (D), the transition state (‡), and the transition state configuration ($y^{‡}$).

sight into reaction mechanisms. This requires knowledge of the full potential energy surface (PES) – the energy landscape on which the nuclear motion takes place – as well as understanding of how the molecules explore the PES during the course of reaction. The microscopic, dynamical approach to calculation of rate constants was pioneered by *Wigner* [2] and has been an active area of research ever since [3–10]. Recently, *Miller*, who has made important contributions to this area, has reviewed various classical, semi-classical, and quantum mechanical methods involved [11][12].

There is still a fundamental question to ask, namely whether the characterization of a given chemical reaction in terms of an overall reaction rate constant is appropriate. Most of the theoretical methods mentioned above are developed for thermal systems, where the internal energy of the reactant molecules is equilibrated through interaction with the surroundings. But what about systems where the internal energy distribution is not maintained in equilibrium by the surroundings? Since the early seventies, a vast amount of theoretical work has been carried out, by *Baer* and *Hase*, and others [1], in order to determine the connection between certain characteristics of nuclear motion and the concept of reaction rate constants.

Nuclear motion takes place on the femtosecond time scale, and the field of femtochemistry deals with monitoring in real time the nuclear motion during a chemical change [13–20]. An ultrafast (femtosecond) pump laser activates the molecular system under consideration by depositing energy into a certain chemical bond or region of the system. A second laser may be used to probe the subsequent nuclear motion as it unfolds on the PES. The key to femtosecond activation is *localization*; in space, time, and energy [20]. The pump pulse creates, on the femtosecond timescale, an ensemble of reactant molecules (a wave packet) with a *de Broglie* wavelength on the order of 0.05–0.5 Å and an uncertainty in energy of a few hundred wave numbers. This should be compared with the length scale of chemical changes, which is *ca.* 2–5 Å, and a total energy of several thousand wave numbers. In a (quasi-)one-dimensional system, that is, a system where the wave packet remains localized along the coordinate it was created, a characteristic time for either vibration or breakage of this bond is obtained by following the motion of the wave packet in time.

In more complex systems, the question arises whether the initially created wave packet maintains its coherence or becomes delocalized over larger parts of the molecule. The answer to this question depends on the time scale for intramolecular vibrational energy redistribution (IVR) compared with the time scalefor motion in the originally activated coordinate. In the limit of a very rapid IVR, the energy will be redistributed uniformly, and, keeping in mind the narrow energy dispersion of the wave packet, the molecules will exhibit RRKM behavior, provided that the TST assumption is valid. On the other hand, if the energy flow between certain degrees of freedom is relatively slow, creating a 'bottleneck' in phase space, the dynamics of one part of the molecule may be nearly de-coupled from the rest of the molecule. This causes a bifurcation of the time scale for reaction into prompt and delayed decays. In this case, a simple statistical picture of the reaction dynamics is not applicable. A theoretical treatment of this bifurcation in quantum mechanical terms has been given recently by *Remacle* and *Levine* [21][22].

Femtosecond activation, targeted at only a part of the molecule, provides a unique opportunity for probing dynamical characteristics of this part of the molecular phase space and for controlling the dynamics in the 'reduced-space' picture of the reaction [23]. Hence, the degree of ergodicity and the extent of localization may become controllable given the unique (coherent) preparation on the femtosecond time scale.

In this paper, we discuss recent experiments and perform classical trajectory calculations on a model system to illustrate these concepts for a unimolecular dissociation reaction. The paper is organized as follows: The following section summarizes the phenomenological treatment of a unimolecular decay. In *Sect. 3*, we outline the theoretical framework for extracting rate-related information from a classical trajectory calculation. A numerical example is presented in *Sect. 4*, where we analyze the effect of varying different characteristic time scales in the system. Further discussion and comparison with experimental results follow in *Sect. 5*, and conclusions are given in *Sect. 6*.

2. Phenomenological Description of a Unimolecular Decay

We consider the unimolecular decay

$$A^* \xrightarrow{k} \text{Products}, \tag{2.1}$$

where A* represents a molecule that is sufficiently activated to react without further input of energy. Since there is no back reaction the kinetics is very simple. The rate equation is given by

$$\frac{dN_{A*}(t)}{dt} = -kN_{A*}(t), \tag{2.2}$$

with the solution,

$$N_{A*}(t) = N_{A*}(t_0)\exp(-kt). \tag{2.3}$$

This rate equation implies a random (uniform) reaction probability, which means that the probability of reaction in any infinitesimal time interval dt is given by $k\,dt$. Alternatively, one can define the 'lifetime distribution' $P(t)$ [1], where $P(t)\,dt$ is the probability that a molecule decays in the infinitesimal time interval form t to $t + dt$. Then,

$$\frac{dN_{A*}(t)}{dt} = -P(t)\,N_{A*}(t_0), \tag{2.4}$$

or

$$P(t) = k\exp(-kt). \tag{2.5}$$

As mentioned in the *Introduction*, even for simple reactions it may not be possible to assign a single overall rate constant since several time scales may be involved in the decay of A* [1][13–24]. A generalization of the simple picture presented above then involves the introduction of several rate constants,

$$A^* \xrightarrow{k_1, k_2, \ldots, k_n} \text{Products}. \tag{2.6}$$

The multitude of rate constants amounts to a multi-exponential decay of the reactants,

$$N_{A*}(t) = N_{A*}(t_0) \sum_n c_n \exp(-k_n t), \tag{2.7}$$

and a lifetime distribution of the form

$$P(t) = \sum_n c_n k_n \exp(-k_n t). \tag{2.8}$$

Thus, the characterization of the reaction by an overall rate constant may not be appropriate. To obtain a compact form for the rate equation in this case we define the function

$$\Pi(t) = -\int dt\, P(t) = \sum_n c_n \exp(-k_n t). \tag{2.9}$$

Then, $N_{A*}(t)/N_{A*}(t_0) = \Pi(t)$, and, from *Eqns. 2.4* and *2.7* we deduce that, in general, the rate equation for a multi-exponential decay can be expressed as

$$\frac{dN_{A*}(t)}{dt} = \frac{\partial \ln[\Pi(t)]}{\partial t} N_{A*}(t). \tag{2.10}$$

Finally, we consider the kinetics of a system with a second channel that deactivates A* without leading to products. In other words, we supply the reaction in *Eqn. 2.6* with a second reaction channel

$$A^* \xrightarrow{k_{deact}} A. \tag{2.11}$$

This deactivation process is commonly considered in the literature as collisional deactivation [1][25–27], where a surrounding inert gas serves as a bath for absorbing energy. Alternatively, we can think of a molecular system, where only a small sub-manifold of modes interact strongly with the reactive coordinate, whereas another manifold of modes merely serve as a bath. The latter manifold can, for instance, consist of vibrational modes in a polyatomic molecule or vibrational levels on a different electronic surface. In *Fig. 2* we show a schematic picture of a reduced (reactive) space containing the reactive coordinate coupled to the manifold $\{R_i\}$. The coupling can, for instance, be of the Golden Rule type such that

$$k_{deact}(E) = \frac{2\pi}{\hbar} |V_{coup}|^2 \varrho_{\{R_i\}}(E), \tag{2.12}$$

where $|V_{\text{coup}}|$ characterizes the coupling strength between the two manifolds and $\varrho_{\{\text{R}_i\}}(E)$ is the density of states in $\{\text{R}_i\}$. The rate equation for the total decay of A* then takes the form

$$\frac{dN_{\text{A}*}(t)}{dt} = \frac{\partial \ln [\Pi(t)]}{\partial t} N_{\text{A}*}(t) - k_{\text{deact}} N_{\text{A}*}(t), \tag{2.13}$$

where $\Pi(t) = \Sigma_n c_n \exp(-k_n t)$ is the function, *Eqn. 2.9*, that characterizes the decay of A* without the deactivating process. This equation is readily rewritten as

$$\frac{dN_{\text{A}*}(t)}{dt} = \frac{\partial \ln [\Pi(t) \exp(-k_{\text{deact}} t)]}{\partial t} N_{\text{A}*}(t), \tag{2.14}$$

which has the solution

$$\frac{N_{\text{A}*}(t)}{N_{\text{A}*}(t_0)} = \Pi(t) \exp(-k_{\text{deact}} t), \tag{2.15a}$$

or

$$N_{\text{A}*}(t) = N_{\text{A}*}(t_0) \sum_n c_n \exp[-(k_n + k_{\text{deact}}) t]. \tag{2.15b}$$

Hence, in the kinetic model, inclusion of a bath has an overall damping effect on the number of reactants as a function of time. The decay of the reactants follows the same single- or multi-exponential behavior as it had without the bath but with each rate constant being the sum of the original and the deactivation rate constant.

The rate equation for the deactivated reactants is

$$\frac{dN_{\text{A}}(t)}{dt} = k_{\text{deact}} N_{\text{A}*}(t), \tag{2.16}$$

with the solution

$$N_{\text{A}}(t) = N_{\text{A}*}(t_0) \sum_n c_n \frac{k_{\text{deact}}}{k_n + k_{\text{deact}}} \{1 - \exp[-(k_n + k_{\text{deact}}) t]\}. \tag{2.17}$$

Assuming that only activated reactants are present at time zero and that the stoichiometry of the reaction is such that one reactant transforms into one product, *i.e.*, $N_{\text{product}}(t) = N_{\text{A}*}(t_0) - N_{\text{A}*}(t) - N_{\text{A}}(t)$, we find that

$$N_{\text{product}}(t) = N_{\text{A}*}(t_0) \left(1 - \sum_n \frac{c_n}{k_n + k_{\text{deact}}} \{k_{\text{deact}} + k_n \exp[-(k_n + k_{\text{deact}}) t]\} \right). \tag{2.18}$$

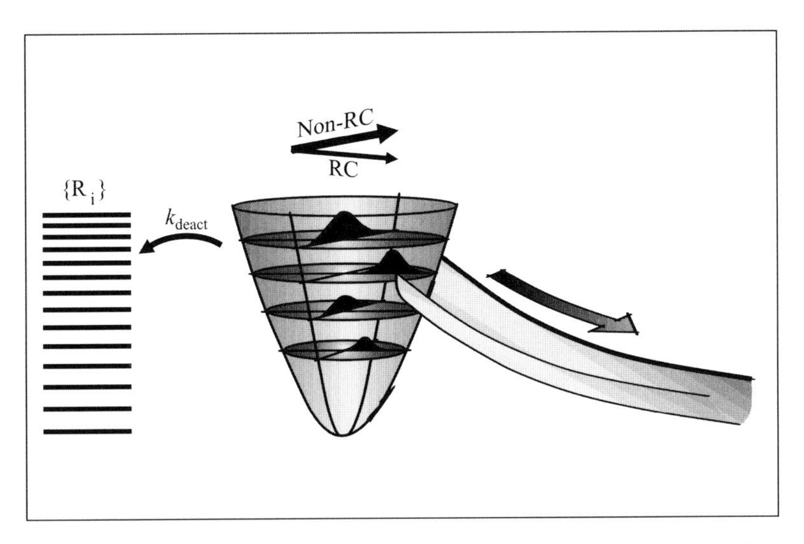

Fig. 2. *Illustration of a system divided into a 'reactive space' and a bath.* The reactive space is coupled irreversibly to the manifold, $\{R_i\}$, of bath states.

Thus, by monitoring the temporal behavior of either the number of reactant molecules themselves or the lifetime distribution for a given system, one can evaluate whether the phenomenological model is applicable and, in the affirmative case, determine the rate constant(s). In the following, we recapitulate the framework for extracting these quantities from classical trajectory calculations and discuss what to expect of their behavior.

3. Theoretical Considerations

3.1. Unimolecular Decay in General

The reactant and product regions in configuration space are separated by the transition-state configuration given by a certain value of the reactive coordinate, $y = y^{\ddagger}$ (see *Fig. 1*). If only reactants are present at time $t = t_0$, the number of reactant molecules at any later time can be expressed as

$$N_{A*}(t) = N_{A*}(t_0) \int d\mathbf{q}\, d\mathbf{p}\, \varrho_0(\mathbf{q}, \mathbf{p})\, H_{A*}[y(t)], \tag{3.1}$$

where $\varrho_0(\mathbf{q}, \mathbf{p})$ is the normalized probability distribution characterizing the initial ensemble of reactants and

$$H_{A*}(y) = \begin{cases} 1, & y \leq y^{\ddagger} \\ 0, & y > y^{\ddagger} \end{cases}. \tag{3.2}$$

The lifetime distribution is defined as $P(t) = -\dot{N}_{A*}(t)/N_{A*}(t_0)$ and we find

$$P(t) = \int d\mathbf{q}\, d\mathbf{p}\, \varrho_0(\mathbf{q}, \mathbf{p})\, \delta[y(t) - y^{\ddagger}]\, \dot{y}(t). \tag{3.3}$$

If the TST assumption is valid, which means that no trajectories cross the transition state more than once, *Eqn. 3.3* can be written as

$$P(t) = \int d\mathbf{q}\, d\mathbf{p}\, \varrho_0(\mathbf{q}, \mathbf{p})\, \delta[t - \tau(\mathbf{q}, \mathbf{p})], \tag{3.4}$$

where $\tau(\mathbf{q}, \mathbf{p})$ is the solution to the equation, $y[\tau(\mathbf{q}, \mathbf{p})] = y^{\ddagger}$.

By considering the case where $\varrho_0(\mathbf{q}, \mathbf{p})$ describes an equilibrium distribution, further insight into the temporal behavior of the quantities can be gained.

3.2. Microcanonical Rate Constants

The dynamical approach to equilibrium rates dates back to *Wigner* [2], whose insight continues to inspire the development of the field [3–12]. The equilibrium rate constant is defined as the flux through the transition state from reactants to products, on the assumption that the possible states of the reactants are populated according to the appropriate equilibrium distribution during the course of reaction.

In the microcanonical case, a classical expression for the rate constant can be given with the form [3][5][7]

$$k_{\mathrm{MIC}}(E) = \frac{1}{\varrho_{\mathrm{R}}(E)\, h^n} \int d\mathbf{q}\, d\mathbf{p}\, \delta[\mathcal{H}(\mathbf{q}, \mathbf{p}) - E]\, \chi_{\mathrm{r}}(\mathbf{q}, \mathbf{p})\, \dot{y}\, \delta(y - y^{\ddagger}), \tag{3.5}$$

where $\mathcal{H}(\mathbf{q}, \mathbf{p})$ is the Hamiltonian for the system, $\varrho_{\mathrm{R}}(E)$ is the density of states of the reactants, and $\chi_{\mathrm{r}}(\mathbf{q}, \mathbf{p})$ is the characteristic function of the set of all phase points lying on reactive trajectories:

$$\chi_{\mathrm{r}}(\mathbf{q}, \mathbf{p}) = \begin{cases} 1, & (\mathbf{q}, \mathbf{p}) \text{ lies on a reactive trajectory} \\ 0, & \text{otherwise} \end{cases}. \tag{3.6}$$

Hence, the rate constant is calculated from the points on the transition state and knowledge of their 'entire history' as expressed in the characteristic function, $\chi_{\mathrm{r}}(\mathbf{q}, \mathbf{p})$. A full trajectory calculation is, therefore, in general, needed to determine its value.

In the TST approximation, which says that the necessary and sufficient condition for a trajectory to be reactive is that it leaves the transition state configuration in the direction of the products, $\chi_r(\mathbf{q}, \mathbf{p})$ is replaced by the step function $\chi_+(\mathbf{q}, \mathbf{p}) = \theta(\dot{y})$:

$$k_{\mathrm{MIC}}^{\mathrm{TST}}(E) = \frac{1}{\varrho_{\mathrm{R}}(E)\, h^n} \int d\mathbf{q}\, d\mathbf{p}\, \delta\,[\mathcal{H}(\mathbf{q}, \mathbf{p}) - E]\, \theta(\dot{y})\, \dot{y}\, \delta(y - y^{\ddagger}). \quad (3.7)$$

By making this substitution no dynamics is involved in the calculation of the rate constant.

As mentioned above, the expression for the rate constant given in *Eqn. 3.5* is valid only when the possible states of the reactants are populated uniformly. The assumption of an uniform population throughout the reaction, together with the TST assumption, make up the core of *RRKM* theory, whence $k_{\mathrm{MIC}}^{\mathrm{TST}}(E)$ is also denoted $k_{RRKM}(E)$ [1].

The question is now how to describe the rate of a process where the population of the reactant states is not *a priori* equilibrated during the course of reaction. One way, of course, is simply to monitor, as a function of time, the decay of reactants using *Eqn. 3.1*, as we shall often do in the following. It is, however, illustrative to derive a general expression for the lifetime distribution in the case where the reactants are prepared in a microcanonical ensemble initially. Such a derivation is given in the *Appendix*, and we find that

$$P(t; E) = \frac{1}{\varrho_{\mathrm{R}}(E)\, h^n} \int d\mathbf{q}\, d\mathbf{p}\, \delta\,[\mathcal{H}(\mathbf{q}, \mathbf{p}) - E]\, \theta[y(t) - y^{\ddagger}]\, \dot{y}\, \delta(y - y^{\ddagger}). \quad (3.8)$$

Thus, the lifetime distribution of an initially prepared microcanonical ensemble is also determined solely by the reactive flux through the transition state. Furthermore, by taking the limit $t \rightarrow t_0^+$, we find

$$\begin{aligned} \lim_{t \rightarrow t_0^+} P(t; E) &= \frac{1}{\varrho_{\mathrm{R}}(E)\, h^n} \int d\mathbf{q}\, d\mathbf{p}\, \delta\,[\mathcal{H}(\mathbf{q}, \mathbf{p}) - E]\, \theta(\dot{y})\, \dot{y}\, \delta(y - y^{\ddagger}) \\ &= k_{RRKM}(E). \end{aligned} \quad (3.9)$$

Hence, by comparison with the phenomenological model of the previous section, *Eqns. 2.3* and *2.5*, we can now conclude that *if an initially prepared microcanonical ensemble decays exponentially, the rate constant is given by* $k_{RRKM}(E)$. In passing, we note that a similar statement has been made in connection with the decay to equilibrium of an initially prepared non-equilibrium mixture of two different isomers [9]. The exponential decay of an initially prepared microcanonical ensemble (or, equivalently, a random dissociation probability) is therefore, termed '*RRKM* behavior'. The dynamical origin of *RRKM* behavior or, rather, of non-*RRKM* behavior has been the subject of intense investigation [1]. This has given rise to two different classes of non-

RRKM behavior: *Intrinsic* and *apparent* non-*RRKM* behavior. It has been shown that *RRKM* behavior is a manifestation of ergodic intramolecular nuclear motion. This means that IVR is rapid compared with the reaction rate. Along these lines, intrinsic non-*RRKM* behavior refers to a situation where an initially prepared microcanonical ensemble does not decay exponentially because a microcanonical ensemble is not maintained during the course of the reaction due to nonergodic intramolecular nuclear motion (the existence of 'bottlenecks' in phase space). In other words, transitions between certain states in the reactant molecules are less likely than transitions leading to product formation.

The term 'apparent non-*RRKM* behavior' covers most other reasons for nonexponential decay – often due to a nonstatistical preparation of the reactant molecules [1]. The observed decay in a unimolecular reaction is, in general, a convolution of the initial preparation and the underlying intramolecular nuclear motion. However, nonstatistical preparation in terms of *coherent, selective activation* may be just the key to highlighting the specifics of the intramolecular nuclear motion, hence providing a tool for determining the ergodic or nonergodic nature of a given molecule.

As mentioned in the *Introduction*, a femtosecond laser pulse can be used to activate a unimolecular process. Such a laser pulse generates, in general, a localized distribution (wave packet) in the above-mentioned sense. Also, a second femtosecond laser pulse may be used in a pump-probe setup to monitor, as a function of time, the passage through a certain nuclear configuration during the course of reaction. Below, we recapitulate a simple classical model describing the activation as well as the probing process and relate it to the quantities discussed above.

3.3. Coherent Activation and Probing of a Unimolecular Reaction

In a classical picture, where we also assume that the femtosecond (pump) pulse is short compared to the time scale of the nuclear motion, the initial distribution of nuclear configurations in phase space, $\varrho_0(\mathbf{q}, \mathbf{p})$, can be written as [28][29]

$$\varrho_0(\mathbf{q}, \mathbf{p}) \propto W_{\mathrm{FC}}(\mathbf{q}; E_{\mathrm{pump}})\, \varrho_{\mathrm{gr}}(\mathbf{q}, \mathbf{p}). \tag{3.10}$$

Here $\varrho_{\mathrm{gr}}(\mathbf{q}, \mathbf{p})$ is a distribution representing the vibrational ground state of the molecule in the electronic ground state, and $W_{\mathrm{FC}}(\mathbf{q}; E_{\mathrm{pump}})$ is the *Franck-Condon* window function. The latter is given by the *Fourier* transform of the temporal envelope of the laser pulse with respect to its peak photon energy, E_{pump}, evaluated at the vertical transition energy between the excited- and ground-state PES, as a function of the nuclear configuration. Other schemes neglect $\varrho_{gr}(\mathbf{q}, \mathbf{p})$ entirely, and depend only on the *Frank-Condon* window function and the temporal envelope of the laser pulse [30].

In any case, the above picture reflects the nature of femtosecond activation, which is to create a spatially as well as energetically localized ensemble of molecules on the excited-state PES. For further discussion of femtosecond activation including more advanced methods using pulse sequences, *e.g.*, the pump-dump scheme, see [13–20].

In a pump-probe setup, a second femtosecond pulse promotes the molecules to a second excited electronic state after a certain time delay with respect to the first pulse. A transient is recorded that reflects the population promoted as a function of the time delay. In a classical picture, where we again assume that the duration of each of each pulse is short compared with the time scale of the nuclear motion, the transient signal can be written as [28][29]

$$S(t) \propto \int d\mathbf{q}\, W_{FC}\,[\mathbf{q}(t); E_{probe}\,]\, \varrho_0\,(\mathbf{q}), \tag{3.11}$$

where $\varrho_0(\mathbf{q}) = \int d\mathbf{p}\, \varrho_0(\mathbf{q}, \mathbf{p})$. The *Franck-Condon* window function picks out a certain nuclear configuration. Hence, on tuning the probe laser to pick out the transition-state configuration, the transient signal would measure the distribution of molecules passing through the transition state as a function of time, *i.e.*, the transient reflects the lifetime distribution. Alternatively, the probe laser can be tuned to pick out the initial configuration. In this case, the transient reflects the time evolution of the auto-correlation function [24]

$$C(t) = \int d\mathbf{q}\, \varrho_0\,[\mathbf{q}(t)]\, \varrho_0\,(\mathbf{q}). \tag{3.12}$$

Therefore, in general, a pump-probe experiment can be used to determine whether the initially prepared ensemble maintains its coherence during the course of reaction.

In the following, we consider a simple numerical simulation of a unimolecular decomposition and elaborate on the origin of non-*RRKM* behavior and how to probe it experimentally.

4. Numerical Simulation of a Unimolecular Reaction

As mentioned above, *RRKM* behavior or deviations therefrom is related to the ergodicity of the intramolecular motion: IVR *vs.* the rate of reaction. Based on a simple model representing a molecular isomerization process, an analysis of this relationship has been carried out by *Berne* and co-workers [25][31][32]. With a simple two-dimensional bound potential, *De Leon* and

Berne [31] have illustrated the transition from non-*RRKM* to *RRKM* behavior as the dynamics change from regular to chaotic. Using this model potential and a stochastic scheme for coupling the two degrees of freedom [25], we study here the unimolecular decay process. To simplify the analysis, we construct a model where the TST assumption is built in, so to speak: when a trajectory reaches the transition state from the reactant side it is stopped and counted as reactive. This enables us to discuss deviations from *RRKM* behavior solely in terms of nonergodic motion. Thus, we consider a reactive coordinate (RC) with $y^{\ddagger} = 0$ and the potential on the reactant side of the transition state given by

$$V_{\mathrm{RC}}(y) = 4y^2(y^2 - 1) + 1\,, \quad y \leq 0\,. \tag{4.1}$$

Details of the potential on the product side are of no importance since trajectories are stopped when they reach $y = y^{\ddagger} = 0$. Hence, the potential along the reactive coordinate is of the general form depicted in *Fig. 1* (see also *Fig. 6*). The dissociation barrier for this potential is $D = V_{\mathrm{RC}}(0) = 1$. If we choose for the nonreactive coordinate (nRC) the same potential as *De Leon* and *Berne* [31], the Hamiltonian without coupling between the two modes becomes

$$\mathscr{H}(x, y, p_x, p_y) = \frac{p_x^2 + p_y^2}{2m} + V_{\mathrm{nRC}}(x) + V_{\mathrm{RC}}(y), \tag{4.2}$$

where the nonreactive mode is the non-linear oscillator,

$$V_{\mathrm{nRC}}(x) = D[1 - \exp(-\lambda x)]^2\,, \tag{4.3}$$

with $\lambda = 1.95$, $D = 10$, and $m = 8$.

To describe coupling between the two modes, we use a stochastic kinetic energy exchange model [25]. In this model random, instantaneous energy-transfer events change the magnitude (but not the direction) of the momenta of the two degrees of freedom according to:

$$\begin{aligned} |p_x'| &= |p_y| \\ |p_y'| &= |p_x|\text{'} \end{aligned} \tag{4.4}$$

where the prime indicate the momenta after the energy-transfer event. The total energy of the system is thereby conserved, and the strength of the coupling is characterized by the mean time between energy-transfer events, τ, where $\tau^{-1}dt$ is the probability of an energy-transfer event taking place in any infinitesimal time interval dt.

This stochastic scheme, which, from a physical standpoint, resembles a collision process, is chosen here for two reasons: *i*) it enables us to discuss the 'coupling strength' between the two modes in terms of a characteristic time rather than a coupling term in the Hamiltonian; *ii*) the coupling is generic. By the latter, we imply that the coupling does not depend on the specifics of the Hamiltonian, such as resonance conditions *etc.* The nonreactive mode can thus be thought of as a single molecular vibration or merely as a 'collective coordinate' representing a multi-dimensional space of vibrational modes exchanging energy with the reaction coordinate.

Before we proceed, a few comments about this model should be made. First, we consider the time scale of the dynamics, that is, the unit of the dimensionless time. In most of our simulations presented below, we consider energies on the order of the dissociation barrier, D. At this energy, one time unit corresponds roughly to half a vibrational period in the nonreactive coordinate. Hence, to obtain a molecular time scale we assert that one unit of time is on the order of 100 fs.

Second, to justify our choices of parameters, we briefly recall the results of *Berne* and co-workers [25][31], who investigated the dynamics in a double-well potential given by *Eqn. 4.1* and its mirror image. Choosing $\tau = 1$, they show that for total energies slightly above D, decay to equilibrium follows a single exponential with the relaxation time given by $1/k_{RRKM}(E)$. This result is in perfect agreement with their finding that at this exchange rate between modes, the dynamics in their system is practically ergodic. We assume that the decay of A* will follow a similar trend.

4.1. Unimolecular Decay of a Microcanonical Ensemble

First, we monitor the decay of the reactants as well as the lifetime distribution for a microcanonical distribution of reactants at time zero with $E = 1.1$ and $\tau = 1$. The decay of reactants is calculated using *Eqn. A.3*, and the normalized lifetime distribution, defined as the lifetime distribution divided by $k_{RRKM}(E)$, is calculated using *Eqn. A.6*.

The results are shown in *Fig. 3*. The number of reactants and the lifetime distribution both decay single exponentially with a relaxation time $\tau_{\mathrm{rxn}} = 55.9$, giving the decay constant, $k = 1/\tau_{\mathrm{rxn}} = 0.018$. In comparison, *Eqn. 3.7* gives $k_{RRKM} = 0.017$ with $E = 1.1$. Note that in this model the *RRKM* rate constant is independent of τ. Thus, the system we consider exhibit *RRKM* behavior under these conditions. This could be expected since the rate of reaction in this case is much slower than IVR, *i.e.*, $\tau_{\mathrm{rxn}} \gg \tau$. Making τ even smaller has no effect on the relaxation time, which is a fur-

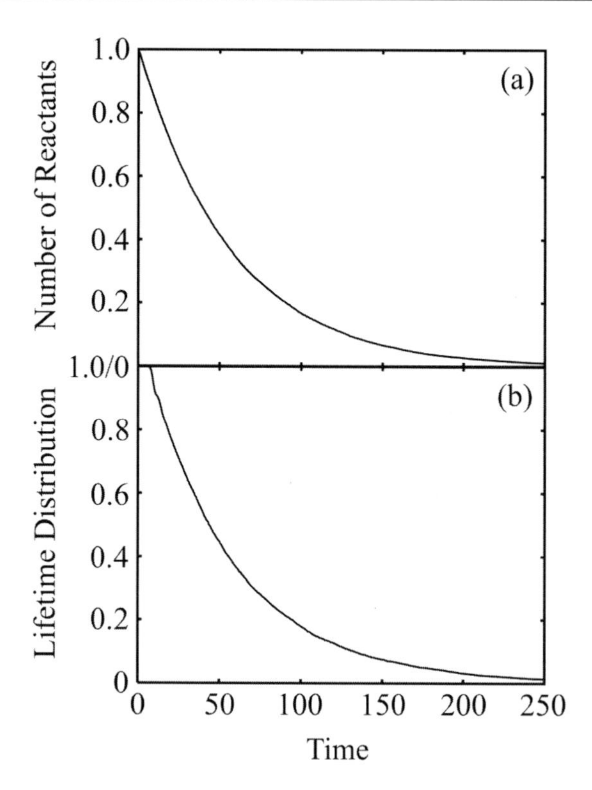

Fig. 3. *Microcanonical results at* E = *1.1 for* $\tau = 1$. *a*) The normalized number of reactants as a function of time, $N_{A*}(t)/N_{A*}(t_0)$; *b*) the normalized life time distribution, $\kappa(t; E) = P(t; E)/k_{RRKM}(E)$. The reader is reminded that the time unit on this and subsequent figures is on the order of half a vibrational period in the nonreactive coordinate (see text); the energy units (E) is in reference to the dissociation energy ($D = 1$).

ther indication of *RRKM* behavior. On the other hand, making τ larger results in non-*RRKM* behavior, as exemplified below, so $\tau = 1$ seems to be the upper limit for statistical dynamics at this energy.

To investigate further the dynamics for a slower energy exchange rate, we now consider the same initial conditions but with $\tau = 10$. The results are shown in *Fig. 4*. Here the decay is no longer single exponential but has two components; a fast decay with $\tau_1 \approx 2.8$ and a slow one with $\tau_2 \approx 121$. The bi-exponential behavior is most clear for the lifetime distribution, which can be understood by comparing *Eqns. 2.7* and *2.8*. The bi-exponential fit to the number of reactants has only a small contribution from the fast component (~10%). This, however, is compensated in the lifetime distribution by multiplication of each term by its respective decay constant, which results in a magnification of the fast component.

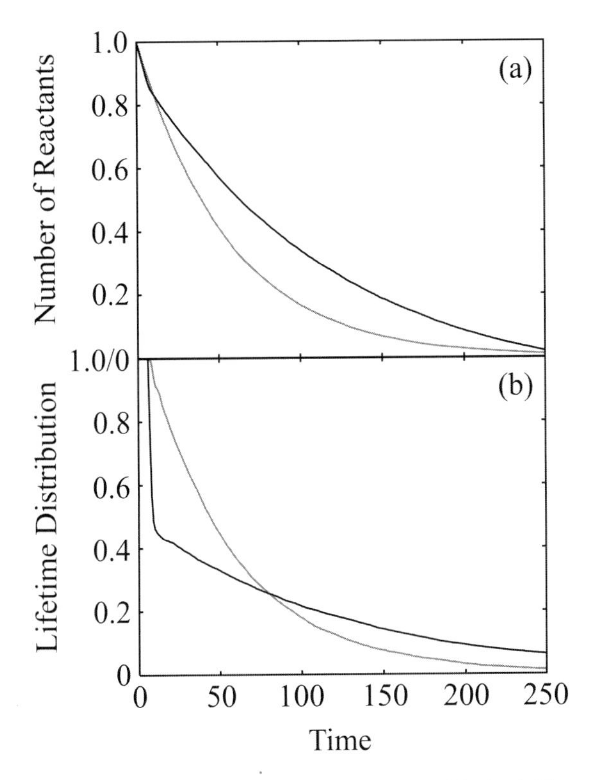

Fig. 4. *Microcanonical results at* E = *1.1 for* $\tau = 10$ *shown in black. a*) The normalized number of reactants as a function of time; *b*) the normalized life time distribution. For comparison, the results of *Fig. 3* are shown in gray.

Up to now, we have discussed only degrees of freedom that can reversibly exchange energy. In the following we allow for the possibility of energy relaxation (deactivation) of the reactants.

4.1.1. *Unimolecular Decay Combined with Energy Relaxation*

In *Sect. 2*, we discussed the kinetics characterizing a system coupled to an energy bath. Here, we include such a bath in our dynamical calculations. The nature of the coupling between the reactive space and the bath naturally depends on the system under consideration. For the model considered here, we make the following choices: *i*) The bath is coupled only to the nonreactive coordinate, *ii*) the coupling is a stochastic energy transfer with the mean time $\tau_{\text{deact}} = 1/k_{\text{deact}}$ between transfer events and an energy transfer resulting in $|p'_x| = 0$. Hence, we can think of this as a hierarchic model for a multi-dimensional molecule where the reaction coordinate is directly (dynamically) coupled to another 'nearby' nuclear coordinate, which, in turn, mediates coupling to more 'distant' vibrational degrees of freedom.

With the inclusion of a bath of this kind, we calculate the decay of reactants with the same initial conditions as above. For the coupling to the bath, we choose an intermediate deactivation time (with respect to τ) of $\tau_{\mathrm{deact}} = 8$. Besides the decay in the number of reactants, A*, we also monitor the rise in the deactivated reactant A and the dissociation product, using $H_{A*} = [1 - \theta(y)]\ \theta[\mathscr{H}(\mathbf{q}, \mathbf{p}) - E]$. The results are shown in *Fig. 5.* Also shown are the kinetic results obtained from *Eqns. 2.15*, *2.17*, and *2.18* with the decay rate constant(s) k_n and their respective weight c_n (for $\tau = 10$) determined from *Figs. 3* and *4*. Both for $\tau = 1$ and $\tau = 10$, there is perfect agreement between the results of the simulation and the kinetic model.

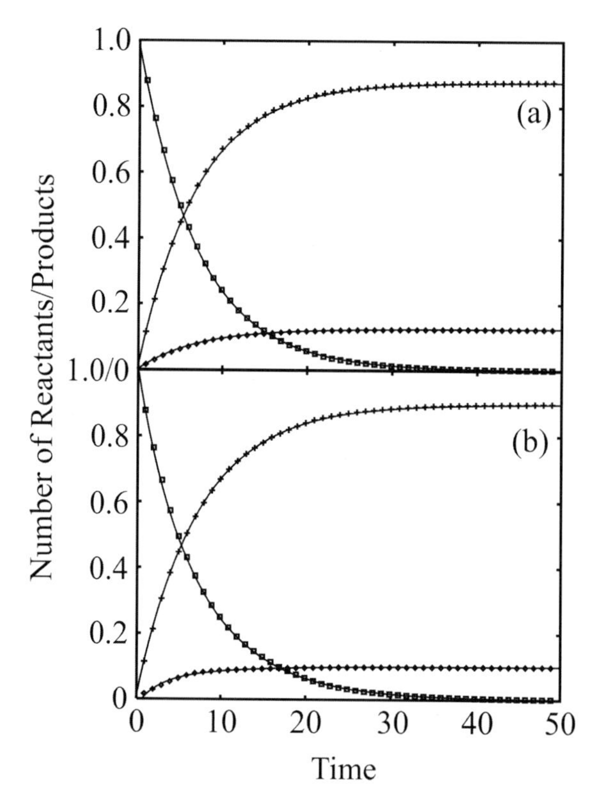

Fig. 5. *Microcanonical results at with* E = *1.1 with a)* $\tau = 1$ *and b)* $\tau = 10$*, including a bath.* Normalized results obtained in trajectory calculations are shown for reactants (squares), deactivated reactants (crosses), and products (diamonds). The solid curves illustrate the results of a kinetic analysis (see text).

In conclusion, we see for the initially prepared microcanonical ensemble: *i*) When the exchange rate between the two degrees of freedom is fast enough, the ensemble (no bath) decays single exponentially with the decay rate constant $k = k_{RRKM}$. *ii*) When the exchange rate between the two degrees of freedom is slowed down, the ensemble does not remain microcanonical during the course of reaction; the time scale bifurcates, and two decay rate constants are observed. *iii*) Finally, adding a bath speeds up all processes without changing their characteristics (single- or multi-dimensional decay), according to kinetic theory.

With these observations in mind, we now deal with the characteristics of a reaction following *selective activation* in order to address questions concerning statistical *vs.* nonstatistical dynamics and how selective activation, in general, can disentangle underlying dynamical features that are often buried by statistical preparation.

4.2. Decay of Selectively Activated Ensembles

By selective activation we mean activation of a spatially localized ensemble. This is feasible in a molecular system using a femtosecond laser, as discussed in connection with *Eqn. 3.10*. Hence, we now assume that the Hamiltonian of *Eqn. 4.2* describes the nuclear dynamics of the system in an excited electronic state. In the model system considered, we do not actually have multiple PESs, but, in order to *mimic* femtosecond activation, we create initial ensembles in the following way. First, we calculate the ground-state wave function, $\psi_{\mathrm{gr}}(\mathbf{q})$, of the Hamiltonian, *Eqn. 4.2*, setting $\hbar = 10^{-3}$. This value is chosen (arbitrarily) to ensure that $\psi_{\mathrm{gr}}(\mathbf{q})$ is localized with respect to the amplitude of the motion at energies above the dissociation barrier. Next, we displace this function to the classical turning point, (x', y'), on the potential determined from the equations:

$$\begin{aligned} V_{\mathrm{nRC}}(x') &= \alpha E \\ V_{\mathrm{RC}}(y') &= (1-\alpha) E \,. \end{aligned} \tag{4.5}$$

The first equation may have two solutions and we choose the positive one. Now, α is a number between zero and one, where $\alpha = 1$ corresponds to activation of the nonreactive coordinate, whereas $\alpha = 0$ corresponds to activation of the reaction coordinate. Finally, we localize the energy by choosing the simple *Franck-Condon* window function:

$$W_{\mathrm{FC}}(\mathbf{q}) = \begin{cases} 1, & |V(\mathbf{q}) - E| \leq 0.05 \\ 0, & \text{otherwise} \end{cases} \,. \tag{4.6}$$

All together, the initial ensembles to be used here are created by *Monte Carlo* sampling of the distribution functions:

$$\varrho_0(\mathbf{q},\mathbf{p}) = W_{\mathrm{FC}}(\mathbf{q})\,|\psi_{\mathrm{gr}}(\mathbf{q}-\mathbf{q}')|^2\,\delta(\mathbf{p}), \tag{4.7}$$

with proper normalization.

4.2.1. *Bond-Selective Activation*

We consider the two limiting cases, namely, activation of either the nonreactive or the reactive coordinate. The two ensembles corresponding to activation of the nonreactive (1) and the reactive (2) coordinate at $E = 1.1$ are illustrated in *Fig. 6*. Since the ensembles are at rest at time zero, their initial motion is directed by the gradient of the potential, as indicated by the arrows on the figure. With initial conditions sampled from these ensembles, we calculate the decay of the reactants for both $\tau = 1$ and $\tau = 10$. The results are shown in *Fig. 7* together with the previous results for the initial microcanonical ensemble.

For $\tau = 1$, there is not much difference between the microcanonical results and the ones obtained by selective activation, they all exhibit *RRKM* behavior. As discussed above, the *RRKM* behavior of initially prepared microcanonical ensemble is due to the fact that the time scale for IVR (τ) is much shorter than the time scale for reaction, which ensures a microcanonical distribution throughout the course of reaction. Here we see, in addition, that the time scale for IVR in this case is sufficiently short to destroy the coherence of the selectively prepared ensembles and to fill phase space in an ergodic fashion prior to reaction. The coherence of the selective prepared ensembles is notable, however, at very short times. The decay of the reactants does not begin until $t \approx 4$. This delay is the so-called *induction time*, which is the time it takes for the fastest trajectory to reach the transition-state configuration [33][34].

The situation for $\tau = 10$ is quite different. Here, we see that the ensemble prepared selectively in the nonreactive coordinate closely follows the behavior of the initially prepared microcanonical ensemble, whereas the ensemble prepared selectively in the reactive coordinate deviates substantially. As mentioned above, the decay of the initially prepared microcanonical ensemble is, in fact, bi-exponential, but the fast component constitutes only *ca.* 10%, which corresponds roughly to the fraction of trajectories that dissociate directly. The decay of the ensemble prepared selectively in the reactive coordinate is also bi-exponential with the same components. Here, however, the fast component constitutes *ca.* 85%. The fast component arises from those

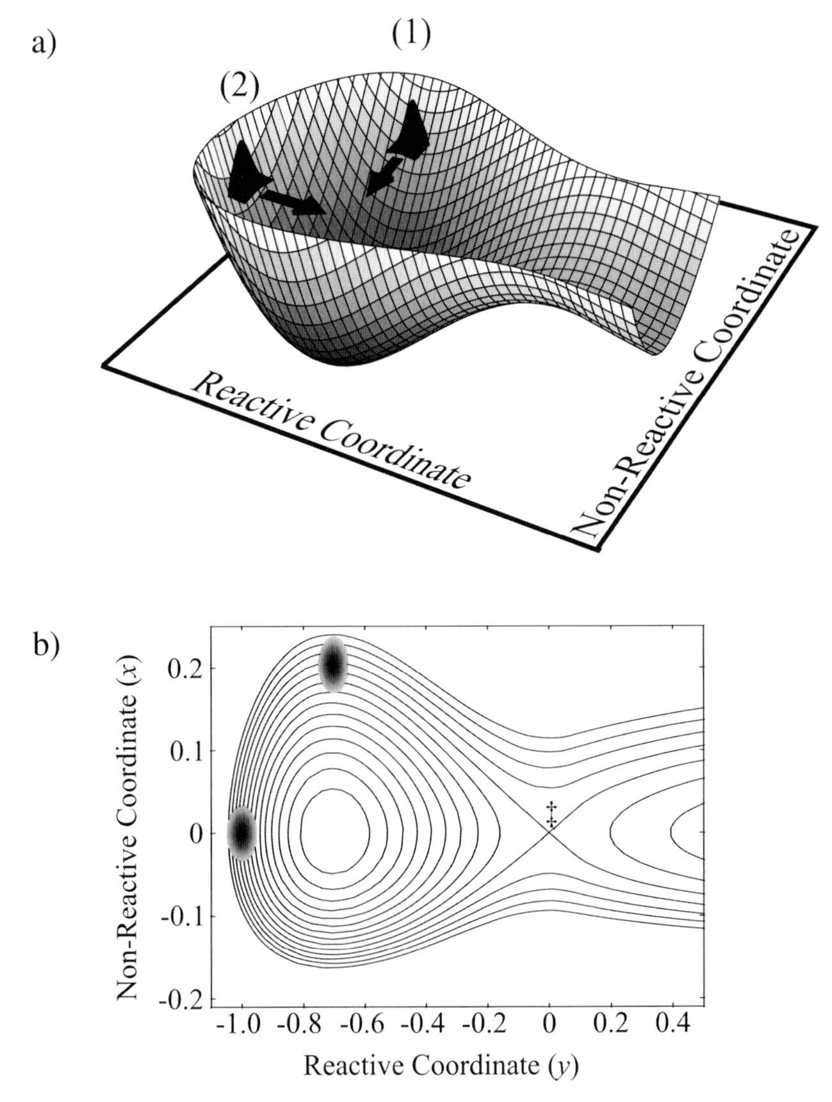

Fig. 6. *Illustration of the two ensembles created by selective activation of either the nonreactive* (1) *or the reactive* (2) *coordinate.* The arrows in *a* indicate the direction of the initial motion. On frame *b*, the contours are drawn for $E = 0.1, 0.2, 0.3, \ldots, 1.4$.

molecules that dissociate directly, in contrast to those that dissociate only after having energy transferred back and forth between the modes. The latter gives rise to the slow component. This also explains why the decay of ensemble prepared selectively in the nonreactive coordinate closely follows the behavior of the initially prepared microcanonical ensemble, since here dissociation can occur only after energy is transferred to the reactive coordinate.

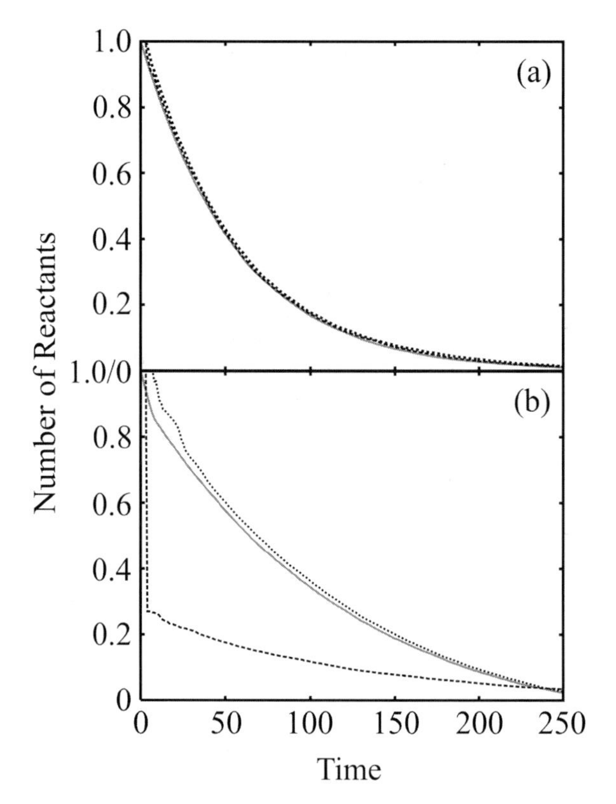

Fig. 7. *Normalized number of reactants for the selective activation of the nonreactive* (dotted) *and the reactive coordinate* (dashed). Microcanonical results are shown for comparison (gray solid lines). *a*) Results for $\tau = 1$; *b*) result for $\tau = 10$.

Thus, by monitoring the decay of selectively prepared ensembles one is able to *magnify* one or the other of the two mechanisms for dissociation. This is further illustrated in *Figs. 8* and *9*, where we show the lifetime distributions in the four cases. Again we see that the results for the ensemble prepared selectively in the nonreactive coordinate follow those obtained from the initial microcanonical ensemble. This, however, is only the case for the ensemble prepared selectively in the reactive coordinate when the energy exchange is fast ($\tau = 1$). For the slow exchange rate ($\tau = 10$), the lifetime distribution has a narrow, dominant peak around $t = 4.1$. This is exactly the time it takes for a trajectory started at the center of this ensemble to reach the transition state, as expected from the interpretation given above. Hence, in this case, the initially prepared ensemble maintains its coherence on the time scale of reaction.

To further address the issue of coherence, we have also calculated the auto-correlation function, *Eqn. 3.12*, for the ensemble initially prepared in

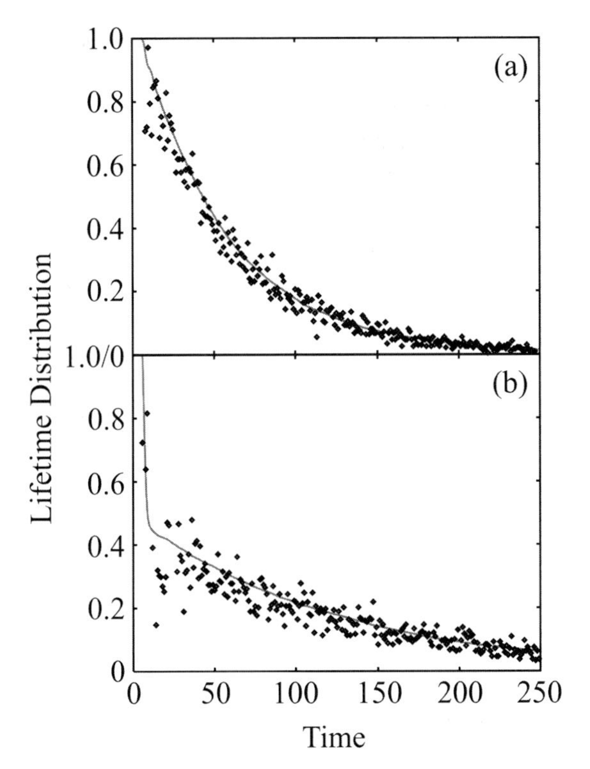

Fig. 8. *Lifetime distributions normalized by* $k_{RRKM}(E)$ *calculated for ensemble* (1) *are shown with diamonds.* The solid gray lines are the microcanonical results. *a*) Results for $\tau = 1$; *b*) result for $\tau = 10$.

the nonreactive coordinate. The results are shown in *Fig. 10* together with the function $\exp(-t/\tau)$ for $\tau = 1$ and $\tau = 10$, respectively. Two features should be noted. The auto-correlation functions show recurring peaks with a separation of $\Delta t = 2.16$, which is exactly the vibrational period in the nonreactive coordinate at $E = 1.1$, and the amplitude of the peaks decays according to the function $\exp(-t/\tau)$. In general, the peak height of the auto-correlation function is expected to decay with a rate equal to the sum of the rate of reaction (energy transfer to the reactive coordinate in this case) and the rate of spreading within the nonreactive coordinate itself [24]. Hence, in this case, the rate of reaction is much higher than the rate of spreading within the non-reactive coordinate.

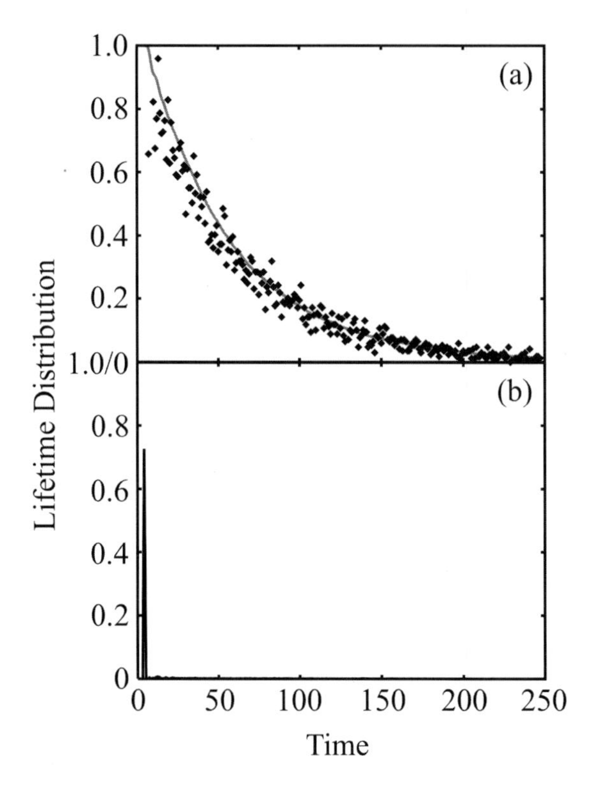

Fig. 9. *Lifetime distributions calculated for ensemble* (2) *with a)* $\tau = 1$ *and b)* $\tau = 10$. The results in frame *a*, shown with diamonds, are normalized by $k_{RRKM}(E)$ and the microcanonical results are shown for comparison. The results in frame *b* are unscaled.

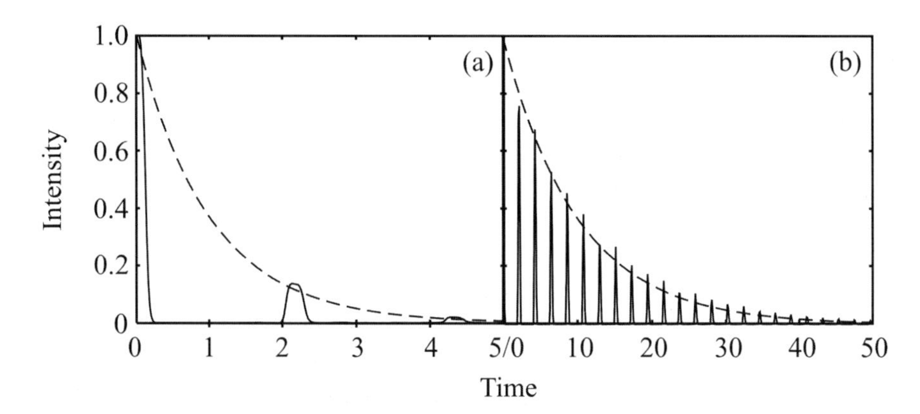

Fig. 10. *The auto-correlation function calculated for ensemble* (1) *with* a) $\tau = 1$ *and* b) $\tau = 10$. Also shown is the function $\exp(-t/\tau)$ for each case (dashed).

4.2.1.1. *Unimolecular Decay Combined with Energy Relaxation*

Now, we add a bath as prescribed above and repeat the simulation with selective activation. For the fast exchange rate ($\tau = 1$), the integrated results, that is, the number of reactants/products as a function of time, follow exactly the results obtained for the initially prepared microcanonical ensemble (*Fig. 5*). This is no surprise since we saw above, without the bath, that the coherence of the initially prepared ensembles was destroyed to give *RRKM* results for $\tau = 1$. Thus, the statistical behavior was expected also with the bath.

For the slow exchange rate ($\tau = 10$), however, the results illustrate nicely how selective activation can disentangle the underlying *fundamental* time scales. In this case, we learned that three time scales for reaction are involved in the decay of the initially prepared microcanonical ensemble: $\tau_1 \approx 2.8$, $\tau_2 \approx 121$, and $\tau_{\text{deact}} = 8$. The fast component arises from direct reaction, whereas the slow component originates from molecules that undergo energy exchange prior to dissociation. Also, we learned that the fastest time for dissociation of molecules from any of the two selectively prepared ensembles (the induction time) is *ca.* 4.1. Now, we see how these time scales show up in the results.

In *Fig. 11* we show the integrated results for both of the initially prepared ensembles, along with fits to the obtained results. In both cases the induction time, $t \approx 4$, is still detectable since no product is produced before that time. However, the activated reactants decay during that time, due to interaction with the bath. This is indeed the case and that part of the decay is, in both cases, single exponential with time constant $\tau_{\text{deact}} = 8$. Around $t = 4$, products appear in both cases. For ensemble (1) this is a smooth single exponential rise with the rate constant $k = k_2 + k_{\text{deact}}$ accompanied by a change in the decay rate constant from k_{deact} to k as well. For ensemble (2), there is an abrupt rise in the number of products, accompanied by a similar fall in the number of reactants, followed by single exponential behavior with the same rate constants as ensemble (1). Thus, the number of reactants essentially follows the behavior depicted in *Fig. 7,b*, multiplied by the damping function $\exp(-t/\tau_{\text{deact}})$, which is the expected result from the kinetic treatment mentioned above.

The induction time plays the role of the dissociation time in a one-dimensional, reduced-space picture of the dissociation process. Although the induction time is detectable in *Fig. 11,a*, it is hugely magnified in *Fig. 11,b*. This shows that, even with inclusion of the bath, a one-dimensional picture, where activation of the reactive coordinate leads to direct dissociation, is applicable. The slow energy-exchange rate between the reactive and the non-reactive coordinate 'shields' the reactive coordinate from the bath, *i.e.*, the nonreactive

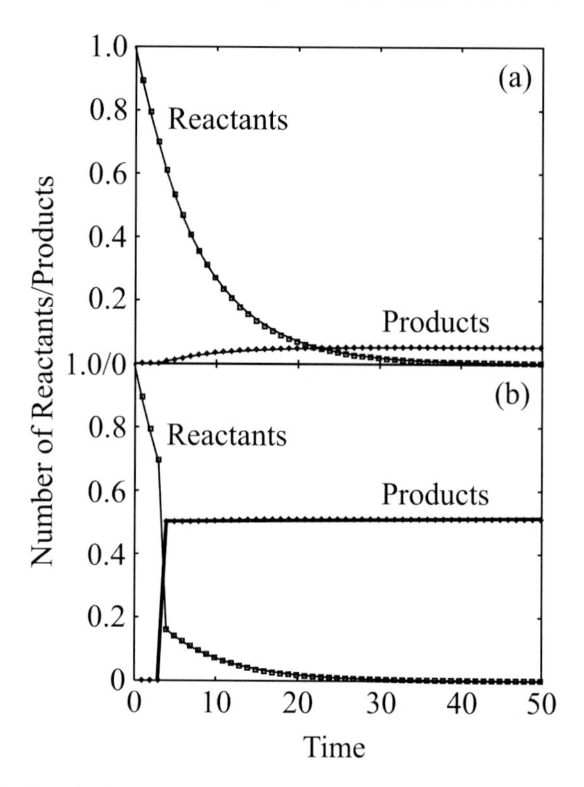

Fig. 11. *Results for selective activation with $\tau = 10$ including a bath.* Normalized results obtained in trajectory calculations are shown for reactants (squares) and products (diamonds). *a)* Ensemble (1); *b)* ensemble (2). The solid curves illustrate the results of a kinetic analysis, see text.

coordinate acts as a bottle neck, as further illustrated in *Fig. 12*, which shows the lifetime distributions as a result of activation of the reactive coordinate. We see that, apart from the amplitude, the lifetime distribution is unaffected by the coupling to the bath. Finally, we show in *Fig. 13* the auto-correlation function following activation of the nonreactive coordinate. The auto-correlation function shows the same behavior as without the bath (*Fig. 10*), namely, recurrences with a period of $\Delta t = 2.16$. However, in this case, there are two mechanisms leading to depletion of the signal: energy exchange with the reactive coordinate, and loss to the bath. Thus, the depletion rate constant is now a sum of $1/\tau$ and $1/\tau_{deact}$, as indicated in the figure. As mentioned above, there is in principle a third mechanism, which could lead to decay of the signal, namely, spreading within the nonreactive coordinate. But, again, this takes place on a much longer time scale for this system.

In summary, we see that for the system considered here, conclusions drawn on the basis of decay rates are of truly dynamical nature, *i.e.*, give in-

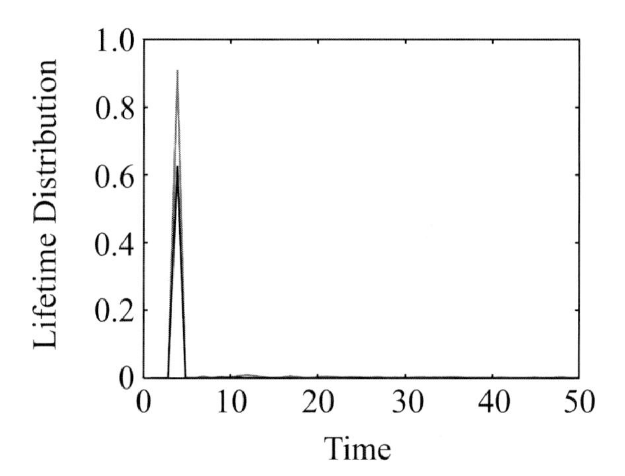

Fig. 12. *Lifetime distribution calculated for ensemble* (2) *with* $\tau = 10$ *including a bath.* The results of *Fig. 9,b*, are shown for comparison.

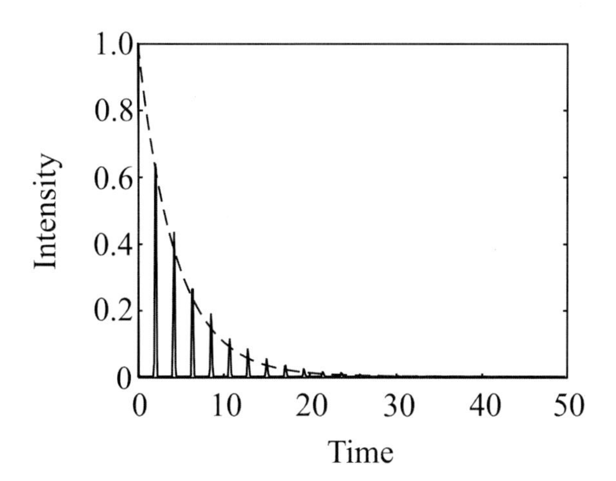

Fig. 13. *The auto-correlation function calculated for ensemble* (1) *with* $\tau = 10$ *including a bath.* Also shown is the function $\exp[-t(1/\tau + 1/\tau_{\mathrm{deact}})]$ (dashed).

sight into the dynamics of the system. In other words, selective activation does not introduce – but rather probes – a certain dynamical behavior.

4.2.2. *Localized Activation in General*

Having considered the two limiting cases, we now turn our attention to a 'mixed' situation, where the initially localized wave packet does not correspond to activation of either the reactive or the nonreactive coordinate; energy is deposited in both coordinates. Especially, we focus on the significance of the amount of energy deposited. It is obvious that the characteristic times for dissociation and vibration depend on the total energy, but the question we now address is the robustness of the trends observed above. For the interesting case where the dynamics is nonergodic ($\tau = 10$), we expect the following: the decay of an initially prepared microcanonical ensemble will continue to follow a bi-exponential decay as the energy is increased. The two time scales might change slightly, but, more importantly, the ratio between the two time components will change, making the short-time component more pronounced since more trajectories will be able to dissociate directly as the total energy is increased. For direct activation of the reactive or nonreactive coordinate, we expect no significant alteration in the profile of the decay, apart from a possible change in the characteristic times. Hence, there seems to be no dramatic dependency on the energy in these cases.

This, however, is not the case for the 'mixed' activation. In *Fig. 14,a* we show the decay of the reactants resulting from localized activation in this way, with $\alpha = 0.5$ in *Eqn. 4.7*, for different energies. For energies $E = 1.1$ and $E = 1.5$, the profile of the decay resembles that for activation of the nonreactive coordinate, whereas the profile at has $E = 2.2$ a dominant contribution from the fast component and thereby resembles that for activation of the reactive coordinate, *cf. Fig. 7,b.* ($E \approx 2$), the results show clear signs of a bifurcating wave packet in the sense that the fast and the slow component each comprise about half of the decay. Hence, for the 'mixed' activation the character of the dynamics changes as the energy is increased. This implies that even if one is not able to target the reactive coordinate directly, localized ac-

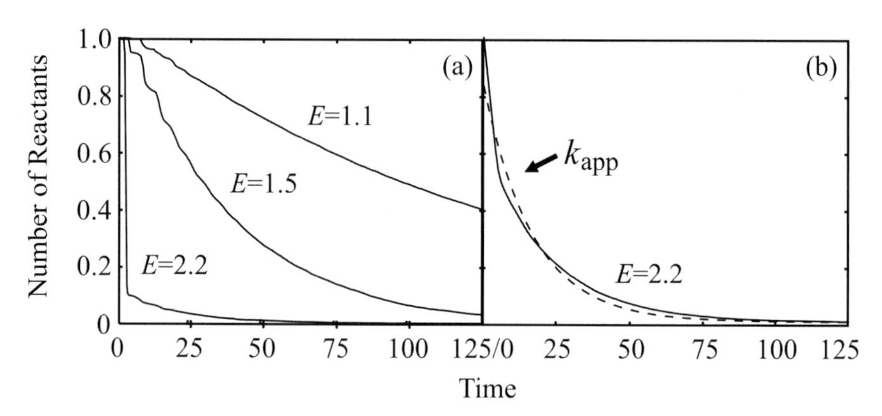

Fig. 14. a) *Decay of reactants following a localized activation depositing energy in both the reactive and the nonreactive coordinate.* b) *Decay of an initially prepared microcanonical ensemble* (solid) *together with the 'best fit' single-exponential decay* (dashed).

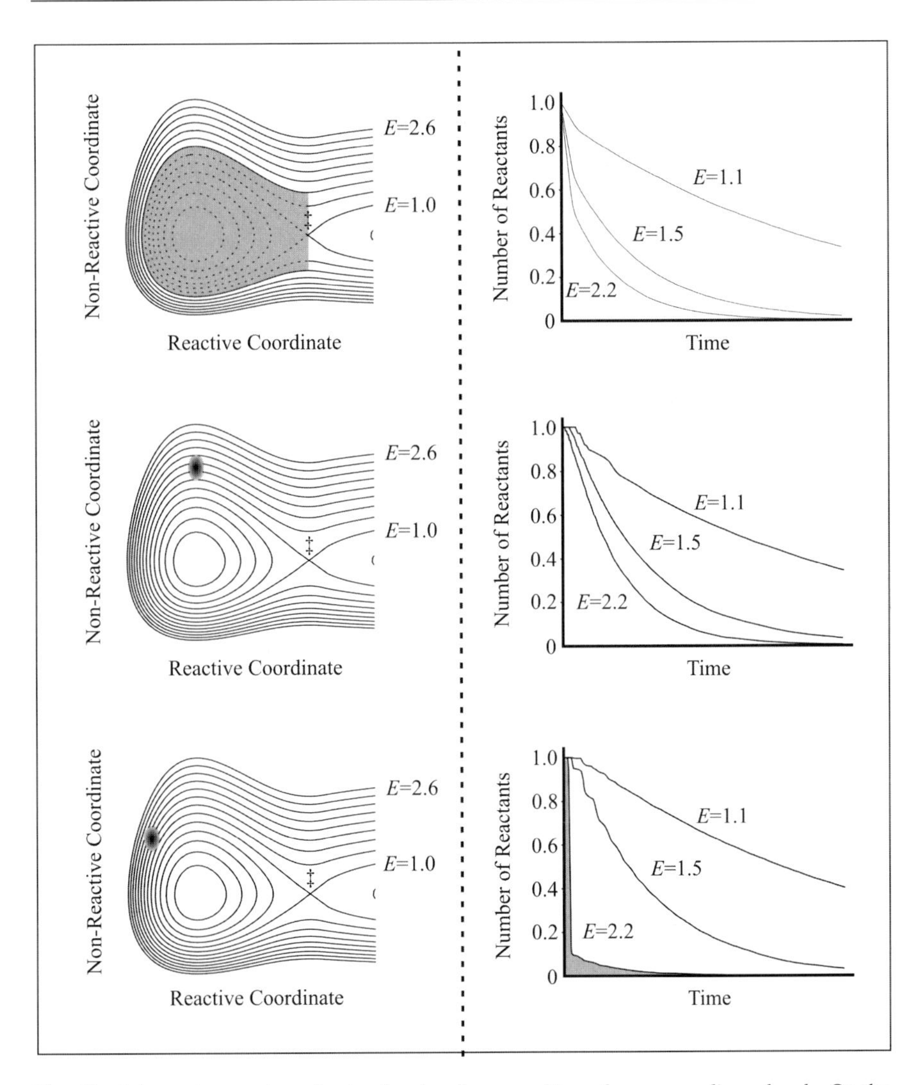

Fig. 15. *Schematic overview of unimolecular decomposition of a nonergodic molecule.* On the left is shown the potential surface with an indication of the initial preparation of the system, and on the right is shown the decay of the reactants. *Top:* Microcanonical preparation. The decay of the reactants shows a bi-exponential behavior reflecting the bifurcation into prompt (direct) and delayed reaction, which can only occur after energy transfer from the nonreactive to the reactive coordinate. The ratio between the two reaction types is shifted towards prompt reaction as the total energy in increased. *Middle:* Localized preparation of the nonreactive coordinate. The reactants decay due to delayed reaction at all energies, since reaction can only occur after energy transfer from the nonreactive to the reactive coordinate. *Bottom:* Localized, mixed preparation. The character of the reaction changes from purely delayed to (almost) purely prompt reaction as the total energy is increased.

tivation may still be used to elucidate the details of the dynamics, provided that the deposited energy is high enough. Finally, we show for comparison the decay of an initially prepared microcanonical ensemble at $E = 2.2$, *Fig. 14,b*. The bi-exponential decay is not nearly as pronounced as in *Fig. 14,a*. In fact, in an actual experiment with a less than infinite signal-to-noise ratio, as we have in the simulation, one could be misled to interpret this decay as a single exponential. In the figure, we show the 'best fit' single exponential with an *apparent* rate constant $k_{app} = 0.06$ compared with the two rate constants from a bi-exponential fit of $k_1 = 0.33$ and $k_1 = 0.046$.

With these remarks and several others throughout this section, we have tried to highlight the experimental implication of the results obtained in this simulation. The essence is summarized in *Fig. 15*, showing the results for the nonergodic system ($\tau = 10$) in three different cases: Microcanonical initial conditions (top), selective activation of the nonreactive coordinate (middle), and 'mixed' activation (bottom).

5. Comparison with Experimental Observations

Here, we make a few remarks on our findings compared with experimental results, with the focus on nonergodic behavior where selective activation results in coherent motion in each mode on a time scale shorter than the rate for energy exchange.

We argued above that this quasi-one-dimensional motion could be monitored in a suitably tuned pump-probe experiment. The lifetime distribution of *Fig. 9,a*, depicts the number of molecules passing through the transition-state configuration as a function of time, showing a simple rise and decay. This behavior compares very well with the observations made in our early femtosecond studies of the dissociation process of ICN [35–37] ($ICN^* \rightarrow I + CN$) and the subsequent analysis [38], which relied exactly on the principle of selective activation and sustained coherence.

The auto-correlation functions depicted in *Fig. 10*, monitoring the coherence in the nonreactive coordinate, bear a close resemblance to the transients observed in another of the early femtosecond experiments, namely, the study of the electronic pre-dissociation of NaI [39–42]. The auto-correlation function measures the fraction of molecules in the initial nuclear configuration as a function of time, just like the transient signal in the experiment, if the probe laser is tuned accordingly [24]. The difference here is that the depletion of the signal in the NaI experiments (which involve a one-dimensional PES) is due to dissociation (as well as wave-packet spreading [24][42]), whereas the depletion in *Fig. 10* is due to energy transfer. However, the principle is the same: One monitors the coherence, as well as loss of population due to 'reaction'. The coherent recurrences provide information about a characteristic vibrational period, and the decay of the signal monitors the loss of coherence and the 'rate of reaction'.

Finally, we address the issue of nonergodic motion in larger molecules. In a recent experiment, *Diau et al.* [43] have measured the rate of CO elimination from a series of cyclic ketones with varying ring size, following femtosecond activation. The measured rates were compared with theoretical predictions from *RRKM* calculations. The measured rates remain almost unaltered as the number of vibrational modes is increased, whereas the rates predicted by *RRKM* theory decrease by more than two orders of magnitude. In light of the above analysis, this indicates nonergodic behavior where the available energy does not redistribute among all degrees of freedom. The observed rate of reaction is, therefore, characteristic for a subset of vibrational modes, which was extracted by coherent preparation. This has implications for possible control of chemical reactions [23].

6. Conclusion

Here, we summarize the numerical analysis with emphasis on the result for selective activation and discuss the experimental implications. When the system is selectively activated it is possible, in the regime of beating IVR, to recover specific dynamical characteristics, that of a directed dynamics along the reactive coordinate, and possibly the control of the outcome of the reaction. At relatively low energy, when IVR is complete, microcanonical and selective excitation will give similar results.

The relevant time scales to compare when evaluating whether a microcanonically prepared system follows *RRKM*-behavior or not are those of the direct dissociation and IVR. Molecules with an initial energy in the reactive coordinate higher than the dissociation barrier may dissociate directly, *i.e.*, without acquiring further energy. To maintain a statistical distribution, IVR should be fast enough that the void after these molecules can be 'filled up' without delay. This is the intrinsic, dynamical requirement for observing *RRKM* behavior.

Non-statistical results arising from femtosecond activation may be a fingerprint of *intrinsic* non-*RRKM* behavior, that is, reflect nonstatistical (nonergodic) nuclear motion on the time scale of the reaction. This is because the initial preparation can be made at a 'point' in phase space such that the propagation of the wave packet will experience the intrinsic forces, including IVR. Hence, femtosecond activation is a tool for experimentally probing nonergodic nuclear motion, as shown in experiments on cyclic ketones [43], and in work from *P.-Y. Cheng*'s group [44].

Femtosecond activation can be used not only to determine whether the nuclear dynamics in a given molecular system is statistical in nature or not, but, in the latter case, also to target the dynamics of specific bonds in the molecule. The key points here are coherence and localization. At high energies this localization gives rise to direct trajectories since the transition state 'opening' is wide and can encompass the initial configuration, while at lower energies the opening is narrow and most trajectories will be indirect.

Nonergodic molecules may exhibit intrinsic coherent motion; statistical (noncoherent) preparation will have a tendency to average this out. Femtosecond activation, on the other hand, is coherent in nature and, therefore, has the capacity to reveal the otherwise 'hidden' coherence, as has been seen in numerous examples.

This work was supported by the *National Science Foundation*, the *Office of Naval Research*, and *US AFOSR*.

Appendix

Here we derive an expression for the lifetime distribution for a unimolecular reaction under the assumption that at *time zero* the states of the reactants are populated according to the microcanonical equilibrium distribution, $\varrho_R(E)$. This is done in analogy with previous treatments of isomerization processes [9][31]. We assert that $y = y^\ddagger$ separates the activated reactants A* from products such that

$$H_{A^*}(y) = \begin{cases} 1, & y \leq y^\ddagger \\ 0, & y > y^\ddagger \end{cases} \tag{A.1}$$

is the characteristic function of the set of all phase points in the reactant region. With no products to begin with we have,

$$\rho_0(\mathbf{q},\mathbf{p}) = \frac{1}{\rho_R(E)\,h^n}\,\delta[\mathcal{H}(\mathbf{q},\mathbf{p}) - E]\,H_{A^*}(y), \tag{A.2}$$

and the expression for the number of reactants, *Eqn. 3.1*, takes the form,

$$N_{A^*}(t;E) = \frac{N_{A^*}(t_0;E)}{\rho_R(E)\,h^n}\int d\mathbf{q}\,d\mathbf{p}\,\delta[\mathcal{H}(\mathbf{q},\mathbf{p}) - E]\,H_{A^*}(y)\,H_{A^*}[y(t)]. \tag{A.3}$$

This expression is recognized as a microcanonical time-correlation function (except for normalization). In evaluating the time derivative of this expression, we may, therefore, rely on the fact that, in an equilibrium system, the correlations among dynamical variables at different times depend only on the separation of these times [9][10]. Thus, for the time derivative of *Eqn. A.2* we find,

$$\begin{aligned}\frac{\rho_R(E)\,h^n}{N_{A^*}(t_0;E)}\,\frac{dN_{A^*}(t;E)}{dt} &= \int d\mathbf{q}\,d\mathbf{p}\,\delta[\mathcal{H}(\mathbf{q},\mathbf{p}) - E]\,H_{A^*}(y)\,\dot{H}_{A^*}[y(t)] \\ &= -\int d\mathbf{q}\,d\mathbf{p}\,\delta[\mathcal{H}(\mathbf{q},\mathbf{p}) - E]\,H_{A^*}(y)\,\dot{H}_{A^*}[y(-t)] \\ &= -\int d\mathbf{q}\,d\mathbf{p}\,\delta[\mathcal{H}(\mathbf{q},\mathbf{p}) - E]\,H_{A^*}[y(t)]\,\dot{H}_{A^*}(y).\end{aligned} \tag{A.4}$$

Using $H_{A*}(y) = 1 - \theta(y - y^{\ddagger})$, we obtain for the lifetime distribution:

$$P(t;E) \equiv -\frac{\dot{N}_{A*}(t;E)}{N_{A*}(t_0;E)} = \frac{1}{\rho_R(E)\,h^n} \cdot \int d\mathbf{q}\,d\mathbf{p}\,\delta[\mathcal{H}(\mathbf{q},\mathbf{p}) - E]\,\theta[y(t) - y^{\ddagger}]\,\dot{y}\,\delta(y - y^{\ddagger}). \tag{A.5}$$

From this equation and its limit as $t \to 0$ (*Eqn. 3.7*), we obtain the normalized lifetime distribution, $\kappa(t;E)$, defined as the ratio between the lifetime distribution and the *RRKM* rate constant, $\kappa(t;E) \equiv P(t;E)/k_{RRKM}(E)$:

$$\kappa(t;E) = \int d\mathbf{q}\,d\mathbf{p}\,[f^{(+)}(E) - f^{(-)}(E)]\,H_{A*}[y(t)], \tag{A.6}$$

where

$$f^{(\pm)}(E) = \frac{\delta[\mathcal{H}(\mathbf{q},\mathbf{p}) - E]\,\dot{y}\,\delta(y - y^{\ddagger})\,\theta(\pm\dot{y})}{\int d\mathbf{q}\,d\mathbf{p}\,\delta[\mathcal{H}(\mathbf{q},\mathbf{p}) - E]\,\dot{y}\,\delta(y - y^{\ddagger})\,\theta(\pm\dot{y})}. \tag{A.7}$$

This expression separates the contributions from trajectories leaving the transition state in opposite directions. In our simulations (*Sect. 4*), the trajectories leaving the transition state in the direction of products never re-enter the reactant region, implying that $\int d\mathbf{q}\,d\mathbf{p}\,f^{(+)}(E)\,H_{A*}[y(t)] = 0$ at all $t > 0$. Thus, these trajectories can be omitted from the calculation.

REFERENCES

[1] T. Baer, W. L. Hase, 'Unimolecular Reaction Dynamics', Oxford University Press, New York, 1996; and refs. therein.
[2] E. Wigner, *Trans. Faraday Soc.* **1938**, *34*, 29.
[3] J. C. Keck, *Adv. Chem. Phys.* **1962**, *13*, 85.
[4] J. C. Keck, *Adv. At. Mol. Phys.* **1972**, *8*, 39.
[5] P. Pechucas, F. J. McLafferty, *J. Chem. Phys.* **1973**, *58*, 1622.
[6] P. Pechucas, in 'Dynamics of Molecular Collisions', Vol. B, Ed. W. H. Miller, Plenum Press, New York, 1976, pp. 269.
[7] W. H. Miller, *J. Chem. Phys.* **1974**, *61*, 1823.
[8] W. H. Miller, S. D. Schwartz, J. W. Tromp, *J. Chem. Phys.* **1983**, *79*, 4889.
[9] D. Chandler, *J. Chem. Phys.* **1978**, *68*, 2959.
[10] D. Chandler, 'Introduction to Modern Statistical Mechanics', Oxford University Press, New York, 1987.
[11] W. H. Miller, *J. Phys. Chem.* **1998,** *A 102*, 793.
[12] W. H. Miller, *Faraday Discuss.* **1998**, *110*, 1.
[13] A. H. Zewail, Femtochemistry: 'Ultrafast Dynamics of the Chemical Bond', World Scientific, Singapore, 1994.
[14] 'Femtosecond Chemistry', Eds. J. Manz and L. Wöste, VCH Publ., New York, 1995.
[15] 'Femtosecond Chemistry', Eds. M. Chergui, World Scientific, Singapore, 1996.
[16] 'Chemical Reactions and Their Control on the Femtosecond Time Scale', Eds. P. Gaspard and I. Burghardt, John Wiley & Sons, New York, 1997.
[17] 'Femtochemistry and Femtobiology: Ultrafast Reaction Dynamics at Atomic-Scale Resolution', Eds. V. Sundström, Imperial College Press, London, 1997.
[18] 'Ten Years of Femtochemistry', Eds. A. W. Castleman Jr. and V. Sundström, *J. Phys. Chem. A*, **1998**, *102*.
[19] E. Scrieber, 'Femtosecond Real-Time Spectroscopy of Small Molecules and Clusters', Springer, New York, 1998.
[20] A. H. Zewail, *J. Phys. Chem., A* **2000**, *104*, 5660.

[21] F. Ramacle, R. D. Levine, *J. Phys. Chem.* **1996**, *100*, 7961.
[22] F. Ramacle, R. D. Levine, *J. Phys. Chem., A* **1998**, *102*, 10195.
[23] A. H. Zewail, *Physics Today* **1980**, *33*, 27.
[24] K. B. Møller, N. E. Henriksen, A. H. Zewail, *J. Chem. Phys.* **2000**, *113*, 10477.
[25] G. Gershinsky, B. J. Berne, *J. Chem. Phys.* **1999**, *110*, 1053.
[26] W. H. Miller, *J. Phys. Chem.* **1995**, *99*, 12387.
[27] K. B. Møller, A. H. Zewail, *Chem. Phys. Lett.* **1999**, *309*, 1.
[28] L. E. Fried, S. Mukamel, *J. Chem. Phys.* **1990**, *93*, 3063.
[29] Z. Li, Y.-L. Fang, C. C. Martens, *J. Chem. Phys.* **1996**, *104*, 6919.
[30] H. Dietz, V. Engel, *J. Phys. Chem., A* **1998**, *102*, 7406.
[31] N. De Leon, B. J. Berne, *J. Chem. Phys.* **1981**, *75*, 3495.
[32] B. J. Berne, *Chem. Phys. Lett.* **1984**, *107*, 131.
[33] D. Zhong, A. H. Zewail, *J. Phys. Chem.* **1998**, *107*, 131.
[34] K. B. Møller, A. H. Zewail, *Chem. Phys. Lett.* **1998**, *295*, 1.
[35] M. Dantos, M. J. Rosker, A. H. Zewail, *J. Chem. Phys.* **1987**, *87*, 2395.
[36] M. J. Rosker, M. Dantus, A. H. Zewail, *J. Chem. Phys.* **1988**, *89*, 6113.
[37] M. Dantos, M. J. Rosker, A. H. Zewail, *J. Chem. Phys.* **1988**, *89*, 6128.
[38] R. Bersohn, A. H. Zewail, *Ber. Bunsen-Ges. Phys. Chem.* **1988**, *92*, 373.
[39] T. Rose, M. J. Rosker, A. H. Zewail, *J. Chem. Phys.* **1988**, *88*, 6672.
[40] M. J. Rosker, T. Rose, A. H. Zewail, *Chem. Phys. Lett.* **1988**, *146*, 175.
[41] T. Rose, M. J. Rosker, A. H. Zewail, *J. Chem. Phys.* **1989**, *91*, 7415.
[42] P. Cong, A. Mokhtari, A. H. Zewail, *Chem. Phys. Lett.* **1990**, *172*, 109.
[43] E. W.-G. Diau, J. L. Herek, Z. H. Kim, A. H. Zewail, *Science* **1988**, *279*, 847.
[44] I.-R. Lee, W.-K. Chen, Y.-C. Chung, P.-Y. Cheng, *J. Phys. Chem., A* **2000**, *104*, 10595.

Photochemistry Meets Natural-Product Synthesis

by **Gerhard Quinkert** and **Knut Eis**

Institut für Organische Chemie, Johann Wolfgang Goethe-Universität, Frankfurt am Main, Marie-Curie-Straße 11, D-60439 Frankfurt am Main

1. Introduction

This essay deals with its subject from a personal point of view. During a timespan, in which photochemistry developed from a phenomenological branch of the natural sciences into a molecular discipline, a change in motivation took place in the rather remote field of chemical synthesis. What had originally been thought of as a scientific method for solving structural problems transmuted in the course of time into a molecular manifestation of *the will to design*. To translate control over photochemical reactions into a lynchpin of total syntheses of complex target structures was only a natural step.

1.1. Photochemistry of Previtamin D

In the early 1960s, opinion took hold that one method of obtaining new molecules might be excitation by light [1] [2]. Molecules existing temporarily after light absorption were known to differ from their nonexcited opposite numbers in their electronic structure[1]), in their topology and, as a rule,

---

[1]) To give semantic weight to this conception, otherwise identical molecules that differed in their electronic structure came to be known as *electronic isomers*, and molecules with the same electronic structure but different multiplicity as *spin isomers* [3]. An example of electronic isomers are molecules of acetone, the electronic structure of which may be formalized as:

- ${}^{1}_{0}(\pi^2 n^2)$ in the ground state,
- ${}^{?}_{1}(\pi^* \pi^2 n)$ in the first excited state, and
- ${}^{?}_{2}(\pi^* \pi n^2)$ in the second excited state.

Electronic excitation from ground state to the first or the second excited state may then be described as $(\pi^* \leftarrow n)$ or as $(\pi^* \leftarrow \pi)$ [4]. Spin isomers such as those of acetone accordingly receive the notation ${}^{3}_{1}(\pi^* n)$ or ${}^{1}_{1}(\pi^* n)$.

also in their reactivity[2]). With this spectroscopically based insight, it would, in principle, be possible to treat every chemical reaction – independently of the electronic structure of the molecules involved and – by also taking photophysical aspects into account – by the same mechanistic study. The initially tolerated contradictory impulse to hide light-induced reactions away in mystic darkness (*vide infra*) was soon felt to be outmoded. *The solving of the problem of the nature of the antirachitic vitamin D* [5] supplies an elegant example.

It had been established that the symptoms of rickets could be eliminated, or even avoided altogether, not only by oral administration of cod-liver oil or by exposing children to UV light, but also even by UV irradiation of the food given to these children. This suggested that both the skin and certain foodstuffs must contain some biologically active substance that was *activated* by exposure to UV light. This substance – which, it was felt, should prove to be applicable therapeutically – was provisionally assigned to a subclass of steroids: the sterols. The further conjecture that *all* sterols would prove activatable by UV light [6][3]) was shown to be untenable, however. Even before its structure was known, the photoactivatable sterol was assigned the function of a provitamin D that underwent photochemical transformation into the – also still unidentified structurally – antirachitic vitamin D. *Adolf Windaus* received the 1928 *Nobel* Prize for Chemistry for his investigations into sterols and the experimentally established relationship between certain sterols and vitamin D. Natural-product chemistry and photochemistry had finally met. Exactly as *Giacomo Ciamician* had hoped for and, in his subsequently famous 1912 New York lecture on *The Photochemistry of the Future* [9], had emphatically demanded.

Work on the structural relationships between provitamin D_3 (**1a**) and vitamin D_3 (**7a**), and between provitamin D_2 (**1b**) and vitamin D_2 (**7b**) had to await the conclusive determination in 1932 of the structure of the parent compound cholesterol[4]) (*Scheme 1*).

There are, therefore, several D provitamins and several D vitamins[5]). *Windaus* developed a view of how the D vitamins arose out of the D provitamins: lumisterols and tachysterols were allocated adjacent places between these two in each series (*Scheme 2*).

---

[2]) In the formal description of light-induced reactions, account should be taken – as in this essay from *Scheme 3* on – of the electronically excited educt and the photophysical changes it undergoes along the reaction coordinate.

[3]) This suspicion, even ten years later, prompted *Adolf Butenandt* to subject steroid hormones to the effect of UV light. 17-Keto-steroids, however, were found in this investigation not to require the slightest light excitation to be biologically active. Estrone [7] and 3β-hydroxyandrostan-17-one [8] are converted into their corresponding 13-α-epimers on UV irradiation. The isolated photoproducts have lost the biological activity of the photoeducts, though.

[4]) On the Odyssey leading ultimately to the correct structure of cholesterol, see [10]: Chapt. 3.

[5]) The subscripts (D_2, D_3...) denote differences in the compounds' side chains, which are not involved in the photochemical process. If an observation applies equally to representatives of both series, no subscript is used.

Scheme 1. *A Plethora of Isomers of the Provitamins D* (**1**)

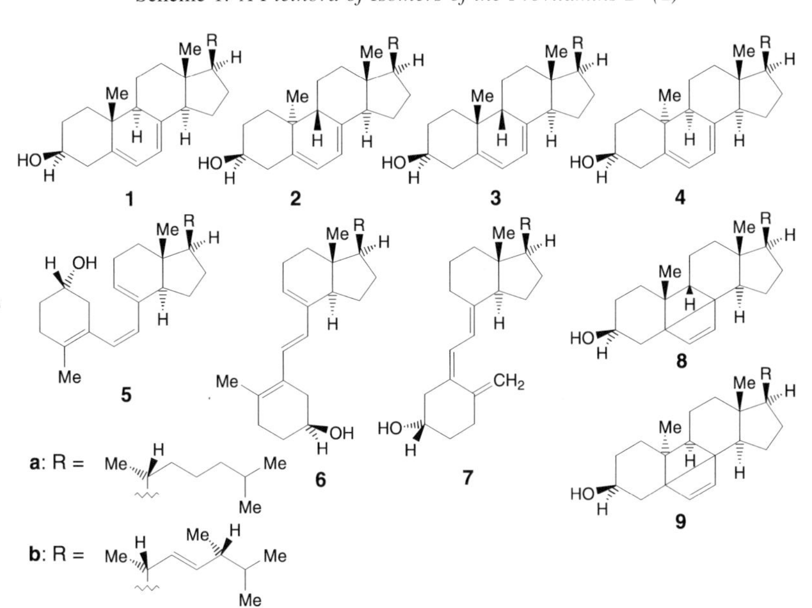

Scheme 2. *The Isomerization Sequence of* Windaus

Provitamin D $\xrightarrow{h\nu}$ Lumisterol $\xrightarrow{h\nu}$ Tachysterol $\xrightarrow{h\nu}$ Vitamin D

The transformations of each isomer to the next within one such reaction series should proceed photochemically. As a matter of fact, the most diverse isomerizations were ascribed to light. To the transformation of provitamin D into lumisterol: selective inversion of one of eight stereogenic centres[6]); to the transformation of lumisterol into tachysterol: opening of ring B with formation of a 9,10-*seco*-sterol; and, finally, to the transformation of tachysterol into vitamin D: rearrangement of a conjugated triene system. Isomerizations altogether, with nothing to do with one another formally, let alone mechanistically. Light was believed capable of everything possible. Only now is it conceivable to properly appreciate the former feeling that all steroids, whether plant, animal, or microorganism in origin, had first to be activated by light in order to become biologically active [6].

Butenandt, close to *Windaus* both personally and in thinking, has said of him [5] that '*he was truly not interested in the chemical structure of any in-*

[6]) A selective inversion, here at C(13), similarly takes place in the previously mentioned[3]) photoisomerization of 17-keto-steroids.

dividual compound, but only in the biological interrelationships in nature'. Given that, at that period, *Hans Meerwein* had long since begun publishing his fundamental ideas about reaction mechanisms involving organic molecules, it is not easy for those born afterwards to comprehend the coexistence of these two wholly different worlds within organic chemistry[7]). Especially not since *Emil Fischer*, decades previously, had already applied himself, with undivided interest, both to the world of natural products (carbohydrates, polypeptides) and the world of chemical reactions (*Walden* inversion, asymmetric reactions). It was, however, not only natural-product chemists who had lost the bigger picture, though. *Rudolf Criegee*, who was close to *Meerwein* both personally and in thinking, has remarked of him [12] that '*natural products chemistry in its particular meaning remained foreign to him all his life*'.

By the 1950s things had reached the stage where *Windaus*'s isomerization scheme could no longer be sustained. Firstly, structural corrections had had to be made. Lumisterol (**2**) differs in configuration from provitamin D (**1**) not only at C(10), but also at C(9) [13]. The 9,10-seco-sterol with a (*Z*)-configured central C=C bond in its conjugated triene system is not tachysterol (**6**), to which the (*E*)-configuration must in reality be ascribed, but the then newly discovered previtamin D (**5**) [14]. Furthermore, previously assumed reaction pathways between some of the participating isomers required modification. Hence, previtamin D (**5**) is transformed into vitamin D (**7**) not photochemically, but thermally, and provitamin D (**1**) cannot be converted directly into lumisterol (**2**) either photochemically or thermally. The new isomerization scheme according to *Léon Velluz* and *Egbert Havinga* (*Scheme 3*) had an appearance completely different to that of *Windaus* (*Scheme 2*).

Scheme 3. *Isomerization Network of* Velluz [14] *and* Havinga [15]

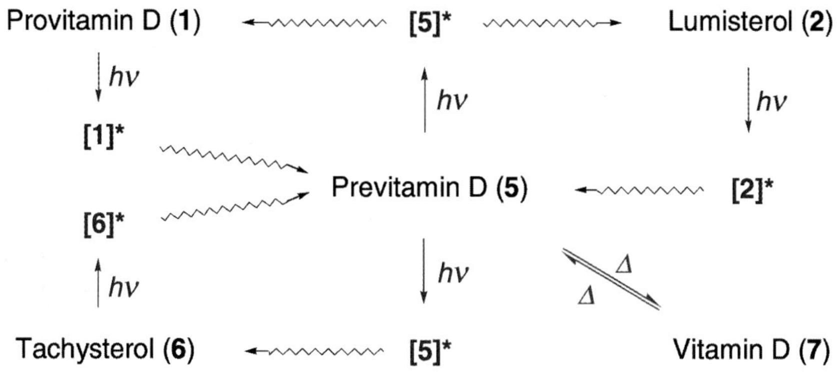

Provitamin D (**1**) ← [5]* → Lumisterol (**2**)

Provitamin D (**1**) —$h\nu$→ [1]*

Previtamin D (**5**) —$h\nu$→ [5]* (upper)

Lumisterol (**2**) —$h\nu$→ [2]*

[1]* → Previtamin D (**5**)

[6]* → Previtamin D (**5**)

[2]* → Previtamin D (**5**)

Tachysterol (**6**) —$h\nu$→ [6]*

Previtamin D (**5**) —$h\nu$→ [5]* (lower)

Previtamin D (**5**) ⇌ Vitamin D (**7**) (Δ / Δ)

Tachysterol (**6**) ← [5]*

7) *Karl Dimroth* told of these two worlds and their effect on him in his last lecture before joining the ranks of the emeriti in February 1979 in Marburg; see [11].

Provitamin D (**1**) and lumisterol (**2**) can both be transformed into previtamin D (**5**). Previtamin D may be converted photochemically into provitamin D (**1**), lumisterol (**2**), and/or tachysterol (**6**). Thermally, it may be converted into vitamin D (**7**), even at body temperature. After the isomers had been assigned their correct structures and put in the right places in the isomerization network, it did not take long any more until the individual reaction steps could be interpreted plausibly.

1.2. The *Woodward-Hoffmann* Rules

Full understanding of the reversible photoisomerization of previtamin D (**5**) and tachysterol (**6**) posed slight difficulty only. After all, light-induced (*E*)/(*Z*)-isomerizations in conjugated, unsaturated molecules had been known for a long time [16], and photoisomerization about a C=C bond belongs to the simplest photoreactions formally. The mechanistic debate was structured more clearly by *R. S. Mulliken* [17]. According to his MO calculations, ${}^1_1$[ethylene]* and ${}^3_1$[ethylene]* each adopt a particular conformation in the energy minimum, with the two CH_2 groups arranged orthogonally to one another. In the transition to ${}^1_0$[ethylene], reversion to the planar atomic arrangement takes place. If dealing with an ethylene derivative capable of existing as an (*E*)- and as a (*Z*)-isomer, the orthogonal conformers can be transformed into either of the planar configurational isomers[8]). Discussion about the '*Present Status of the Photoisomerization about Ethylenic Bonds*' [18], still incessant even today, is a good example of how an increasing number of individual cases makes it all the more difficult to propose a comprehensive mechanistic picture. Provided that only a few examples of a particular reaction type are known, the picture appears simple and clear. With deeper investigation into a growing number of cases, the picture gets complex and blurred. After taking additional experimental findings into account, and evaluating the conclusions drawn from them, the view of the reaction type under study remains complex but becomes clear again. The place of a mechanistic singularity, however, has now been replaced by an intergrading array of mechanistic variants.

The light-induced ring openings and cyclizations depicted in *Scheme 3* abruptly became interpretable once *R. B. Woodward* and *R. Hoffmann* had put forward their rules concerning preservation of orbital symmetry during pericyclic reactions [19]. Previtamin D (**5**) and Vitamin D (**7**) may be converted one into the other *via* antarafacial 1,7-sigmatropic H-shift. Provitamin D (**1**)

[8]) The features of one-way and conventional two-way isomerizations, and the factors controlling their modes are discussed in [18].

and lumisterol (**2**) are each connected with previtamin D (**5**) through conrotatory electrocyclic reactions (*Scheme 4*).

Now it could not only be explained why previtamin D (**5**) cyclizes to the two stereoisomers of *anti*-configuration at the two newly produced stereogenic centres C(9) and C(10) (**1** and **2**), but also why previtamin D (**5**) cyclizes thermally to the two stereoisomers (isopyrovitamin D (**3**) and pyrovitamin D (**4**)) of *syn*-configuration at these centres[9]).

The interpretation of the chemical reactions of *Scheme 4* did not merely represent a solution to a longstanding star number in problem seminars at leading universities. As well as this, the veil over the *no mechanism reactions* had been blown aside. But even more importantly, there was from this time on no longer any convincing justification for special treatment for electronically excited molecules and their chemical transformations in organic chem-

Scheme 4. *The Four Cyclo-Isomers, Accessible by Ring Closure of Previtamin D* (**5**)*, and Their Photoproducts*

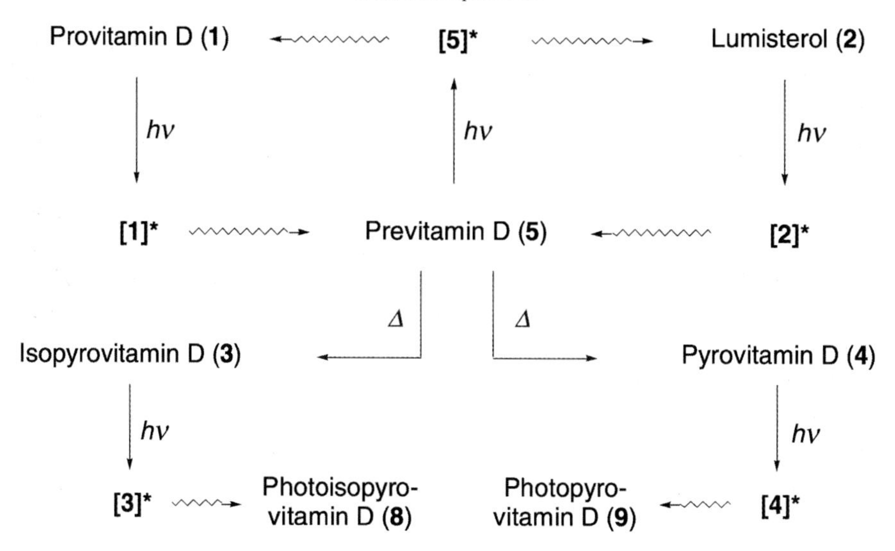

[9]) *Woodward* himself wrote [20] about how, while working out a strategic concept for the synthesis of vitamin B_{12} in Cambridge (USA), he came upon the *Woodward-Hoffmann* rules as the solution to fundamental questions about the process of those reactions described, not without irony, by *W. von E. Doering* as *no mechanism reactions* [21]. For a psychological analysis by his daughter, implying that *Woodward*'s mind was preadapted for the orbital-symmetry rules, see [22]. In a painstaking analysis, *J. A. Berson* [23] has demonstrated how *E. Havinga* and *L. J. Oosterhoff* came close to the interpretation of what were to become the *Woodward-Hoffmann* rules, without appreciating the generality of the reaction principle, with their speculative derivation [24] of the stereostructure of the photochemical or thermal cyclization products of previtamin D (**5**) from the symmetry of the highest occupied (by electrons) π-MO of the conjugated, unsaturated triene system.

istry [3a]. To put it unequivocally: the *Woodward-Hoffmann* rules brought unity – irrespective of the electronic structure of participating molecules – to the understanding of chemical reactions. Moreover, thanks to the discussion of photoprocesses they offered a clear rejection to the separation of chemistry and physics[10]).

Photochemical reactions begin with an electronically excited educt and end with a product in the electronic ground state. Somewhere between electronically excited educt and any electronically unexcited product, there must be a transition from a higher- to the lowest-potential-energy hypersurface. If electronic activation and deactivation are interpreted as physical processes, deactivation may occur in an electronically excited educt, an electronically exited product, or somewhere in between. Experimentally, the majority of photochemical reactions appear to belong to the last-named group, in which the reacting molecules change from a higher- to the lowest-energy hypersurface during the actual chemical process (with radiationless deactivation). It is not possible to truly understand the *Woodward-Hoffmann* rules without taking account of the available hypersurfaces and the possible transitions between them [26]. That the *Woodward-Hoffmann* rules are also capable of accurately predicting[11]) the stereoselectivity of the light-induced cyclization of conjugated, unsaturated trienes can be seen as a convincing argument for the beginning of this reaction elsewhere than on the ground-state hypersurface[12]).

While the $9\alpha,10\beta$- and $9\beta,10\alpha$-cyclohexadienes **1** and **2**, respectively, react photochemically to give the open-chain triene **5** (*Scheme 4*), the $9\beta,10\beta$- and $9\alpha,10\alpha$-cyclohexadienes **3** and **4**, respectively, are not capable of this. Conrotatory ring-opening in the latter case would give not **5**, but a highly strained stereoisomer of **5** with a (*E*)-configured C=C bond in ring A or ring C. Hence, instead of reacting by conrotatory ring opening to a highly strained triene, **3** and **4** undergo disrotatory ring closure to the at least somewhat less strained cyclobutene derivatives **8** and **9**, respectively. That conjugated, unsaturated trienes follow the prediction of the *Woodward-Hoff-*

---

[10]) If, in the past, those more interested in photophysical aspects, and those more interested in photochemical aspects had increasingly diverged, they altered their behavior after becoming converts to the *Principles of Preservation of Orbital Symmetry*. The 1965 *Solvay Conference* on *Reactivity of the Photoexcited Organic Molecule* [25] marked the change: under the banner of the *Woodward-Hoffmann* rules, the newly respectable photochemists successfully gained admission to the sanctuary of the astonished photophysicists.

[11]) The word 'predicting' is used here to mean the deducing of a known fact as well as the forecasting of a new fact [27].

[12]) This, therefore, disallows the reaction of vibrationally energetically excited molecules existing in the electronic ground state and which might arise after *internal conversion* from a higher singlet hypersurface or after *intersystem crossing* from a higher triplet hypersurface.

mann rules in every case, sometimes reacting as 1,3,5-trienes, sometimes as 1,3-dienes, to the corresponding less strained cyclo-isomer, is also seen in the following example (*Scheme 5*).

The purple *ortho*-quinodimethane-derived seco-isomer *rac*-**10** can only cyclize in conrotatory manner, on grounds of strain[13]). It, therefore, reacts photochemically as a triene, *via* [*rac*-**10***], with ring closure to give the yellowish cyclohexadiene derivative *rac*-**12**, and thermally under the mildest conditions at which reaction takes place at all, as a diene, with ring closure to give the colorless cyclobutene derivative **11**[14]).

Scheme 5. *The Kinetically and Thermodynamically Determined Cyclo-Isomers* **11** *and* rac-**12** *Accessible from* rac-**10** [3a][28a]

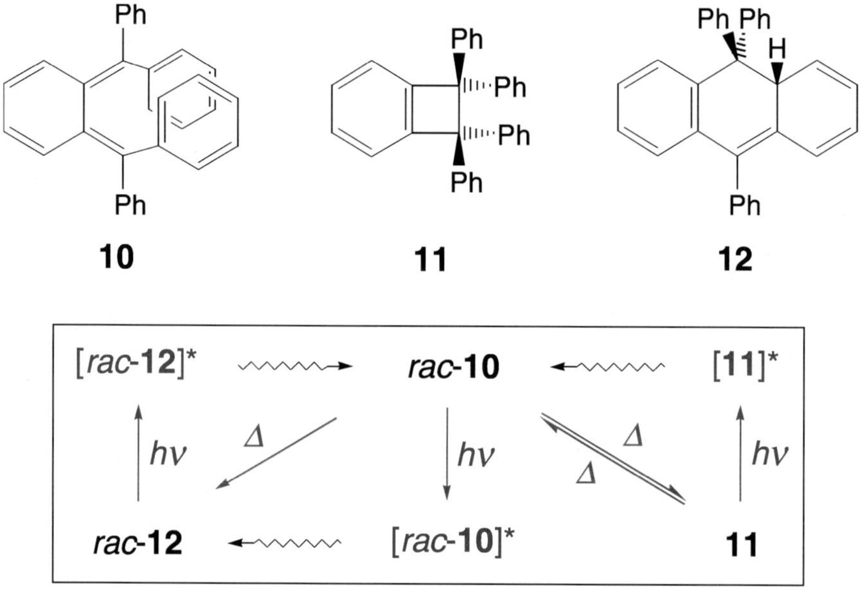

[13]) In a conrotatory motion, the two inner Ph residues avoid one another, in a disrotatory motion they collide.

[14]) Compound **11** rearranges at room temperature (*via rac*-**10**) into *rac*-**12**. Compound **11**, with its newly formed four-membered ring, is the kinetically determined cyclization product of *rac*-**10**; *rac*-**12**, with newly formed six-membered ring, the thermodynamically favored one. The conrotatory isomerization at room temperature of *rac*-**10** into *rac*-**12** is – like the disrotatory isomerization of **8** or **9** into **3** or **4** at higher temperature – not allowed. *Not allowed* does not mean that the *forbidden* reaction cannot take place. It has more the character of a traffic regulation (one way street) and means that reaction in the *forbidden* direction is less likely as a predictable event.

1.3. Commercial Preparation of Vitamin D

1.3.1. *Partial Synthesis with Photochemical Formation of Previtamin D*

1.3.1.1. *UV Irradiation at 0° with Light from a Medium-Pressure Hg Lamp*

Vitamin D_3 (**7a**) is obtained commercially by partial synthesis. This partial synthesis begins with inexpensive cholesterol and its transformation into provitamin D_3 (**1a**). It ends with light-induced ring opening to previtamin D_3 (**5a**) and its thermally induced 1,7-H-shift to vitamin D_3 (**7a**, see *Scheme 3*). Since **5** reacts back to **1** and onwards to **2** and **6**, the result is a photostationary mixture. UV Irradiation of **1** is expediently carried out at 0°, to largely exclude isomerization of **5** into **7**. This is important for the prevention of further reaction of **7** into the so-called *over-irradiation* products[15]), the components of which would appreciably raise the diversity of the irradiated mixture.

Table 1,a, gives some details about chemical yields of **7** obtained by irradiation of **1** at 0° with light from a medium-pressure Hg lamp until the stated compositions of **1**, **5** and **6** are reached. The crude irradiation mixtures were chromatographed at 0°, and those fractions consisting substantially of **5** and **6** were converted in boiling EtOH into a mixture of **6** and **7**. Compound **6** could be largely separated out by fractionalizing *Diels-Alder* reaction with maleic anhydride[16]). After further chromatography, 9% of crystalline **7** was obtained [34].

All four participants, **1**, **2**, **5**, and **6**, in the photoisomerization process absorb light, albeit with differing intensities in the region between 254 and 350 nm (*vide infra*), and six different photoreactions with different quantum yields happen in concert. Taking account of this fact, it is not surprising to learn that the most diverse efforts have been undertaken to raise the chemical yield of **7**.

---

[15]) See [29][30] for the structures of the various components in the *over-irradiation* mixtures. Since **5** is also converted to a secondary photoproduct on prolonged UV irradiation, the above-mentioned mixture represents not a photostationary state but a quasi-photostationary one.

[16]) The rapid reaction of tachysterol with maleic anhydride had convinced *Kurt Alder* [32] to assign it the structure of a triene system of (*E*)-configuration at the central C=C bond. This assumption was confirmed by comparing the UV and IR data of three triene pairs of (*E*)- or (*Z*)-configuration at the central C=C bond [33].

Table 1. *How To Improve the Chemical Yield of* **7** *by Photochemical Sophistication* [31]

	Conditions	Products
a)	$h\nu$, Hg medium pressure, 0°C →	**1** (25%) + **5** (25%) + **6** (50%)
	chromatography at 0°C →	**5** + **6**
	EtOH, reflux →	**6** + **7**
	chemical separation chromatography →	**7** (9% yield)
b)	$h\nu$, Hg medium pressure, 0°C →	**1** (25%) + **5** (25%) + **6** (50%)
	chromatography at 0°C →	**5** + **6**
	$h\nu$, fluorenone, 0°C →	**5** + **6**
	EtOH, reflux, separation, chromatography →	**7** (28% yield)
c)	254 nm, 0°C →	**1** (6%) + **5** (28%) + **6** (64%) + **2** (2%)
	350 nm, 0°C →	**1** (8%) + **5** (73%) + **6** (9%) + **2** (10%)
	chromatography at 0°C →	**5** (66%)
	EtOH, reflux →	**5** + **7**
	chromatography →	**7** (50% yield)

1.3.1.2. *Use of Photosensitizers*

The photochemical configurational isomerization of previtamin D (**5**) into tachysterol (**6**) (and back) differs from the photochemical constitutional isomerization of provitamin D (**1**) or lumisterol (**2**) into previtamin D (**5**) (and back; see *Scheme 3*): not only in the reaction type, but also in which of the $\pi^*\pi$-spin isomers is involved. Unlike the constitutional isomerization, the configurational isomerization can proceed not only under direct light excitation (*singlet mechanism*), but also by triplet-triplet energy transfer (*triplet mechanism*), with the aid of photosensitizers.

In triplet-triplet energy transfer, a photosensitizing triplet donor transfers its energy to an acceptor in the electronic ground state. The overall process begins with the light absorption of the donor and may be described as follows: ${}^{1}_{0}[D] \xrightarrow{h\nu} {}^{1}_{1}[D]^*$; ${}^{1}_{1}[D]^* \rightarrow {}^{3}_{1}[D]^*$; ${}^{3}_{1}[D]^* + {}^{1}_{0}[A] \rightarrow {}^{1}_{0}[D] + {}^{3}_{1}[A]^*$. By this way, ${}^{3}_{1}[A]^*$ is formed indirectly after a series of physical events: transfer of ${}^{1}_{0}[D]$ into ${}^{1}_{1}[D]^*$ by light absorption, intersystem crossing from ${}^{1}_{1}[D]^*$ into ${}^{3}_{1}[D]^*$, and energy transfer of the latter into ${}^{3}_{1}[A]^*$.

For an ideal donor-acceptor pair for triplet-triplet energy transfer, ${}^{1}_{1}[A]^{*}$ should lie higher than ${}^{1}_{1}[D]^{*}$ and ${}^{3}_{1}[A]^{*}$ lower than ${}^{3}_{1}[D]^{*}$, while intersystem-crossing efficiency should be high. Ketone photosensitizers and conjugated diene or triene acceptors meet these requirements for a molecular donor-acceptor pair for triplet-energy transfer. Unlike ketones, with small energy differences between ${}^{1}_{1}[D]^{*}$ and ${}^{3}_{1}[D]^{*}$, that between ${}^{1}_{1}[A]^{*}$ and ${}^{3}_{1}[A]^{*}$ in conjugated polyenes is considerable. Since ketones (aromatic ones in particular) absorb in the long-wavelength range ($\pi^{*}n$ absorption) where dienes and trienes are optically transparent, and since they jump from ${}^{1}_{1}[D]^{*}$ to ${}^{3}_{1}[D]^{*}$ in high quantum yield, and do not change chemically under the irradiation conditions, they are frequently used as photosensitizers [35].

If provitamin D (**1**) is subjected to UV irradiation and subsequent chromatography, and the triplet-sensitizer fluorenone is then added to the primary mixture (see *Table 1,b*) and irradiation performed a second time, the result is a secondary mixture consisting mainly of previtamin D (**5**) and, to a small degree, of tachysterol (**6**). Boiling-EtOH treatment leads to a tertiary mixture of mainly vitamin D (**7**) and of slightly tachysterol (**6**). Fractionalizing *Diels-Alder* reaction, chromatography, and recrystallization finally affords **7** in 28% yield [34][17]).

1.3.1.3. *Consecutive UV Irradiations with Light of Varying Wavelength*

An influence of excitatory light wavelength on the component composition of the provitamin D (**1**) irradiation product occasionally has been observed [38]. A glance at the molar extinctions at various wavelengths between 254 and 350 nm (*Table 2,a*) of the four components in the irradiation mixture prompts the suspicion that relatively-long-wavelength light extensively converts the already enriched tachysterol (**6**) back to previtamin D (**5**). At 308 nm, the proportion of **6** in the irradiated mixture is extremely high (*Table 2,b*). As might be expected, this is almost completely overcome by subsequent irradiation with laser light of longer wavelength [29].

Irradiation of an analytical sample at room temperature with 248-nm light gave a primary product of the composition given in *Table 2,b, Line 1*. Subsequent irradiation with 337-nm or 353-nm light led to the secondary mix-

[17]) To enable photosensitized isomerization of tachysterol (**6**) into previtamin D (**5**) for the industrial preparation of vitamin D (**7**), use has been made of H_2O-soluble photosensitizers [36], easier to remove in the workup of the crude irradiation product. Noteworthy is the efficient photoisomerization of **1** into **5** in the hydrophobic microdomains of aqueous poly(sodiumstyrene sulfonate-co-2-vinylnaphthalene) [37]. The reaction is initiated by the singlet-singlet energy transfer ${}^{1}_{1}$[Naphthalene]* + ${}^{1}_{0}$[**1**] → ${}^{1}_{0}$[Naphthalene] + ${}^{1}_{1}$[**1**]*. That **6** is practically absent is due to a triplet-triplet energy transfer ${}^{3}_{1}$[Naphthalene]* + ${}^{1}_{0}$[**6**] → ${}^{1}_{0}$[Naphthalene] + ${}^{3}_{1}$[**6**]*.

Table 2. a) *Molar Extinctions for Compounds* **1**, **2**, **5**, *and* **6** *at Various Wavelengths* [39].
b) *Composition of the Photostationary State at Various Wavelengths* [40].

a)

λ [nm]	ε			
	1	**2**	**5**	**6**
254	4500	4130	7250	11450
300	1250	1320	930	11250
330	25	30	105	2940
340	20	25	40	242
350	10	20	25	100
at λ max [nm]	11900 (282)	850 (280)	9000 (262)	24600 (281)

b)

λ [nm]	**1**	**2**	**5**	**6**
248	2.9	n.d.	25.8	71.3
254	1.5	2.5	20	75
302	3.4	17	43	26
308	13.3	35.5	3.41	42.3
248+337	8.8	9.8	79.8	1.5
248+353	0.1	8.7	80.1	1.0

tures listed in *Lines 5* and *6* [40]. In a preparative irradiation of **1** at 0° in a *Rayonet* reactor with 254-nm light [39], a primary mixture of **1** (6%), **2** (2%), **5** (28%), and **6** (64%) was obtained. After application of 350-nm light, a secondary mixture of **1** (8%), **2** (10%), **5** (73%), and **6** (9%) was found. The chemical yield of **5** after MPLC at 0° was 66%. After equilibration in boiling EtOH, a tertiary mixture of **5** and **7** in a 83:17 ratio was obtained, from which 50% vitamin D (**7**) could be isolated by MPLC (see *Table 1,c*).

The case study of the *preparation of vitamin D* (**7**) *with the aid of electrocyclic ring cleavage of provitamin D* (**1**) *to previtamin D* (**5**) (*Scheme 3*) makes it clear that a change in the wavelength of excitatory light may be able to influence the composition of the many-component photoproduct. An effect of this kind may have many causes. So far, the more *trivial* instance has been considered: that a photoeduct (here: **1**) reacts in a photochemical primary reaction to give a primary photoproduct (here: **5**), and this reacts in a photochemical secondary reaction to give a secondary photoproduct (here: with the components **1**, **2**, and **6**). If no other chemical change takes place, a photostationary mixture[15]) is obtained, its composition (of **1**, **2**, **5**, and **6**) is determined by the absorptions of the participant molecules and the quantum yields of the reactions taking place (*Scheme 6*).

Scheme 6. *Quantum Yields of Photoisomerizations That Start or End with Previtamin D (**5**)* [41b]

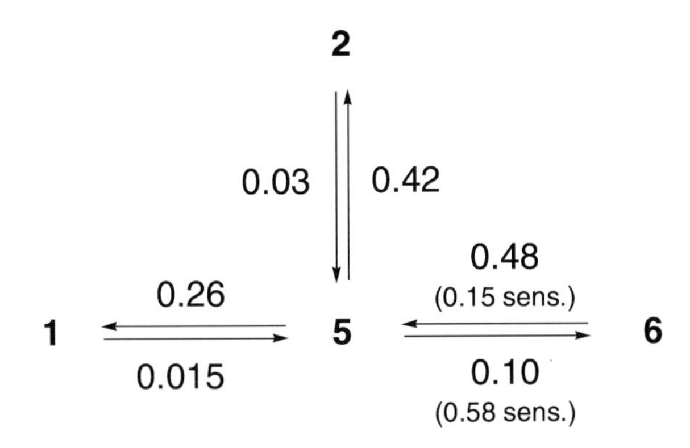

A second possible cause of wavelength dependency in the composition of a photoproduct ($\mathbf{P}^1+\mathbf{P}^2$) is selective light absorption of two coexisting conformers $\mathbf{E}^1$ and $\mathbf{E}^2$ in the photo-educt (*Scheme 7*)[18]).

Assuming that equilibration of the electronically excited conformers $[\mathbf{E}^1]^*$ and $[\mathbf{E}^2]^*$ is slower than that of the unexcited conformers $\mathbf{E}^1$ and $\mathbf{E}^2$, and is also slower than the chemical reaction of $[\mathbf{E}^1]^*$ specifically to $\mathbf{P}^1$ and from $[\mathbf{E}^2]^*$ specifically to $\mathbf{P}^2$, the composition of the photoproduct ($\mathbf{P}^1+\mathbf{P}^2$) will be determined by the composition of the photoeduct ($\mathbf{E}^1+\mathbf{E}^2$). An example of this is given by the light-induced isomerization of previtamin D (**5**) into tachysterol (**6**) at two different temperatures. At the higher temperature, equilibration takes place between $\mathbf{E}^1$ and $\mathbf{E}^2$. At the lower temperature, however, it does not.

The product obtained at 125 K (or above) does not change if cooled to 92 K, but that obtained at 92 K is transformed into the first product on raising the temperature to 125 K. *Havinga* and co-workers [41] interpreted this finding by describing **5** as two conformation types (*cZc*)-**5.1** and (*tZc*)-**5.2**, and assuming that (*cZc*)-**5.1** transforms *via* $^1_1$[(*cZc*)-**5.1**]* into product conformation (*cEc*)-**6.1**, while (*tZc*)-**5.2** is transformed *via* $^1_1$[(*tZc*)-**5.2**]* into product conformation (*tEc*)-**6.2**. This interpretation requires that the barrier to equilibration between $^1_1$[(*cZc*)-**5.1**]* and $^1_1$[(*tZc*)-**5.2**]* is higher in all instances than that between (*cZc*)-**5.1** and (*tZc*)-**5.2**[19]).

A third possible cause of wavelength dependency in the composition of a photoeduct ($\mathbf{P}^1+\mathbf{P}^2$) is wavelength-dependent photoactivation of a photoeduct ($^1_0[\mathbf{E}]$) both to the electronically isomeric $^1_1[\mathbf{E}]^*$ and to the electronical-

[18]) See [29b][38][42] and refs. cit. therein. For the limitation of the photochemical applicability of the *Curtin-Hammett* principle, see [43].

[19]) $\pi^*\pi$-Excitation raises the rotational barrier about the C(5)–C(6) bond.

Scheme 7. *The Two Conformers* **5.1** *and* **5.2** *of Previtamin D at Low Temperature Photo-isomerize Stereospecifically into the Tachysterol Conformers* **6.1** *and* **6.2,** *Respectively* [29b]

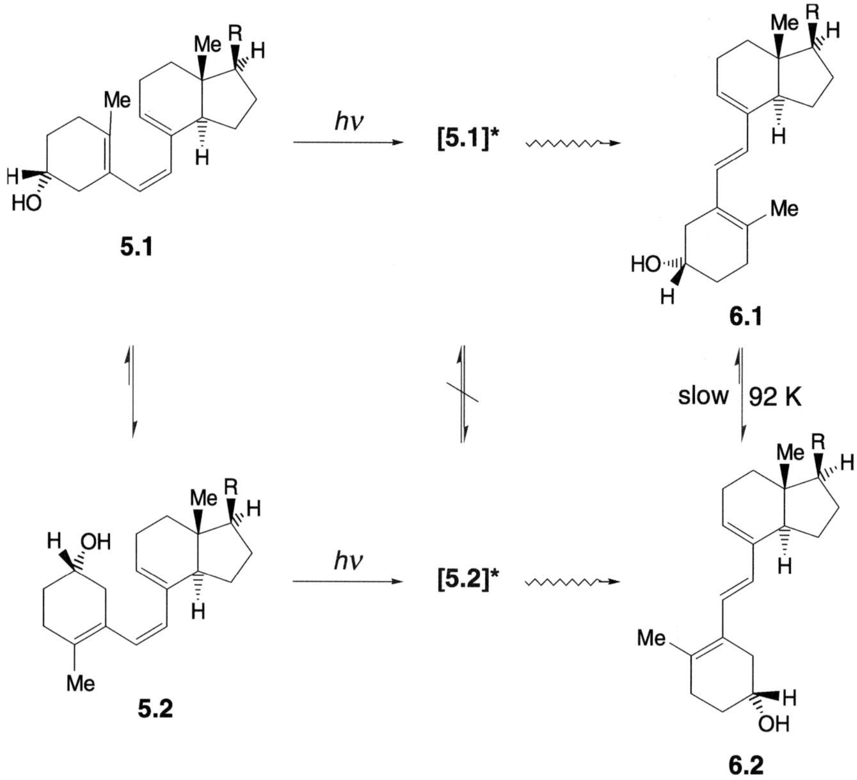

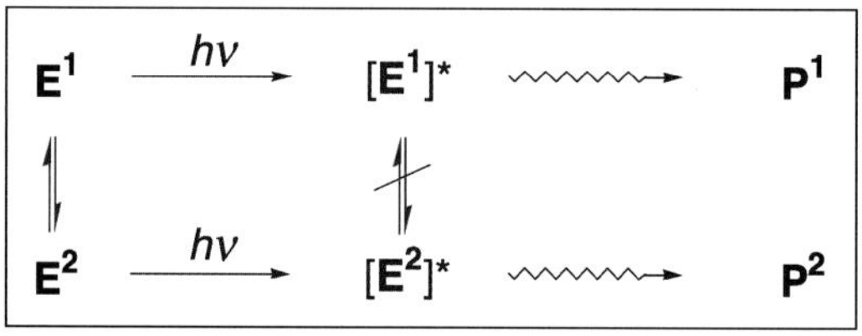

ly isomeric $^{1}_{2}$[**E**]*, followed by the specific reaction of $^{1}_{1}$[**E**]* into $\mathbf{P}^1$ and of $^{1}_{2}$[**E**]* into $\mathbf{P}^2$ (*Scheme 8*).

As an example of this, it is shown that a sudden change becomes apparent in the UV irradiation of previtamin D (**5**) between 302 and 305 nm. The quantum yield of the photocyclization of **5** into provitamin D (**1**) and lumisterol (**2**) increases, while that of the light-induced isomerization of **5** into tachysterol (**6**) shrinks. *Dauben*, *Kohler*, and co-workers [44] make use of a

Scheme 8. *Wavelength-Dependent Access to Different Hypersurfaces Determining Specific Product Composition* [44]

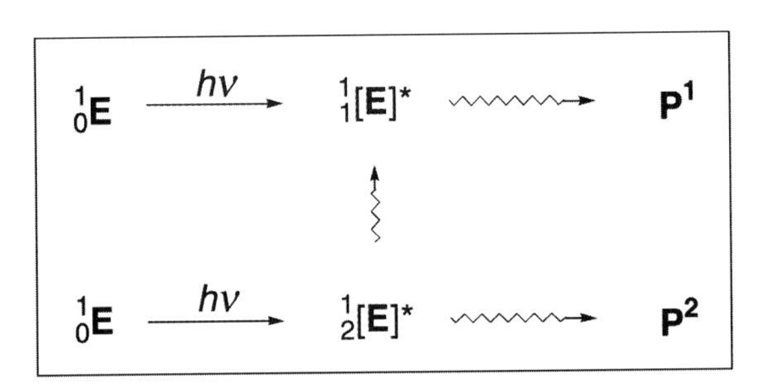

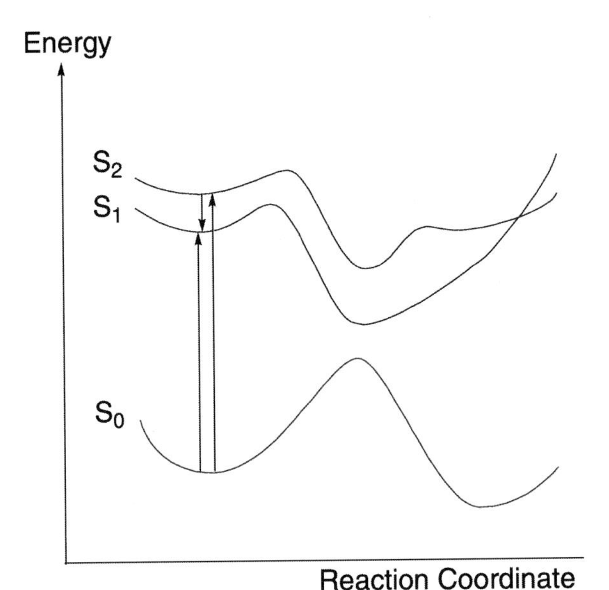

section through the polydimensional hypersurface[20]), in which the process for conrotatory cyclization to the corresponding cyclohexadiene is evident.

Two very close-lying hypersurfaces (S_1 for the $^1_1$[triene]*-electron isomer and S_2 for the $^1_2$[triene]*-electron isomer) are available to the conjugated trienes. The somewhat lower-lying hypersurface S_1 has a trough, where the ground-state hypersurface S_0 exhibits a barrier. In this region, a quantum jump from S_1 to S_0 has a high probability of occurring, assuming that the trough of S_1 is sufficiently populated. This prerequisite is *a priori* fulfilled

[20]) A two-dimensional cut through a set of poly-dimensional energy hypersurfaces leads [45] to a simplified picture of a reactive pathway to be expected.

for molecules elevated by light of wavelength of <303 nm from S_0 to S_2. Molecules promoted from S_0 to S_2 by absorption of light of wavelength of <303 nm can enter into (*E*)/(*Z*)-isomerization while in this state, or – jumping to the S_1-hypersurface – they can cyclize.

The assumption of different reaction behavior in two excited electronic isomers begs the question of why electron isomer ${}^1_2$[**E**]* on the S_2-hypersurface converts more rapidly into the original educt ${}^1_0$[**E**] or the product $\mathbf{P}^2$, instead of relaxing, in the beginning partially through internal conversion to electronic isomer ${}^1_1$[**E**]* and then completely by transformation of ${}^1_1$[**E**]* into ${}^1_0$[**E**] or into $\mathbf{P}^1$. The interpretation in [44] is an exception to a rule established by *Kasha* [4] originally for light emission and later extended to photoreactions. According to this, transitions in the condensed phase to ${}^1_0$[**E**] by fluorescence emission or to ${}^1_0$[**P**] by photochemical reaction can also occur from ${}^1_1$[**E**]*, if a higher hypersurface than S_1 has been reached through light absorption.

Surprised, the photochemist is going to mull this information over. As long as the relative weights and individual UV absorptions of two photoeduct conformations are not known, the idea about abrupt change of light absorption within a narrow wavelength region remains mere guesswork. Perhaps two different types of photochemistry initiated at different hypersurfaces might really be a solution, notwithstanding the fact that such an assumption would conflict with *Kasha*'s extended rule. Either way, the evidence is uncertain, and the case remains unsettled still.

During all the years while discussion of the mechanisms set out in *Schemes 3* and *4* were taking place, the view of the biological role of the D vitamins (**7**) had shifted. Firstly, it had been established experimentally that endogenic photoproduction of previtamin D_3 (**5a**) from provitamin D_3 (**1a**) by sunlight on the skin represented the major source of the vitamin, rather than intake of **7a** in food [46]. Insofar as this is the case, it is no longer possible to describe **7a** as a vitamin. In fact, **7a** functions as a prohormone, its metabolites **13** and **14** being biologically active. With the aid of synthetic membranes, serving as model epidermal production sites for **7a**, it had been shown that conformational restraints essentially restricted debate to the photochemical primary product (**5.1** in *Scheme 7*) and excluded any photochemical secondary products derived from it [47]. It had then been found [48] that, in humans, it was not **7a**, thermally generated from previtamin D_3 (**5a**) that circulated, but the 25-hydroxy compound **13**.

Metabolites, 25-hydroxyvitamin D (**13**) and 1α,25-dihydroxyvitamin D (**14**), are produced enzymatically, mainly in the liver and kidneys [49]. Their existence and hormonal activity were discovered after the successful synthesis of radioactively labelled vitamin D_3. The synthesis of vitamin D (and its metabolites) is the next item on the agenda.

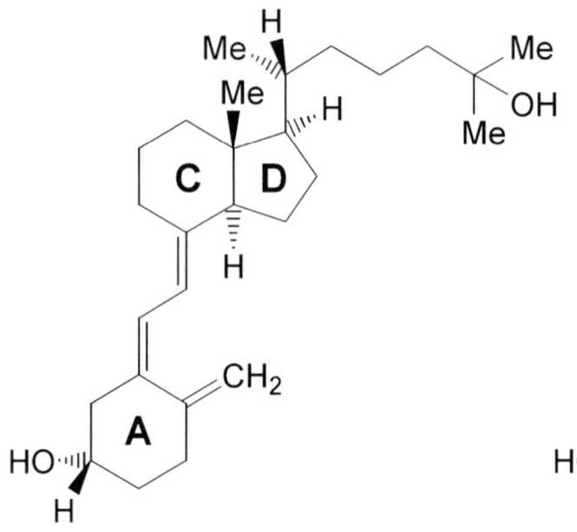

13 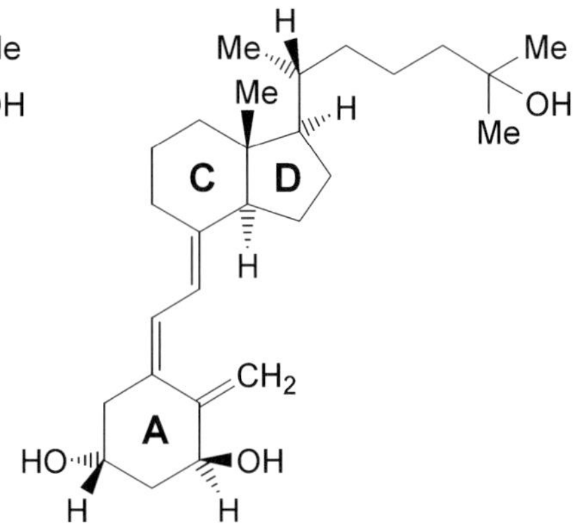 **14**

1.4. Total Syntheses of Vitamin D

1.4.1. *The* Braunschweig *Synthesis*

In the autumn of 1952, one of the authors had received a topic in the area of vitamin-D_3 synthesis for a thesis to be done. In a *Memorial Lecture* later to be held for *Hans Herloff Inhoffen*, it was said [50]: '*Professor Inhoffen had risen from his chair as though some third person, to whom respect should be shown, had come into his office. The author had followed suit, of course, and so learned – eye to eye – how the total synthesis of the D vitamins was to be seen as a reverence to* Inhoffen*'s mentor,* Windaus, *and, as well as this, was understood in the community of natural product synthetic chemists as a duty to be fulfilled. Because, only by successful total synthesis could the underlying structure of the envisaged target compound be proven unequivocally. This pleased the author greatly since, after all, by this way the chemical topic was epistemologically based and (*vide infra*) strategically determined.*'

The *Braunschweig* synthesis of vitamin D_3 (**7a**) [51] was undertaken with the intention of confirming the underlying target structure[21]). It was planned

[21]) Ever since *W. H. Perkin*, Jr. had published his experiments on the synthesis of terpenes [52], total synthesis was considered to provide the final proof of structure in every case. This '*would require, however, that a synthesis might be devised of such a simple kind that there would be no longer room for doubt as to the structure of these important substances*'. When, in the 1960s, structural determination by X-ray structure analysis became superior in effectiveness to the previous practice of ascertaining the structure of a natural product by chemical degradation and logical piecing together of the degradation products in a chemical jigsaw puzzle, organic natural-product chemistry ran into crisis [53]. *In the past organic chemists had spent half the time on degradation and half on synthetic work. Now a turning of the talent formerly spent on degradative work to synthetic work took place* (see [54]: Sect. 3; for a detailed discussion see [55]: Chapt. 7).

as a *relay synthesis*: organized around a degradation product accessible from the naturally occurring target compound by chemical degradation as a *relay compound*. For effective division of labor, the relay compound would be synthesized from the starting compound by, say, half of the synthetic team, while the other half converted the relay compound into the target compound by chemical means. At the end, there would exist, for each individual synthetic step, a detailed experimental procedure, the sum of which would be usable with sufficient effectiveness in a chemical laboratory or in a chemical plant for the production of the required target compound. The opportunity for time saving by a relay synthesis, compared with a regular from A to izzard total synthesis, is obvious.

The *Braunschweig* vitamin-D_3 synthesis did not operate about one relay compound, but with *seven* of them (*Scheme 9*).

Without exception, the relay compounds contain rings C and D of the vitamins, and greater or lesser parts of the so-called side chains on ring D and

Scheme 9. Braunschweig *Synthesis of Vitamin D_3* (**7a**). Relay compounds **15** through **21** (*blue field*) obtained by degradation of vitamin D_3 (**7a**) and vitamin D_2 (**7b**) (*yellow field*), respectively.

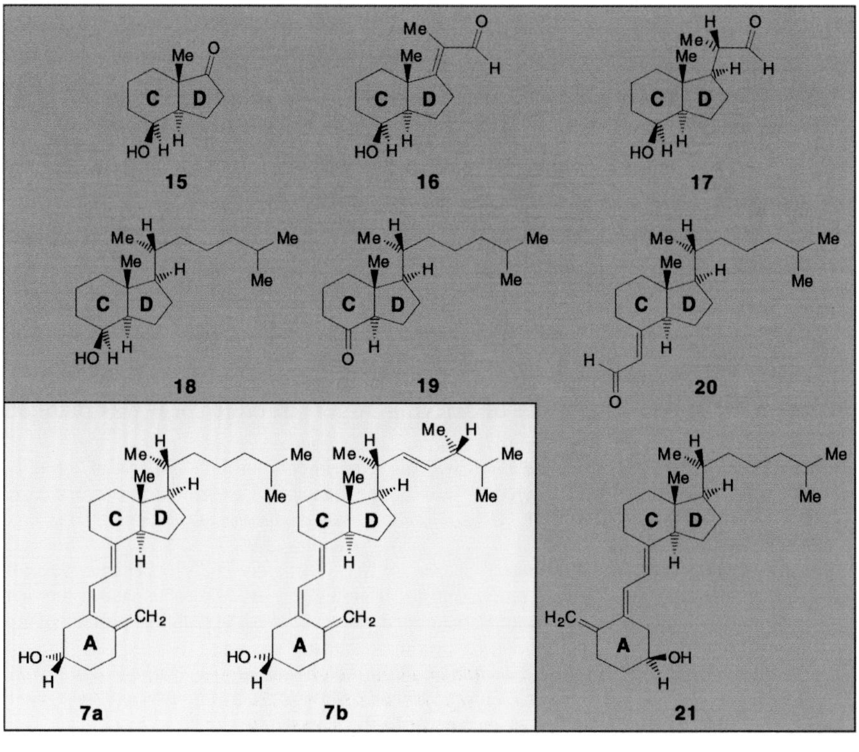

the bridge between rings C and A. They appear, one after the other, in the second half of the synthesis which closes with the endgame shown in *Scheme 10.*

An aldol condensation fixes the easily accessible ring-A unit **22** onto the ring-C/D unit relay compound **20**. The resulting condensation product **23/24** 1:1 is separated by chromatography. *Wittig* reaction[22]) transfers epimer **23** into 5,6-*trans*-vitamin D_3 (**21**). The latter, on photoisomerization, affords vitamin D_3 (**7a**).

The *opening act* is dedicated to the construction of aldehyde **20** *via* the bicyclic hydroxy ketone **15** (*Scheme 11*). It starts with the so-called *Hagemann* ester *rac*-**25**, which, *via* **26** and *rac*-**27**, is converted into *rac*-**28**/*rac*-**29**.

Scheme 10. Braunschweig *Synthesis of Vitamin D_3* (**7a**): *the End Game* [51c]

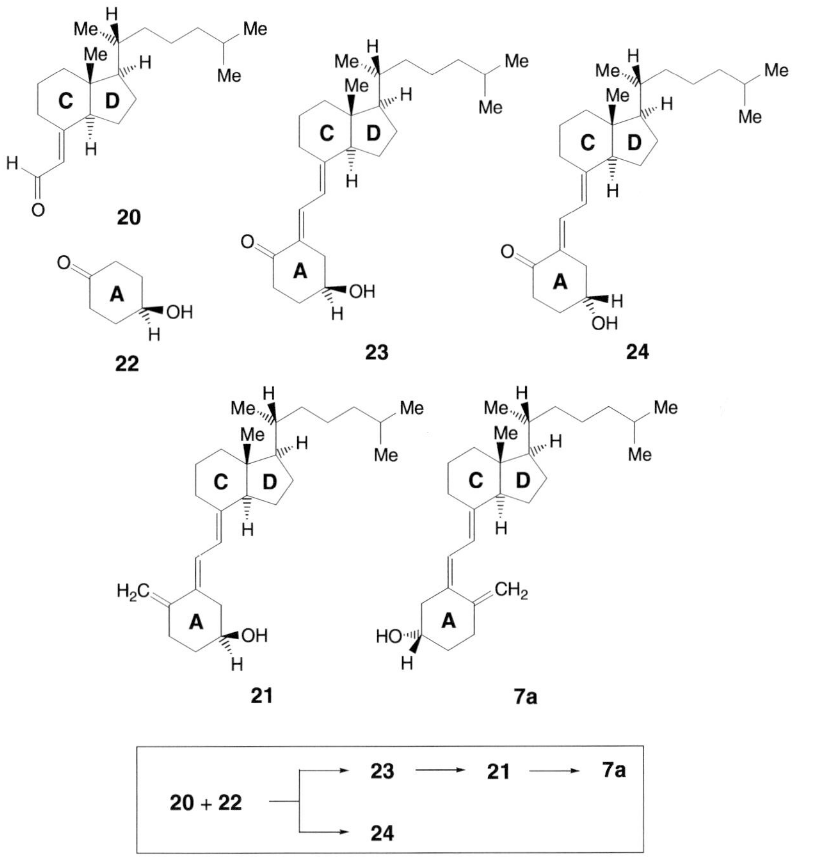

20 + 22 → **23** → **21** → **7a**; **20 + 22** → **24**

[22]) The *Wittig* reaction published in 1954 [56] played a fundamental role in the *Braunschweig* synthesis of vitamin D_3. '*It seems appropriate to point out that, without the* Wittig *reaction, no completed vitamin-D synthesis would lie before us today*' [57].

Scheme 11. Braunschweig *Synthesis of Vitamin* D_3 (**7a**): *the Opening* [51d–g]

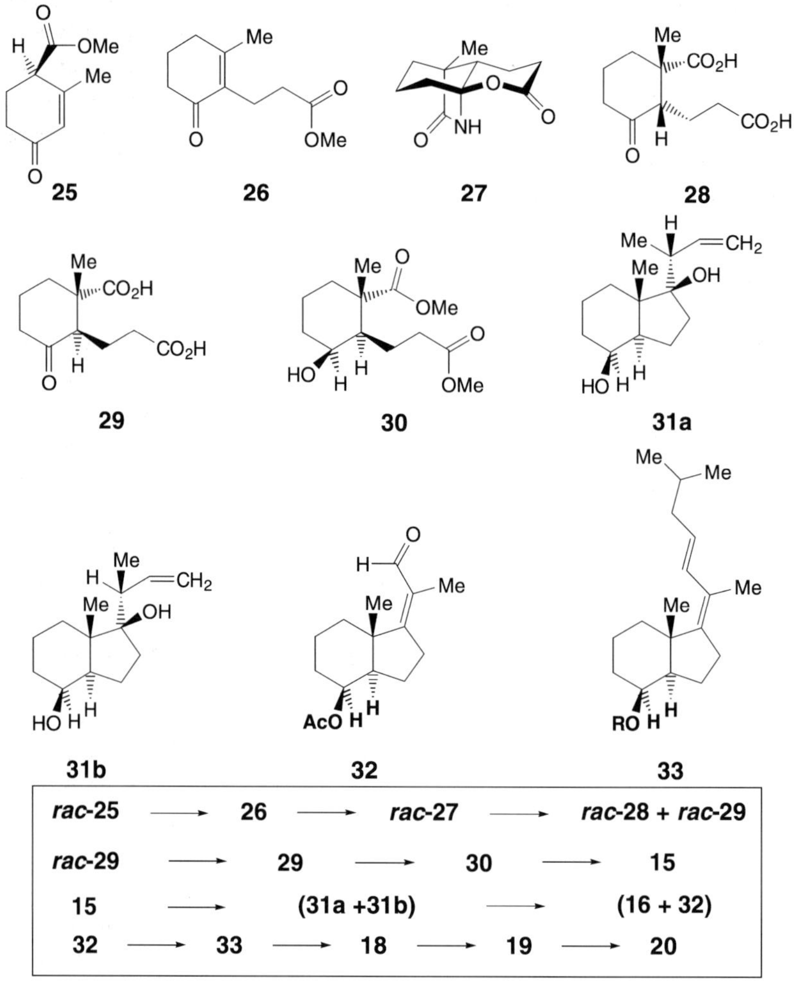

25 **26** **27** **28**

29 **30** **31a**

31b **32** ***33***

***rac*-25** ⟶ **26** ⟶ ***rac*-27** ⟶ ***rac*-28 + *rac*-29**

***rac*-29** ⟶ **29** ⟶ **30** ⟶ **15**

15 ⟶ **(31a + 31b)** ⟶ **(16 + 32)**

32 ⟶ **33** ⟶ **18** ⟶ **19** ⟶ **20**

Resolution of crystalline *rac*-**29** (by means of brucine) gives **29**, which, after conversion to **30**, affords **15** (*Scheme 9*) by *Dieckmann* condensation. The bicyclic hydroxy ketone **15** reacts with crotylmagnesium bromide to furnish **31a** + **31b**. On successive treatment with osmium tetroxide and lead tetraacetate the diastereoisomeric aldehydes arise which can be dehydrated to afford **16** + **32**. The latter compound, obtainable by chromatography, on *Wittig* reaction leads to **33** which, on catalytic hydrogenation, gives **18** (*Scheme 9*). Oxidation converts **18** to **19**, and by successive *Wittig* reaction, followed by selective ozonolysis, unsaturated aldehyde **20** finally is formed.

The pathway adopted in the *Braunschweig* synthesis, to be sure, is too elaborate for a practically applicable preparation of vitamin D_3.

1.4.2. *The* Leeds *Synthesis*

The Leeds synthesis of vitamin D is also a relay synthesis. The relay compounds employed here are given in *Scheme 12*.

Scheme 12. Leeds *Synthesis of Vitamin D_3* (**7a**). The relay compounds (*blue field*) obtained by degradation of vitamin D_3 (**7a**) and vitamin D_2 (**7b**), respectively [58a].

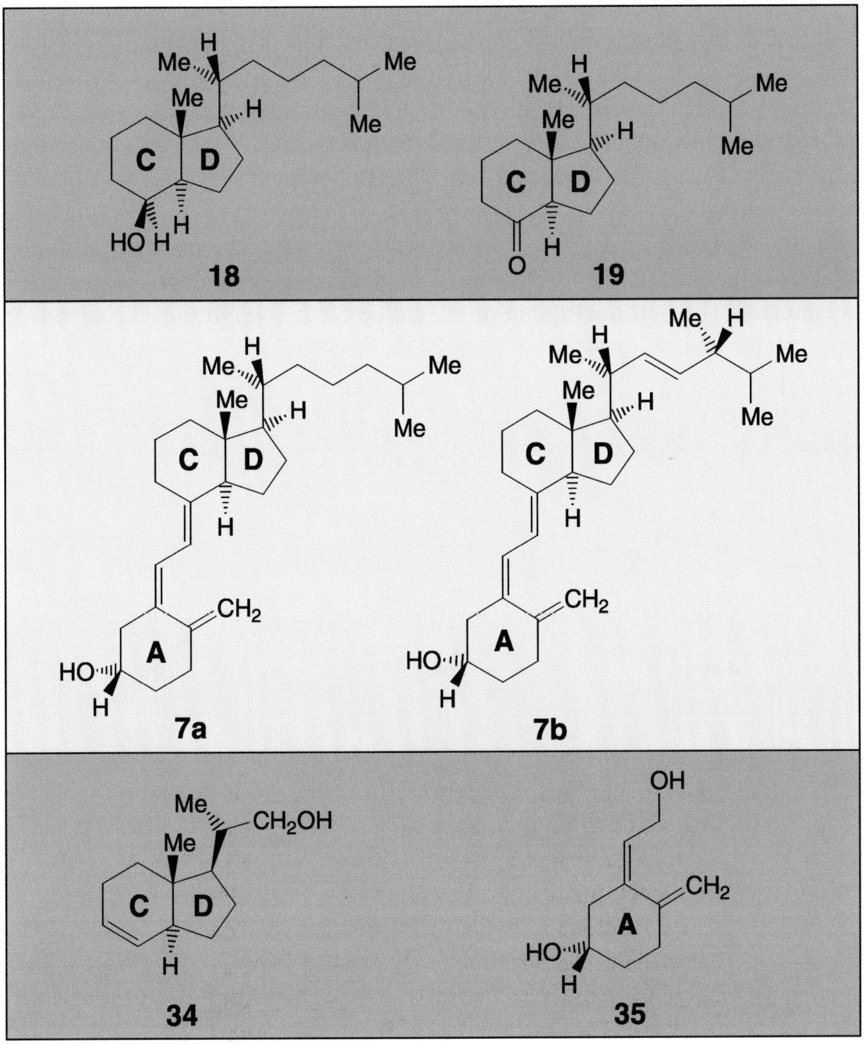

Lythgoe [58] and co-workers took some logical steps from the *Braunschweig* synthesis, making use of the Ring-A building block **35**, so as to ensure from the outset the proper (*S*) configuration at C(3) and the (*Z*)-configuration at the C=C bond between C(5) and C(6). Compound **19** was the putative C/D-building block, with the *Horner-Wittig* procedure as the connecting reaction.

The preparation of **35** started from the chiral, non-racemic *Diels-Alder* adduct **36a**, obtainable by treatment of butadiene with dimenthyl fumarate, followed by hydrolysis or by resolution of *rac*-**36a** with quinine as a resolving agent. Subsequent reduction afforded **36b**. From there, the further route *via* **36c**, **37a**, and **37b** led to **35** (*Scheme 13*).

Ring-C/D-building block **19** was obtained in a multistep sequence. The chain of events began with a *Diels-Alder* reaction of (*E*)-penta-2,4-dienoic acid and methyl methacrylate furnishing, *rac*-1-methylcyclohex-3-exe-1,2-dicarboxylic acid, resolution of which, by help of quinine as resolving agent, gave 1-methylcyclohex-3-ene-1,2-dicarboxylic acid. The latter compound, after successive dihydroxylation (*via* the intermediate epoxide) and bis-decarboxylation ($Pb(OAC)_4$) was converted to a derivative of the *trans*-diol **42**. This derivative on *Claisen* rearrangement (using the *ortho*-ester of (*R*)-dihydrocitronellic acid in the *Johnson* variant), followed by another *Claisen* rearrangement (with 1-(dimethylamino)-1-methoxyethylene in the *Eschenmoser* variant) afforded **43** or **44**, respectively (*Scheme 14*).

Scheme 13. Leeds *Synthesis of Vitamin D_3* (**7a**): *Route Leading to Ring-A Building Block* **35** [58b]

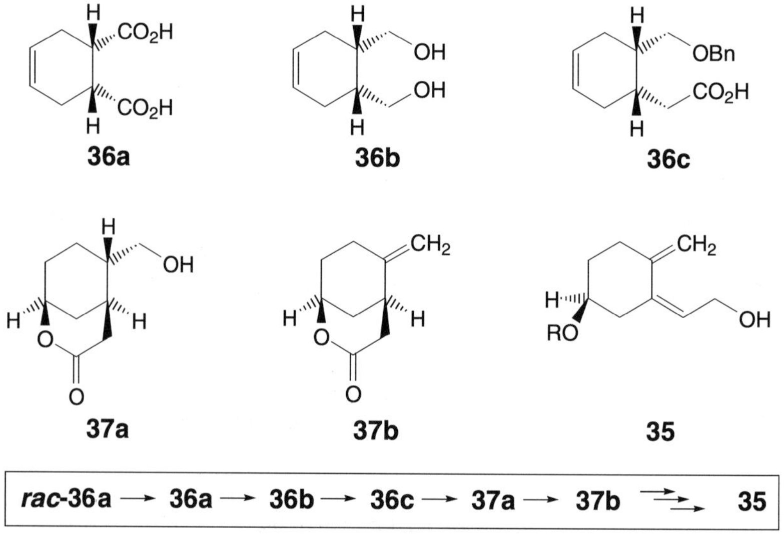

rac-**36a** → **36a** → **36b** → **36c** → **37a** → **37b** → → **35**

Scheme 14. *Leeds Synthesis of Vitamin D_3* (**7a**): *Route Leading to Ring-A Building Block* **19** *Containing Rings C and D* [58c–d]

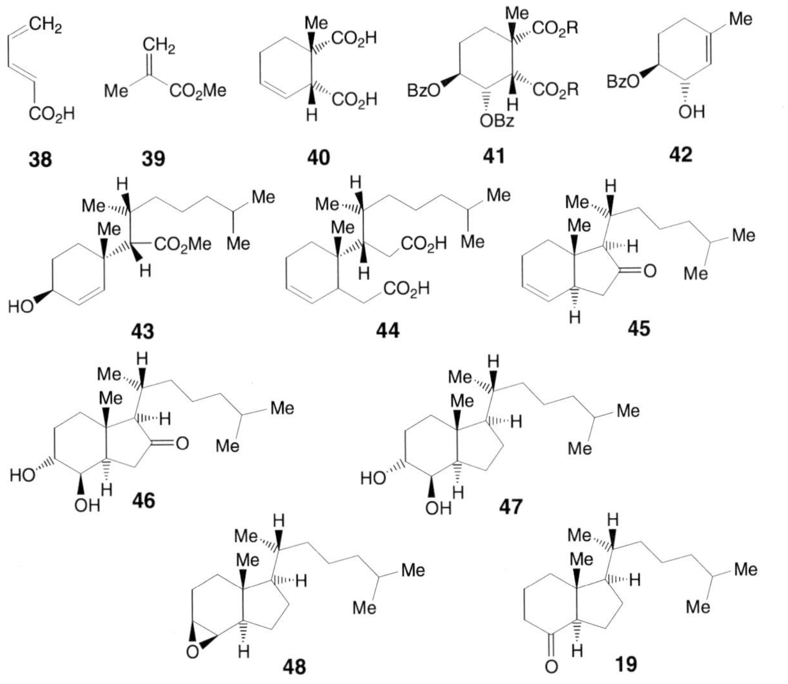

38 + **39** → *rac*-**40** → **40** → **41** → **42**
→ **43** → **44** → **45** → **46** → **47**
→ **48** → **19**

Cyclization of the dimethyl ester afforded ketone **45**, which, by way of the crystalline $8\alpha,9\alpha$-epoxide, was converted to the $8\beta,9\alpha$-diol **46**. The corresponding diacetate, *via* the ethylene thioacetal, was deoxygenated to give **47**. The $8\beta,9\beta$-epoxide **48**, accessible from **47**, after treatment with $LiAlH_4$ and oxidation of the resulting hydroxy compound, finally furnished relay ketone **19**.

The Leeds synthesis of vitamin D_3 requires too many steps to be commercially interesting. All the same, the *Horner-Wittig* connection of the two building blocks containing ring A, or rings C and D, respectively, played a not inconsiderable role in the total synthesis of the vitamin-D metabolites **13** and **14**[23]). *Scheme 15* [31] summarizes various synthetic approaches in this area.

[23]) On the frenzied synthesis activity up until the mid-1990s, see [31][59].

Scheme 15. *An Array of Synthetic Approaches to 1,25-Dihydroxyvitamin D_3* (**14**) [31]

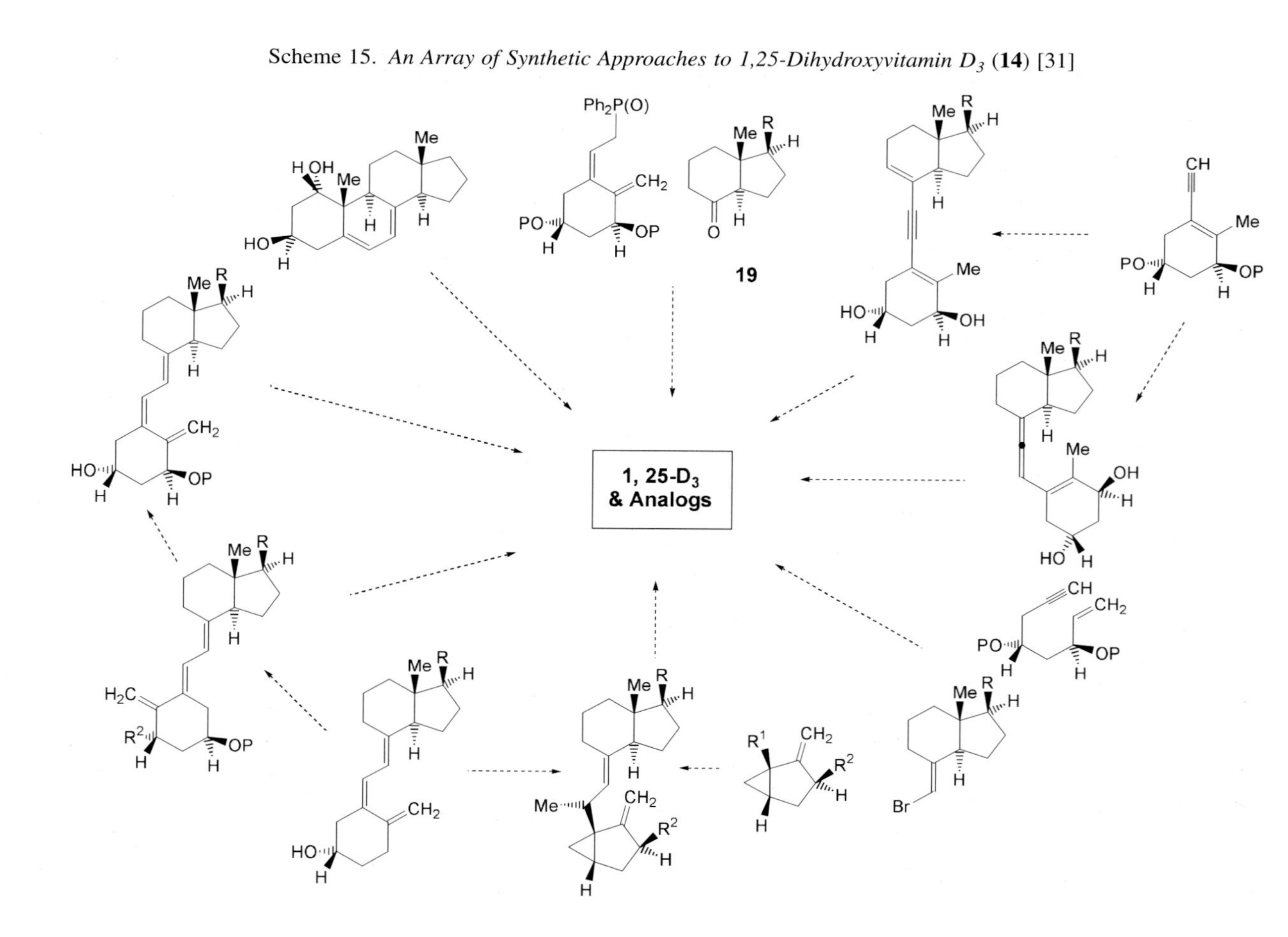

1.5. Open Questions

The completion of the *Braunschweig* Vitamin-D_3 synthesis [51g] did not meet only with praise. Doubts were voiced[24]) as to whether a total synthesis that included a photochemical reaction in its execution would meet the standards of a structural determination (*vide supra*): photochemical reactions, after all, were perceived as less comprehensible than those reactions not dependent on involvement of light. The announcement, on the other hand, of having achieved the first total synthesis of a D vitamin by non-photochemical methods [61] might easily be misunderstood as a bias against the use of light-induced reactions in a natural-product synthesis.

One of the authors, after experience in the vitamin-D area, saw in the study of mechanisms of light-induced reactions a future field of activity [62]. To understand photochemical reactions well enough to be able to use them reliably as key reactions for synthetic aims (in particular, in the field of natural-product synthesis) was the declared [63] motif of the years to come. First, however, the apprenticeship begun in Braunschweig was to be continued with *D. H. R. Barton* in London.

Translating a chapter on *Conformational Analysis of the Steroid Nucleus* into German for a monograph on steroids [64] had made the translator acquainted with the dynamic stereochemistry pioneered by *Barton* [65]. What was more fitting than the wish to take part in the solving of problems of conformational analysis? A biographical note [66] states:

'*When I arrived at Imperial College* Barton *quizzed me about my plans. I told him that I was looking forward to the opportunity to work in his laboratory on a conformational analysis theme, and that afterwards, to qualify as a lecturer at Technische Hochschule Braunschweig, I was thinking of a project in organic photochemistry. He inquired after some particulars, but mentioned in passing that, since the principles of conformational analysis had now been formulated, detailed elaboration was best left to other laboratories. Oh, and by the way, he was already investigating organic photochemistry and would I perhaps like to work together with him studying the photochemistry of 2,4-cyclohexadienones?*

The photochemistry of 2,4-cyclohexadienones was to prove fruitful [67]. *A ring-opening takes place, which was believed to furnish a dieneketene that, under the right conditions, adds protic nucleophiles. As the photoeducts are easily accessible from phenols, these phenols may hence be converted in this manner into open-chain compounds. This structural alteration proceeds in high chemical yield and should lend itself to synthetic application. For ap-*

---

[24]) *A. Lüttringhaus*, personal communication; see [60].

propriate design, though, deeper insights into the reaction mechanism were desirable.

Having returned to Braunschweig, I turned to other problems in organic photochemistry. The photochemistry of 2,4-cyclohexadienones – so I believed – was to be developed further at Imperial College. Months later I met Barton *again, and he inquired after new results in 2,4-cyclohexadienone photochemistry. My expressed belief – that this topic was his field – was emphatically rebutted and I was implored to carry on those studies regardless'.*

2. Photochemical Ring-Opening in Cyclohexa-2,4-dienones: Key Reactions in Natural-Product Syntheses

2.1. UV Absorption in Cyclohexa-2,4-dienones

Normally, a photoreaction begins with light absorption[25]). Cyclohexa-2,4-dienones absorb light in two closely adjacent wavelength regions. The shorter wavelength, and more intensely absorbing, region is ascribed to the $\pi^* \leftarrow \pi$ excitation, while that of longer wavelength and less intense absorption is ascribed to $\pi^* \leftarrow$n excitation[26]). The influence of the polarity[27]) of the solvent used on the location of the two regions of absorption is regarded as a diagnostic criterion for determination of the electronic excitation concerned [69]. A change in solvent from one of low polarity to one of high polarity results in a shift of the $\pi^*\pi$-absorption region towards red, and of the π^*n-absorption region towards blue[28]). Comparison of the absorption spectrum of (±)-6-acetoxy-6-methylcyclohexa-2,4-dien-1-one (*rac*-**49**; *Fig. 1*) with that of 2,3,4,5,6,6-hexamethylcyclohexa-2,4-dien-1-one (**50**; *Fig. 2*) begs the question of whether, at least at high levels of dienone ring substitution[29]) and with use of a polar solvent, the energetic sequence of the $\pi^*\pi^2$n and $\pi^*\pi$n^2 electronic isomers, and hence their characteristic type of photochemistry, might be invertable[30]).

[25]) *Normal photoreactions* include transformations of an electronically excited starting material, arising either out of direct light excitation or from energy transfer from a light-excited sensitizer. *Abnormal photoreactions* (*photoreactions without light*) are those transformations that commence with an electronically excited starting material or sensitizer arising from a thermally induced reaction.

[26]) See [4] concerning the notation regarding $\pi^*\pi$- and π^*n-absorption regions.

[27]) A series of empirical values of polarity parameters for a wide range of solvents can be found in [68].

[28]) The blue shift of the π^*n-absorption region in carbonyl compounds had been observed as early as 75 years ago [69]. For a detailed discussion, see [70].

[29]) The location of the maximum of the $\pi^*\pi$-absorption region of cyclohexa-2,4-dienones in MeOH follows rules [71] that take into account the number and positions of the Me substituents present.

[30]) On the answer to this point, *vide infra*.

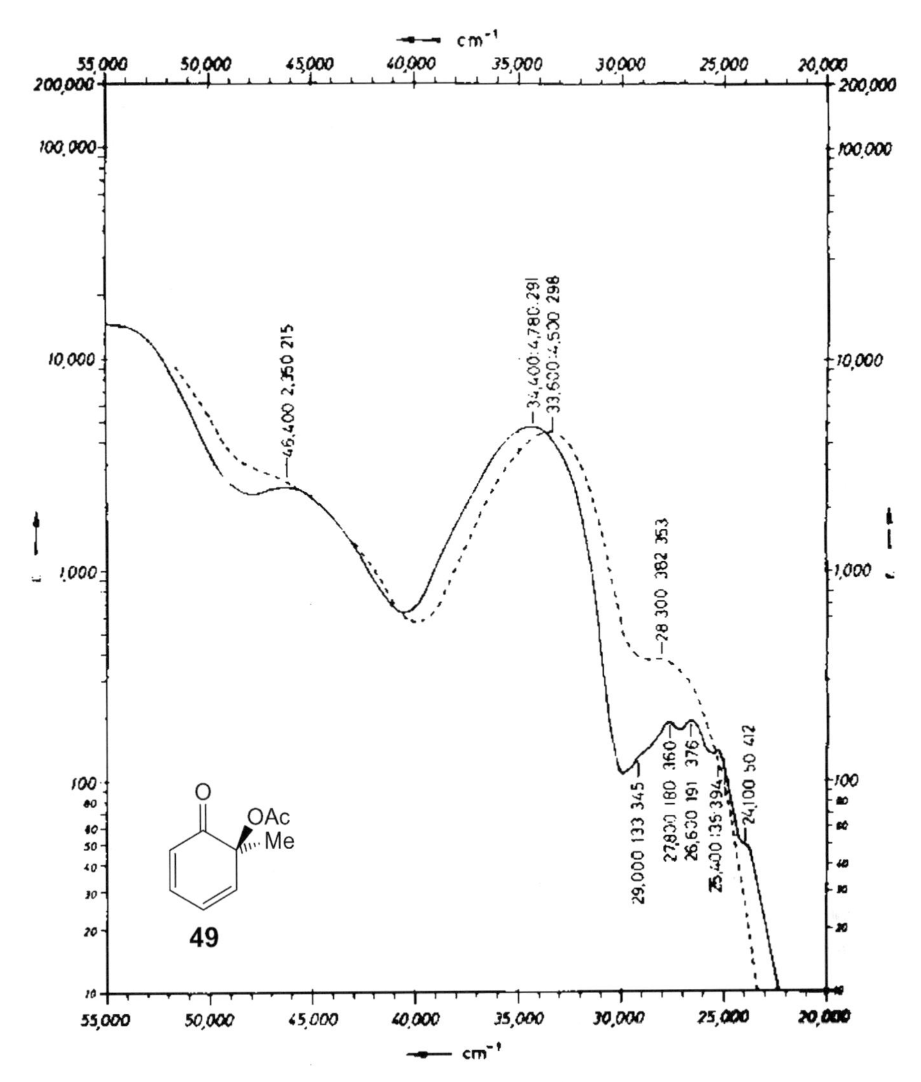

Fig. 1. *UV Spectra of 6-acetoxy-6-methylcyclohexa-2,4-dien-1-one* (**49**) *at room temperature in heptane* (——) *and MeOH* (----)

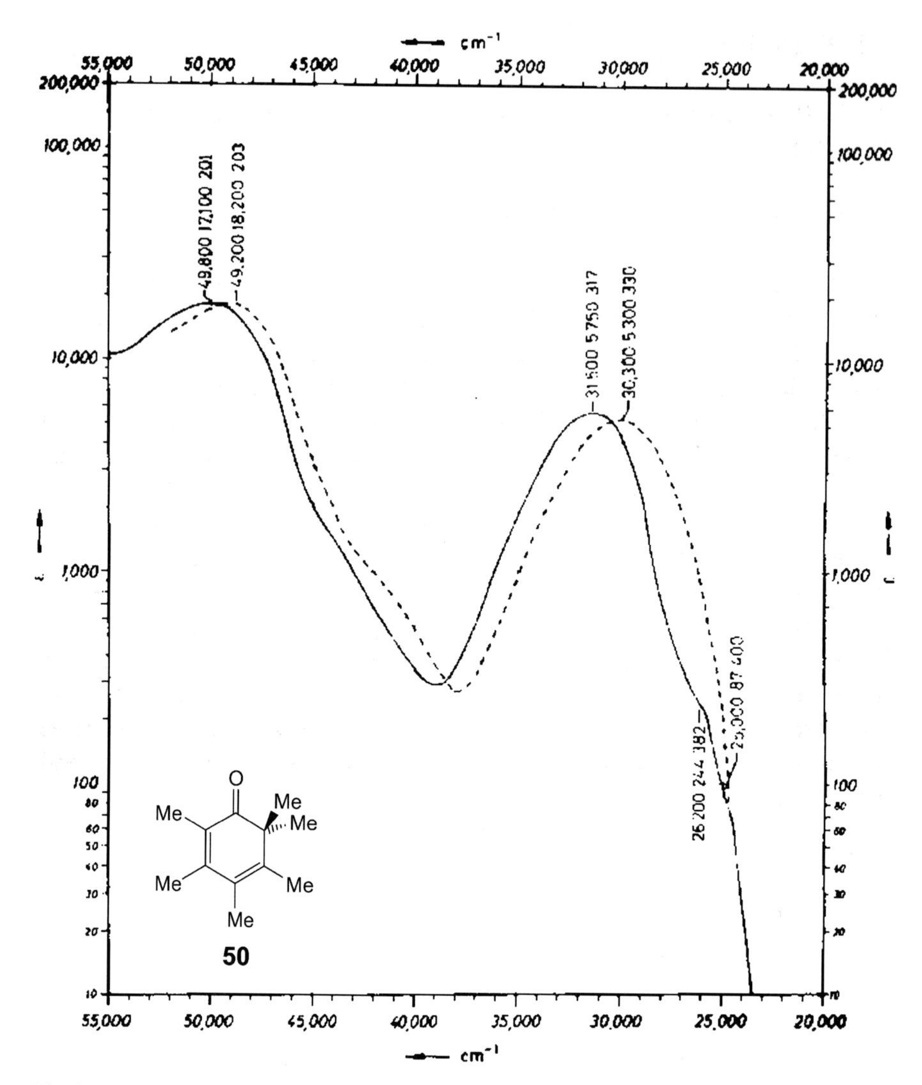

Fig. 2. *UV Spectra of 2,3,4,5,6,6-hexamethylcyclohexa-2,4-dien-1-one* (**50**) *at room temperature in heptane* (——) *and MeOH* (----)

2.2. Photochemistry of Cyclohexa-2,4-dienones

2.2.1. *Primary Photochemical Products from $^1_1$[Cyclohexa-2,4-dienones]*: Dieneketenes*

2.2.1.1. *Direct Detection of Dieneketenes*

Carrying out photochemical reactions at low temperature makes it possible to investigate primary products that are too short-lived at room temperature to be detectable by the sort of conventional spectrometers found in every organic chemistry laboratory. Working at low temperature, it proved possible to detect spectroscopically the photochemical seco-isomerizations of countless cyclohexa-2,4-dienones to the corresponding dieneketenes, while at higher temperature, the thermal cyclo-isomerization of these dieneketenes to the original dienones could be observed [72] [73][31]) (*Scheme 16*).

With the aid of flash spectroscopy, it also proved possible to detect ketene transients at room temperature [72d] [72f]. The fact that these transients are indeed ketenes is confirmed unequivocally by the ketene IR band at *ca.* 2100 cm^{-1}: this would appear whenever low-temperature irradiation was performed on a given cyclo-isomer and disappear once more after warming to room temperature [72]. In a series of examples, it was possible to observe several ketene bands; this could be put down to the involvement of several dieneketene *conformations* [72d] [72f].

Scheme 16. *Thermo-reversible Photoisomerization of Cyclohexa-2,4-dienones* [72]

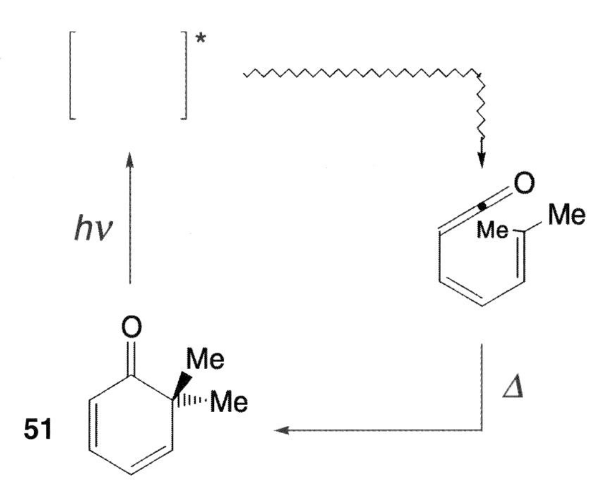

[31]) For technical details about the equipment used for low-temperature spectroscopy, see [28b] [72b–d].

Scheme 17. *The Four Types of Idealized Dieneketene Conformations*. For designation of conformations, see Footnote 6 in [72e]

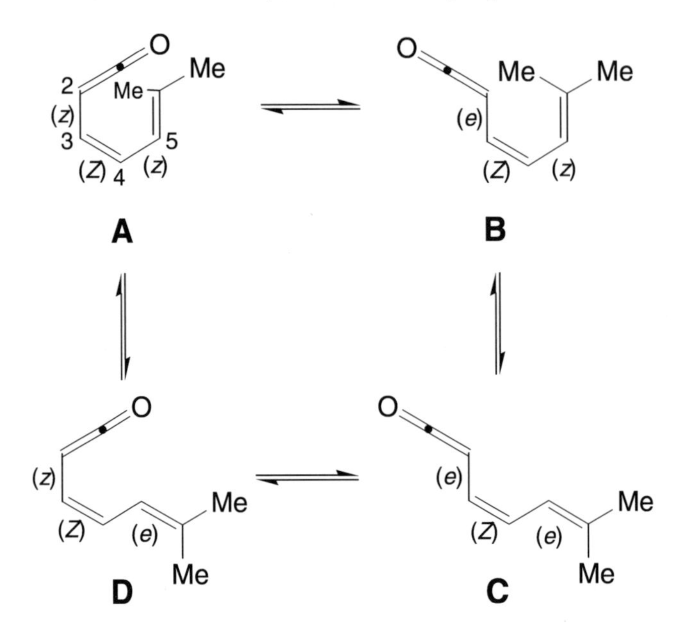

Two dieneketene *configurations* are to be expected, if C(6) of the related cyclohexa-2,4-dienone is a stereogenic center. It would hardly be worth mentioning this fact without striking exceptions at hand. *ortho*-Quinol acetates (*rac*-**49** for example) only provide the dieneketene with (3*Z*,5*E*)-configuration (*Scheme 18*).

Scheme 18. *Each of the Two* ortho-*Quinolacetate Conformers Undergoes Stereospecific Ring Opening* [72e]

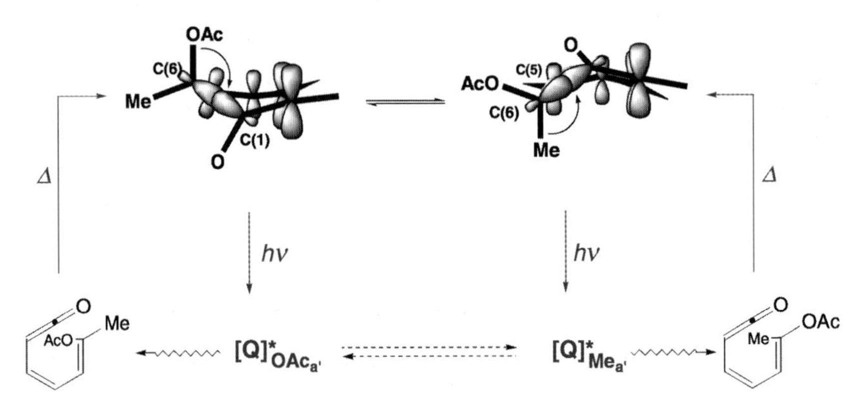

The observed stereoselection can be explained provided two premises are agreed on (see Fig. 15 in [72e], and Scheme 21 in [72f]). Firstly, the equilibrium between the two *ortho*-quinolacetate conformers is assumed to be placed completely on the side of that conformer with the AcO group in the pseudoequatorial position[32]). Secondly, each of the two conformers is supposed to undergo stereospecific ring opening[33]).

2.2.1.2. *Secondary Reactions of Initially Generated Dieneketenes*

Both intramolecular and intermolecular reaction pathways are available to the conjugated dieneketenes. Which product components emerge victoriously out of the competing processes depends on the substitution pattern and the reaction conditions.

2.2.1.2.1. *Cyclization*

Conjugated dieneketenes of (*Z*)-configuration about the central C=C bond convert through electrocyclic reaction into the cyclohexa-2,4-dienones from which they, as a rule, originated. To derive structure-reactivity relationships for the thermally induced cyclo-isomerization of a dieneketene, both the light-induced preparation of the dieneketene and the quantitative monitoring of its thermal secondary reactions were carried out at the same temperature. It was hence possible, by kinetic spectrophotometry in cyclohexane, to obtain the corresponding rate constants and, from these, the associated activation parameters for an entire temperature range (see Table 3 in [72e]). An approximate picture of foreseeable reaction behavior was thus obtained, if the steric influence of substituents on the population of the four idealized conformations **A** to **D** (*Schema 17*) was estimated. Substituents at C(5), C(2), and C(4) promote cyclization, in order of diminishing effectiveness. All in all, low activation energies (low activation enthalpies) and low *Arrhenius* factors (significantly negative activation entropies) are typical. The cyclization-rate constants vary only negligibly with solvent polarity (see Figs. 16 and 17, and Table 6 in [72e]).

[32]) This conformer is promoted as well by through-bond interaction between the σ-orbital of the C(6)–C(7) bond and the π^*-orbital of the C=O group (*stereoelectronic effect*) as by through-space interaction between the negative end of the AcO group and the positive end of the ring C=O group (*electronic effect*).

[33]) Ring opening proceeds stereospecifically in accordance with the least-motion principle (see Sect. 3.3 in [72e] and Sect. 5 in [72f]).

If a stereogenic center is created on ring closure, then the original and the *reconstructed* dienones do not have to be configurationally identical. This is the case, for example, when **52** or *ent*-**52** are transformed *via* configurationally isomeric transients **E** and **F** into *rac-52* (*Scheme 19*).

The binary composition of the ketene transients was proven by low-temperature ^{1}H-NMR spectroscopy [62] [72b] [74a]. Formal analysis of the reaction kinetics[34]) indicates that the two ketene components were generated concomitantly and not successively [74b]. The racemization, easily followed by monitoring the CD spectra (*Fig. 3*), gives indirect information about the temporary ring opening taking place[35]).

Androsta-2,4-dien-1-one (**53**) may be transformed into the epimeric 10α-androsta-2,4-dien-1-one (**54**) *via* the seco-isomeric dieneketene **55** (*Scheme 20*). The CD spectra of the two cyclo-isomers (*Fig. 4*) are practically mirror images of each other, facilitating in-depth investigation of the cyclo/seco-isomerizations.

The thermally reversible photoisomerizations [62] lead with sufficiently long irradiation times to a photostationary state, in which the thermodynamically favored 10β-steroid predominates. The quantitative composition of the epimer mixture depends on the wavelength of the exciting light and on the nature of the solvent used [72d]. The influence of wavelength is the result of

Scheme 19. *Photoracemizations of* **52** *and* ent-**52** [72f]

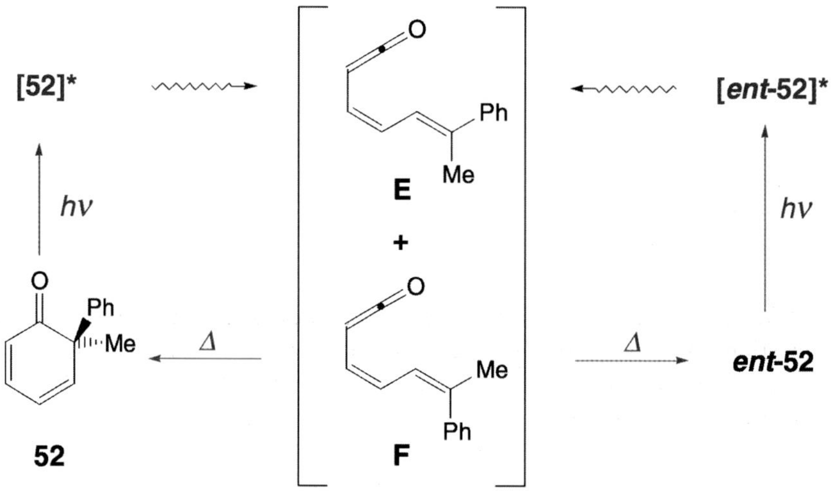

[52]* → [E + F] ← [ent-52]*

hν, Δ

52, E, F, ent-52

[34]) See [75] regarding formal kinetic examination, interpretation, and characterization of light-induced reactions; see [62] for application to cyclohexa-2,4-dienones.

[35]) See Sect. 3.3.2 in [76] for the racemization of the antibiotic (+)-geodin and the lichen compound (+)- or (–)-usnic acid *via* seco-isomers.

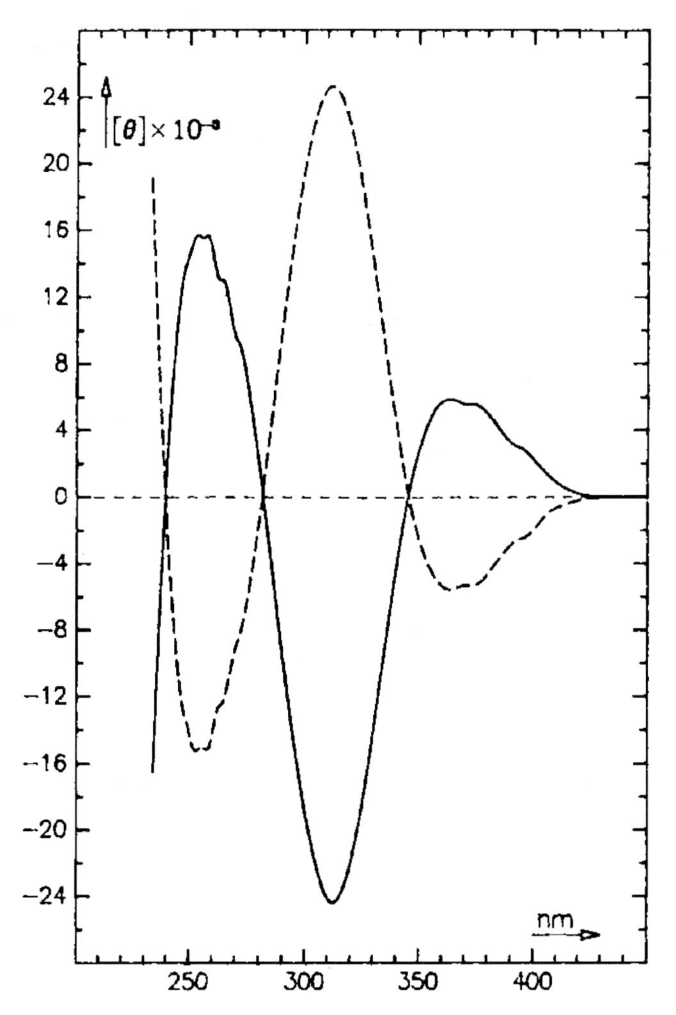

Fig. 3. *CD Spectra of the cyclohexa-2,4-dienones* **52** (----) *and* ent-**52** (——) *at room temperature in MeOH*

the differences in absorption shown by the two cyclo-isomers in the region in question. More powerful than the differences in extinctions, though, is the fact that the quantum yield of the transformation of **54** into **53** is 3.7 times higher than that of **53** into **54**. Or, put another way, **53** is also favored over **54** kinetically.

The mixture of *rac*-**56** and the *meso*-bis-dienone **57** are also involved in a photostationary state. Here, the effect of circularly polarized light on the optically inactive fractions of *rac*-**56** or **57** was studied, in order to be able to distinguish between the racemic mixture and the *meso*-compound by observation which of the two has become optically active (see *Sect. 2.3.1*).

Scheme 20. *Photoepimerizations of* **53** *and* **54** [72d]

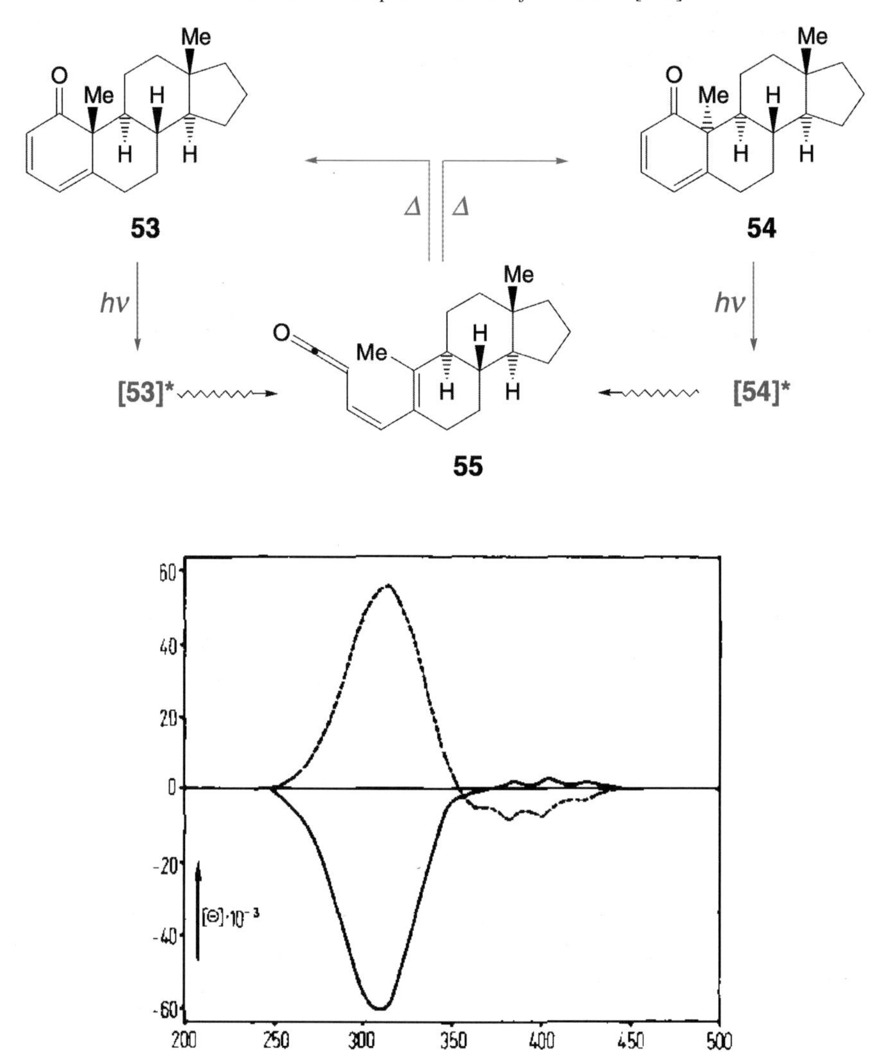

Fig. 4. *CD Spectra of the epimeric cyclohexa-2,4-dienones* **53** (——) *and* **54** (----) *at room temperature in hexane*

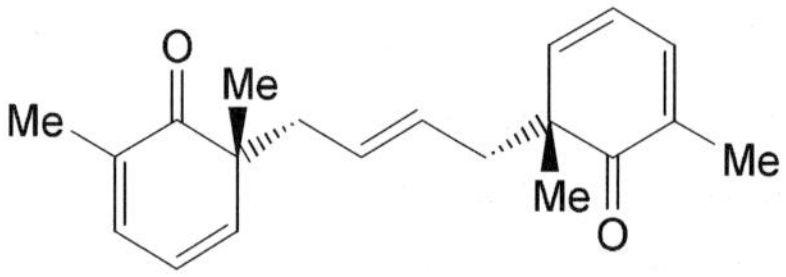

Me O Me Me Me O

56

Me O Me Me Me O

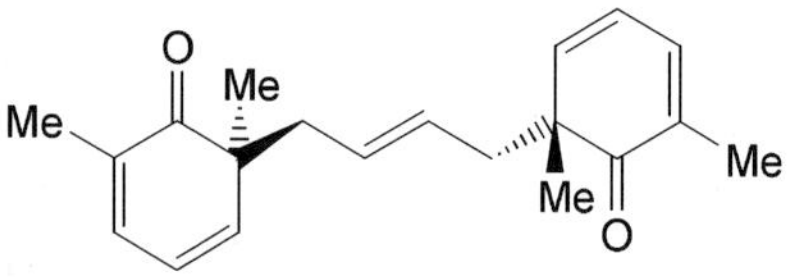

57

Cyclization products of a completely different type are formed if C(8) of an extended dieneketene plays the role of an electrophilic center capable of taking part in an intramolecular aldol addition. *Scheme 21* illustrates what happens in such cases.

Scheme 21. *Intramolecular Aldol Addition of an Extended Dieneketene Where C(8) Plays the Role of an Electrophilic Center* [72f]

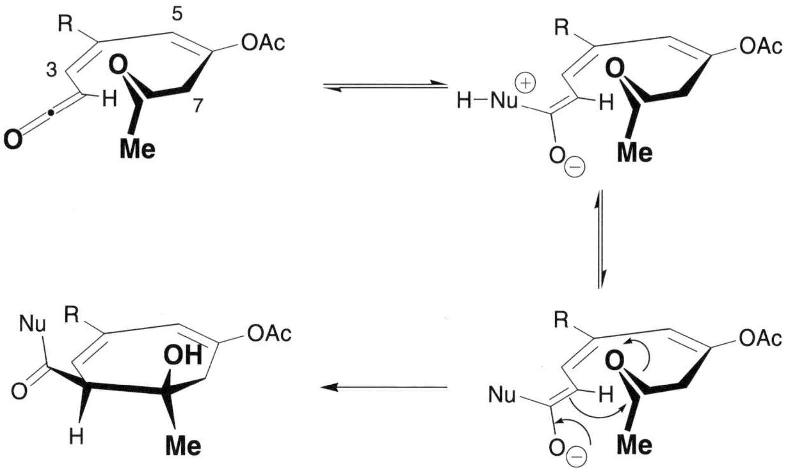

R 5 OAc 3 O H 7 O Me

H–Nu R OAc O H O Me

Nu R OAc O H O Me

Nu R OAc OH O H Me

2.2.1.2.2. *Addition of Protic Nucleophiles*

Protic nucleophiles add to dieneketenes. *Scheme 22* shows the course of reaction, as could be determined step by step by conventional spectroscopy at low temperature and by flash spectroscopy around room temperature [72f].

In the case of UV irradiation of *rac*-**52** in either EtOH or Et_2O containing cyclohexylamine, a relationship has been observed between the wavelength of the exciting light and the relative proportions in the isolated reaction product [74c] of components of type **58** (with a UV maximum in Et_2O of 279 nm) and of type **59** (with a UV maximum in Et_2O of 255 nm) [77]. For 365-nm light, the information given in *Scheme 22* applies. When 313-nm light was used, this reaction outcome was eclipsed by a photoisomerization of the primary product (in a determined ratio of **58/59**) into a secondary product (with an altered ratio to the detriment of the configurational isomer **58** absorbing at longer wavelength). A sequence of two photoisomerizations, independent of one another, can be established easily with the help of *Mauser* diagrams[34]) [72b] [72f] [74c]. It may be avoided if cyclohexa-2,4-dienones are excited in the long-wavelength UV or, even better, in the visible range.

Scheme 22. *Spectroscopically Monitored Complex Chemistry, Leading to Adducts* **58** *and* **59**, *of the Addition of a Protic Nucleophile* (Nu–H) *to Configurationally Isomeric Dieneketenes* (**G+H**), *Photochemically Accessible from* rac-**52** [72f]

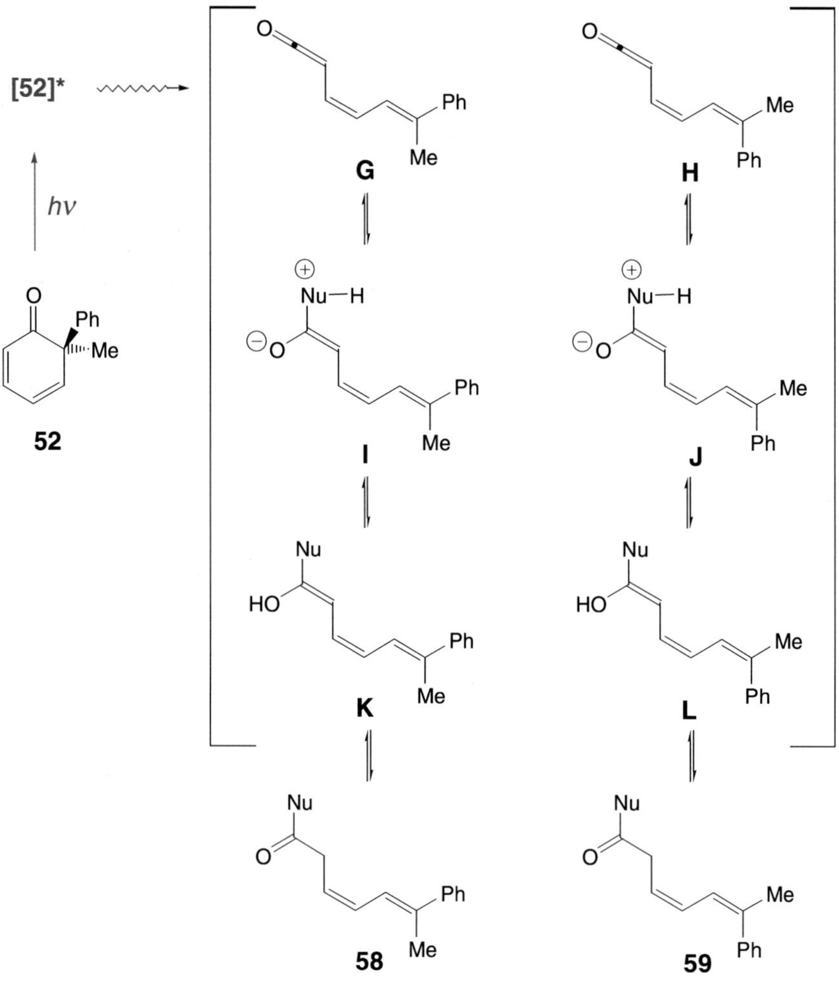

[52]* hν 52 G H I J K L 58 59 Ph Me O Nu–H Nu HO

Under these conditions, the photochemical transformation of cyclohexa-2,4-dienones into doubly unsaturated carboxylic acids or their derivatives is an ideal photoreaction for the synthetic chemist, since the products created absorb at significantly shorter wavelengths than the starting materials employed. What this implies does not require further laboring, in the light of the previtamin-D (**5**) photochemistry outlined earlier (*Sect. 1.1*). In-depth study of the addition of protic nucleophiles to photochemically generated dieneketenes was to result in insights into reaction mechanisms that enable the synthetic chemist to avoid a dead end, when trying to achieve intramolecular nucleophilic addition.

The topic of the addition of protic nucleophiles to dieneketenes is not exhausted in *Scheme 22*. It also turns out that concentration and nucleophilicity of the addend can influence the course of reaction. This also holds good for the quantum yield, which determines the degree to which absorbed light is used for the desired conversion. Since the previously mentioned cyclization (*Sect. 2.2.1.2.1*) competes generally with adduct formation, everything needs to be done to ensure that every ketene molecule produced is captured by a nucleophile. The quantum yield then reaches its maximum value. On the other hand, the maximum quantum yield indicates the threshold value for the concentration of the protic nucleophile, that must be reached or exceeded if the *reagent light* is not to be wasted. This threshold value decreases with increasing nucleophilicity.

In *Scheme 22*, it was tacitly assumed that the ketonization of the conjugated, unsaturated ketene *O,Nu*-semiacetals **K** and **L** led regioselectively to the 1,2-adducts **58** and **59**. In principle, 1,4- and 1,6-adducts also need to be allowed for. The consequences of this can be seen in *Scheme 23* – an extend-

Scheme 23. *The Dieneketene* **55**, *Photochemically Accessible from the Epimeric Cyclo-Isomers* **53** *or* **54** (*Scheme 20*), *on Addition of a Protic Nucleophile* (Nu–H) *is Transferred into 1,6- and/or 1,2-Adducts of Types* **60** *or* **61**, *Respectively* [72d]

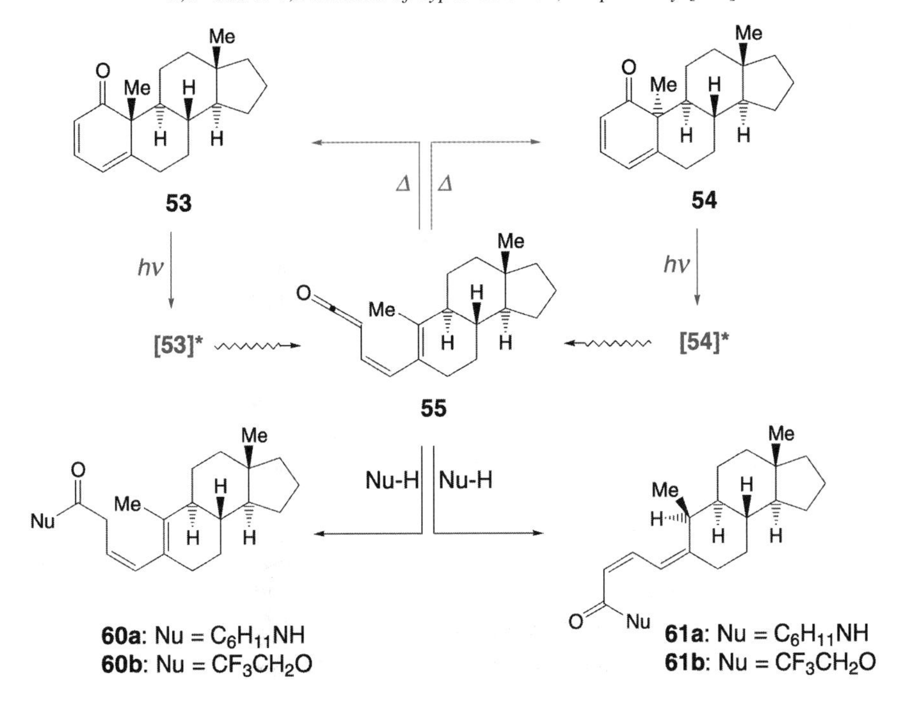

ed version of *Scheme 20*. Here, in the case of the androsta-2,4-dien-1-one configurational isomers **53** and **54**, the 1,6-adduct **61a** was also observed, in addition to the 1,2-adduct **60a**, only, though, if the concentration of the strong nucleophile cyclohexylamine had been kept low. If, under these conditions, the wavelength of the light was changed from 365 to 313 nm, the complexity of the product composition increased, as the (2*Z*,4*Z*)-configured primary component underwent a secondary photoisomerization and then reappeared together with the other three isomers: those of (2*Z*,4*E*)-, (2*E*,4*Z*)-, and (2*E*,4*E*)-configurations.

1,6-Adducts – and also 1,4-adducts – have been observed in other instances too. Only, however, when no strong nucleophile was present in high concentration, and when the substitution pattern of the cyclohexa-2,4-dienone promoted the quasi-cyclic conformation in the resulting dieneketene (**A** in *Scheme 17*). The widely accepted overall reaction scheme [72d–f], developed with great demands on physicochemical methodology, was probably of still greater significance for the decades-old debate over nucleophilic addition to (conjugated, unsaturated) ketenes [78] than for the synthetic chemist interested in convenient preparative access to substituted hexadienoic acids from cyclohexa-2,4-dienones.

If weakly nucleophilic 2,2,2-trifluoroethanol (TFE) is used as solvent (and hence in high concentration), photoisomerizations from **53** to **54** (and back) are indeed observed, but with no adduct formation whatsoever[36]). If, however, 1,4-diazabicyclo[2.2.2]octane (DABCO) is added to the irradiated solution, the 1,2-adduct **60b** is obtained in 92% chemical yield [79]. The addition of DABCO was done with the intention [72e] of increasing the relatively slight reactivity of the ketene group **B** by converting it to the highly reactive acylammonium group **O** [80] (*Scheme 24*).

The catalytic effect of DABCO on the addition of protic nucleophiles to dieneketenes is not limited to the case at hand. The *ortho*-quinol acetates *rac*-**62** and *rac*-**63** are additional examples; the corresponding 1,2-adducts (*rac*-**64** and **65**) can be obtained in this manner in high chemical yield (*Scheme 25*) [72e].

The catalytic properties of DABCO (or other tertiary amines) attain preparative significance in the *intramolecular version* of the addition of a protic, nucleophilic group to an activated dieneketene to produce macrolides or macrolactams: by photolactonization or photolactamization, respectively.

[36]) Of the alternative conjectures, either that no ketene is produced in TFE or that the ketene produced is not capable of addition because of rapid cyclization due to TFE, only the latter merits consideration. The existence of the ketene transients has been affirmed not only indirectly, by the observed photoepimerization, but also directly, by detection by flash spectroscopy [72d].

Scheme 24. *How DABCO Promotes the Conversion of* **B** *to* **P** [72e]

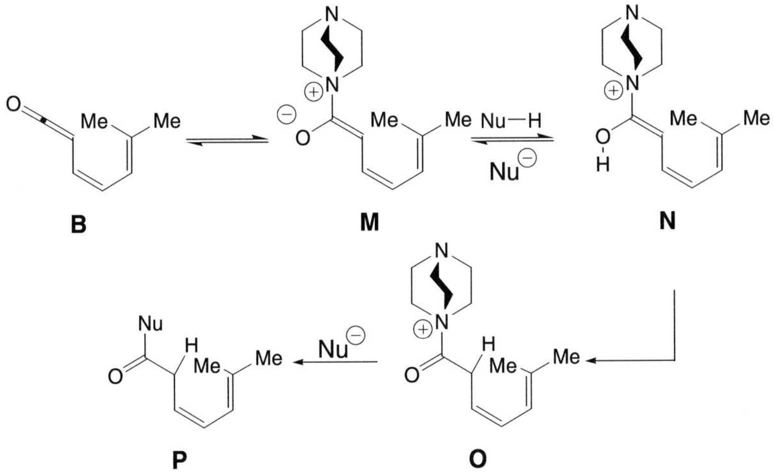

Scheme 25. ortho-*Quinol Acetates* rac-**62** *and* rac-**63** *to Furnish Trifluoroethyl Esters* rac-**64** *and* **65**, *Respectively, on Irradiation in the Presence of DABCO Only* [72e]

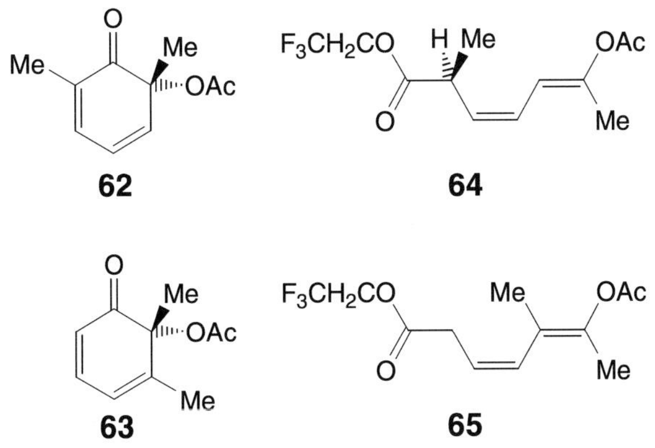

The term *photolactonization* refers to a new preparative approach by means of which type-**S** lactones of large ring size may be obtained photochemically from hydroxyalkylated *ortho*-quinol acetates of type **Q** *via* transient dieneketenes of type **R** (*Scheme 26*) [81][37]).

Zwitterion **U** is a doubly activated derivative of hydroxy ketene **R**, tried and tested as an intermediate in the preparation of 16- to 20-membered ring lactones of type **S** [81] (*Scheme 27*).

[37]) Similarly, *photolactamization* defines a new preparative way by means of which lactams of large ring size are obtainable photochemically from aminoalkyl-substituted *ortho*-quinol acetates *via* transient dieneketenes [82].

Scheme 26. *Formal Description of Photolactonization* [81]

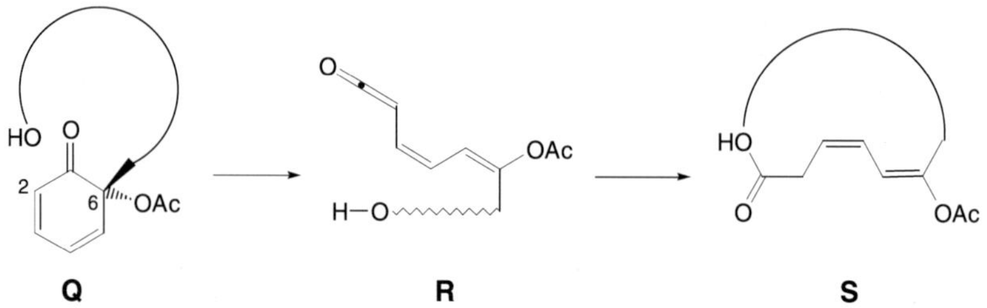

Scheme 27. *How DABCO Promotes the Conversion of* **R** *to* **S** [81c]

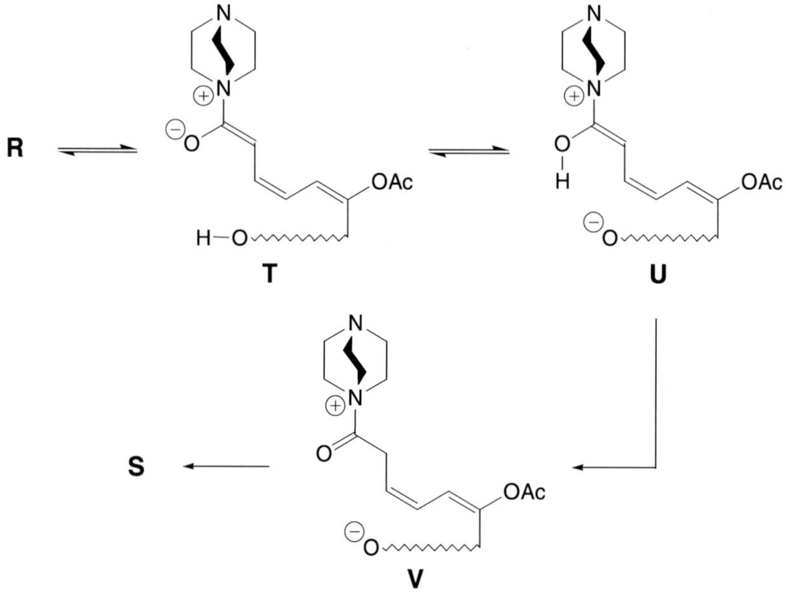

The total syntheses of the 16-membered macrolide antibiotic (–)-A 26771 B (*Sect. 2.3.4*) and of the 18-membered lichen macrolide (+)-aspicillin (*Sect. 2.3.3*) were accomplished by photolactonization as the key reaction step.

2.2.1.2.3. *Bicyclization*

Table 3 contains *ortho*-quinol acetates that, in TFE or in some cases even in MeOH, undergo photoisomerization to bicyclo[3.1.0]hex-3-en-2-ones (see *Scheme 28*).

Table 3. *Conditions under Which Irradiation of the Given* ortho-*Quinol Acetates Leads to Adducts* (yellow fields), *Bicyclo[3.1.0]hex-3-en-1-ones* (green fields), *or Reforms Educt* (blue field) [72e]. a: In the presence of DABCO: adduct formation.

o-Chinol-acetate	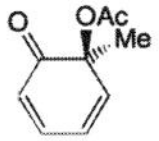 49	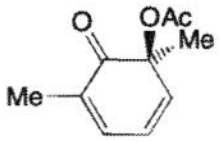 62	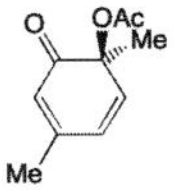 69	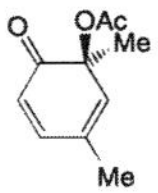 70	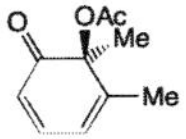 63	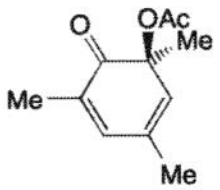 67	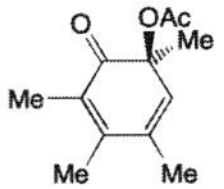 71	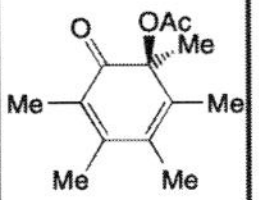 72
Cyclo-hexylamine								
MeOH								
TFE		a			a			

Scheme 28. ortho-*Quinol Acetates* rac-**62** *and* rac-**67** *to Furnish Bicyclo[3.1.0]hex-3-en-1-ones* rac-**66** *and* rac-**68**, *Respectively, on Irradiation in TFE* [72e]

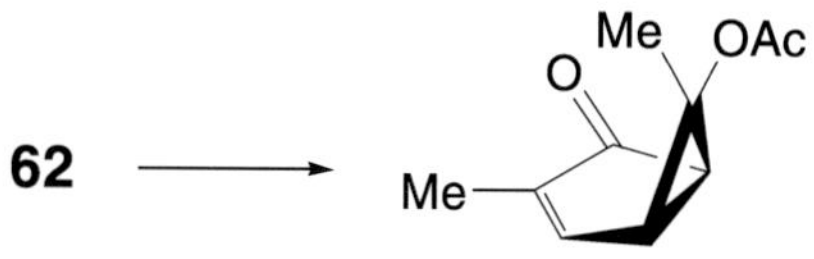

The reaction was first observed by *Harold Hart* [83]. The $\pi^*\pi n^2$-structure was ascribed to the electronically excited starting material[38]). On the grounds of the absorption spectra, it was viewed as possible that, in TFE, the $\pi^*\pi n^2$-electronic isomer might lie lower energetically than the $\pi^*\pi^2 n$-electronic isomer. One might, therefore, conclude that the $\pi^*\pi^2 n$-electronic isomers would undergo ring opening to afford the seco-isomer of the cyclohexa-2,4-dienone involved while the $\pi^*\pi n^2$-electronic isomer would rearrange to the related bicyclohexenones (*Scheme 29*).

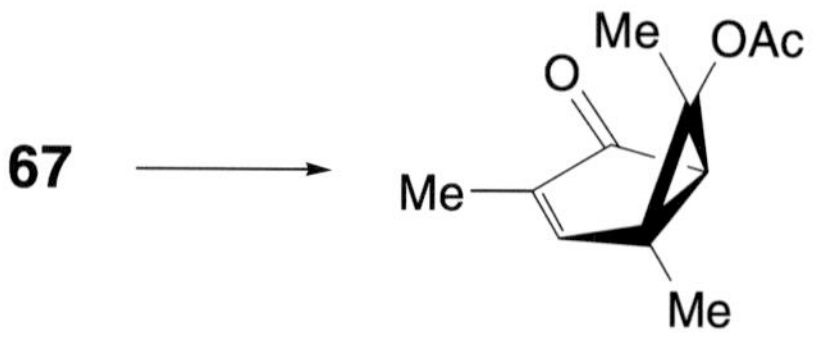

Such a view would imply that it might be possible to achieve mastery of control over photoproducts by straightforward manipulation of the reaction medium. It was taken up as such, as a challenge for the developing picture of the chemistry of electronically excited molecules in general. At the beginning of the 1970s, *W. G. Dauben*, *L. Salem*, and *N. J. Turro* attempted to develop the widely accepted view [72b][84] of the dichotomy into the two types of cyclohexa-2,4-dienone photochemistry into the basis for future classification of photochemical reactions [85a][85b], or even for the foundation of a theory of photochemical reactions [85c].

Fruitful though this approach might have been for general accessibility to multihypersurface chemistry [3a], their setting up of qualitative correlation

[38]) If this were true, only the singlet-spin isomer would need to be taken into account. The triplet-spin isomer of an *ortho*-quinol acetate, obtainable by use of photosensitizers of triplet energy as low as 42 kcal/mol, reacts to give the product of a dienone-phenol rearrangement [72e].

Scheme 29. *Previous view: $\pi^*\pi^2 n$ Photochemistry Gives Dieneketenes* (DK) *Which Cyclize to Furnish Cyclohexa-2,4-dienones* (CHD) *and/or Add Protic Nucleophiles to Afford Related Adducts* (AD); *$\pi^*\pi n^2$ Photochemistry Gives Bicyclo[3.1.0]hex-3-en-1-ones* (BC)

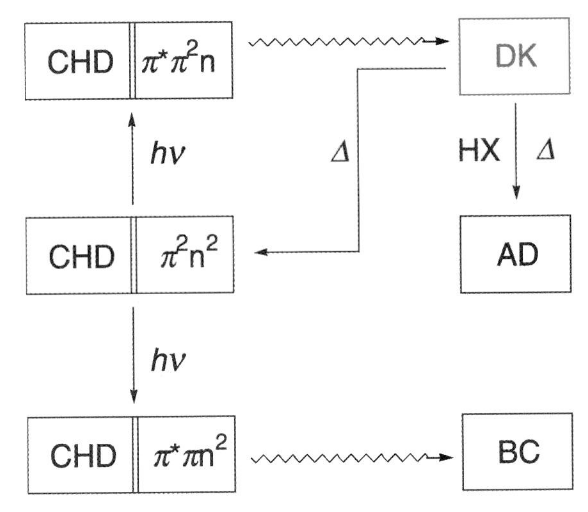

diagrams suffered from the outset from the fact that the test case predicted[11]) a *photochemical* dichotomy that did not in truth exist at all[39]). In-depth studies [3b], in fact, revealed that photoexcited cyclohexa-2,4-dienones react in every instance to give the dieneketene seco-isomers. The branching into the various reaction pathways, manifesting as *cyclization*, *bicyclization*, or *adduct formation*, only takes place – *thermally* induced – at the ketene transient stage (*Scheme 30*).

Photoisomerizations of cyclohexa-2,4-dienones to bicyclo[3.1.0]hexenones must be reckoned with if the dieneketene substitution pattern affects the population of the idealized conformation types (see *Scheme 17*), to the benefit of **A** or **D**. The electronic structure of a dieneketene in its ground state resembles electronically excited conjugated trienes[40]).

The putative polar transition structure finds supporting evidence in the easily ascertainable acceleration of the reaction (see Figs. 16 and 17, and Table 7 in [72e]) with increasing solvent polarity [68], as well as by the structure of monocyclic five-membered ring compounds of type **71** that appear *via* dipolar intermediates of type **W** as side-products (*Scheme 31*).

[39]) There is one insignificant reaction pathway that leads from $\pi^*\pi n^2$-excited *ortho*-quinol acetates to phenols. Together, and in negligible quantities, there appear the corresponding resorcin derivative (as a photo-dienone-phenol rearrangement product), and the phenol that originally served as the starting material for the *Wessely* acetoxylation leading to the *ortho*-quinol acetate in question. Since the phenol mixture is also accessible through triplet sensitization, light-induced reaction must start with the triplet-spin isomer (see *Footnote 38*).

[40]) The latter meet twisting with charge separation and, interestingly enough, undergo conversion into bicyclo[3.1.0]hex-3-enes [86].

Scheme 30. *Present View: $\pi^*\pi^2n$ Photochemistry Gives Dieneketenes* (DK) *Which React Thermally to Cyclohexa-2,4-dienones* (CHD), *Bicyclo[3.1.0]hex-3-en-1-ones* (BC), *or Form Adducts* (AD) *on Reaction with Protic Nucleophiles* [3b]

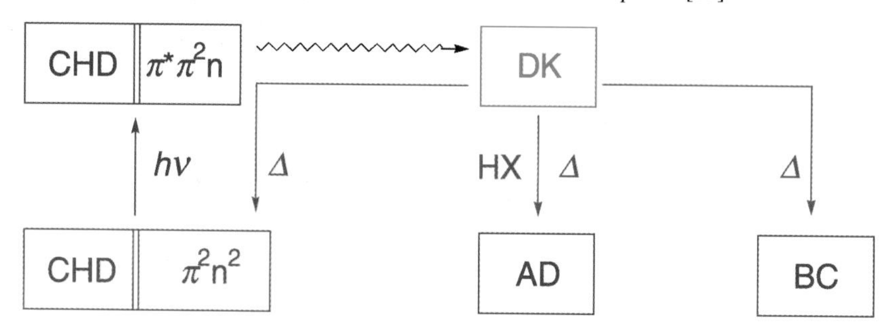

Scheme 31. *Monocyclic By-product* **71** *Uncovers the Polar Character of the Transition State that Separates Dieneketenes from Bicyclo[3.1.0]hex-3-en-1-ones* [72e]

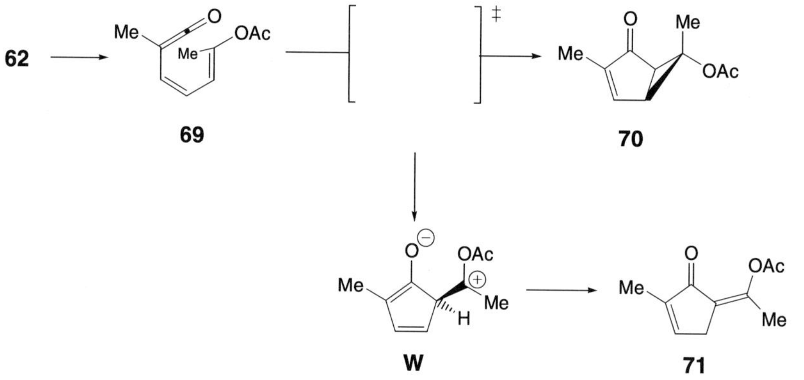

Over-irradiation products [72e], formed by light-induced bicyclo[3.1.0]-hexenone-phenol rearrangement, have hampered the preparative use of the bicyclization of *ortho*-quinol acetates.

Scheme 32, in an extended *Jablonski* diagram, summarizes photophysical and photochemical processes that in the case study of *ortho*-quinol acetates have been observed.

The photochemical ring opening to furnish seco-isomeric dieneketenes takes place, whether 365-nm ($\pi^*\leftarrow$n absorption) or 313-nm-light ($\pi^*\leftarrow\pi$ absorption) will be used. This holds true as well for photoproduct composition[41]) as for quantum yield of consumed educt molecules [72c]. In passing, it ought to be mentioned that, in the presence of the triplet-spin isomer

[41]) Except for the trivial case where the primary photoproduct absorbs light and is converted to a secondary photoproduct.

Scheme 32. *Extended* Jablonski *Diagram Uncovering the Photoprocesses of* ortho-*Quinol Acetates* (wavy lines: radiationlesss transitions) [72e]

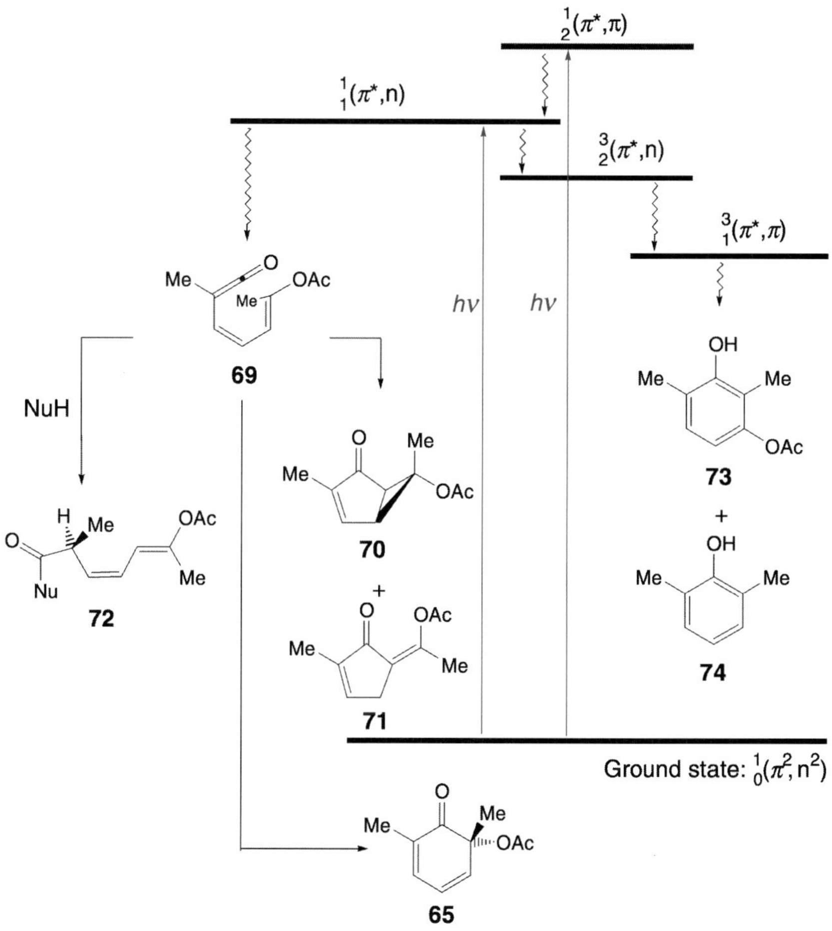

quencher piperylene ((*E*)- and/or (*Z*)-penta-1,3-diene), quantum yield remains unchanged. All that fits the view that ring opening of a cyclohexa-2,4-dienone starts at the $\pi^*\pi^2$n-electronic isomer's hypersurface, reached either directly by light absorption of the ground state $\pi^2 n^2$ electronic isomer or indirectly by internal conversion from the $\pi^*\pi n^2$ one, which had been reached by light absorption before.

2.3. Total Syntheses of Natural Products

Having now surveyed the essentials of cyclohexa-2,4-dienone photochemistry, with the aid of numerous representatives distributed widely across a common structure space, it is next appropriate to evaluate some individual cyclohexa-2,4-dienones from the point of view of their utility in natural product synthesis. To assess the value of cyclohexa-2,4-dienone photochemistry to the synthetic chemist, it is also necessary to take account of the preparative accessibility of this class of compounds and the pathways onwards from them to the chosen target compound. The fact that *ortho*-quinol acetates are the subject of particular attention in this essay has to do with their ease of preparation by means of *Wessely* acetoxylation of 2-alkylated phenols. The chemical yields here are in most cases acceptable. That is not so, however, for cyclohexa-2,4-dienones, obtainable from 2-alkylated phenols through *C*-alkylation or abnormal *Reimer-Tiemann* reaction. Here, as a rule, yields leave something to be desired. In the syntheses of the natural products dimethylcrocetin (*Sect. 2.3.1*) and (–)-variotin (*Sect. 2.3.2*), the rather restricted accessibility of cyclohexa-2,4-dienones from 2-alkylated phenols, by means of alkylation or abnormal *Reimer-Tiemann* reaction, cautions an appropriate degree of modesty. Without the effective *Wessely* approach, the photochemical syntheses of the macrolides (–)-aspicillin (*Sect.* 2.3.3) and (+)-A 26771 B (*Sect. 2.3.4*) would not pass the evaluation test. Either way, photochemical ring opening in cyclohexa-2,4-dienones calls out to be put to the test of natural-product synthesis.

2.3.1. *Total Synthesis of Dimethylcrocetin*

Esters of crocetin (= 8,8′-diapocarotene-8,8′-dioic acid; **75b**) are naturally occurring food-coloring agents of lemon yellow hue [87a]. Their structure suggests a symmetrical synthesis strategy. Depending on retrosynthetic analysis used to disconnect the target molecule (of, for example, dimethylcrocetin (**75c**))[42]) into suitable building blocks, the result is either a class of syntheses in which disconnection has been made at one or two C=C bonds (*Scheme 33,A*), or one in which two C–C bonds have been taken apart (*Scheme 33,B*).

The abstract construction scheme ($C_8 + C_4 + C_8$) may be demonstrated with the building blocks of 2,6-dimethylphenol (**76a** = **74**) and (*E*)-1,4-dibromobut-2-ene (**77a**; *Scheme 34*).

As well as the *O,O*- and the *O,C*-bis-alkylation components **78a** and *rac*-**79a**, respectively, it is necessary to reckon with the stereoisomeric *C,C*-bis-alkylation components **80a** and *rac*-**81a**. These components should, on irradiation in MeOH, be transformed into a mixture of the stereoisomers *rac*-

[42]) For the synthesis of **75c**, see [87b]. For the fact that no commercially viable synthesis of **75c** exists, see [87a].

Scheme 33. *Symmetrical Synthesis Strategy for Target Molecules of Type* **75** (*A:* disconnections at C=C bonds; *B:* disconnections at C–C bonds)

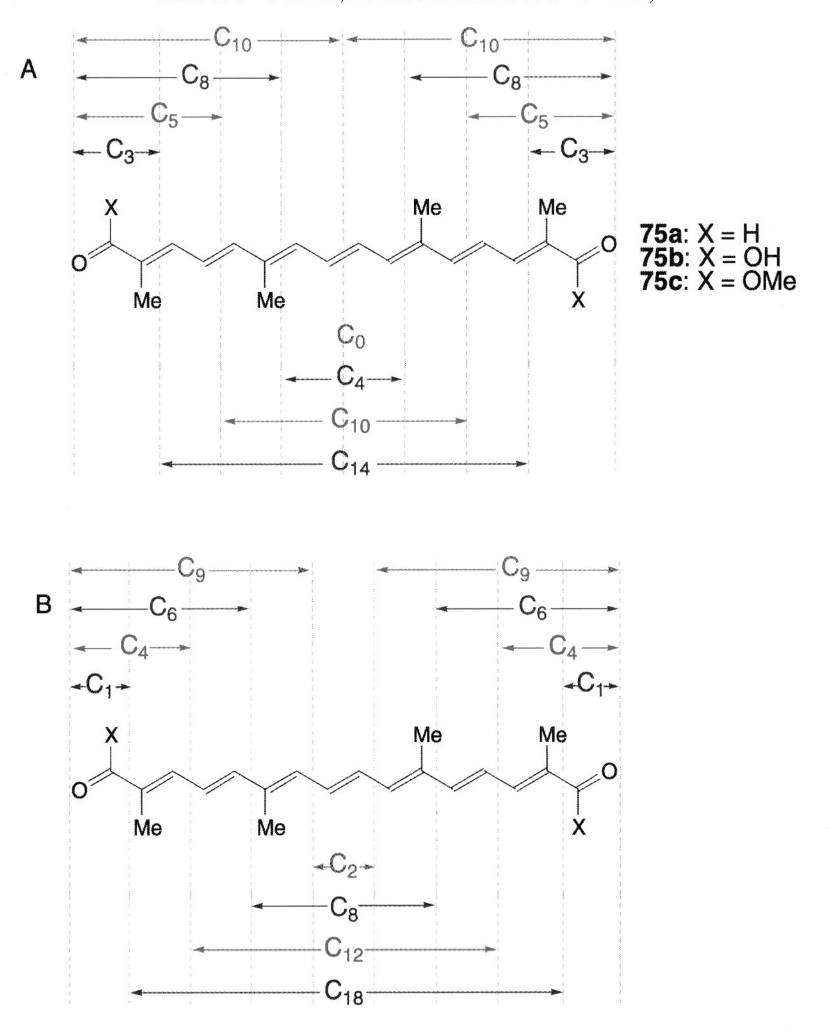

82a, **83a**, *rac*-**84a**, *rac*-**85a**, *rac*-**86a**, and **87a**, and, after exhaustive dehydrogenation of the mixture, into **75c**.

The three-step synthesis of **75c** from **76a** and **77a** is characterized by the lack of any need for onerous protecting-group manipulation. From the starting phenol comes a dienone, from that a dieneketene, from that a doubly unsaturated methyl ester. The *C*,*C*-bis-alkylated components **80a** and *rac*-**81a**, formed in the introductory synthetic step, already contain the complete set of C-atoms needed for crocetin (**75b**), albeit still arranged differently constitutionally. In the second step, the photochemical ring opening takes place, af-

Scheme 34. *Synthesis of Dimethylcrocetin* (**75c**) *Following the Pattern* $C_8 + C_4 + C_8$ [88]

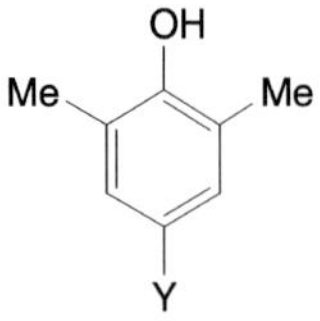

76a = **74** : Y = H
76b: Y = Br

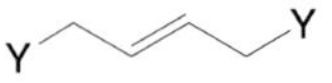

77a: Y = Br
77b: Y = Cl

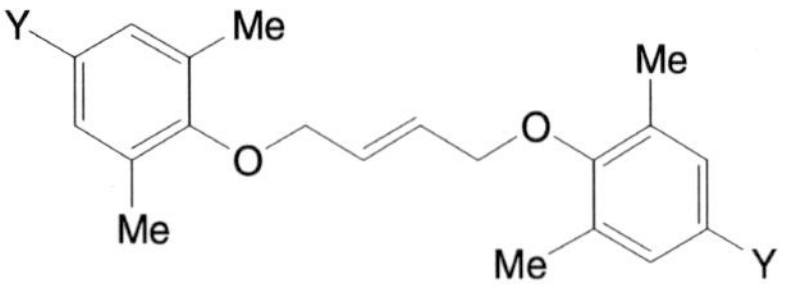

78a: Y = H
78b: Y = Br

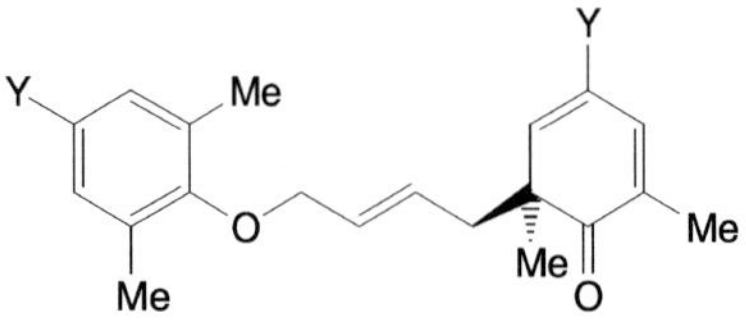

79a: Y = H
79b: Y = Br

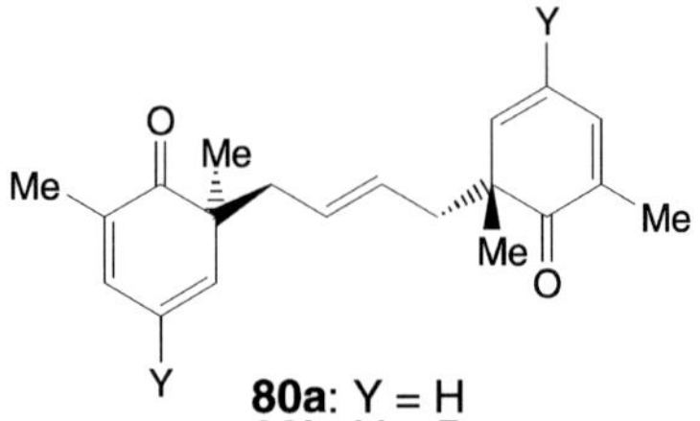

80a: Y = H
80b: Y = Br

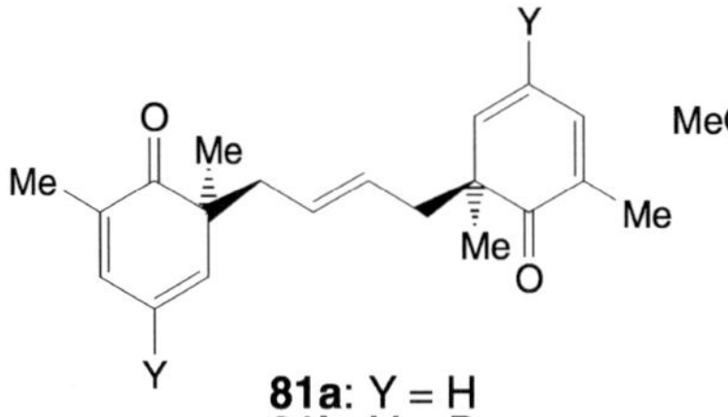

81a: Y = H
81b: Y = Br

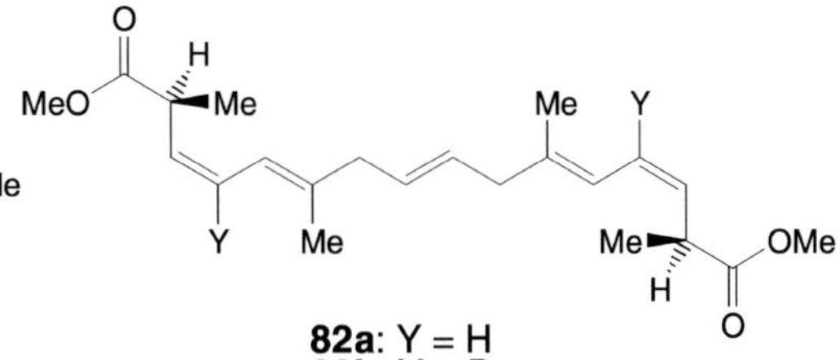

82a: Y = H
82b: Y = Br

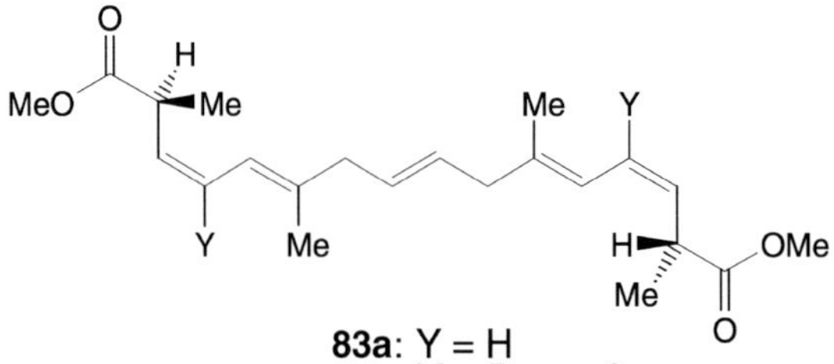

83a: Y = H
83b: Y = Br

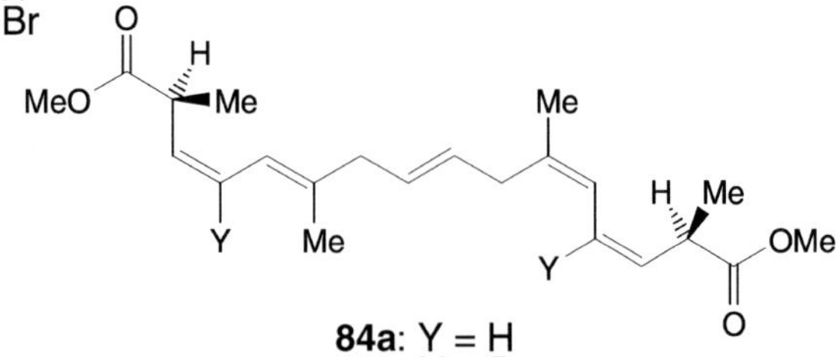

84a: Y = H
84b: Y = Br

Scheme 34 (cont.)

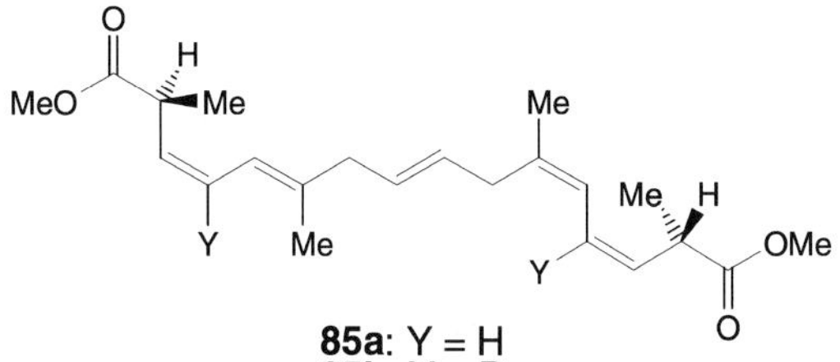

85a: Y = H
85b: Y = Br

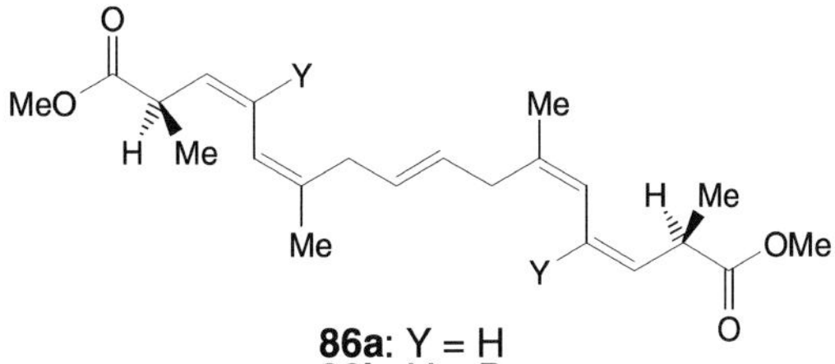

86a: Y = H
86b: Y = Br

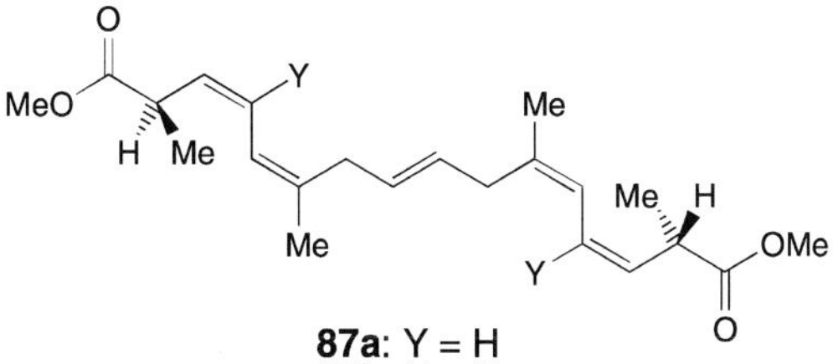

87a: Y = H
87b: Y = Br

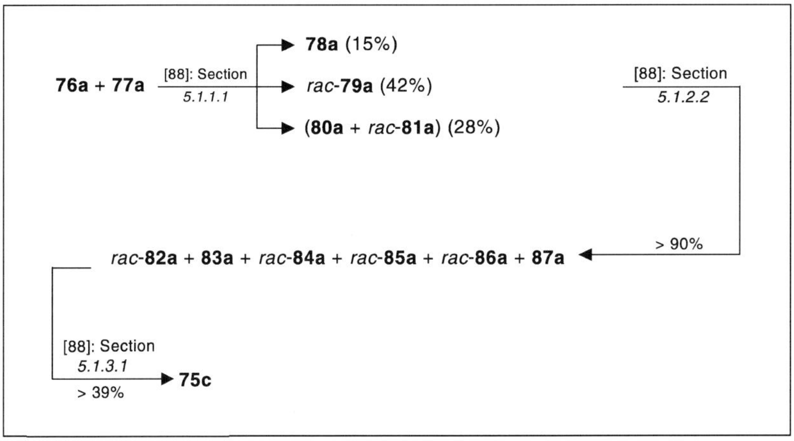

76a + **77a** —[88]: Section *5.1.1.1*→ **78a** (15%); *rac*-**79a** (42%); (**80a** + *rac*-**81a**) (28%)

—[88]: Section *5.1.2.2*, > 90%→ *rac*-**82a** + **83a** + *rac*-**84a** + *rac*-**85a** + *rac*-**86a** + **87a**

—[88]: Section *5.1.3.1*, > 39%→ **75c**

fording the seco-isomeric bis(dieneketenes), which add MeOH to furnish the mixture of *rac*-**82a**, **83a**, *rac*-**84a**, *rac*-**85a**, *rac*-**86a**, and **87a**. These stereoisomers each have the complete chain of 16 C-atoms, with the four Me groups in their correct positions, and with the terminal methyl-ester groups. In the final step, the two missing C=C bonds are introduced. From previous experimentation [89] with stereoisomers of **75c**, it may be assumed that, under the dehydrogenation conditions, the desired target compound with (all-*E*)-configuration, as the thermodynamically most stable stereoisomer, will be produced either predominantly or even exclusively.

The overall amount of separation work necessary is small. At the bis-alkylation product stage, **80a** and *rac*-**81a** may easily be separated from **78a** and *rac*-**79a**. To separate the two *C,C*-bis-alkylation components **80a** and *rac*-**81a**, formed in equal proportions, from one another would not be necessary for the conduct of the synthesis, but was undertaken once, for assignment of configurations. To establish experimentally which of the two fractions was the racemic mixture, the enantioselective photorearrangement of *rac*-**81a** into **81a**/*ent*-**81a** $\neq$ 1 and of (*rac*-**81b** into **81b**/*ent*-**81b** $\neq$ 1) was performed with the aid of circularly polarized light[43]). In the photochemical ring opening of *rac*-**81**, the two enantiomers **81** and *ent*-**81** react at different rates, thus bringing about a kinetic racemate resolution (see *Scheme 35*).

As dehydrobrominations usually can be performed with higher chemical yields than dehydrogenations, an improvement is to be expected if the photochemical synthesis of **75c** begins with **76b** instead of **76a**. As a matter of fact, by this synthetic variant, the overall yield, the conversion of **76a** to **76b** included, amounts to 17%, at least [88]. In passing, it should be mentioned that the presence of a Br-atom does nor prevent the photochemical ring opening of the cyclohexadienone moiety to occur. Intersystem crossing, caused by heavy-atom effect, obviously does not take place. The quantum yield for educt molecules (**80b** + *rac*-**81b**) consumed is independant of the exciting light (313- or 365-nm light; see Sect. 5.2.2.1.2 in [88]) and practically the same ($\Phi \approx 0.55$) as in the case of 6,6-dimethylcyclohexa-2,4-dien-1-one [72c].

2.3.2. *Total Synthesis of (–)-Variotin* [79][91a]

The synthesis of the fungicidal antibiotic (–)-variotin (**88**), also known by the name of (–)-pecilocin, was underpinned by efforts to study synthetic building blocks accessible photochemically from products of abnormal *Reimer-Tiemann* reaction in the context of their utility in natural-product total

[43]) For photochemistry with circular polarized light, see [90].

Scheme 35. *Guide Providing Orientation in a Complex System of Light-Induced Isomerizations Hereby Converting* rac-**81a** *to* **81a**/ent-**81a** ≠ 1 (for details, see [88]: Sect. 5.1.2.3)

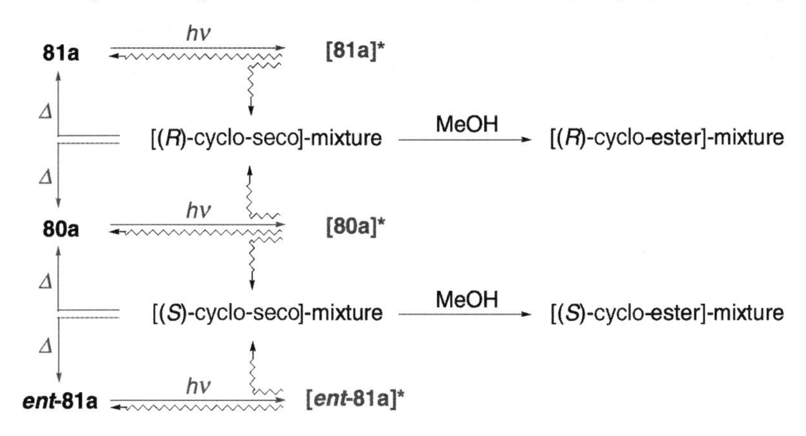

[(S)-cyclo-seco]-mixture [(R)-cyclo-seco]-mixture

synthesis. It is not necessary to delve for very long into the constitutional formula of **88** to arrive at the disconnection of the bonds between N and C(1), and between C(7) and C(8), giving pentanal (**89**), and the synthons **X** and **Y** (*Schemes 36* and *37*).

The need to translate the hypothetical synthon **Y** into the actually existing chemical compounds **92a** and **93a** (as progenitors of vinyl carbanion equivalents) results in a reaction sequence beginning with the photochemical ring opening of the abnormal *Reimer-Tiemann* compound *rac*-**90**, giving the dieneketene seco-isomer of constitution **91**. Nucleophilic addition of 2-pyrrolidone (**X–H**) onto **91** might produce (*via* **Z**) a mixture of **92a** and **93a**, and this might be converted to metallo-organic compounds of types **92b** and **93b**, matching the reactivity profile of the synthon.

Scheme 36. *Synthesis of (–)-Variotin* (**88**) [91a]: *the Molecules*

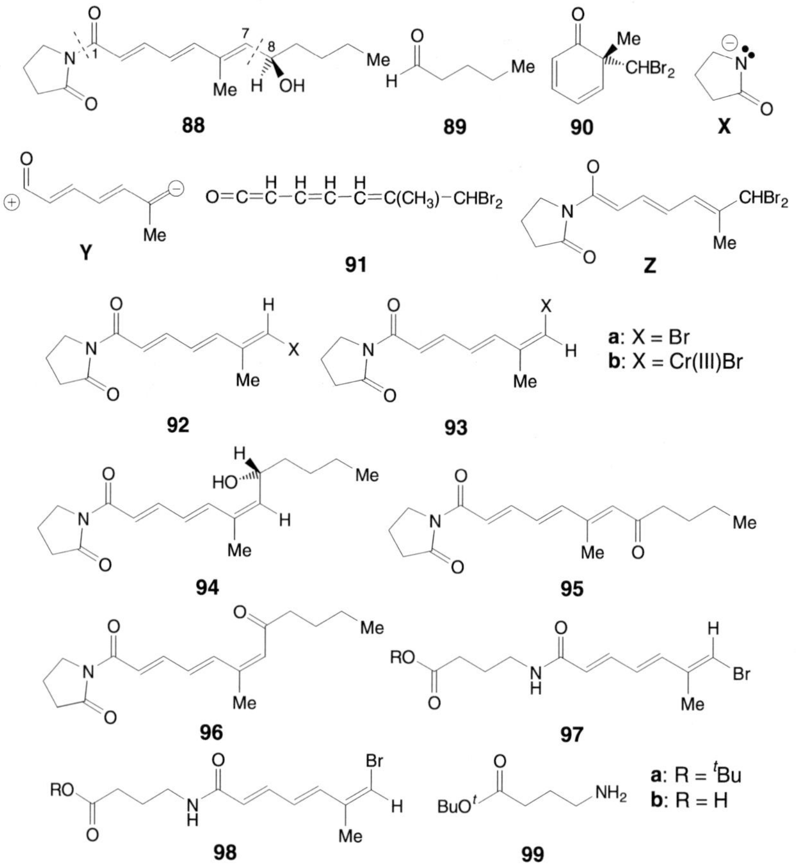

The alkenylchromium(III) compounds **92b** and **93b**, obtained under Ni catalysis conditions according to the methodology of *Takai* and co-workers [93a], participate in pentanal addition to afford the easily separable coupling components (±)-variotin (*rac*-**88**) and (±)-isovariotin (*rac*-**94**). The chirogenic[44]) coupling reaction took place with an insufficient degree of enantioselectivity[45]), and so an alternate route *via* didehydrovariotin (**95**) and didehydroisovariotin (**96**) was adopted. Oxazaborolidine-catalyzed borane reduction [93b] was used to convert **95** and **96** enantioselectively into **88**[46]) or **94**, re-

---

[44]) *'A chirogenic reaction step'*, as defined by *A. Eschenmoser* [92] *'is one in which a chiral product is formed from an achiral starting material, any necessary reagent being also achiral'*.

[45]) See [91a] on catalyses that resulted in only moderate enantiomeric excesses.

[46]) The spectroscopic properties are identical with those of *rac*-**88** reported in the literature [94a–c]. The optical rotation ($[\alpha]_D = -20.4$ in EtOH), deviates considerably from that reported for the natural product ($[\alpha]_D = -5.68$ in EtOH) [94d].

Scheme 37. *Synthesis of (–)-Variotin* (**88**) [91a]: *the Conditions*

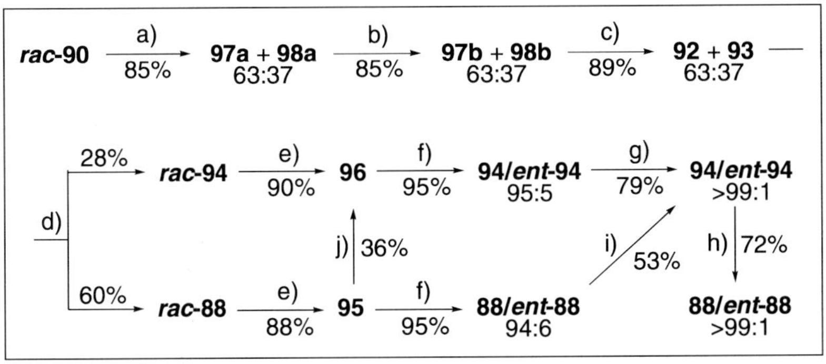

***rac*-90** —a) 85%→ **97a + 98a** 63:37 —b) 85%→ **97b + 98b** 63:37 —c) 89%→ **92 + 93** 63:37 —d)→

28% → ***rac*-94** —e) 90%→ **96** —f) 95%→ **94/*ent*-94** 95:5 —g) 79%→ **94/*ent*-94** >99:1

60% → ***rac*-88** —e) 88%→ **95** —f) 95%→ **88/*ent*-88** 94:6

95 —j) 36%→ **96**; **88/*ent*-88** 94:6 —i) 53%→ **94/*ent*-94** >99:1; **94/*ent*-94** >99:1 —h) 72%→ **88/*ent*-88** >99:1

a) 10 equiv. DABCO, 3 equiv. *tert*-butyl 4-aminobutanoate (**99**) $h\nu > 340$ nm, 4 h. b) 10 equiv. TFA, r.t., 16 h. c) 1. 2,6-Dichlorobenzoyl chlorid, Et_3N, THF, 0°, 1 h, 2. DMAP, benzene, 70°, 1 h. d) 1. Pentanal, CrC_2, 1% $NiCl_2$, DMSO, r.t., 2 h; 2. HPLC. e) DDQ, $CHCl_3$, 40°, 0.5 h. f) CBS Reduction, toluene, –78°, 35 h. g) Crystallization ($AcOEt/Et_2O$). h) $h\nu > 380$ nm, Thioxanthen-1-one, CH_2Cl_2/Et_2O, 5 h, 3 cycles of radiation and chromatography. i) $h\nu > 340$ nm 16 h. j) 1. $h\nu > 340$ nm, 0.8 h, 2 cycles of radiation and chromatography, 2. crystallization.

spectively. The mixture of bromides **92a/93a** was obtained in a three-step sequence, by irradiation of *rac*-**90** in the presence of **99**, *via* the mixtures (**97a/98a**) and **97b/98b**[47]).

The conversion of *rac*-**90** to carboxylic-acid derivatives of constitutional type **AB** (*Scheme 38*), *via* a sequence involving ketenes of constitutional type **91** and intermediates of constitutional type **AA**, is reminiscent of a fragmentation designed by *Eschenmoser* and *Frey* [95]. However, in the absence of light, we were unable to perform this in the case at hand.

Scheme 38. *Stepwise Conversion of* rac-**90** *to Unsaturated Carboxylic-Acid Derivatives of Constitutional Type* **AB**

rac*-90** —$h\nu$→ **[*rac*-90] ⇝ O=C=CH–CH=CH–CH=C(CH₃)–CHBr₂ (**91**)

91 —NuH→ ⁻O–C(=CH–CH=CH–CH=C(CH₃)–CHBr₂)–Nu⁺–H (**AA**)

AA —HBr→ O=C(Nu)–CH=CH–CH=CH–C(CH₃)=CHBr (**AB**)

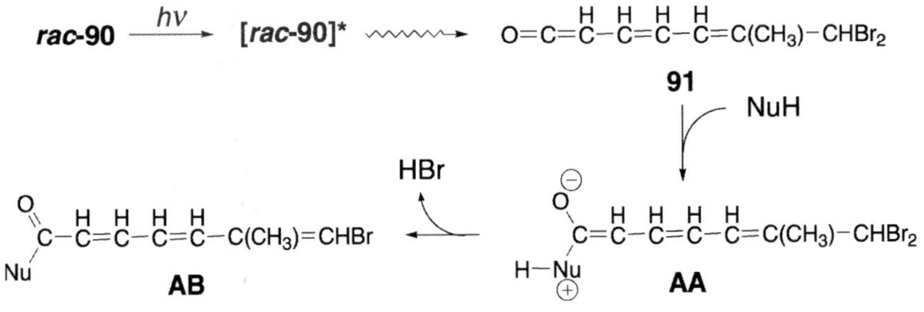

[47]) This detour was made since previous irradiation experiments with *rac*-**49** in Et_2O, containing 2-pyrrolidone, had furnished the desired pyrrolidone derivative only in poor yield; see [91b].

Syntheses of *rac*-**88** have been reported in the literature [94]. The synthesis of **88** (*Schemes 36* and *37*), to the best of our knowledge, is the first one completed.

2.3.3. *Total Syntheses of Aspicillin*

X-Ray crystal-structure analysis [97a] of the 18-membered lichen macrolide (+)-aspicillin (**100**, *Fig. 5*) was able to confirm the proposed constitution [96] and to determine the relative configuration [96a] [97a]. The absolute configuration was established by (cycloocta-1,5-diene)copper(I)-chloride-catalyzed treatment of (*S*)-2-methyloxirane[48]) with the *Grignard* compound obtained from the mixture of THP-ethers of 10-bromodecan-1-ol to give the C_{13}-diol degradation product of **100** [97b], after which a broadly structured synthetic project began in Frankfurt [97] [98]. Its goal was to test the synthetic usefulness of the photolactonization, selecting a suitable macrolide as a test case. Syntheses of (+)-aspicillin (**100**) [99] (and of (–)-aspicillin (*ent*-

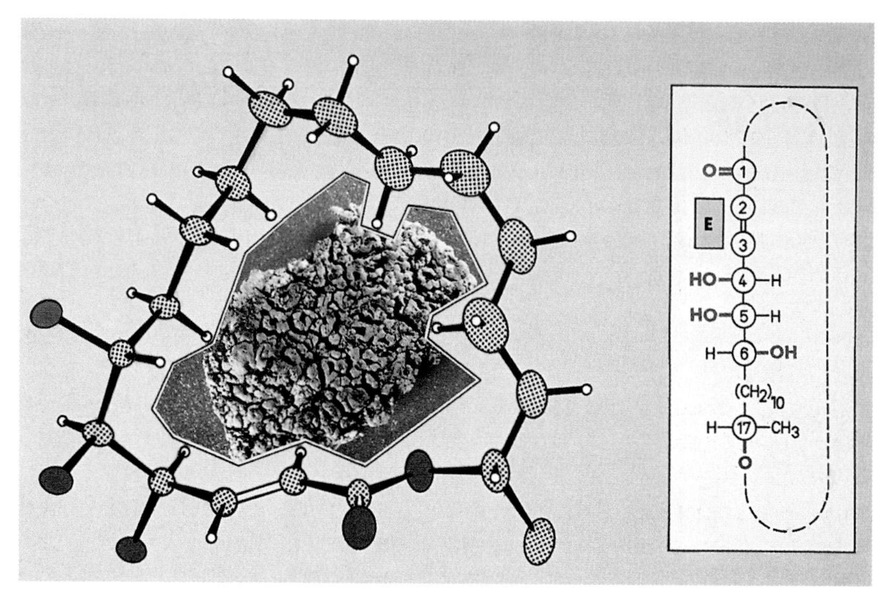

Fig. 5. *Cutout of a photographed sample of the crustose lichen* (*Aspicilia caesiocinerea*) *from the southern part of the Black Forest with conformation observed in the crystalline state and* Fischer *projection of* (+)-*aspicillin* (**100**)

[48]) (*S*)- and (*R*)-2-methyloxiranes have been used in several total syntheses of **100** and *ent*-**100**, functioning as building blocks incorporating the future C(17).

100) [100]) were soon also being reported from other laboratories, and so today it is possible to undertake a comparative examination of different synthetic strategies. The accent in the following discussion lies on the relevant synthetic strategy. Tactical measures and protecting-group manipulation (more tiresome than anything) that aided a given strategy to success, and detailed knowledge of which clearly serves to add color to a sober report, should be taken from the relevant original publications.

From the strategic point of view, macrolide syntheses may be divided into two classes. In syntheses of one class, ring closure takes place fairly early on in the course of synthesis. Functional groups and stereogenic centers may then selectively be introduced onto the cyclic skeleton (*ring strategy*). It is known that the restriction in conformational space associated with the transition from an open- to a closed-chain system is probably the most important topological means of promoting selectivity[49]). The long-standing objection that structural alterations on a lactone skeleton are inadvisable because of the ease of ring opening proved on closer scrutiny to be mere prejudice after all. The magnitudes-greater reactivity, at least towards nucleophilic reagents, of a lactone group compared to an ester does indeed apply for lactones of conventional ring size and *synperiplanar* arrangement of the lactone group, but is, however, invalid for lactones with a greater number of ring members and *antiperiplanar* arrangement of the lactone group – as *Rolf Huisgen* had shown as early as 1959 [102]. These latter lactones consistently behave like acyclic carboxylates under hydrolytic conditions.

In the second class of macrolide syntheses, functional groups and stereogenic centers are introduced into the acyclic precursors (*open-chain strategy*). Not until the end – or shortly before – does ring closure take place. The objection brought against this, that '*the construction, in a desired stereochemical sense, of asymmetric arrays in flexible, open-chain systems, is a relatively little-known art, in which stereoselectivity is rare, or little understood when observed, and generalizations are dangerous*' [101], wholly accurately reflected the state of synthetic capabilities at the time, but fortunately no longer applies today[50]).

Before comparing several syntheses of (+)-aspicillin (**100**) and its non-natural enantiomer *ent*-**100** with each other, we should say a few words about characterization of total syntheses aimed at one and the same target compound. In our particular case, in the synthesis of **100** (or *ent*-**100**) by a route that can be specified by the building blocks used and by its characteristic intermediates (green field of *Scheme 39*), one wants to know the constitution-

[49]) '*The creation of new asymmetric centers within rigid systems is best exemplified by cyclic – and especially fused polycyclic – systems*' [101].

[50]) The progress in aldol methodology [103] developed in numerous laboratories provides an eloquent testimony of this.

Scheme 39. *A System for Characterization of Total Syntheses of* **100** (*ent*-**100**) *with Blanks to Be Filled to Indicate* a) *Constitutional Design* (sequence of building-block addition); b) *Configurational Matrix* (to indicate in which order and by what type of stereoselection the stereogenic centers are introduced; for details, see text); c) *Numerical Overall Data*; d) *Quick Overview* (of building blocks and characteristic intermediates).

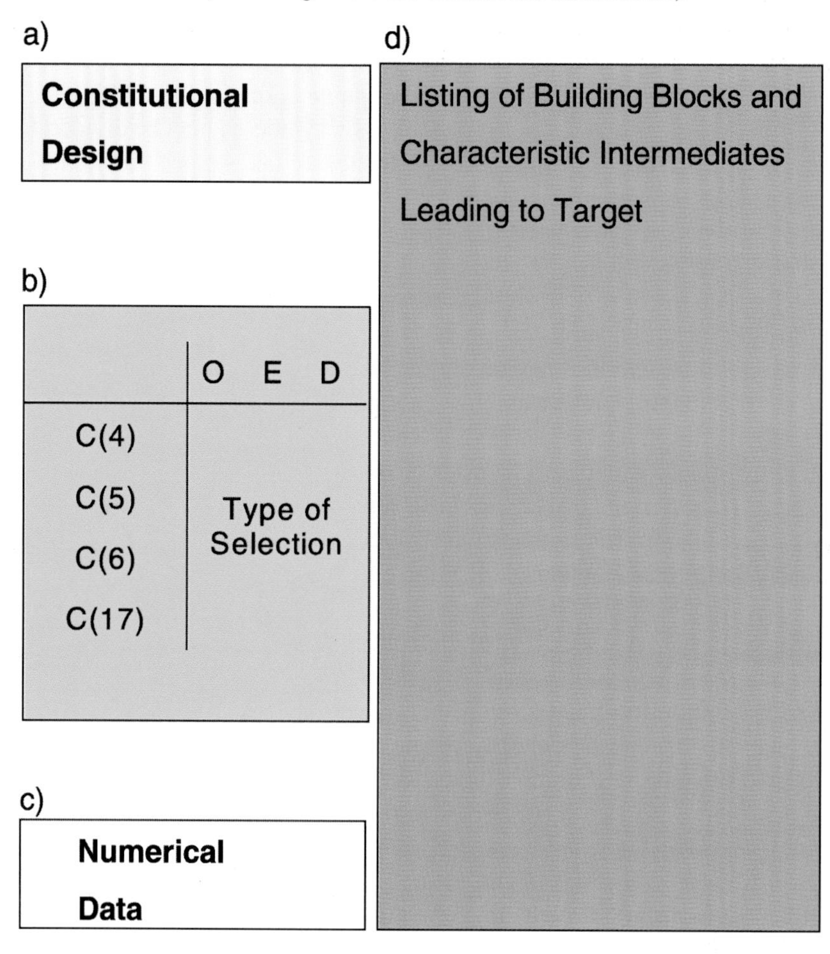

al pattern, *i.e.*, which building blocks (yellow fields) have been used. Numerical data (number of steps and overall yield, in the white field) may give some hint as to the practicability of the particular synthesis[51]). Another important piece of information concerns the various stereogenic centers (at C(4), C(5), C(6), and C(17)) in the target structure. The way and order in which these centers have been introduced describe the configurational design and are con-

[51]) In those syntheses where a compound of arbitrary availability was chosen as reference compound for calculating the number of synthetic steps and overall yield, these data have been left out.

veniently expressed by the orange matrix. Such a matrix denotes the type of selection that has been at work: *diastereoselection* (D), *enantioselection* (E), or *no stereoselection* (O). No stereoselection here means that one or several stereogenic centers have been introduced by means of chiral, nonracemic building blocks (*e.g.*, available from the chiral C pool). In this case evolution, and not the synthetic chemist, has done the selection.

2.3.3.1. *Activities in Frankfurt*

2.3.3.1.1. *Synthetic Variants with Photolactonization as the Key Reaction*

Three mutually contrasting phases of an overall synthesis are evident in the synthetic variants under discussion here. They agree in their *middle game*, but differ in their *openings* and *end games* (*Scheme 40*).

Scheme 40. *Combinatorial Set of the Three Phases of the Photochemical Synthesis of* **100**. Figures in the white fields indicate overall yields or, in parentheses, overall number of steps.

Opening	Middle Game	End Game
(S)-Methyl-oxirane **61.0%** Steps: 9	OH (CH$_2$)$_{10}$ OH H Me → **33.6%** Steps: 6 → HO H O H Me O	Osmylation **40.0%** Steps: 1
Alkylation $ZnMe_2$ **53.5%** Steps: 7		Epoxidation/ Epoxide-cleavage **73.5%** Steps: 2
Yeast-reduction **39.8%** Steps: 11		

Combinatorial Set
8% (16)
15% (17)
7% (14)
13% (15)
5% (18)
10% (19)

2.3.3.1.1.1. *The Constant Middle Game*

The middle game involves the conversion of a hydroxyalkyl-substituted phenol (*Scheme 40*) to diastereoisomeric C_{18}-cyclohexadienones (*Scheme 41*) by *Wessely* acetoxylation. These are subjected to UV irradiation to give

Scheme 41. *Characterization of the Photochemical Synthesis of* **100** (composed of the yellow opening, the yellow middle game, and the blue end game of *Scheme 40* [97a][97b])

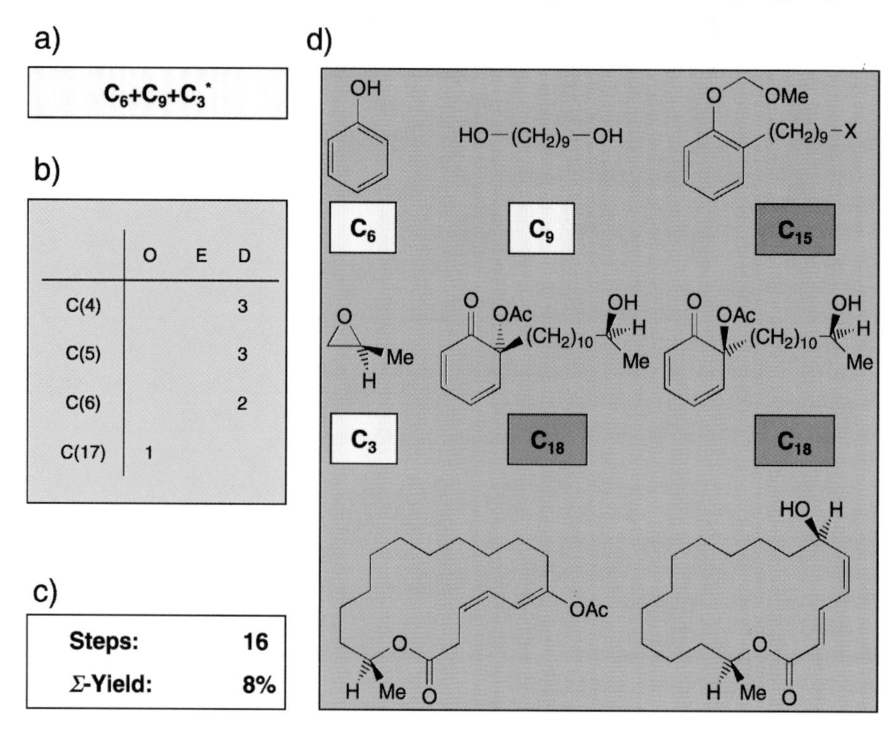

the corresponding photolactone, diastereoselective reduction of which by the *Yamamoto* reagent [104] provides the hydroxydiene macrolide shown.

2.3.3.1.1.2. *Variations in the Opening*

For the preparation of the hydroxyalkyl-substituted C_{18} phenol with (*S*)-configuration at C(17) (*Scheme 40*), one variant involving no stereoselection and two using enantioselection have been developed.

2.3.3.1.1.2.1. *Without Stereoselection*

Ring opening of (*S*)-2-methyloxirane (C_3^* building block in *Scheme 41*), accessible from (*S*)-lactic acid [105]) by the *Grignard* compound of the C_{15} phenol derivative leads to the hydroxyalkyl-substituted phenol of *Scheme 40*.

2.3.3.1.1.2.2. *Enantioselection by TADDOLate-Mediated Methylation*

The C_{17} aldehyde of *Scheme 42*, accessible by reaction of the C_{11} building block with the C_6 unit, is transformed into the (*S*)-configured hydroxyalkyl-substituted phenol by treatment with the C_1 unit in the presence of the chiral, nonracemic auxiliary **AX**[52]).

Scheme 42. *Characterization of the Photochemical Synthesis of* **100** (composed of the orange opening [97e] , the yellow middle game, and the blue end game of *Scheme 40* [97c][97d])

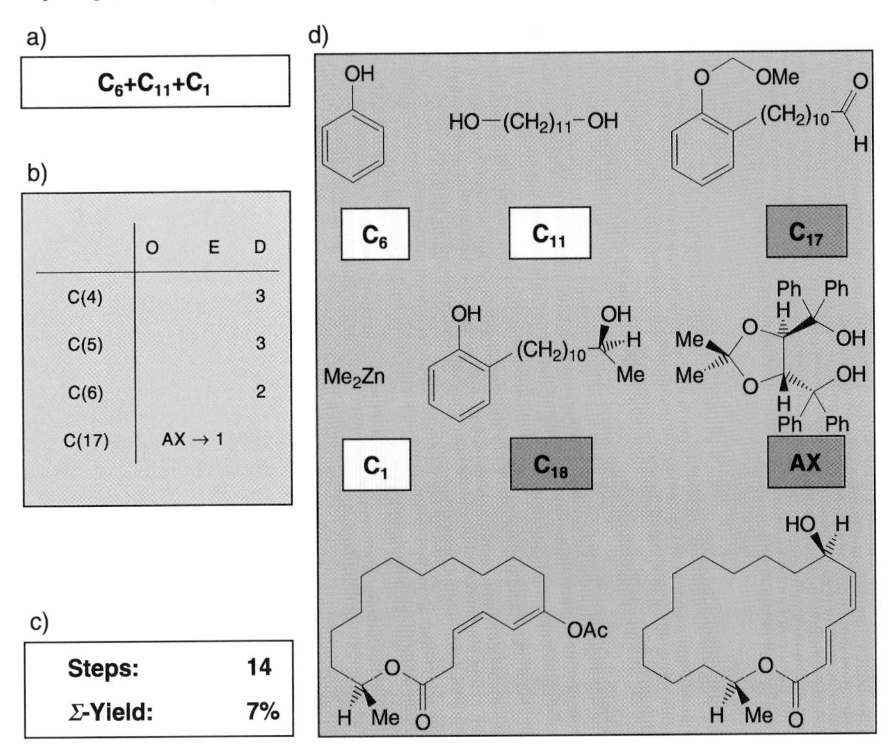

2.3.3.1.1.2.3. *Enantioselective Reduction by Yeast*

The C_{14} intermediate (X = Br) resulting from treatment of the C_6 building block with the C_8 unit (*Scheme 43*) reacts with the C_4 unit, obtained by reduction of ethyl acetoacetate by yeast (**Y**) to give the C_{18} intermediate. This is then transformed into the hydroxyalkyl-substituted phenol of *Scheme 40*.

52) For TADDOLate-mediated enantioselective alkylation of aldehydes, see Scheme 9 in [106].

Scheme 43. *Characterization of the Photochemical Synthesis of* **100** (composed of the green opening [97c], the yellow middle game, and the blue end game of *Scheme 40* [97b])

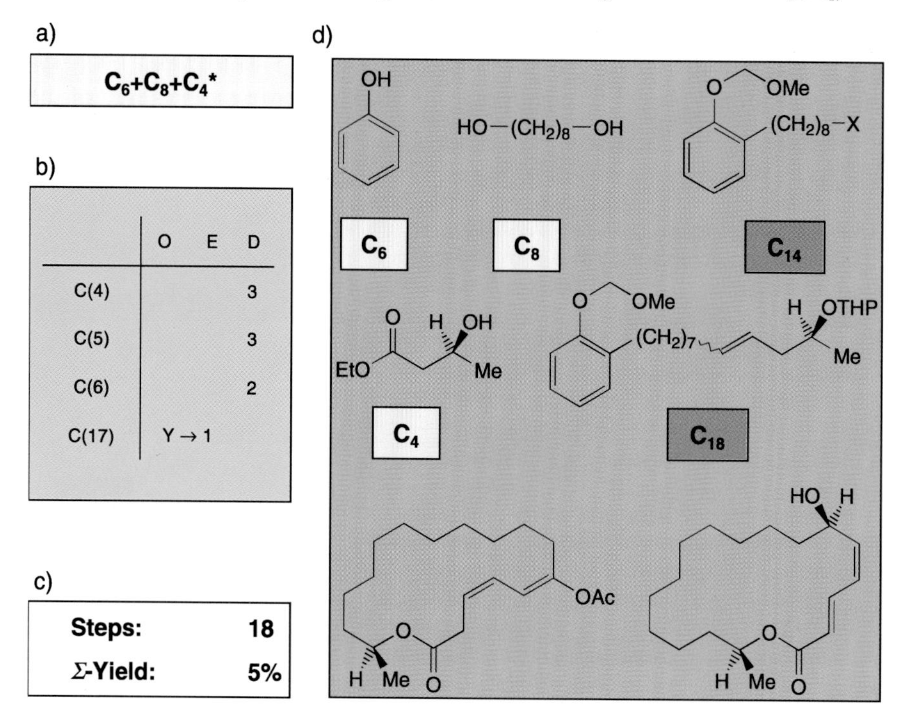

2.3.3.1.1.3. *Variations in the End Game*

The previously outlined synthetic variants of **100** have only differed from one another in their openings. The cases still outstanding differ in their respective end game. In place of a one-step dihydroxylation (by OsO_4), a two-step reaction sequence (an epoxidation and opening of the epoxide ring to a 1,2-*cis*-diol) may occur. Although the number of reaction steps grows by one in each case, the overall yield of **100** in the $C_6 + C_9 + C_3^*$ synthetic variant (*Scheme 41*) increases from 8 to 15% (*Scheme 44*).

The overall yield for the $C_6 + C_{11} + C_1$ synthetic variant increases from 7 (*Scheme 42*) to 13% (*Scheme 45*).

And finally, the overall yield for the $C_6 + C_8 + C_4^*$ variant goes up from 5 (*Scheme 43*) to 10% (*Scheme 46*).

Scheme 44. *Characterization of the Photochemical Synthesis of* **100** (composed of the yellow opening [97e], the yellow middle game, and the yellow end game of *Scheme 40* [97b][97d])

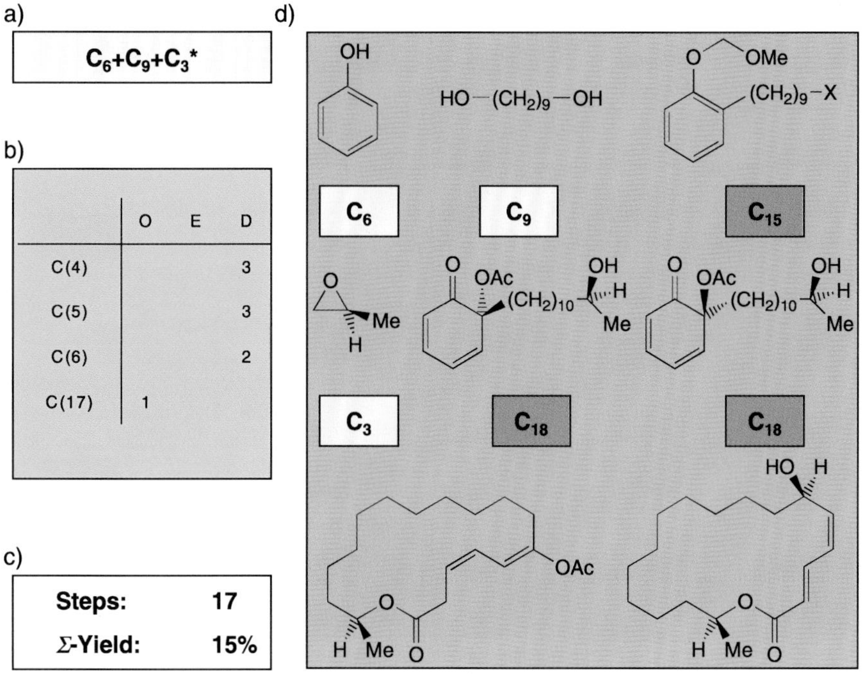

a)

C_6+C_9+C_3*

b)

	O	E	D
C(4)			3
C(5)			3
C(6)			2
C(17)	1		

c)

Steps:	**17**
Σ-Yield:	**15%**

d)

2.3.3.1.2. *Synthetic Variant with Two Chiral, Nonracemic Building Blocks from the Chiral C Pool*

Before we can declare the photolactonization-based synthesis of lichen macrolide **100** a success, we need a reference case. One synthetic reference leading to **100**, in which the complete set of stereogenic centers was introduced with no stereoselection, *i.e.*, using chiral, nonracemic building blocks from the chiral C pool [107] (in this case: D-mannose and D-lactic acid), was designed and carried out in Frankfurt, too (*Scheme 47*).

With its $C_3^* + C_7 + C_6^* + C_2$ synthetic plan (yellow field), it follows the *open-chain strategy* (green field) and may be characterized by the configuration matrix (orange field), numerical data for overall yield (13%), and total number of steps in the synthesis (white field). The relatively high number of steps can be traced back to the required protecting group manipulation; this is encountered as a rule with the use of sugar building blocks. Measured against the reference synthesis, the photochemical synthesis of **100**, with its opening with no stereoselection, with its end game *via* an intermediate epoxide, and with little use of protecting-group manipulation (*Scheme 44*; overall yield 15%) can be appreciated.

Scheme 45. *Characterization of the Photochemical Synthesis of* **100** (composed of the orange opening [97e], the yellow middle game [97b], and the yellow end game [97d] of *Scheme 40;* compare with *Scheme 42)*

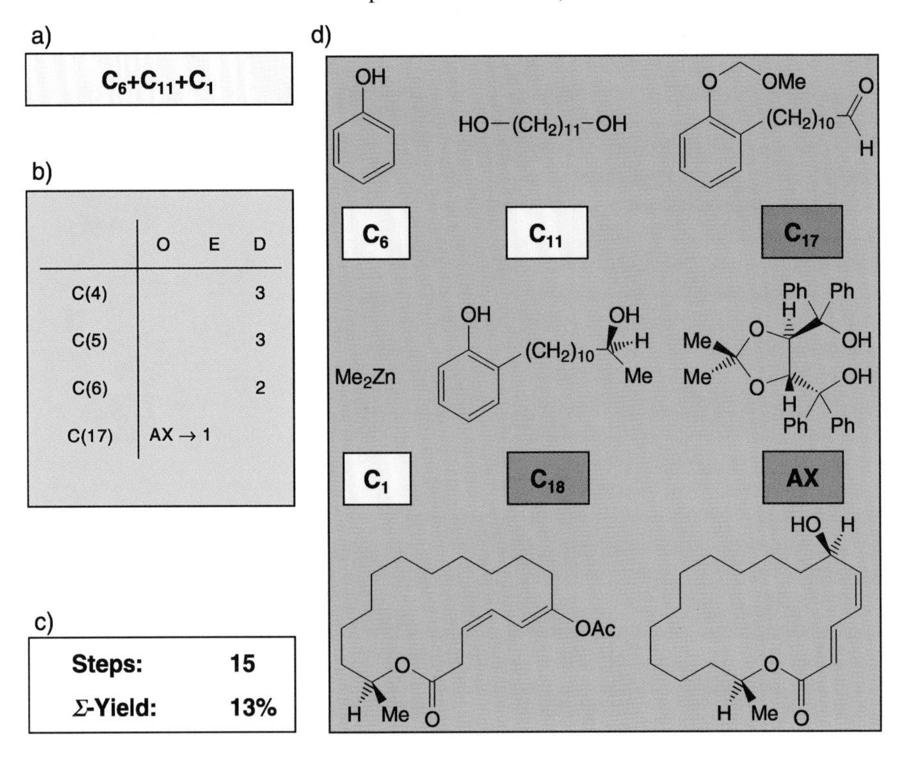

2.3.3.2. *Activities outside Frankfurt*

2.3.3.2.1. *Application of the Open-Chain Strategy*

2.3.3.2.1.1. *Ring Closure by Lactonization*

2.3.3.2.1.1.1. *Synthesis of* **100**: *with a* $C_5^{\ddagger}$ *Building Block from the Chiral C Pool and an Enantioselective Enzymatic Reduction of a* C_8 *Building Block*

S. C. Sinha and *E. Keinan* have put forward a $(C_5^{\ddagger} + C_2 + C_1) + C_8^{\ddagger} + C_2$ synthesis of **100**, in which the three adjacent stereogenic centers were taken from D-arabinose ($C_5^{\ddagger}$ building block) and the fourth introduced into the C_8 building block by enzymatic reduction of a preceding carbonyl group (*Scheme 48*).

Once more, the large number of steps of the synthesis can be attributed to the use of sugar building blocks and the protecting-group manipulation that this demands.

Scheme 46. *Characterization of the Photochemical Synthesis of* **100** (composed of the green opening [97c], the yellow middle game [97b], and the yellow end game [97d] of *Scheme 40*)

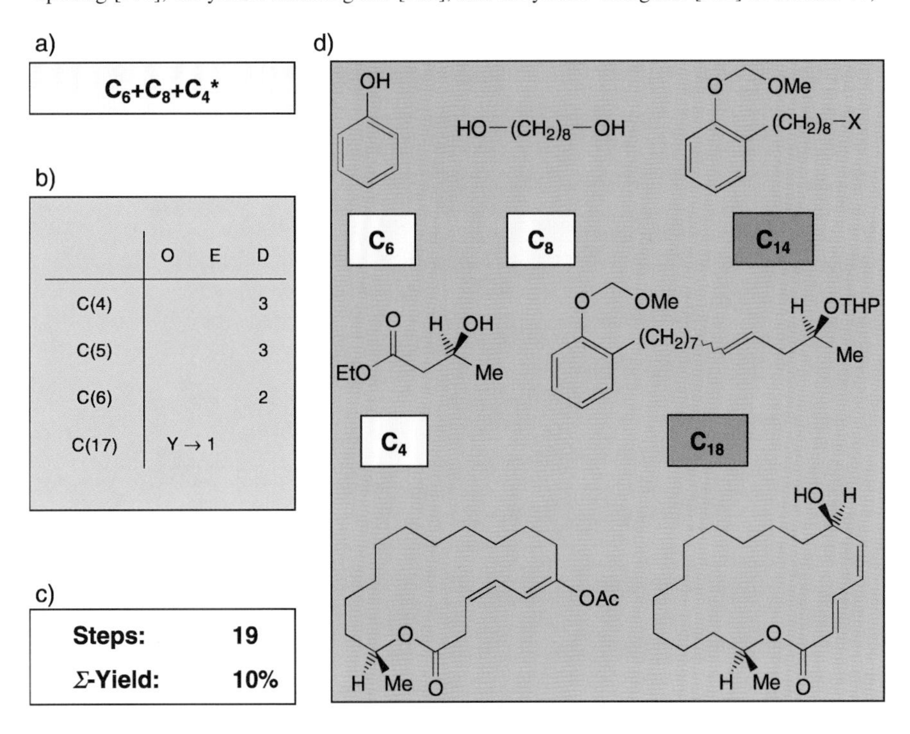

2.3.3.2.1.1.2. *Synthesis of* **100**: *Introduction of All of the Stereogenic Centers into a Naked C Skeleton by* Sharpless *Asymmetric Dihydroxylations in Serial Order*

Sinha and *Keinan* have reported a second synthesis of **100**, which makes use not of building blocks, but of auxiliaries from the chiral C pool (**AX** as well as *ent*-**AX**). The synthesis begins with a reaction sequence that follows the $C_{13} + C_2 + C_1$ pattern to convert the C_{13} aldehyde into the C_{16} hydrocarbon. This is then transformed by a threefold *Sharpless* asymmetric dihydroxylation (and partial acetalization) into the highly functionalized C_{16} intermediate. The latter compound, after another chain elongation by means of the same C_2 building block as used earlier, furnishes the C_{18} intermediate, and this then provides **100** by a three-step route (*Scheme 49*).

Scheme 47. *Characterization of a Synthesis of* **100**, *in which Two Building Blocks from the Chiral C Pool Provide All the Required Stereogenic Centers* [98]

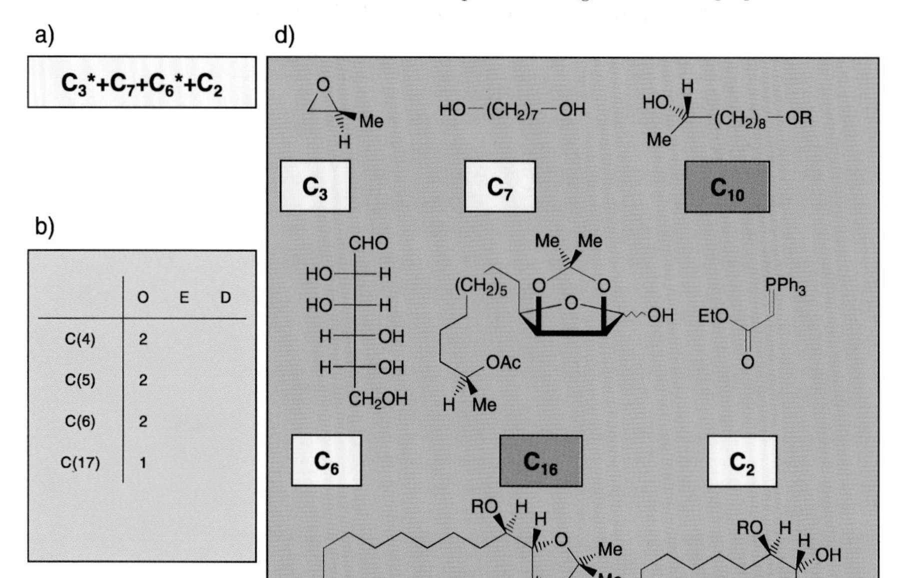

2.3.3.2.1.1.3. *Synthesis of* **100**, *in Which an Alkenyl-Substituted Furan Intermediate Plays a Pivotal Role*

The characteristic features of the *Kobayashi* approach (*Scheme 50*) are: the construction of the C_{18} alkenyl-furan by a $C_3^* + C_{11} + C_4$ sequence, the *Sharpless* asymmetric dihydroxylation of the olefinic C=C bond, and the oxidative transformation, efficiently and under mild conditions, of the substituted furan ring into a 4-oxobut-2-enoic-acid moiety, which is diastereoselectively reduced by $Zn(BH_4)_2$.

Scheme 48. *Characterization of a Synthesis of* **100**, *in Which a* C_5^* *Building Block from the Chiral C Pool Provides Three of the Four Required Stereogenic Centers* [99c], *While the Missing One Is Introduced into a* C_8 *Building Block by Enantioselective Reduction*

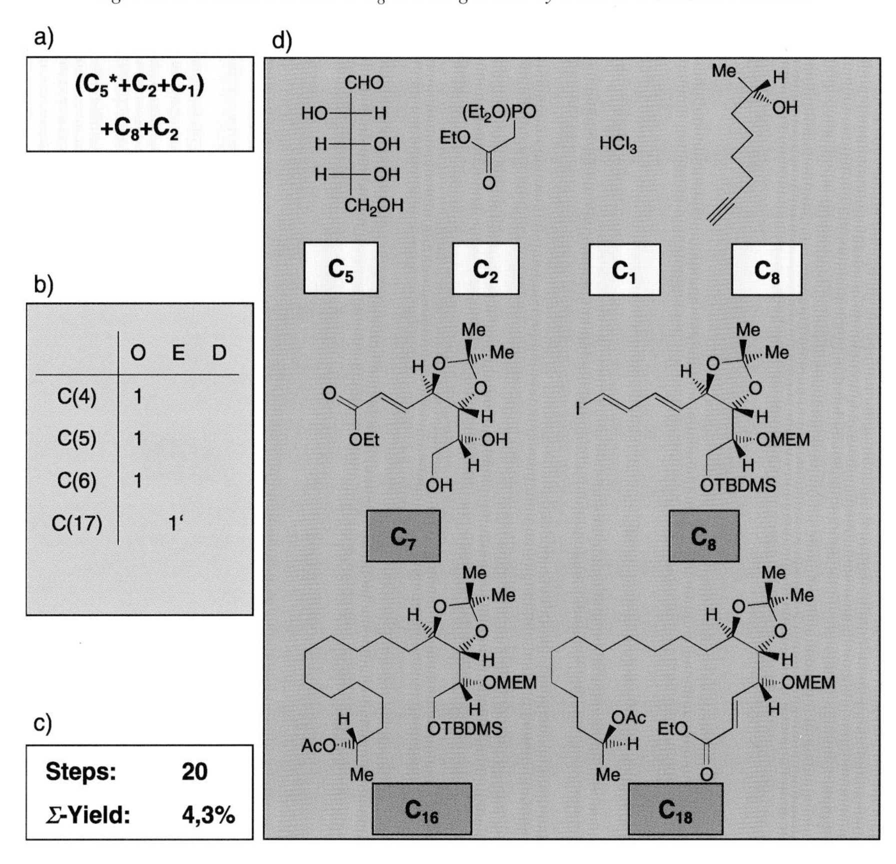

2.3.3.2.1.1.4. *Synthesis of* **100** *that Proves the Usefulness of* Enders'*s RAMP-SAMP Hydrazone Method* [108]

The SAMP-hydrazone (use of **AX**) of the C_2 building block propiophenone (*Scheme 51*) is alkylated with the C_{10} building block and further transformed by a six-step procedure into the protected C_{12} intermediate. The RAMP-hydrazone (use of *ent*-**AX**) of the derivative of the C_3 building block 2,2-dimethyl-1,3-dioxan-5-one is stereoselectively α,α'-dialkylated both with the C_1 building block and the C_{12} intermediate, with subsequent diastereoselective reduction of the carbonyl group by L-*Selectride*. *Horner* olefination, with the other C_2 building block, elongated the C chain. Here, again, the need for extensive protecting-group manipulation raises the total number of steps.

Scheme 49. *Characterization of a Synthesis of* **100**, *in Which an Intermediate* C_{16} *Hydrocarbon Is Provided with Functional Groups and All the Stereogenic Centers by Three* Sharpless *Asymmetric Dihydroxylations in Serial Order* [99c]

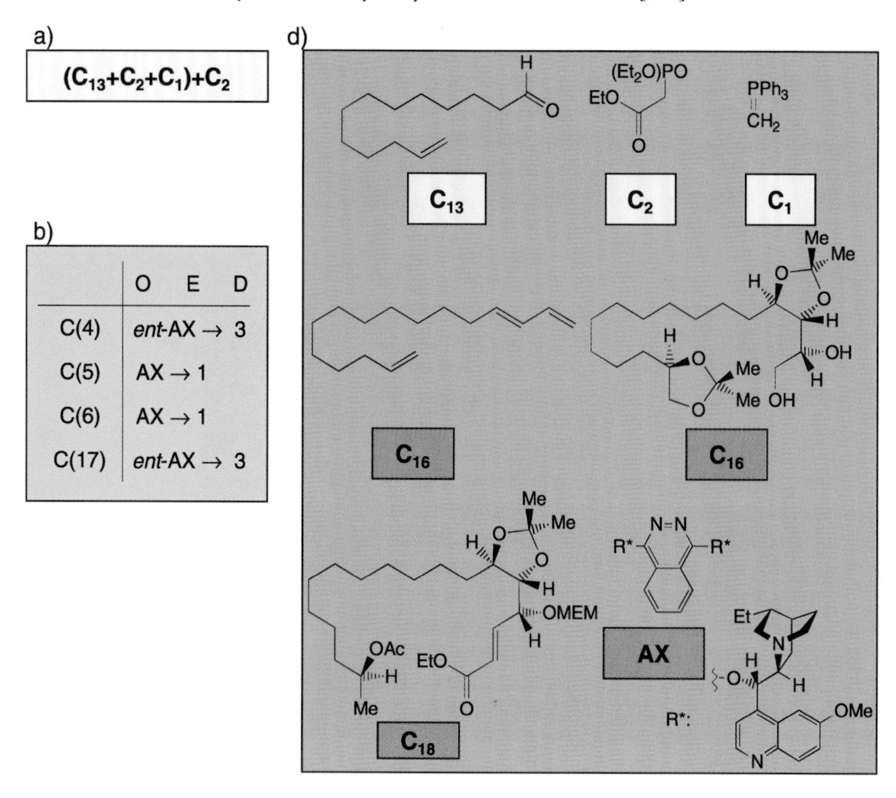

	O E D
C(4)	*ent*-AX → 3
C(5)	AX → 1
C(6)	AX → 1
C(17)	*ent*-AX → 3

2.3.3.2.1.1.5. *Synthesis of* ent-**100** *According to the Pattern* $[(C_8 + C_3^*) + (C_4^* + C_2)] + C_1$

Dimethyl (2*R*,3*R*)-tartrate provides the central C_4 building block. *Horner* olefination is used to attach the C_2 building block at one end. Asymmetric *Sharpless* epoxidation converts the resulting olefin to the C_6 epoxide. The C_{11} chain, produced from the C_8 and C_3 building blocks, is attached at the other end. The missing C_1 atom is provided by CH_2N_2; an *Arndt-Eistert* reaction leads, *via* the C_{18} diazo ketone, to the C_{18} hydroxycarboxylic acid. Cyclization of the seco-isomer gives *ent*-**100** (*Scheme 52*).

Scheme 50. *Characterization of a Synthesis of* **100**, *in Which the* C_3^* *Building Block Supplies the First Stereogenic Center.* The other stereogenic centers are diastereoselectively introduced by *Sharpless* asymmetric dihydroxylation of the alkenyl-substituted furan or by $Zn(BH_4)_2$ reduction of the C_{18} furan cleavage product, respectively [99d]

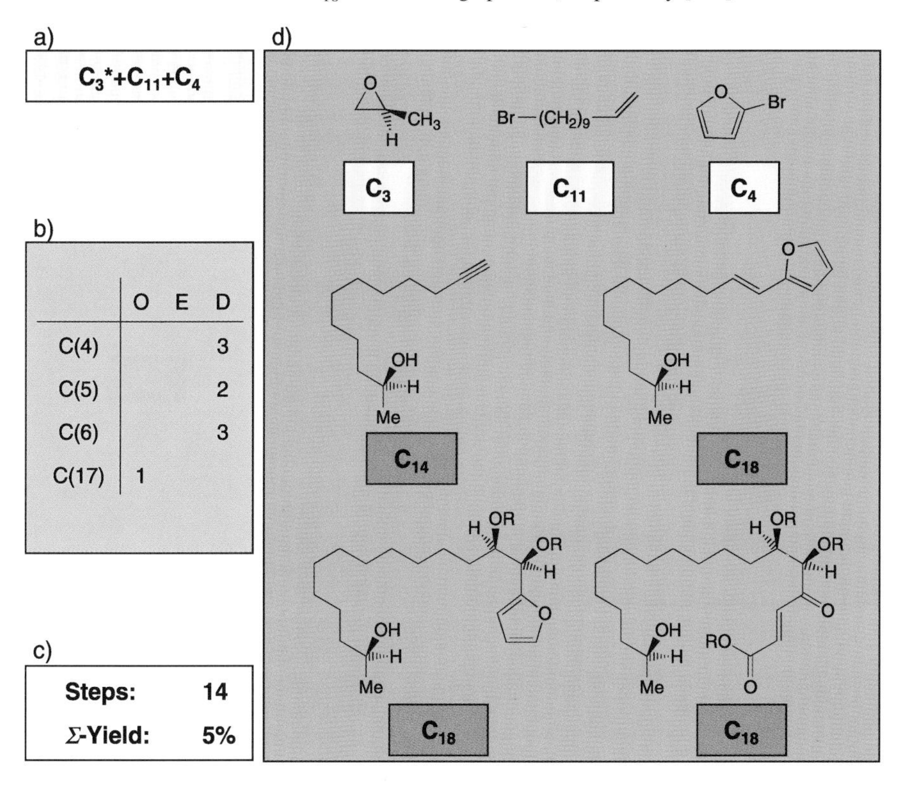

2.3.3.2.1.1.6. *Synthesis of* ent-**100** *According to the Pattern* $[(C_8 + C_1^*) + (C_4 + C_2 + C_1^* + C_2)]$

G. Solladie's synthesis (*Scheme 53*) exhibits two distinctive marks. One is the characteristic use of the (*R*)-configured C_1 building block on two occasions; firstly, to provide C(17) both with a Me group and with (*R*)-configuration, and, secondly, to put in place the C_7 β-keto sulfoxide (by treatment of C_1 with the olefination product from a C_4 building block and a C_2 unit). After conversion of the β-keto sulfoxide to the C_7 aldehyde, this is attached to the phosphonium halide of the C_9 building block by means of a *Wittig* reaction. The resulting olefination product can be transformed into the C_{18} hydroxycarboxylic acid and, after subsequent cyclization of this C_{18} seco-isomer, protecting groups are removed. The conditions selected to achieve it are especially mild, and this represents the second notable characteristic of the synthesis presently considered.

Scheme 51. *Characterization of a Synthesis of* **100**, *in Which Three of the Four Stereogenic Centers Are Diastereoselectively Introduced by Manipulation of* Enders'*s RAMP-SAMP Hydrazones* [99b]

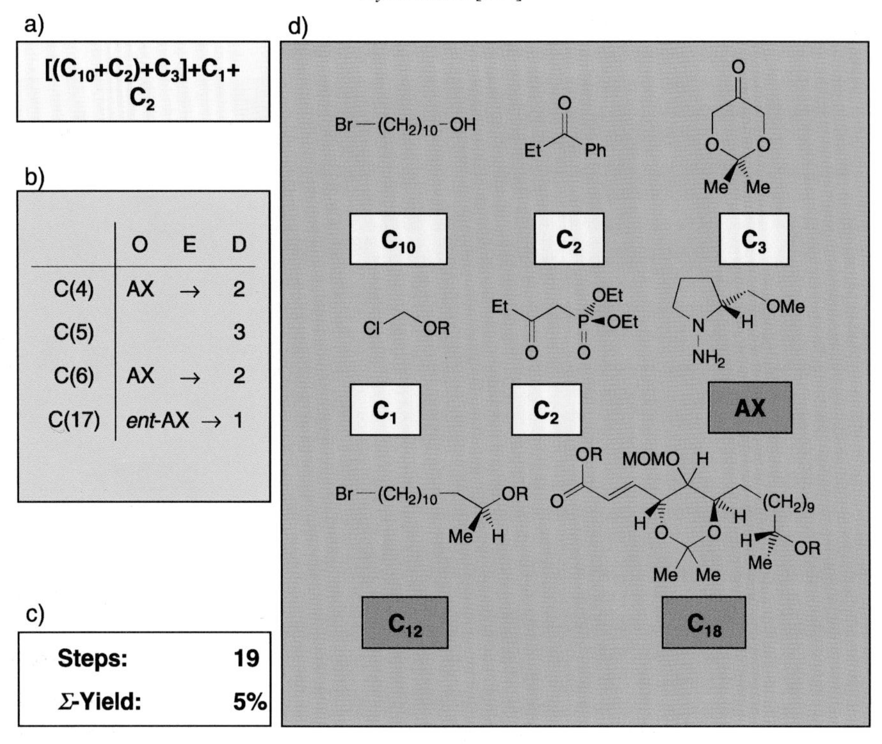

2.3.3.2.1.2. *Ring Closure by Olefin Metathesis*

2.3.3.2.1.2.1. *Synthesis of* **100** *According to the Pattern* $[C_7 + (C_3^{\ddagger} + C_8 + C_2) - C_2]$

Metathetic macrocyclization of a C_{20} reaction product obtained by *Horner* olefination from a C_7 aldehyde and a C_{13} phosphonate (*Scheme 54*). The phosphonate is accessible by transesterification [109a] of the C_2 building block with the alcohol produced by treatment of the *Grignard* compound of C_8 with the C_3 building block. The C_7 aldehyde is made by means of a series of transformations, beginning with the asymmetric *Sharpless* epoxidation of the achiral C_7 building block (with the participation of D-diisopropyl tartrate) [99e] [109b]. According to a sophisticated analysis [110], the (5*E*)-unsaturated epoxy (2*S*,3*R*,4*R*)-alcohol is produced in large enantiomeric excess, because a kinetic enantiomer resolution operates on the selection of one of the two enantiotopic vinyl groups (by rapid further reaction of the minor enantiomer created).

Scheme 52. *Characterization of a Synthesis of* ent-**100**, *in Which Three Stereogenic Centers Are Obtained from the Chiral C Pool.* The fourth stereogenic center is introduced by asymmetric *Sharpless* epoxidation [100a]

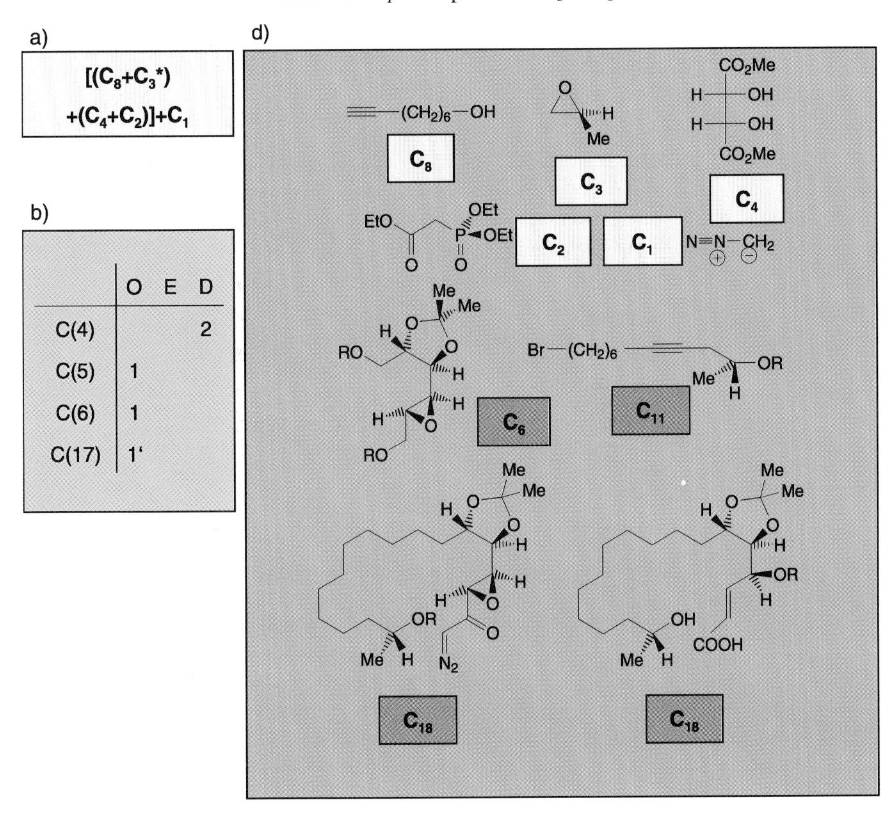

2.3.3.2.1.2.2. *Synthesis of* **100** *According to the Pattern* *[(*C_4^* *+* C_3*) + (*C_3^* *+* C_8 *+* C_2*) –* C_2*]*

S. Hatakeyama's C_{13} phosphonate (*Scheme 54*) is treated in *S. V. Ley*'s synthesis (*Scheme 55*) with a different C_7 aldehyde. The resulting C_{20} triene can be transformed into the 18-membered macrolide by intramolecular metathesis. The C_7 aldehyde is accessible by a fairly lengthy series of reaction steps [111] starting from dimethyl (2*R*,3*R*)-tartrate.

Scheme 53. *Characterization of a Synthesis of* ent-**100**, *in Which C(4) and C(17) Are Diastereoselectively Formed* via *Chiral, Nonracemic β-Keto Sulfoxides* [100b]

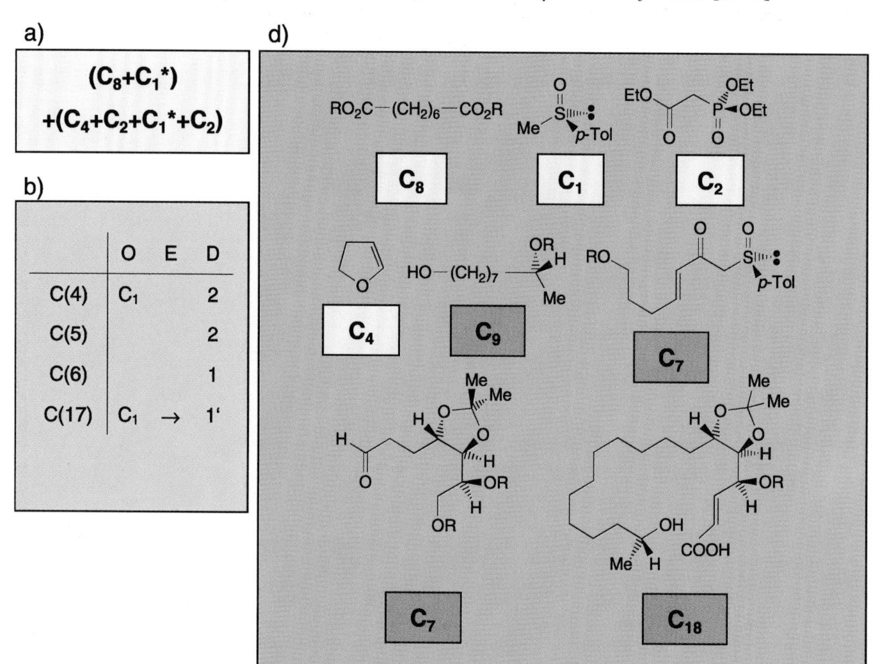

2.3.3.2.2. *Application of the Ring Strategy: Asymmetrically Catalyzed Macrocyclization of a ω-Formylalkyl Alkynate*

In *W. Oppolzer*'s synthesis of **100** (*Scheme 56*), a key role is played by DAIB-catalyzed[53]) macrocyclization of a C_{18} *ω-formylalkyl alkynate*, *via* an intermediate (1-alkenyl)(alkyl)zinc intermediate.

Stereoselective catalyst-promoted formation of the C_{18} (6*R*,17*S*)-hydroxy-ene-lactone is increased by the preexisting C(17) center of chirality. Successive introduction of an (*E*)-C(2)=C(3) bond, epoxidation of the C(4)=C(5) bond, and cleavage of the oxirane ring leads to the target compound.

---

[53]) DAIB =((–)-3-*exo*-(Dimethylamino)isoborneol) was introduced as a chiral auxiliary (**AX**) for the enantioselective addition of dialkylzincs to aldehydes [112]. It has been applied for both intermolecular [113a] and intramolecular [113b] formation of (*E*)-allyl alcohols from acetylenes and aldehydes *via* (1-alkenyl)(alkyl)zinc intermediates.

Scheme 54. *Characterization of a Synthesis of* **100**, *with Formation of a* C_{20} *Open-Chain Intermediate with Terminal Vinyl Groups, Accessible by* Horner *Olefination of a* C_{13} *Phosphonate and a* C_7 *Aldehyde.* The aldehyde is formed from an achiral educt by asymmetric *Sharpless* epoxidation [99e]

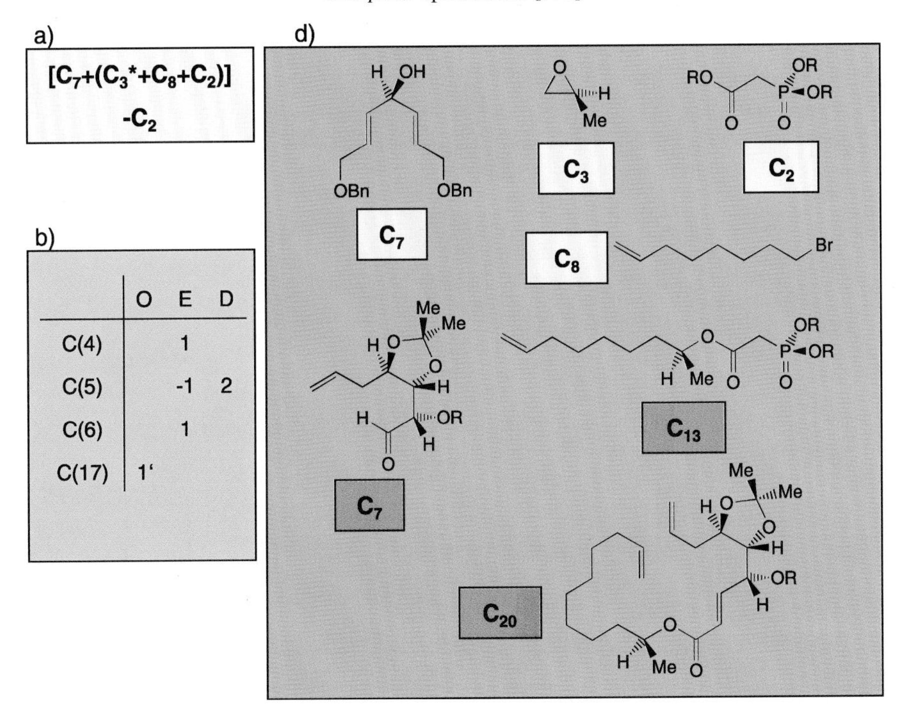

2.3.4. *Total Synthesis of the Antibiotic A 26771 B*

Photolactonization had played a fruitful role in the synthesis of (+)-aspicillin (**100**) because it had been possible to introducc thc functional O group at C(6) and, to take measures well-adapted for the future, the C(4)=C(5) and C(2)=C(3) bonds at the same time as the skeleton was being formed. Would it also prove its worth in the synthesis of the antibiotic A 26771 B (**101**), in which it would be necessary to remove the O-function at C(6), resulting unavoidably from *Wessely* oxidation? Comparison of *Schemes 57* and *58* shows how the strategy used for the photochemical synthesis of **100** was also retained for the photochemical synthesis of **101**.

In both cases, the synthetic pathway leads first of all from the building blocks to the corresponding *ortho*-hydroxyalkyl-substituted phenols (*in the blue fields*), and from these to the cyclohexa-2,4-dienones by means of *Wessely* acetoxylation (*in the orange fields*). These then undergo light-induced transformation into the photolactones (*in the green fields*). Continuing

Scheme 55. *Characterization of a Synthesis of* **100**, *with Formation of a* C_{20} *Open-Chain Intermediate with Terminal Vinyl Groups, Accessible by* Horner-*Olefination of a* C_{13} *Phosphonate* (see *Scheme 54*) *and a* C_7 *Aldehyde*. The aldehyde is produced starting from (2*R*,3*R*)-dimethyl tartrate [99f]

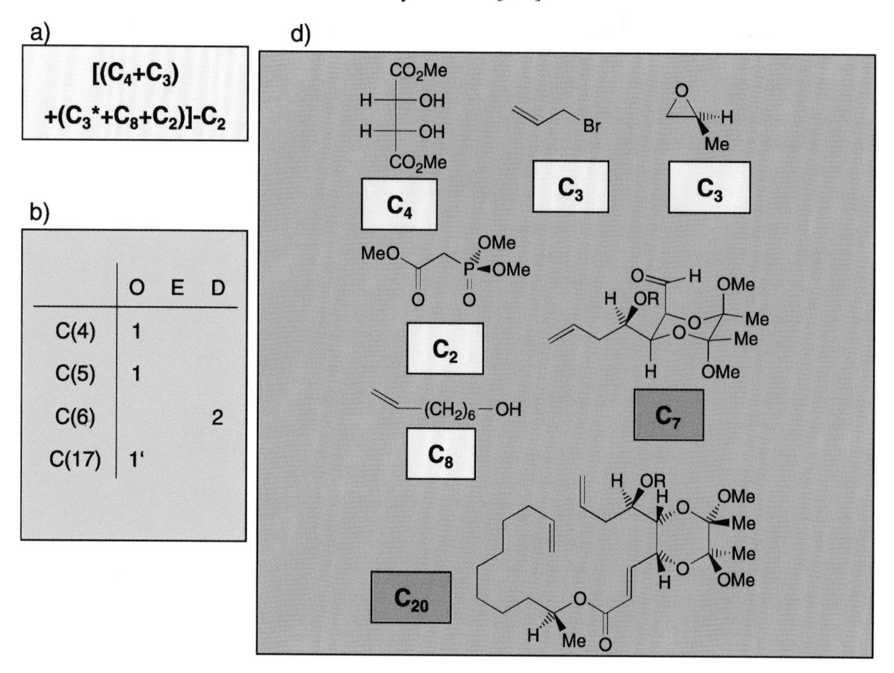

	O	E	D
C(4)	1		
C(5)	1		
C(6)			2
C(17)	1'		

onwards (*in the yellow fields*) through oxo-diene-lactones and hydroxy-diene-lactones finally brings us to the respective title compound. Differences in the target structures were allowed for by selection of different synthetic building blocks (C_7 instead of C_9 unit, C_3 unit with (*R*)- rather than (*S*)-configuration), as well as by certain tactical measures during the execution of each synthesis. Thus, the leaving groups ($PhSO_2^-$ *en route* to the 16-membered lactone, Br^- to the 18-membered one) that must be eliminated to effect the transformation of the photolactones into the oxo-diene-lactones are introduced at different stages of the syntheses: before the *Wessely* acetoxylation of the appropriate phenol in the former case, and after the photolactonization in the latter. The decision of which leaving group is introduced when, and at what stage it should be eliminated, was arrived at from more general studies involving preparative access to macrolides of varying ring size [81c]. While the conversion from the oxo-diene-lactones to the hydroxy-diene-lactones still takes place in the same manner (with *Yamamoto* reagent), the final stretches of the syntheses leading to the different target compounds naturally diverge.

Scheme 56. *Characterization of a Synthesis of* **100**, *in Which a C_{18} Formylalkyl ω-Alkynate, Formed Straightforwardly from Building Blocks C_3, C_{10}, and C_5, Is Diastereoselectively Cyclized, Affording a C_{18} Macrolide, Which Is Smoothly Transformed into* **100** [99a]

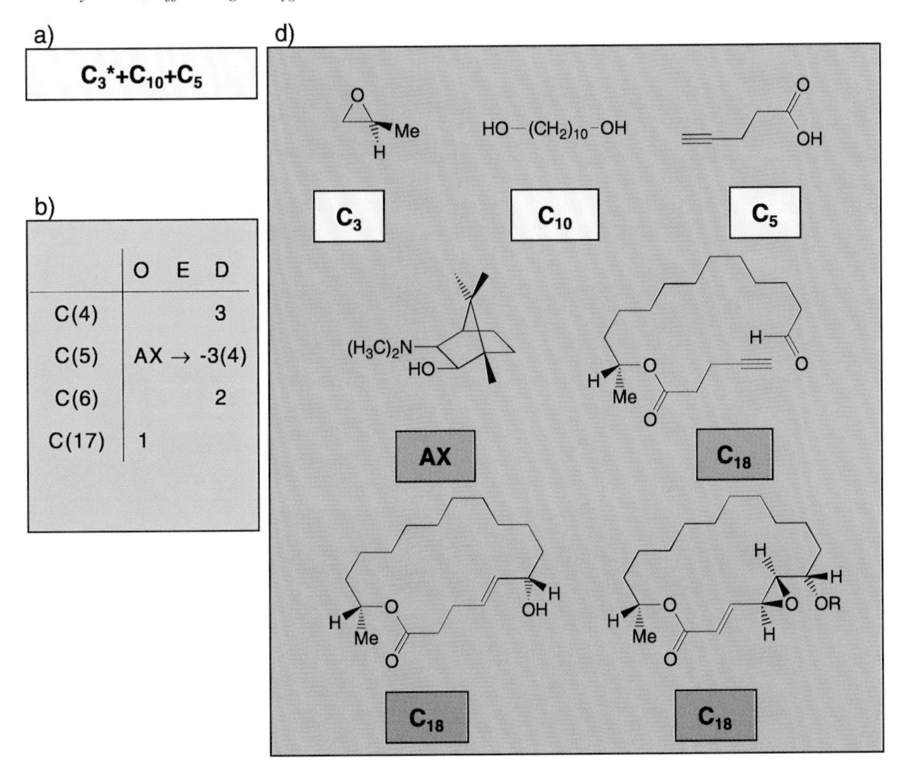

The 16-membered (2*E*,4*Z*,15*R*)-oxo-diene-lactone **106** is smoothly and predominantly converted into the (2*E*,4*Z*,6*S*,15*R*)-hydroxy-diene-lactone **107** (*Scheme 59*). The 18-membered (2*E*,4*Z*,17*S*)-oxo-diene-lactone **102**[54]) similarly goes over almost exclusively into the (2*E*,4*Z*,6*R*,17*S*)-hydroxy-diene-lactone **103**. Computer-assisted conformational analysis [115] convincingly shows that the stereochemical bias caused by the Me substituent (see *Fig. 6*), in each case, is conformationally transmitted to the remote carbinol center formed during reduction.

It moreover proves to be of predictive[11]) value in selecting optimal conditions for the further transforrmations to be done (see *Schemes 59* and *60*).

Having accomplished the synthesis of the antibiotic A 26771 B (**101**), we are now able to answer the question raised earlier, whether, in retrospect, the

[54]) The conformation of the oxo-diene-lactone of *Scheme 57* is thermodynamically preferred. Conformation **102** of *Scheme 59*, however, is kinetically favored. For detailed discussion, see [97b][114].

Scheme 57. *Strategic Route from Building Blocks to (+)-Aspicilin* (**100**) [97b–d] (*cf.* *Scheme 44*)

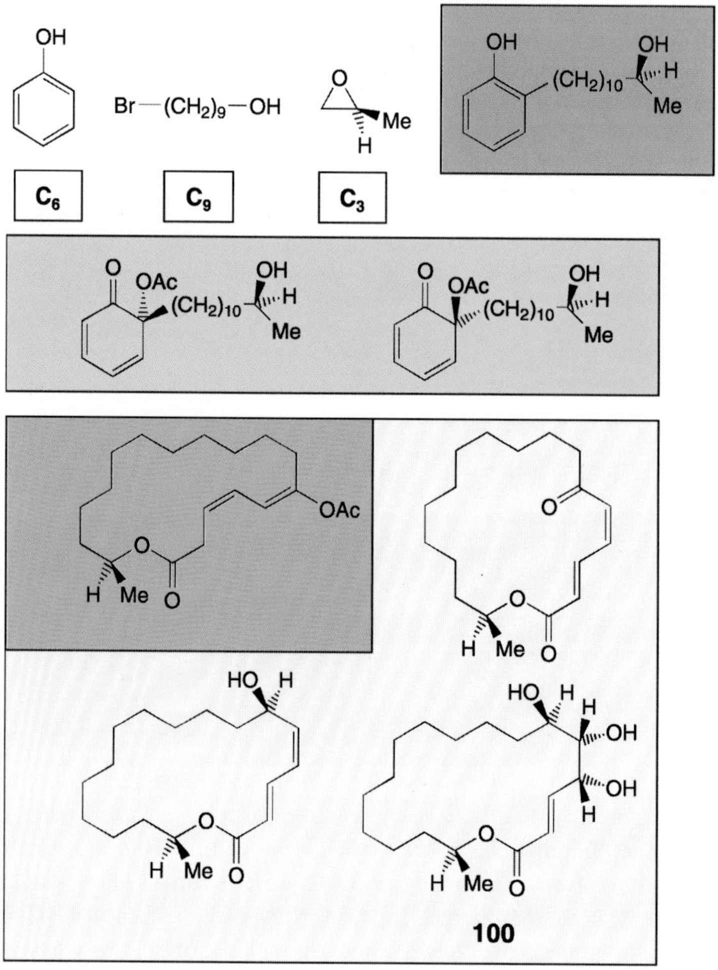

photochemical strategy did actually lead to success. The functionality at C(6), not present in the target structure, proved useful for oxygenation at C(4) and generation of the stereogenic center C(5). So far so good. The answer would have been more in the affirmative, if deoxygenation of **108** in the desired way had been possible. The *Barton-McCombie* reaction provided a product deoxygenated at C(6), but only after the adjacent oxirane ring had been opened between the two C-atoms, affording 17-membered, unsaturated ether-lactones. At the cost of an increase in the number of steps, a detour had to be made to reach **110a** which could be deoxygenated properly.

Scheme 58. *Strategic Route from Building Blocks to Antibiotic (–)-A 26771B* (**101**) [114] (*cf. Scheme 57*)

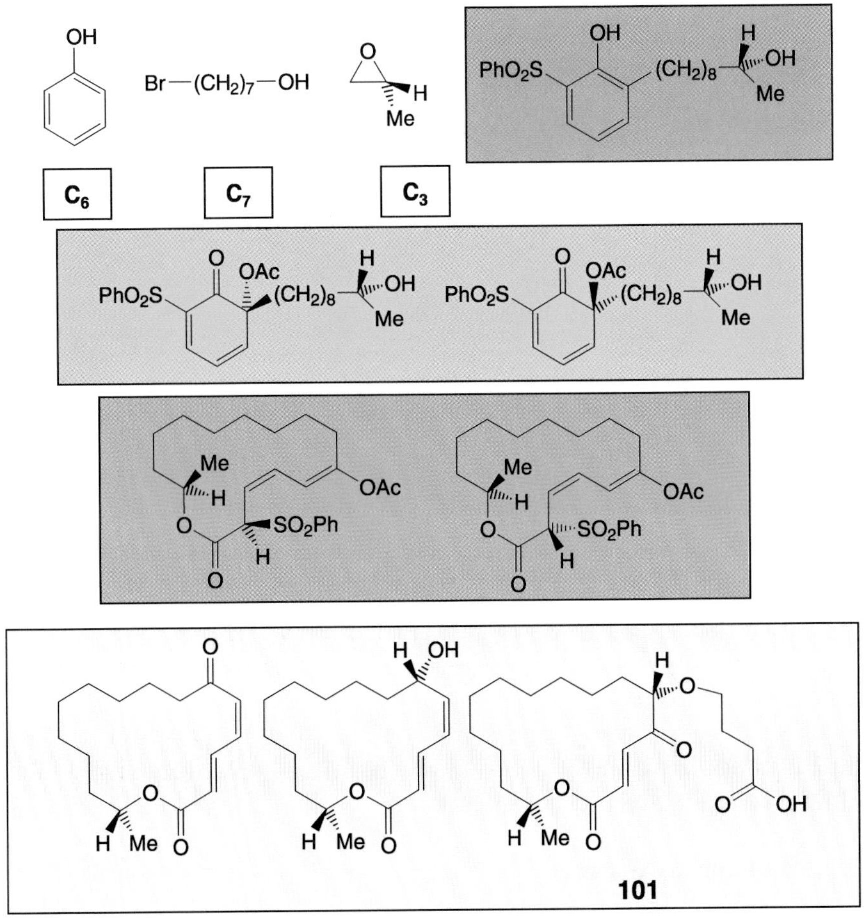

Among the published [114][116] syntheses of **101**, *Keinan*'s synthesis, [116c] making use of *Sharpless*'s asymmetric dihydroxylation, is straightforward and efficient and merits special consideration. *Kobayashi*'s approach [116c] should be mentioned here, too.

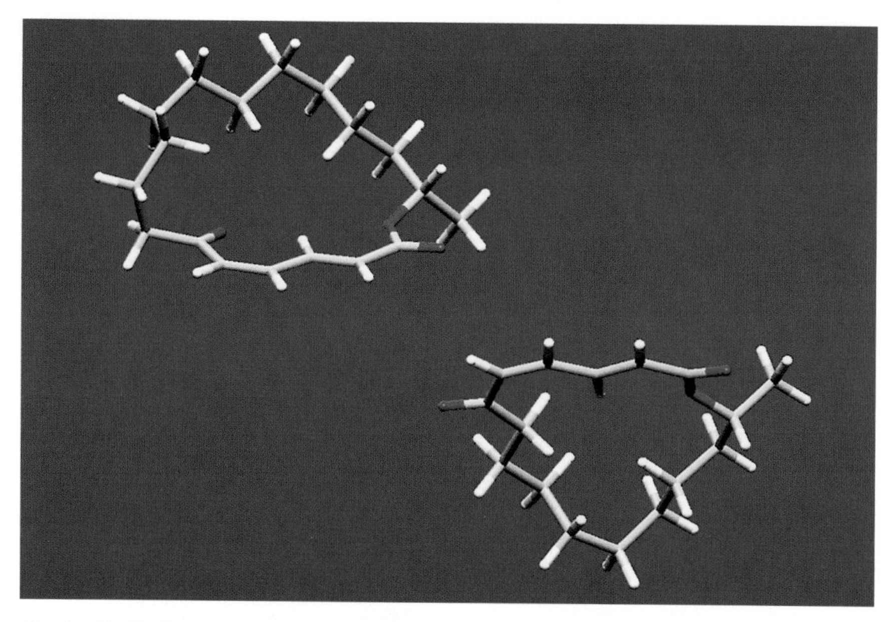

Fig. 6. *Idealized molecular models of (*2E,4Z,17S*)-oxo-diene-lactone* **102** *and (*2E,4Z,15R*)-oxo-diene-lactone* **106** *revealing unhindered access to the C=O groups within the* (Re)*- or* (Si)*-half-space, respectively* [114]

3. Photoenolization of *ortho*-Me-Substituted Acetophenone Derivatives in Steroid Synthesis

3.1. Mechanistic Aspects

Along the road from photoenolization as a reaction principle [117] to photoenolization as a key reaction in the total syntheses of (+)-estrone [118] and (–)-norgestrel [119], some questions have been put, and answers found, from which emerges the prospect of deliberately planned application of photochemical reactions in sophisticated syntheses of target compounds. No more talk of photochemistry shrouded in mystery (see *Sect. 1.1*).

Scheme 61 expresses, suitably briefly, current mechanistic understanding of photoenolization for cases in which all of the transient species between the conformations of the ketone educt (**113s**+**113a**) and those of the photoenol product (**114s**+**114a**) are identifiable [120].

The two conformers **113s** and **113a**, in their electronic ground states, are promoted by light absorption to their related singlet electronic isomers $^1_1$[**113s**]* and $^1_1$[**113a**]*. The latter rapidly decay by intersystem crossing into

Scheme 59. *Tactical Means to Reach* **100** *Starting from* **102** [97b,c], *or* **101** *Starting from* **106** [114]: *the Molecules*

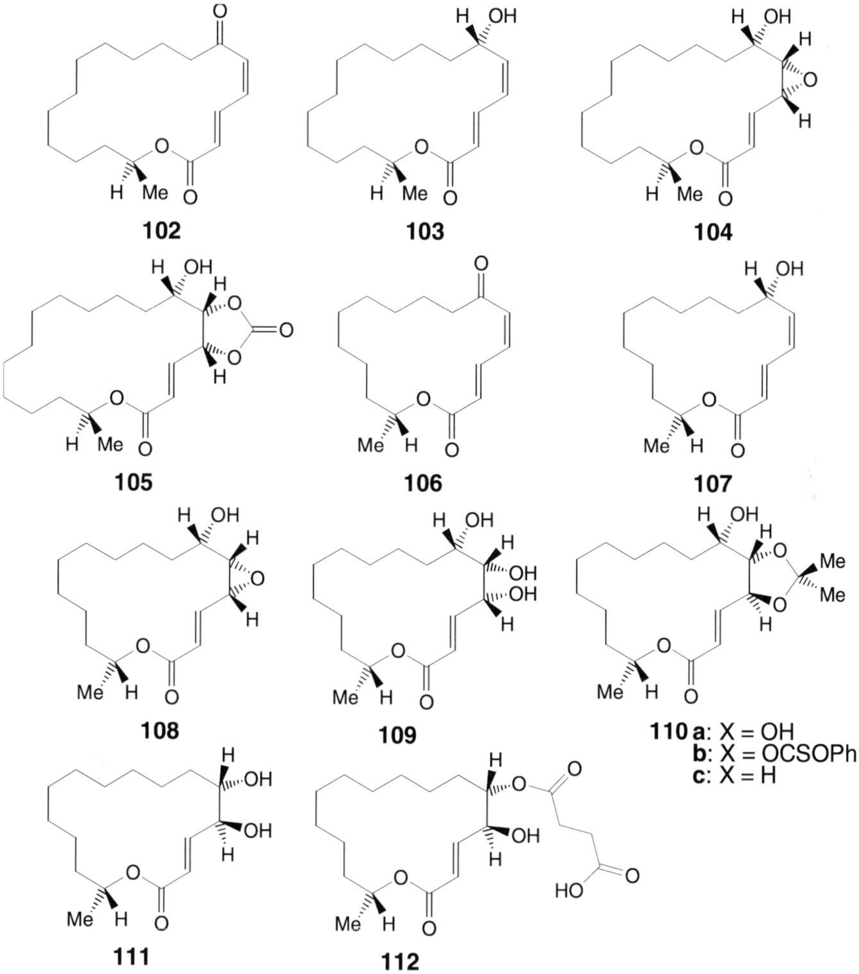

102 **103** **104**

105 **106** **107**

108 **109** **110 a**: X = OH
b: X = OCSOPh
c: X = H

111 **112**

their triplet-spin isomers ${}^{3}_{1}$[**113s**]* and ${}^{3}_{1}$[**113a**]*. For ${}^{3}_{1}$[**113s**]*, an adiabatic[55]) H-shift now takes place to furnish ${}^{3}_{1}$[**114s**]*, which equilibrates with ${}^{3}_{1}$[**114a**]*. Both ${}^{3}_{1}$[**114s**]* and ${}^{3}_{1}$[**114a**]* are further deactivated by intersystem crossing to afford the two photoenol conformers **114s** and **114a** in their electronic ground states. Reketonization of **114s** leads back to original educt, with a

[55]) In adiabatic reactions, the chemical change occurs on the same energy hypersurface (see *Sect. 1.2*). *Most adiabatic photoreactions are proton transfer reactions. All reactions of that sort are reversible insofar as, following deactivation, very fast reverse reactions drive the system back to its original composition so that no permanent change occurs* [26b].

Scheme 60. *Tactical Means to Reach* **100** *Starting from* **102**, *or* **101** *Starting from* **106**: *the Conditions*

102 —a) 96%→ **103** —b) 93%→ **104** —c) 73%→ **105** —d) 60%→ **100**

a) [97b]: Section 2.2.1.1; b) [97c]: Section 2.1.1
c) [97c]: Section 2.3.3.1; d) [97c]: Section 2.3.4.1

106 —e) 92%→ **107** —f) 86%→ **108** —g) 81%→ **109** —h) 78%→ **110a**;

110a —i) 85%→ **110b** —j) 89%→ **110c** —k) 86%→ **111** —l) 69%→ **112** —m) 52%→ **101**

e) [114]: Section 1.3.11; f) Section 1.4.1.2; g) Section 1.5
h) Section 1.6.1; i) Section 1.6.2; j) Section 1.6.3
k) Section 1.6.4; l) Section 1.7.1; m) Section 1.7.2

substantial contribution from the tunnel effect [120e]. In the overall procedure, the (light) energy is wasted[56]) as long as no compound is present able to trap one of the transients irreversibly to furnish a photoproduct. *Scheme 62* shows how photoenols may be trapped efficiently by an appropriate dienophile.

These examples demonstrate the stereospecificity of the intermolecular *Diels-Alder* reactions between the indicated dienophiles and the assumed photoenols. On irradiation in the presence of fumaric acid (**118**) or maleic acid (**120**), **115** stereospecifically affords *rac*-**119** and *rac*-**121**, respectively [121a]. This result indicates rapid reversion to the ketone by the short-lived photoenol configuration of type **114s**, and participation of the long-lived photoenol configuration of type **114a** in the transition structure, accessible by approaching the dienophile in normal *endo*-orientation [121f].

[56]) Waste of this sort may be desirable, to dissipate UV light in an otherwise photodegradable system [120f].

Scheme 61. *Sequence of All the Transients Involved in the Photochemistry of* ortho-*Methylacetophenone* (**s** and **a** denote *syn* and *anti*, respectively)

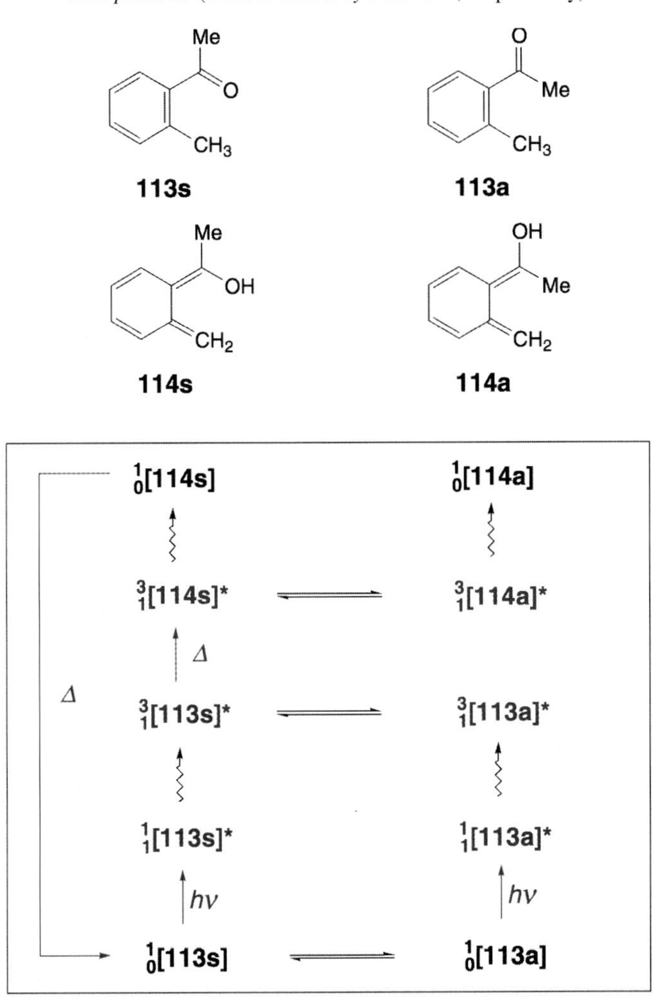

3.2. Total Syntheses of (+)-Estrone and (–)-Norgestrel

3.2.1. *A Photochemical Synthesis by an Intramolecular* Diels-Alder *Reaction*

(+)-Estrone (**122a**) has developed into a yardstick for total synthesis planning and execution. A flood of synthetic approaches to this steroid hormone was the natural consequence. For the commercial preparation of **122a**, partial syntheses based on structural manipulation of easily accessible, natural-

Scheme 62. *Adducts Formed on Irradiation of* ortho-*Methylbenzophenone in the Presence of Dienophiles* (*a:* see [117a], *b* and *c:* see [121a]). The empty space placed within brackets should be filled with the processes of *Scheme 61* enclosed in a frame.

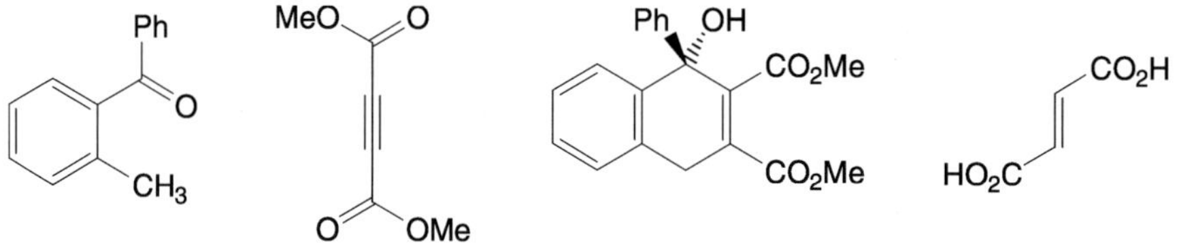

115 **116** **117** **118**

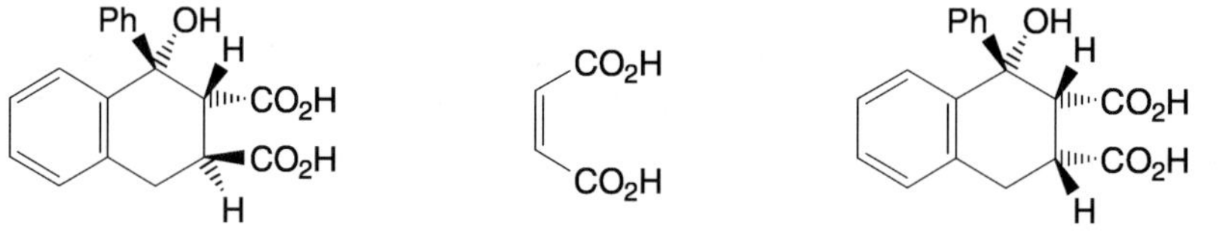

119 **120** **121**

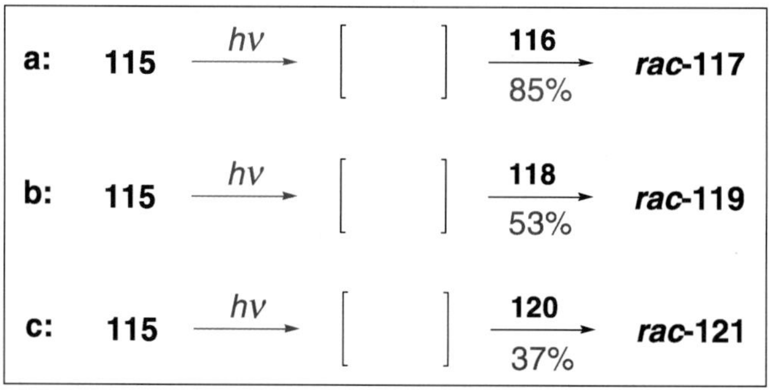

a: **115** $\xrightarrow{h\nu}$ [ ] $\xrightarrow[85\%]{\mathbf{116}}$ ***rac*-117**

b: **115** $\xrightarrow{h\nu}$ [ ] $\xrightarrow[53\%]{\mathbf{118}}$ ***rac*-119**

c: **115** $\xrightarrow{h\nu}$ [ ] $\xrightarrow[37\%]{\mathbf{120}}$ ***rac*-121**

ly occurring steroids still command attention. The situation is different for (–)-norgestrel (**123b**), a compound that fulfills the function of the gestagenic component in oral contraceptives. Not occurring in nature, it possesses an Et group at C(13), and so is unobtainable by partial synthesis from easily available steroids, all of which have a Me group at C(13). Its preparation on an industrial scale is of historical significance in total synthesis generally, and in the development of asymmetric synthesis in particular (see [119c,d]).

Even to the novice synthetic chemist, the functional groups in rings A and D of **123b** suggest the latent precursor **122b**. The substituted anisole ring in **122b** may be transformed into the α,β-unsaturated cyclohexenone ring of **123b** with the aid of a *Birch* reduction. Nucleophilic addition of acetylide anion to the cyclopentanone ring in **122b** provides the substituted cyclopentane ring of **123b**.

122 a: R^1 = H, R^2 = Me
b: R^1 = Me, R^2 = Et

123 a: R = Me
b: R = Et

The problem of synthesis hence essentially reduces to the construction of the tetracyclic skeleton. The assistance of combinatorial analysis[57]) massively open up the breadth of scope of the search for alternative synthetic pathways. A synthetic chemist desiring to prepare a given target molecule selects that variant that appears most suitable from the great variety of virtual[58]) pathways. In the present context, this only involves those synthetic routes in which photoenolization plays a key role. The total synthesis of **123b**, characterized in *Scheme 63*, presents the derivative of the *ortho*-methylacetophenone **AD** and the associated photoenol **AD′**.

Scheme 63. *Characterizatioin of the Photochemical Synthesis of* **123b** *Following the Constitutional Pattern* **A + D → AD → AD′ → ABCD → 123b** (including the results of *Schemes 64* and *65* [119b])

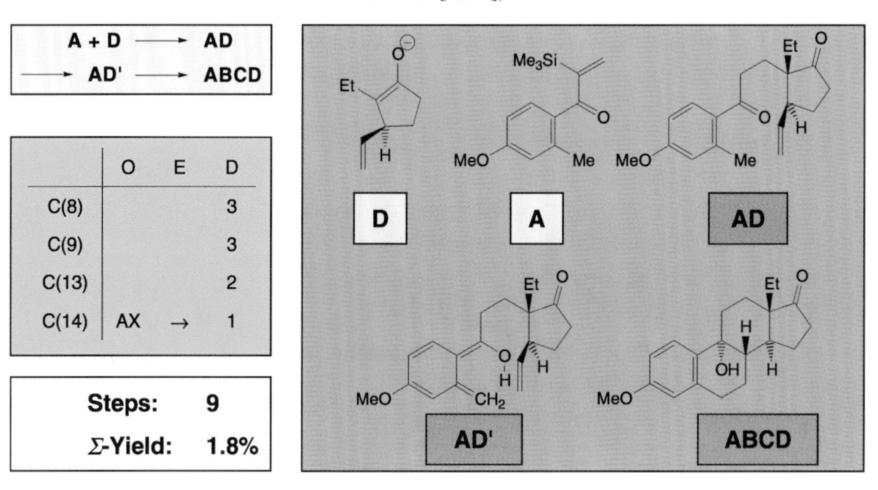

57) See [122] on combinatorial analysis in the case of **123b.**

58) The term *virtual* is used here: '*not physically existing but made to appear from the point of view of the user* (*The Oxford Dictionary of New Words*, Edition 1997). The word virtual has been employed in a sense other than that given here in *J.-M. Lehn*'s essay [123].

The photoenol, being an *ortho*-quinodimethane derivative[59]), should be capable of completing the steroid skeleton by intramolecular, entropy-favored *Diels-Alder* reaction. Identification of optimal reaction conditions (see Sect. 3.4 in [118d]) starts with critical analysis of the reaction potential involved. The electron density on the dienophilic vinyl group is relatively high, and hence would tend to slow reactions. The electron density on the conjugated diene system is likewise high, thanks to the OH groups present, but acts here to accelerate reactions. It can even be increased further by deprotonation of the dienol[60]). Under the optimal conditions (see [118d]), which include:

- *light of wavelength > 340 nm* (brings about selective absorption by the benzoyl chromophore keto group).
- *reaction temperature*: 98° (since the cycloaddition fails otherwise to keep in step with reketonization[61])),
- *addition of pyridine* (for the purpose of deprotonating diene **AD′**, thus enhancing the lifetime of the resulting anion and the reactivity of the diene in the cycloaddition),
- *addition of 2,4,6-trimethylphenol* (impedes photochemical decomposition of the type **ABCD** benzyl alcohols produced),
- *action of oxalic acid in refluxing benzene on the crude irradiation product* (for the dehydration of the primarily formed benzyl alcohols),

the cycloadduct **ABCD** is produced (as the main product), together with its 9β-epimer (minor product)[62]). Their constitution and configuration were confirmed (in the racemic series) by comparison with an authentic sample of **ABCD** (in the series with Me rather than Et at C(13); see Sect. 5.8.2 in [118d]) and by NMR spectroscopy [125]. Because of the instability of the cycloadduct under the conditions used, isolation of the epimeric benzyl alcohols was usually refrained from. Instead, the crude irradiation product was dehydrated into a mixture of styrene derivatives (mostly containing a C(9)=C(11) bond, with lesser quantities possessing a C(8)=C(9) bond. The overall yield for the three-step reaction sequence: *intramolecular* Diels-Alder *reaction – photoenolization – dehydration* amounted to 47% (relative to **AD**). Using standard techniques, it was possible to convert the olefin mixture to **123b** in 45% chemical yield [119b].

[59]) We had met a member of the *ortho*-quinodimethane family earlier (see *Scheme 5*).

[60]) Powerful rate enhancement can be achieved with the assistance of an oxy-anionic substituent effect [124].

[61]) Deuterium incorporation on irradiation of **AD** in the presence of CD_3OD at room temperature shows that photoenolization has taken place. For light-induced H/D exchange in *ortho*-benzylbenzophenone, see [117a].

[62]) For stereoselective cycloaddition, see Sect. 2.3 in [118d] and Scheme 30 in [118f].

The intermediate **AD** is a 1,5-diketone and as such accessible by means of a *Michael* addition of **D** (as an enolate anion) to **A** (in a yield of 52%). It is a characteristic of the *photochemical synthesis* of **123b** that the kinetically favored *cis*-orientation of the ethyl and vinyl groups on the five-membered ring of the *Michael* adduct **AD** ensures the thermodynamically disfavored *trans*-fusion of rings C and D in the *Diels-Alder* adduct of type **ABCD**. The overall yield of **123b**, based on **D**, amounts to 11%[63]). The achiral building block **A** is accessible by conventional means [118d].

It is in the chiral, nonracemic component **D** that the strengths and weaknesses of the total synthesis characterized in *Scheme 63* are manifested. *Scheme 64* outlines the reliable way in which (*R*)-configured, three-membered-ring compound **128** [118e] can be obtained by diastereoselective *Linstead* cyclopropanation [126].

Scheme 64. *Diastereoselective Formation of the Three-Membered Ring Compound* **128** *from the* C_2*-Symmetrical Diester* **125** (for a discussion of how the absolute configuration of the (1*R*,3*R*,4*S*)-8-phenylmenthyloxy residue of the auxiliary **AX** in **125** determines the absolute configuration of **128,** see Fig. 2 in [118e], and Scheme 36 in [118f])

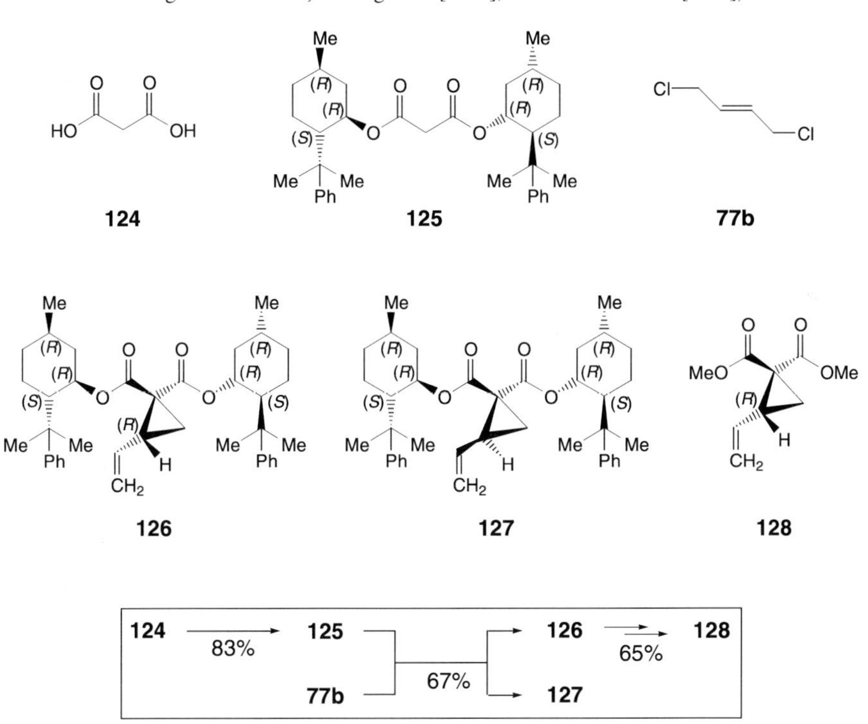

[63]) The overall yields of **122a** and **123a** (both relative to the appropriate Ring D building block) are 23% [118e] and 17% [118f], respectively.

Scheme 65 shows the stereospecific ring expansion [127] of **128** on treatment with the enolate anion of ethyl methylmalonate, furnishing a mixture of the equilibrating five-membered-ring compounds **131** and **132** (= **D** in *Scheme 63*).

These stereostructural achievements displayed in *Schemes 64* and *65* demonstrate the power of stereocontrol exerted here. The moderate overall yield of the multistep procedure, however, does call for a better solution.

Some improvement might be possible by means of enantioselective conjugate addition of a chiral organocuprate [128] to 2-ethylcyclopent-2-enone (**133b**) as *Michael* acceptor (*Scheme 66*).

As the *Michael* donor (here, the ethenide anion) itself is achiral, the chirogenic[44]) addition has to be controlled by use of a chiral auxiliary. While it was possible, operating in conjunction with the proline-derived (*S*)-2-(ethoxymethyl)pyrrolidine (**136**) and CuSCN, to achieve 1,4-addition of lithiopropene to **133a**[64]) furnishing a mixture of the equilibrating isomers **134**/*ent*-**134**

Scheme 65. *Stereospecific Ring Expansion Converting* **128** *to the Mixture of Equilibrating Five-Membered-Ring Compounds* **131** *and* **132** [119b] (see also Fig. 9 in [119d])

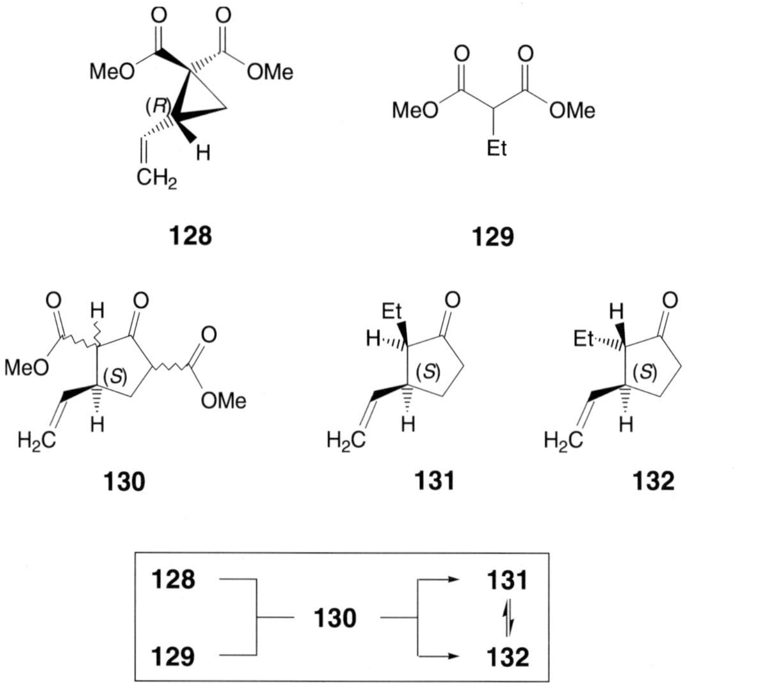

[64]) This greatly improved the total synthesis of (+)-confertin [129b], in which *Linstead* cyclopropanation originally [129a] had been used, too.

Scheme 66. *How* **133a** *or* **133b** *May Be or Has Been* [128] *Converted to a Mixture of Equilibrating Isomers* (**134** + **135**) *or* (**131** + **132**), *Respectively*

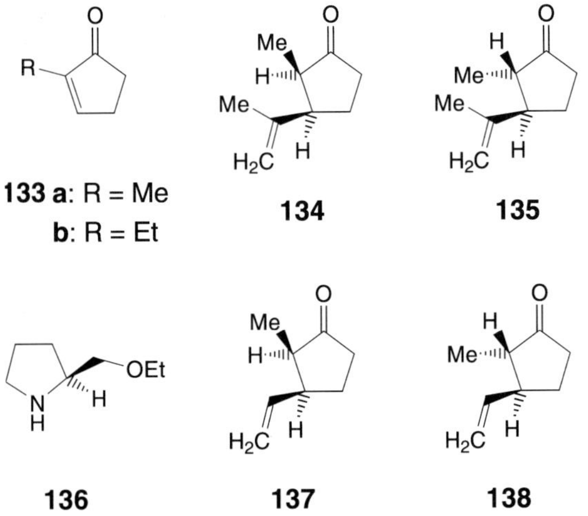

133 a: R = Me
b: R = Et

134 **135**

136 **137** **138**

≫1 and **135**/ent-**135**≫1 in high chemical yield (88%) and with high optical purity (88%), compounds **137**/*ent*-**137**≫1 and **138**/*ent*-**138**≫1 could be obtained from **133a** under similar conditions with only 24% optical purity. Enantioselective preparation of (**131**/*ent*-**131**≫1) + (**132**/*ent*-**132**≫1) by conjugate addition of the ethenide anion to **133b** still awaits solution.

3.2.2. *A Non-Photochemical Synthesis of* **123b** *with an Intermolecular* Diels-Alder *Reaction*

Understanding of photochemical reactions was an enterprise of the first half of the twentieth century. This essay addresses the question of whether, in the period since, they have reached a sufficiently advanced stage of development to be in a position to take on the role of a plannable and reliable key reaction in the strategic approach behind a total synthesis of a complex natural product. To ascertain this, if the question is not to be meant rhetorically, then it is necessary to establish some practices by which *photochemically based total syntheses* are as standard judged systematically against alternative total syntheses, not involving electronically excited molecules, of the same target compounds. Consequently, news of a new total syntheses of, for example, (–)-norgestrel (**123b**) is bound to be received with particular attentiveness. Especially if the competing syntheses have been planned and executed in the same laboratory, without partiality for or against photon-driven reactions. In Frankfurt, compound **123b** was produced not only by *photochemical synthesis* (*Scheme 63*) but also by a pathway containing no photochemical steps.

The *Diels-Alder* reaction plays a pivotal role in both instances; *intramolecularly* after the photochemical preparation of the conjugated diene in *Scheme 63*, and in the [4+2] cycloaddition's *intermolecular* variant, without any involvement of light, in *Scheme 67*. The carrying out of the intermolecular modification had had an additional incentive, as the reaction between the building blocks **AB** (the *Dane* diene **139**) and **D** (the *Dane* dienophile **140a**) had been examined [131] as early as the end of the 1930s, in the hope of obtaining a cycloadduct incorporating the steroid skeleton (*rac*-**141a**; *Scheme 68*).

Even when, twenty years later [132], it had been established that, although the major product component *rac*-**142a** did not have a steroid skeleton, the side product *rac*-**141a** did indeed possess one, *Elisabeth Dane*'s concept still had not been built upon. In 1988, in a lecture entitled '*Five Decades of Steroid Synthesis*' [119c], this fact was remembered with the thoughtful reflection that efforts might be made to change the direction of regioselection or even to achieve enantioselection.

How might the chirogenic[44]) reaction between **139** and a dienophile of type **140** be steered in the desired direction by *molecular reaction mediators*? Is there any kind of general procedure for systematically searching for such mediating agents? The answer is YES, provided that the search is carried out in hierarchically structured testing steps.

Scheme 67. *Characterization of the Synthesis of* **123b** *Following the Constitutional Pattern* **AB + D → ABCD → 123b** *with the Help of Auxiliary* **AX** (R = phenanthren-9-yl [130])

AB + D ⟶ ABCD

	O	E	D
C(8)			3
C(9)			3
C(10)			5
C(13)	AX → 1		
C(14)	↳ -1		2
C(17)			4

Steps: 9

Σ-Yield: 11%

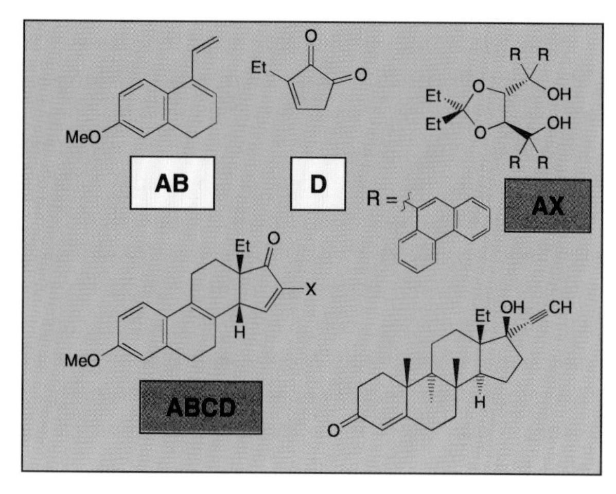

Scheme 68. Diels-Alder *Reactions of Diene* **139** *with Dienophiles of Type* **140** (for conditions, see *Table 4*).

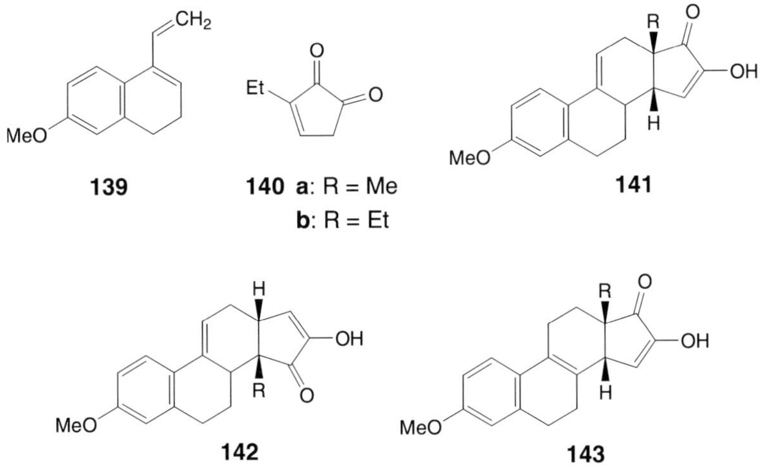

To pass the first test, hopefuls from the throng of candidate molecules must be capable of accelerating the reaction under discussion. Here, it has been known for forty years [133] that *Lewis* acids are capable of accelerating *Diels-Alder* reactions that involve polar dienophiles. *Table 4* shows that this also applies in the present case.

The second test picks out those successful candidates from the first round that are capable of differentially accelerating the parallel reactions leading to the adduct components of types *rac*-**141** and *rac*-**142**[65]). *Table 4* shows that the

Table 4. Diels-Alder *Reactions of Diene* **139** *with Dienophiles of Type* **140**. Conditions and results: [130d]

Entry	Dienophile	Reaction conditions	Adduct	Yield [%]
1	**140a**	Dioxane/reflux	*rac*-**141a** (1) + *rac*-**142a** (3)	82
2	**140b**	Dioxane/reflux	*rac*-**141c** (1) + *rac*-**142c** (1.9)	40
3	**140a**	$BF_3 \cdot OEt_2/Et_2O/-20°$	*rac*-**141a** (49)+*rac*-**142a** (3)	75
4	**140b**	$BF_3 \cdot OEt_2/Et_2O/-20°$	*rac*-**141c** (32)+*rac*-**142c** (1)	53
5	**140a**	$(i\text{-}PrO)_2TiCl_2$, CH_2Cl_2, –30° CH_2Cl_2, conc. HCl, r.t.	*rac*-**143a**	80
6	**140b**	$(i\text{-}PrO)_2TiCl_2$, CH_2Cl_2, –30° CH_2Cl_2, conc. HCl, r.t.	*rac*-**143b**	80
7	**140b**	$(i\text{-}PrO)_3TiCl_2$, CH_2Cl_2, –20°	*rac*-**141b**	70

[65]) A positive sign for reaction acceleration and regioselection in the right direction was provided by *Valenta et al.* [134], who treated the *Dane* diene **139** with the unsymmetrically substituted dienophile 2,6-dimethyl-1,4-benzoquinone to furnish the racemic mixture analogous to *rac*-**141a** and *rac*-**142a**.

applied *Lewis* acids do indeed act regioselectively. It seems reasonable to suppose that the candidates successful so far should also pass the next test, that they should promote the formation of a product of type **141**/*ent*-**141** ≫ 1 when acting in conjunction with a chiral auxiliary under conditions promoting regioselection and enantioselection. A tried and tested auxiliary is **AX** (*Scheme 67*), one of *Seebach*'s TADDOLs [106]. Enantioselection in the chirogenic reaction was achieved by suitably chosen conditions (ee 93% for **141a** and 89% for **141b**; *Table 5*) and perfected by fractional crystallization (ee > 99.7%).

Table 5. *Enantioselective Diels-Alder Reactions of Diene* **139** *with Dienophiles of Type* **140** *Mediated by Ti-TADDOLate Complexes, Formed by Reactions of TADDOL* **AX** *(Scheme 67;* R = phenanthrene-9-yl) *with* $(i\text{-}Pr)_2TiCl_2$ [130d]

Entry	Dienophile	Equiv. of **A**	Temp. [°C]	Time	Yield [%]	$[\alpha]^{20}_{589}$	ee [%]
1	**140a**	2	−25	15 h	35	+191	92
2	**140a**	2	−80	2 d	65	+194	93
3	**140a**	0.25	−80	7 d	78	+172	85
4	**140b**	2	−80	7 d	50	+193	88
5	**140b**	0.2	−80	7 d	77	+196	89

The transformation of **141b** (with a *cis*-fusion of rings C and D) into (–)-norgestrel (**123b**; with rings C and D *trans*-fused) was carried out using methods known from the literature. In both of the syntheses of **123b** discussed here, the overall yield (relative to the appropriate D building block; see *Schemes 63* and *67*) was 11%. The contest, meanwhile, was decided by the more easily accessible D building block and the more elegant way in which the required absolute configuration of the target compound could be synthesized.

4. Conclusion

From the perspective of *comparative chemical synthesis*, emphatically represented in this essay, a photoreaction is just as deserving of consideration as a thermal reaction in designing a total synthesis of a complex target structure[66]). Important for both is that they are thoroughly understood mech-

[66]) Cases exist, such as the natural or commercial preparation of previtamin D from provitamin D (*Sect. 1*) or the ring opening of cyclohexa-2,4-dienones to their dieneketene seco-isomers (*Sect. 2*), in which the photoreaction has the advantage. There are other instances such as the preparation of reactive *ortho*-quinodimethane derivatives as intermediates in steroid synthesis[67]) (*Sect. 3*), in which either photoenolization or thermal seco-isomerizations of appropriate benzocyclobutene derivatives may be used equally well.

[67]) For a summary of syntheses of 19-norsteroids by intramolecular *Diels-Alder* reactions of *ortho*-quinodimethanes, see [135].

anistically. The alternative of whether to use a *Bunsen* burner or a UV lamp for a particular step in a synthesis has at best practical significance. How important it is to keep pace with development in mechanistic photochemistry, however, is exemplified by the fact that those early producers of vitamin D_3 who did not give up the archaic Mg spark as a light source have been out of business for some time.

We are most grateful to Dr. *Andrew Beard* for having translated the German version and to *Deutsche Forschungsgemeinschaft*, *Fonds der Chemischen Industrie*, and *Bundesministerium für Wirtschaft* for support.

REFERENCES

[1] G. W. Robinson, '*Electronic Excited States of Simple Molecules*', in 'Light and Life', Eds. W. D. McElroy, B. Glass, The John Hopkins Press, Baltimore, 1961, p. 11.
[2] G. Quinkert, *Pure Appl. Chem.* **1964**, *9*, 607.
[3] a) G. Quinkert, *Angew. Chem., Int. Ed.* **1975**, *14*, 790; b) G. Quinkert, F. Cech, E. Kleiner, D. Rehm, *Angew. Chem., Int. Ed.* **1979**, *18*, 557.
[4] M. Kasha, *Discuss. Faraday Soc.* **1950**, *9*, 14.
[5] A. Butenandt, *Angew. Chem.* **1960**, *72*, 645.
[6] A. Windaus, '*Nobel Lecture 1928*', in 'Nobel Lectures Chemistry 1922–1941', Elsevier Publ. Co., Amsterdam, 1966, p. 105.
[7] A. Butenandt, A. Wolff, P. Karlson, *Ber. dtsch. chem. Ges.* **1941**, *74*, 1308.
[8] A. Butenandt, L. Poschmann, *Ber. dtsch. chem. Ges.* **1944**, *77*, 394.
[9] G. Ciamician, *Science* **1912**, *36*, 385.
[10] L. F. Fieser, M. Fieser, 'Steroids', Reinhold Publ. Corp., New York, 1959.
[11] C. Reichardt, *Liebigs Ann. Chem./Recl.* **1997**, XXVIII.
[12] R. Criegee, *Angew. Chem., Int. Ed.* **1966**, *5*, 333.
[13] J. Castells, E. R. H. Jones, G. D. Meakins, R. W. J. Williams, *J. Chem. Soc.* **1959**, 1159.
[14] L. Velluz, G. Amiard, B. Goffinet, *Bull. Soc. Chim. France* **1955**, 1341.
[15] M. P. Rappoldt, E. Havinga, *Recl. Trav. Chim. Pays-Bas* **1960**, *79*, 369.
[16] G. M. Wyman, *Chem. Rev.* **1955**, *55*, 625.
[17] R. S. Mulliken, C. C. J. Roothaan, *Chem. Rev.* **1947**, *41*, 219; A. J. Merer, R. S. Mulliken, *Chem. Rev.* **1969**, *63*, 639.
[18] T. Arai, K. Tokumaru, *Adv. Photochem.*, **1995**, *20*, 1.
[19] R. B. Woodward, R. Hoffmann, *Angew. Chem., Int. Ed.* **1969**, *8*, 781.
[20] R. B. Woodward, 'Aromaticity', The Chemical Society, London, 1967.
[21] S. J. Rhoads, in 'Molecular Rearrangements, Part 1', Ed. P. de Mayo, Interscience Publ., New York, 1963, p. 655.
[22] C. E. Woodward, in 'Creative People at Work', Eds. D. B. Wallace, H. E. Gruber, Oxford University Press, New York, 1989, p. 227.
[23] J. A. Berson, 'Chemical Creativity', Wiley-VCH, Weinheim, 1999.
[24] E. Havinga, J. L. M. A. Schlatmann, *Tetrahedron* **1961**, *16*, 146.
[25] '*13th Chemistry Conference of the Solvay Institute*', Reactivity of the Photoexcited Organic Molecule, Interscience Publ., New York, 1967.
[26] a) T. Förster, *Pure Appl. Chem.* **1970**, *24*, 443; b) T. Förster, *Pure Appl. Chem.* **1973**, *34*, 225; c) W. Th. A. M. Van der Lugt, L. J. Oosterhoff, *J. Chem. Soc., Chem. Commun.* **1968**, 1235.
[27] S. G. Brush, *Science* **1989**, *246*, 1124.
[28] a) G. Quinkert, W.-W. Wiersdorff, M. Finke, K. Opitz, F.-G. von der Haar, *Chem. Ber.* **1968**, *101*, 2302; b) G. Quinkert, M. Finke, J. Palmowski, W.-W. Wiersdorff, *Mol. Photochem.* **1969**, *1*, 433; c) K. H. Grellmann, J. Palmowski, G. Quinkert, *Angew. Chem., Int. Ed.* **1971**, *10*, 196.

[29] a) F. Boomsa, H. C. J. Jacobs, E. Havinga, A. Van der Gen, *Recl. Trav. Chim. Pays-Bas* **1977**, *96*, 104; 113; b) H. C. J. Jacobs, *Pure Appl. Chem.* **1995**, *67*, 63.
[30] A. G. M. Barrett, D. H. R. Barton, R. A. Russell, D. A. Widdowson, *J. Chem. Soc., Perkin Trans. 1*, **1977**, 631.
[31] G.-D. Zhu, W. H. Okamura, *Chem. Rev.* **1995**, *95*, 1877.
[32] K. Alder, M. Schumacher, *Fortschritte Chem. Org. Naturstoffe* **1953**, *10*, 1.
[33] H. H. Inhoffen, K. Brückner, R. Gründel, G. Quinkert, *Chem. Ber.* **1954**, *87*, 1407; H. H. Inhoffen, G. Quinkert, *Chem. Ber.* **1954**, *87*, 1418.
[34] S. C. Eyley, D. H. Williams, *J. Chem. Soc., Chem. Commun.* **1975**, 858.
[35] A. A. Lamola, in 'Energy Transfer and Organic Photochemistry', Eds. A. A. Lamola, N. J. Turro, Interscience Publ., New York, 1969.
[36] K.-H. Pfoertner, *J. Chem. Soc., Perkin Trans. 2*, **1991**, 523; K.-H. Pfoertner, M. Voelker, *J. Chem. Soc., Perkin Trans. 2*, **1991**, 527.
[37] M. Nowakowska, V. P. Foyle, J. E. Guillet, *J. Am. Chem. Soc.* **1993**, *115*, 5975.
[38] a) K. Pfoertner, *Helv. Chim. Acta* **1972**, *55*, 921; b) T. Kobayashi, M. Yasumura, *J. Nutr. Sci. Vitaminol.* **1973**, *19*, 123; c) H. J. C. Jacobs, J. W. H. Gielen, E. Havinga, *Tetrahedron Lett.* **1981**, 4013.
[39] W. G. Dauben, R. B. Phillips, *J. Am. Chem. Soc.* **1982**, *104*, 355.
[40] V. Malesta, C. Willis, P. A. Hackett, *J. Am. Chem. Soc.* **1981**, *103*, 6781.
[41] a) P. A. Maessen, H. J. C. Jacobs, J. Cornelisse, E. Havinga, *Angew. Chem., Int. Ed.* **1983**, 22, 718; *Angew. Chem., Suppl.* **1983**, 994; b) H. J. C. Jacobs, E. Havinga, *Adv. Photochem.* **1979**, *11*, 305.
[42] W. G. Dauben, P. E. Share, R. R. Ollmann Jr., *J. Am. Chem. Soc.* **1988**, *110*, 2548.
[43] J. I. Seeman, *Chem. Rev.* **1983**, *83*, 83.
[44] W. G. Dauben, B. Disanayaka, D. J. H. Fumhoff, B. E. Kohler, D. E. Schilke, B. Zhou, *J. Am. Chem. Soc.* **1991**, *113*, 8367.
[45] P. E. Share, K. L. Kompa, S. D. Peyerimhoff, M.C Van Hermert, *Chem. Phys.* **1988**, *120*, 411; W. Th. A. M. van der Lugt, L. J. Oosterhoff, *J. Am. Chem. Soc.* **1969**, *91*, 6042; D. Grimbert, G. Segal, A. Devaquet, *J. Am. Chem. Soc.* **1975**, *97*, 6629.
[46] P. C. Beadle, *Photochem. Photobiol.* **1977**, *25*, 519.
[47] R. M. Moriarty, R. N. Schwartz, C. Lee, V. Curtis, *J. Am. Chem. Soc.* **1980**, *102*, 4257.
[48] J. G. Haddad, Jr., T. J. Hahn, *Nature* **1973**, *244*, 515.
[49] S. J. Marx, U. A. Liberman, C. Eil, *Vitamines and Hormones* **1983**, *40*, 235.
[50] G. Quinkert, Memorial Lecture held for *H. H. Inhoffen*, Braunschweig, 3. Dec. 1993.
[51] a) H. H. Inhoffen, K. Irmscher, *Fortschr. Chem. Org. Naturstoffe* **1959**, *17*, 70; b) K. Irmscher, in 'Über Sterine, Gallensäuren und verwandte Naturstoffe', Ed. H. H. Inhoffen, Bd. 2, p. 652. F. Enke Verlag, Stuttgart, 1959; c) H. H. Inhoffen, K. Irmscher, H. Hirschfeld, H. Stache, A. Kreutzer, *Chem. Ber.* **1958**, *91*, 2309; d) H. H. Inhoffen, G. Quinkert, S. Schütz, G. Friedrich, E. Tober, *Chem. Ber.* **1958**, *91*, 781; e) H. H. Inhoffen, E. Prinz, *Chem. Ber.* **1954**, *87*, 684; f) H. H. Inhoffen, S. Schütz, P. Rossberg, O. Berges, K.-H. Nordsiek, H. Plenio, E. Höroldt, *Chem. Ber.* **1958**, *91*, 2626; g) H. H. Inhoffen, H. Burkhardt, G. Quinkert, *Chem. Ber.* **1959**, *92*, 1564.
[52] W. H. Perkin, Jr., *J. Chem. Soc.* **1904**, 654.
[53] D. H. R. Barton, *Chem. Br.* **1973**, *9*, 149.
[54] J. D. Dunitz, '*Looking Backwards, Glancing Sideways: Half a Century of Chemical Crystallography*', in 'Essays in Contemporary Chemistry: From Molecular Structure towards Biology', Eds. G. Quinkert, M. V. Kisakürek, Verlag Helvetica Chimica Acta, Zürich, 2001.
[55] J. D. Dunitz, 'X-Ray Analysis and the Structure of Organic Molecules', Verlag Helvetica Chimica Acta, Basel, 1995.
[56] G. Wittig, U. Schöllkopf, *Chem. Ber.* **1954**, *87*, 1318.
[57] H. H. Inhoffen, *Angew. Chem.* **1960**, *72*, 875.
[58] a) B. Lythgoe, *Chem. Soc. Rev.* **1980**, *9*, 449; b) B. Lythgoe, R. S. Manwaring, J. R. Milner, T. A. Moran, M. E. N. Nambudiry, J. Tideswell, *J. Chem. Soc., Perkin Trans. 1*, **1978**, 387; c) I. J. Bolton, R. G. Harrison, B. Lythgoe, R. S. Manwaring, *J. Chem. Soc. (C)* **1971**, 2944; d) I. J. Bolton, R. G. Harrison, B. Lythgoe, *J. Chem. Soc. (C)*, **1971**, 2950; e) P. S. Littlewood, B. Lythgoe, A. K. Saksena, *J. Chem. Soc. (C)*, **1971**, 2955.

[59] a) G. Quinkert (Ed.), *Synform* **1985**, *2*, 41; **1986**, *3*, 131; **1987**, *5*,1; b) H. Dai, G. H. Posner, *Synthesis* **1994**, 1383.
[60] G. Quinkert, *Angew. Chem.* **1967**, *79*, 730.
[61] T. M. Dawson, J. Dixon, P. S. Littlewood, B. lythgoe, A. K. Saksena, *J. Chem. Soc. (C)* **1971**, 2960.
[62] G. Quinkert, *Angew. Chem., Int. Ed.* **1972**, *11*, 1072.
[63] G. Quinkert, Dechema-Monographie Nr. 871, in 'Chemische Reaktionstechnik', S. 239, Verlag Chemie, Weinheim, 1964.
[64] D. H. R. Barton, in 'Über Sterine, Gallensäuren und verwandte Naturstoffe', Ed. H. H. Inhoffen, Bd. 2, p. 618. F. Enke Verlag, Stuttgart, 1959.
[65] D. H. R. Barton, *Experientia* **1950**, *6*, 316.
[66] G. Quinkert, in 'The Bartonian Legacy', Eds. A. I. Scott, P. Potier, Imperial College Press, London, 2000, p. 138.
[67] D. H. R. Barton, G. Quinkert, *Proc. Chem. Soc.* **1958**, 197; D. H. R. Barton, G. Quinkert, *J. Chem. Soc.* **1960**, 1; D. H. R. Barton, *Helv. Chim. Acta* **1959**, *42*, 2604.
[68] C. Reichardt, 'Solvents and Solvent Effects in Organic Chemistry', 2nd. Ed., VCH-Verlag, Weinheim, 1988.
[69] G. Scheibe, *Ber. dtsch. chem. Ges.* **1925**, *58*, 586; **1926,** *59*, 2619.
[70] J. N. Murrell, 'The Theory of the Electronic Spectra of Organic Molecules', Methuen & Co., London, 1963.
[71] G. Quinkert, G. Dürner, E. Kleiner, F. Adam, E. Haupt, D. Leibfritz, *Chem. Ber.* **1980**, *113*, 2227.
[72] a) G. Quinkert, *Photochem. Photobiol.* **1968**, *7*, 783; b) G. Quinkert, *Pure Appl. Chem.* **1973**, *33*, 285; c) G. Quinkert, B. Bronstert, D. Egert, P. Michaelis, P. Jürges, G. Prescher, A. Syldatk, H.-H. Perkampus, *Chem. Ber.* **1976**, *109*, 1332; d) G. Quinkert, H. Englert, F. Cech, A. Stegk, E. Haupt, D. Leibfritz, D. Rehm, *Chem. Ber.* **1979**, *112*, 310; e) G. Quinkert, E. Kleiner, B.-J. Freitag, J. Glenneberg, U.-M. Billhardt, F. Cech, K. R. Schmieder, C. Schudok, H.-C. Steinmetzer, J. W. Bats, G. Zimmermann, G. Dürner, D., Rehm, E. F. Paulus, *Helv. Chim. Acta* **1986**, *69*, 469; f) G. Quinkert, S. Scherer, D. Reichert, H.-P. Nestler, H. Wennemers, A. Ebel, K. Urbahns, K. Wagner, K.-P. Michaelis, G. Wiech, G. Prescher, B. Bronstert, B.-J. Freitag, I. Wicke, D. Lisch, P. Belik, T. Crecelius, D. Hörstermann, G. Zimmermann, J. W. Bats, G. Dürner, D. Rehm, *Helv. Chim. Acta* **1997**, *80*, 1683.
[73] O. L. Chapman, J. D. Lassila, *J. Am. Chem. Soc.* **1968**, *90*, 2449; O. L. Chapman, Proceedings XXIIIrd Int. Congr. Pure Appl. Chem. Spec. Lect. 1971, 1, 311.
[74] a) G. Quinkert, B. Bronstert, P. Michaelis, U. Krüger, *Angew. Chem., Int. Ed.* **1970**, *9*, 241; b) H. H. Perkampus, G. Prescher, B. Bronstert, G. Quinkert, *Angew. Chem., Int. Ed.* **1970**, *9*, 241; c) G. Quinkert, M. Hintzmann, P. Michaelis, P. Jürges, *Angew. Chem., Int. Ed.* **1970**, *9*, 238.
[75] H. Mauser, 'Formale Kinetik', Bertelsmann Universitätsverlag, Düsseldorf, 1974.
[76] G. Quinkert, E. Egert, C. Griesinger, 'Aspects of Organic Chemistry', Verlag Helvetica Chimica Acta, Basel, 1996.
[77] G. Quinkert, H. Hintzmann, P. Michaelis, P. Jürges, H. Appelt, U. Krüger, *Liebigs Ann. Chem.* **1971**, *748*, 38.
[78] a) R. Samtleben, H. Pracejus, *J. Prakt. Chem.* **1972**, *314*, 1576, and refs. cit. therein; b) D. P. N. Satchell, R. S. Satchell, *Chem. Soc. Rev.* **1975**, *4*, 231, and refs. cit. therein; c) J. D. Morrison, H. S. Mosher, 'Asymmetric Organic Reactions'; 2. Printing, Chapt. 6. American Chemical Society, Washington, 1976; e) K. Sung, T. T. Tidwell, *J. Am. Chem. Soc.* **1998**, *120*, 3043.
[79] Unpublished results.
[80] M. L. Bender, R. J. Bergeron, M. Komiyama, 'The Bioorganic Chemistry of Enzymatic Catalysis', Chapt. 7, John Wiley and Sons, New York, 1984.
[81] a) G. Quinkert, G. Fischer, U.-M. Billhardt, J. Glenneberg, U. Hertz, G. Dürner, E. F. Paulus, J. W. Bats, *Angew. Chem., Int. Ed.* **1984**, *23*, 440; b) G. Quinkert in 'Organic Synthesis: An Interdisciplinary Challenge', Eds. J. Streith, H. Prinzbach, G. Schill, Blackwell Scientific Publ., Oxford, 1985, p. 131; c) G. Quinkert, U.-M. Billhardt, H.

Jakob, G. Fischer, J. Glenneberg, P. Nagler, V. Autze, N. Heim, M. Wacker, T. Schwalbe, Y. Kurth, J. W. Bats, G. Dürner, G. Zimmermann, H. Kessler, *Helv. Chim. Acta* **1987**, *70*, 771.

[82] G. Quinkert, H. P. Nestler, B. Schumacher, M. Del Grosso, G. Dürner, J. W. Bats, *Tetrahedron Lett.* **1992**, *33*, 1977.

[83] J. Griffiths, H. Hart, *J. Am. Chem. Soc.* **1968**, *90*, 5296.

[84] H. Hart, *Pure Appl. Chem.* **1973**, *33*, 247.

[85] a) L. Salem, W. G. Dauben, N. J. Turro, *J. Chim. Phys.* **1973**, *70*, 694; b) W. G. Dauben, L. Salem, N. J. Turro, *Acc. Chem. Res.* **1975**, *8*, 41; c) L. Salem, *Science* **1976**, *191*, 822.

[86] V. Bonacic-Koutecky, P. Bruckmann, P. Hiberty, J. Koutecky, C. Leforestier, L. Salem, *Angew. Chem.* **1975**, *14*, 575.

[87] a) O. Isler, R. Rüegg, U. Schwieter, *Pure Appl. Chem.* **1967**, *14*, 245; b) 'Carotenoids', Ed. O. Isler, Birkhäuser Verlag, Basel, 1971.

[88] G. Quinkert, K. R. Schmieder, G. Dürner, K. Hache, A. Stegk, D. H. R. Barton, *Chem. Ber.* **1977**, *110*, 3582.

[89] L. Zechmeister, 'Cis-Trans Isomeric Carotenoids, Vitamin A and Arylpolyenes', p. 114, Springer-Verlag, Wien 1962.

[90] a) W. Kuhn, E. Knopf, *Z. Phys. Chem.* **1930**, *7B*, 292; b) O. Buchardt, *Angew. Chem. Int. Ed.* **1974**, *13*, 179.

[91] a) D. Reichert, Dissertation, Universität Frankfurt am Main, 1992; b) P. Eckes, Diplomarbeit, Universität Frankfurt am Main, 1987; c) A. Ebel, Dissertation, Universität Frankfurt am Main, 1994; d) O. Schultheis, Dissertation, Universität Frankfurt am Main, 1991.

[92] S. Drenkard, J. Ferris, A. Eschenmoser, *Helv. Chim. Acta* **1990**, *73*, 1373.

[93] a) N. A. Saccomano, in B. M. Trost, I. Fleming, S. L. Schreiber (Eds.): Comprehensive Organic Synthesis, Vol. 1. Pergamon Press, Oxford, 1991, P. 193. b) S. Itsuno in 'Comprehensive Asymmetric Catalysis', Eds. E. N. Jacobsen, A. Pfaltz, H. Yamamoto, Vol. 1. Springer-Verlag, Berlin, 1999. P. 289.

[94] a) M. Sakakibara, M. Matsui, *Agric. Biol. Chem.* **1973**, 37, 911; b) A. Ishida, T. Mukayama, *Bull. Chem. Soc. Jpn.* **1978**, *51*, 2077; c) H. Braun, Ph. D. Thesis, Universität Frankfurt am Main, 1991; d) S. Takeuchi, H. Yonehara, *J. Antibiot., Ser. A* **1969**, *22*, 179.

[95] a) A. Eschenmoser, A. Frey, *Helv. Chim. Acta* **1952**, *35*, 1660; b) C. A. Grob, *Angew. Chem., Int. Ed.* **1969**, *8*, 535; c) P. Weyerstahl, H. Marschall in 'Comprehensive Organic Syntheses', Eds. B. M. Trost, I. Fleming, E. Winterfeldt, Vol. 6, P. 1041. Pergamon Press, Oxford, 1991.

[96] a) G. Quinkert, N. Heim, J. W. Bats, H. Oschkinat, H. Kessler, *Angew. Chem., Int. Ed.* **1985**, *24*, 987; b) S. Huneck, K. Schreiber, W. Steglich, *Tetrahedron* **1973**, *29*, 3687.

[97] a) G. Quinkert, N. Heim, J. Glenneberg, U.-M. Billhardt, V. Autze, J. W. Bats, G. Dürner, *Angew. Chem., Int. Ed.* **1987**, *26*, 362; b) G. Quinkert, N. Heim, J. Glenneberg, U. Döller, M. Eichhorn, U.-M. Billhardt, C. Schwarz, G. Zimmermann, J. W. Bats, G. Dürner, *Helv. Chim. Acta* **1988**, *71*, 1719; c) G. Quinkert, U. Döller, M. Eichhorn, F. Küber, H. P. Nestler, H. Becker, J. W. Bats, G. Zimmermann, G. Dürner, *Helv. Chim. Acta* **1990**, *73*, 1999; d) G. Quinkert, H. Becker, G. Dürner, *Tetrahedron Lett.* **1991**, *32*, 7397; e) A. Leonhardt, Ph. D. Thesis, Universität Frankfurt am Main, 1992.

[98] G. Quinkert, E. Fernholz, P. Eckes, D. Neumann, G. Dürner, *Helv. Chim. Acta* **1989**, *72*, 1753.

[99] a) W. Oppolzer, R. N. Radinor, J. De Brabander, *Tetrahedron Lett.* **1995**, *36*, 2607; b) D. Enders, O. F. Prokopenko, *Liebigs Ann. Chem.* **1995**, 1185; c) S. C. Sinha, E. Keinan, *J. Org. Chem.* **1997**, *62*, 377; d) Y. Kobayashi, M. Nakano, G. B. Kumar, K. Kishihara, *J. Org. Chem.* **1998**, *63*, 7505; e) T. Nishioka, Y. Iwabuchi, H. Irie, S. Hatakeiyama, *Tetrahedron Lett.* **1998**, *39*, 5597; f) D. J. Dixon, A. C. Foster, S. V. Ley, *Org. Lett.* **2000**, *2*, 123.

[100] a) P. P. Waanders, L. Thijs, B. Zwanenburg, *Tetrahedron Lett.* **1987**, *28*, 2409; b) G. Solladié, I. Fernandez, C. Maestro, *Tetrahedron Lett.* **1991**, *32*, 509; G. Solladié, I. Fernandez, C. Maestro, *Tetrahedron: Asymmetry* **1991**, 2, 801.

[101] R. B. Woodward, in 'Frontiers in Bioorganic Chemistry and Molecular Biology', Ed. Y. A. Ovchinikov, M. N. Kolosov, Elsevier, Amsterdam, 1979.

[102] R. Huisgen, H. Ott, *Tetrahedron* **1959**, *6*, 253.
[103] D. A. Evans, J. V. Nelson, T. R. Taber, *Top. Stereochem.* **1982**, *13*, 1; C. H. Heathcock, 'Asymmetric Synthesis', Vol. 3, Ed. J. D. Morrison, Academic Press, Orlando, 1984; Comprehensive Asymmetric Catalysis I – III, Eds. E. N. Jacobsen, A. Pfaltz, H. Yamamoto, Springer, Berlin, 1999.
[104] S. Iguchi, H. Nakai, M. Hayashi, H. Yamamoto, *Bull. Chem. Soc. Jpn.* **1981**, *54*, 3033; S. Iguchi, H. Nakai, M. Hayashi, H. Yamamoto, *J. Org. Chem.* **1979**, *44*, 1363.
[105] B. Seuring, D. Seebach, *Helv. Chim. Acta* **1977**, *60*, 1175.
[106] D. Seebach, A. K. Beck, A. Heckel, '*TADDOL and Its Derivatives – Our Dream of Universal Chiral Auxiliaries*', in 'Essays in Contemporary Chemistry: From Molecular Structure towards Biology', Eds. G. Quinkert, M. V. Kisakürek, Verlag Helvetica Chimica Acta, Zürich, 2001.
[107] S. Hanessian, 'Total synthesis of natural products: the chiron approach', Pergamon Press, Oxford, 1983; J. W. Scott, in 'Asymmetric Synthesis', Vol. 4, Eds. J. D. Morrison, J. W. Scott, Academic Press, Orlando, 1984.
[108] D. Enders, in 'Stereoselective Synthesis', Eds. E. Ottow, K. Schöllkopf, B.-G. Schulz, Springer-Verlag, Berlin, 1993.
[109] a) S. Hatakeyama, K. Satoh, K. Sakurai, S. Takano, *Tetrahedron Lett.* **1987**, *28*, 2713; b) S. Hatakeyama, K. Satoh, S. Takano, *Tetrahedron Lett.* **1993**, *34*, 7425.
[110] S. L. Schreiber, T. S. Schreiber, D. B. Smith, *J. Am. Chem. Soc.* **1987**, *109*, 1525.
[111] a) J. S. Barlow, D. J. Dixon, A. C. Foster, S. V. Ley, D. J. Reynolds, *J. Chem. Soc., Perkin Trans. 1* **1999**, 1627; b) D. J. Dixon, A. C. Foster, S. V. Ley, D. J. Reynolds, *J. Chem. Soc., Perkin Trans. 1* **1999**, 1631; c) D. J. Dixon, A. C. Foster, S. V. Ley, D. J. Reynolds, *J. Chem. Soc., Perkin Trans. 1* **1999**, 1635.
[112] a) M. Kitamura, S. Suga, K. Kawai, R. Noyori, *J. Am. Chem. Soc.* **1986**, *108*, 6071; b) R. Noyori, M. Kitamura, *Angew. Chem.* **1991**, *103*, 34.
[113] a) W. Oppolzer, R. N. Radinov, *Helv. Chim. Acta* **1992**, *75*, 170; b) W. Oppolzer, R. N. Radinov, *J. Am. Chem. Soc.* **1993**, *115*, 1593.
[114] G. Quinkert, F. Küber, W. Knauf, M. Wacker, U. Koch, H. Becker, H. P. Nestler, G. Dürner, G. Zimmermann, J. W. Bats, E. Egert, *Helv. Chim. Acta* **1991**, *74*, 1853.
[115] a) W. C. Still in 'Current Trends in Organic Synthesis', Ed. H. Nozaki, Pergamon Press, Oxford, 1983; p. 233; b) W. C. Still in 'Selectivity – a Goal for Synthetic Efficiency', Eds. W. Bartmann, B. M. Trost, Verlag Chemie, Weinheim, 1984, p. 263.
[116] a) G. Quinkert (Ed.), *Synform* **1984**, *3*, 125 (for literature in and before 1984); b) S. C. Sinha, A. Sinha-Bagchi, E. Keinan, *J. Org. Chem.* **1993**, *58*, 7789; c) Y. Kobayashi, H. Okui, *J. Org. Chem.* **2000**, *65*, 612.
[117] a) N. C. Yang, C. Rivas, *J. Am. Chem. Soc* **1961**, *83*, 2213; b) E. F. Zwicker, L. I. Grossweiner, N. C. Yang, *J. Am. Chem. Soc* **1963**, *85*, 2671; c) K. R. Huffman, M. Loy, E. F. Ullman, *J. Am. Chem. Soc* **1965**, *87*, 5417.
[118] a) G. Quinkert, *Chimia* **1977**, *31*, 225; G. Quinkert, *Affinidad* **1977**, *34*, 42; b) G. Quinkert, W.-D. Weber, U. Schwartz, G. Dürner, *Angew. Chem., Int. Ed.* **1980**, *19*, 1027; c) G. Quinkert, U. Schwartz, H. Stark, W.-D. Weber, H. Baier, F. Adam, G. Dürner, *Angew. Chem., Int. Ed.* **1980**, *19*, 1029; d) G. Quinkert, W.-D. Weber, U. Schwartz, H. Stark, H. Baier, G. Dürner, *Liebigs Ann. Chem.* **1981**, 2335; e) G. Quinkert, U. Schwartz, H. Stark, W.-D. Weber, F. Adam, H. Baier, G. Frank, G. Dürner, *Liebigs Ann. Chem.* **1982**, 1999; f) G. Quinkert, H. Stark, *Angew. Chem., Int. Ed.* **1983**, *22*, 637.
[119] a) G. Quinkert, Workshop Conferences Hoechst, Eds. W. Bartmann, B. M. Trost, Verlag Chemie, Weinheim, 1984, Vol. 14, p. 213; b) G. Dürner, H. Baier, G. Quinkert, *Helv. Chim. Acta* **1985**, 68, 1054; c) G. Quinkert, Vorlesungsreihe Schering, Heft 19, Schering AG, Berlin, 1988; d) G. Quinkert in 'Stereoselective Synthesis', Eds. E. Ottow, K. Schöllkopf, B. G. Schulz, Springer-Verlag, Berlin, 1993.
[120] a) P. J. Wagner, *Pure Appl. Chem.* **1977**, *49*, 259; b) R. Haag, J. Wirz, P. Wagner, *Helv. Chim. Acta* **1977**, 60, 2595; c) P. K. Das, M. V. Encinas. R. D. Small, Jr., J. C. Scaiano, *J. Am. Chem. Soc.* **1979**, *79*, 6965; d) J. C. Scaiano, *Chem. Phys. Lett.* **1980**, *73*, 319; e) K.-H. Grellmann, H. Weller, E. Tauer, *Chem. Phys. Lett.* **1983**, *95*, 195; f) J. P. Bays, M. V. Encinas, J. C. Scaiano, *Macromolecules* **1979**, *12*, 348.

[121] a) F. Nerdel, W. Brodowski, *Chem. Ber.* **1968**, *101*, 1398; b) P. G. Sammes, *Tetrahedron* **1976**, *32*, 405; c) W. Oppolzer, *Angew. Chem., Int. Ed.* **1977**, *16*, 10 ; e) W. Oppolzer, *Synthesis* **1978**, 793; f) R. M. Wilson, in 'Organic Photochemistry', Ed. A. Padwa, Vol. 7, Marcel Dekker, New York, 1985, p. 339.

[122] G. Quinkert, 'Verhandl. der 120. Versammlung der Ges. Dtsch. Naturforscher und Ärzte 1998 in Berlin', S. Hirzel Verlag, Stuttgart, 1999, p. 163.

[123] J.-M. Lehn, 'Dynamic Combinatorial Chemistry and Virtual Combinatorial Libraries', in 'Essays in Contemporary Chemistry: from Molecular Structure towards Biology', Eds. G. Quinkert, M. V. Kisakürek, Verlag Helvetica Chimica Acta, Zürich, 2001.

[124] M. L. Steigerwald, W. A. Goddard III, D. A. Evans, *J. Am. Chem. Soc* **1979**, *101*, 1994.

[125] D. Leibfritz, E. Haupt, M. Feigel, W. E. Hull, W.-D. Weber, *Liebigs Ann. Chem.* **1982**, 1971.

[126] R. W. Kierstead, R. P. Linstead, B. C. L. Weedon, *J. Chem. Soc.* **1952**, 3610; 3616.

[127] S. Danishefsky, *Acc. Chem. Res.* **1979**, *12*, 66.

[128] N. Krause, *Kontakte (Darmstadt)* **1993**, 3.

[129] a) G. Quinkert, H.-G. Schmalz, E. Walzer, S. Gross, T. Kowalczyk-Przewloka, C. Schierloh, G. Dürner, J. W. Bats, H. Kessler, *Liebigs Ann. Chem.* **1988**, 283; b) G. Quinkert, T. Müller, A. Königer, O. Schultheis, B. Sickenberger, G. Dürner, *Tetrahedron Lett.* **1992**, *33*, 3469.

[130] a) G. Quinkert, M. Del Grosso, A. Bucher, J,W. Bats, G. Dürner, *Tetrahedron Lett.* **1991**, *32*, 3357; b) G. Quinkert, M. Del Grosso, A. Bucher, M. Bauch, W. Döring, J. W. Bats, G. Dürner, *Tetrahedron Lett.* **1992**, *33*, 3617; C) G. Quinkert, H. Becker, M. Del Grosso, G. Dambacher, J. W. Bats, G. Dürner, *Tetrahedron Lett.* **1993**, *34*, 6885; d) G. Quinkert, M. Del Grosso, A. Döring, W. Döring, R. I. Schenkel, M. Bauch, G. T. Dambacher, J. W. Bats, G. Zimmermann, G. Dürner, *Helv. Chim. Acta* **1995**, *78*, 1345.

[131] a) E. Dane, J. Schmitt, *Liebigs Ann. Chem.* **1938**, *536*, 197; b) E. Dane, J. Schmitt, *Liebigs Ann. Chem.* **1939**, *537*, 207.

[132] G. Singh, *J. Am. Chem. Soc* **1956**, *78*, 6109.

[133] P. Yates, P. Eaton, *J. Am. Chem. Soc* **1960**, *82*, 655.

[134] J. Das, R. Kubela, A. MacAlpine, Z. Stojanac, Z. Valenta, *Can. J. Chem.* **1979**, 3308.

[135] a) M. B. Groen, F. J. Zeelen, *Recl. Trav. Chim. Pays-Bas* **1986**, *105*, 465; b) F. J. Zeelen, *Nat. Prod. Rep.* **1994**, *11*, 607.

TADDOL and Its Derivatives – Our Dream of Universal Chiral Auxiliaries[1])

by **Dieter Seebach***, **Albert K. Beck**, and **Alexander Heckel**

Laboratorium für Organische Chemie der Eidgenössischen Technischen Hochschule, ETH Zentrum, Universitätstrasse 16, CH-8092 Zürich

Dreams are there to be realized and
in order to do so, we have to wake up![2])

After 15 years of work in the field of organolithium chemistry [1–5], we had decided in the early eighties to look for higher selectivities with derivatives of the early transition metals Ti, Zr, and V [6–10]. It turned out that

1) Manuscript of lectures held by *D. S.* at the University of Notre Dame (Indiana, USA), at Washington University (St. Louis, USA), at the IUPAC Congress (D-Berlin), at the Hungarian Academy of Sciences (H-Budapest), at the Slowak University of Technology (SK-Bratislava), at the COST Symposium (CH-Basel), at the University of Barcelona (Spain), and at the Symposium on 'Chirality in Chemistry' (DK-Copenhagen), 1999/2000.

2) The background of the picture above shows the spiral galaxy NGC 4603 as seen by the *NASA Hubble Space Telescope* in May 1999 (Photo No. STScI-PR99-19).

organotitanium compounds, indeed, react much more selectively with different functional groups, and that they are able to better differentiate diastereotopic groups and faces than their, at the time, common Li and Mg counterparts. Thus, it was obvious to also test chiral derivatives in enantioselective reactions. Being fans of tartaric acid, we soon came up with Ti-complexes of the diols obtained from (*R,R*)- or (*S,S*)-tartrate acetonide and aromatic *Grignard* reagents, threitol derivatives with the horrible systematic name $\alpha,\alpha,\alpha',\alpha'$-**t**etr**a**aryl-2,2-dimethyl-1,3-**d**ioxolane-4,5-**d**imethan**ols**, for which we proposed the abbreviation TADDOLs [6][11][12]. In the meantime, TADDOL is a synonym for all the structures with X = Y = OH, as indicated in the center of *Scheme 1*, and its derivatives have been employed for a myriad of approaches to 'introduce chirality' in products of chemical reactions, in materials, and in analysis, as will become evident in the following sections.

The preparation of TADDOLs and analogs, and thus the number of structures accessible, is subject to combinatorial variation: almost any aldehyde, ketone, and aryl halide from the catalogs provide the R^1, R^2, and Aryl substituents in the structure in *Scheme 1*, and just a small collection is presented in *Scheme 2*, together with an analog derived from (*R,R*)-cyclohexane-1,2-dicarboxylate.

Scheme 1. *TADDOLs, Analogs, and Derivatives Thereof Constructed from (+)- or (–)-Tartrate, Aldehydes or Ketones, and a* Grignard *Reagent* (shown here is the structure derived from (+)-(*R,R*)-tartrate). Instead of tartrate acetals (R^2 = H) or ketals (R^1,R^2 ≠ H) [12], other chiral nonracemic dicarboxylates may be employed in the *Grignard* reaction [13] [14], Aryl = Ph can be hydrogenated to C_6H_{11} [15] [16], and the diastereoisomeric derivatives with *geminal* Ph/Me instead of Aryl/Aryl groups are available through the 4,5-dibenzoyl-dioxolane [14]. For examples, for OH derivatizations, and for OH/Y and OH/Z substitutions, see *Schemes 2* and *3*.

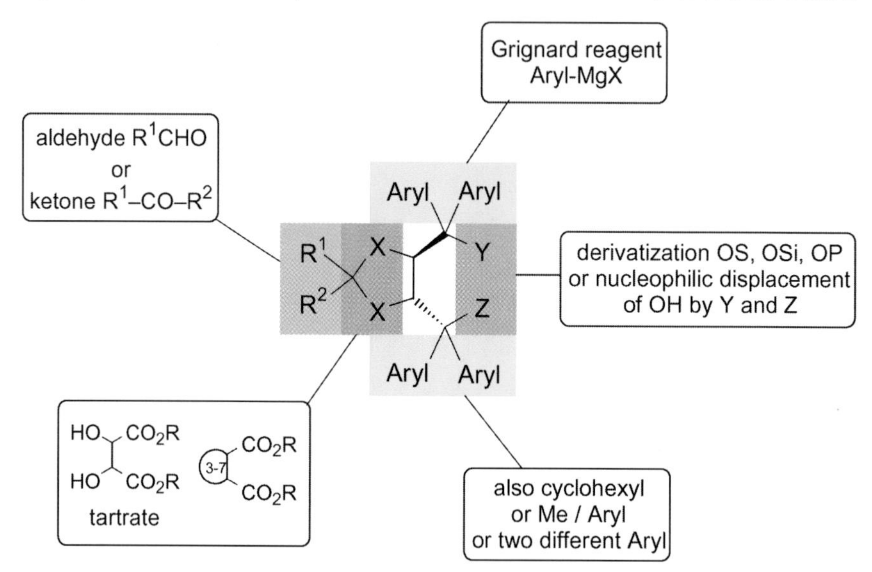

Scheme 2. *Some TADDOLs Prepared from Tartrate Acetals and Ketals (or from cyclohexane-dicarboxylate) and an Aromatic* Grignard *Reagent* (or a combination of two different organometallic nucleophiles). For preparation of the compounds containing a benzylic bromide or four *para*-OH groups (shown in the fourth line of the *Formulae*), suitable precursors or protecting groups have been used in the *Grignard* addition step. The references given underneath the *Formulae* refer to those papers in which the corresponding compounds have been described and characterized, or at least mentioned.

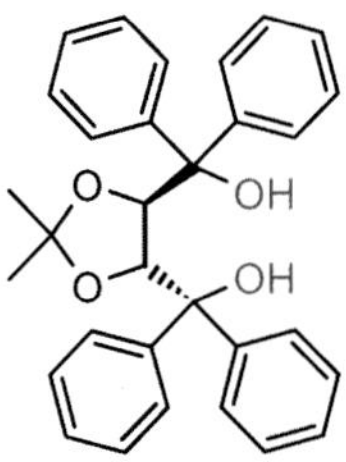

[12] 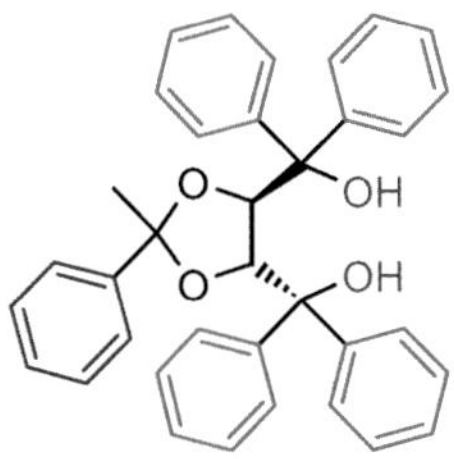 [16][17] 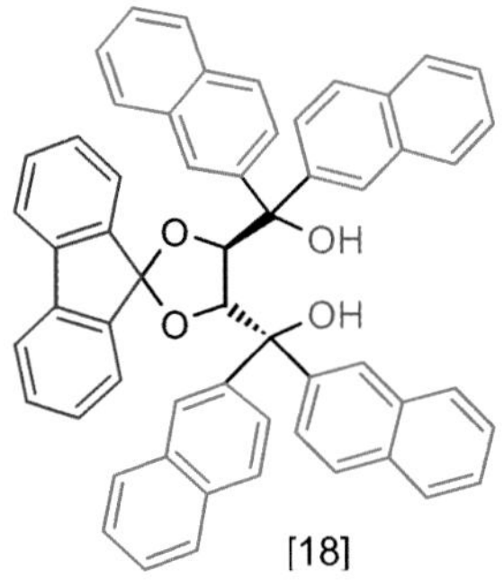 [18]

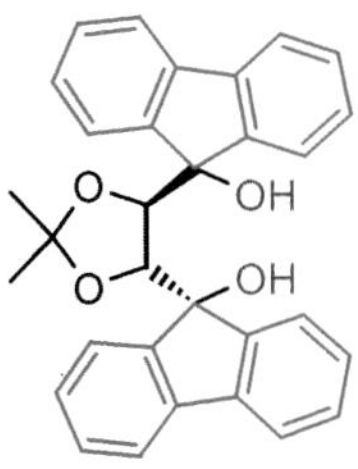

[19] 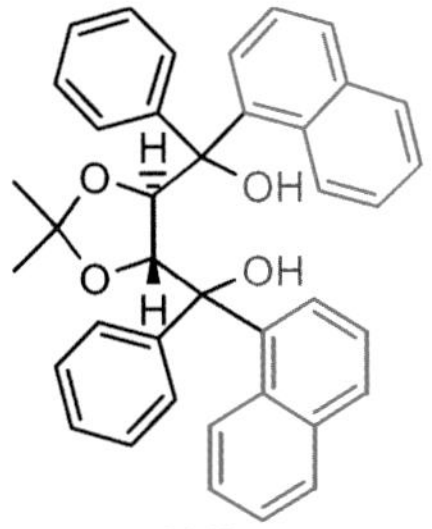 [14] 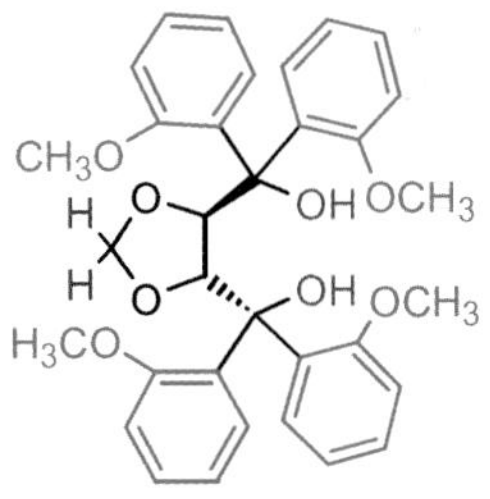 [20]

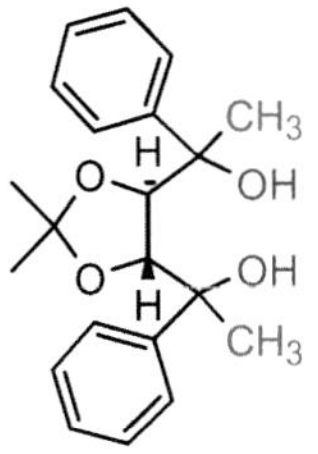

[14] 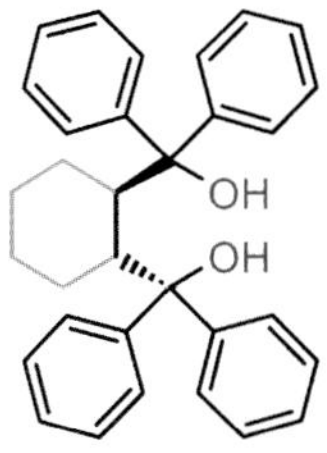 [13][14] 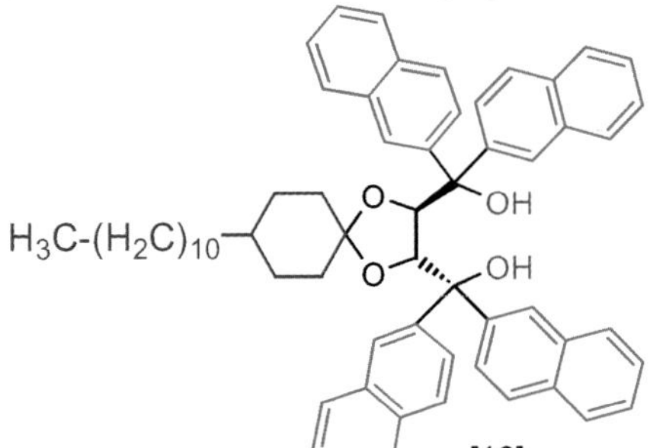 [18]

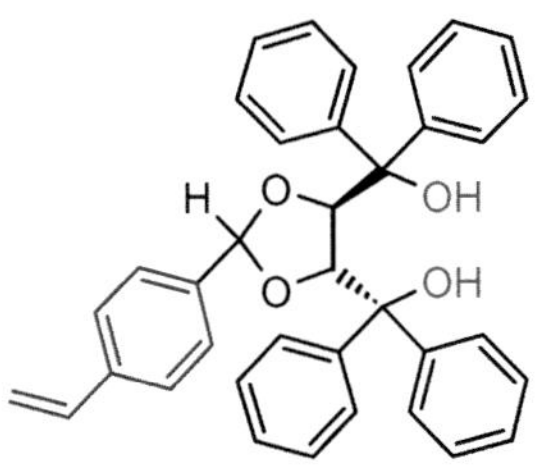

[21] 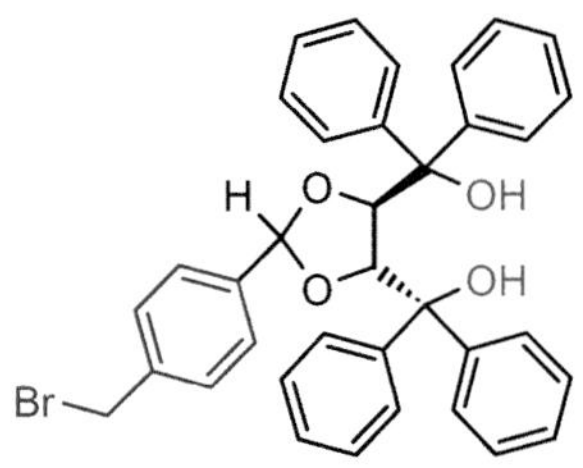 [21] 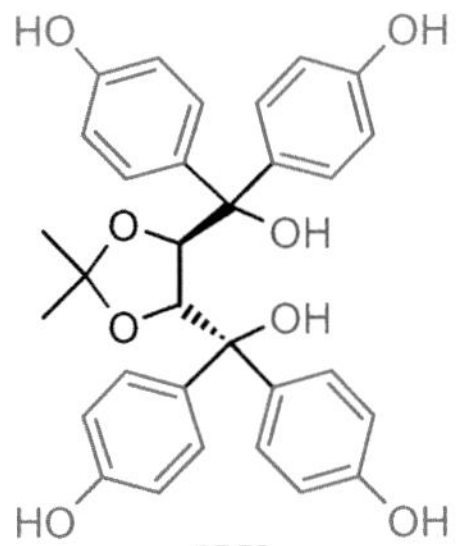 [22]

This variability is increased even more, if we include the derivatives of C_2- or C_1-symmetry which are accessible by derivatization of the OH groups (silylation → $OSiR_3$ or O–SiR_2–O; phosphinylation → OPR_2 or O–PR–O; sulfinylation → O–SO–O) or by substitutions, with the key steps being an *Appel* reaction, providing the monochloride [23], or conversion to the dichloride with thionyl chloride [24][25]. We had long hesitated to try such substitutions which should occur through cations (S_N1 reactions) at the diphenyl-substituted *tertiary* C-atoms of TADDOL, and which could have been competed by elimination, fragmentation or rearrangement [20]. In fact, most reactions leading to the products shown in *Scheme 3* have been optimized to occur with high yields (at least with the parent TADDOL).

If we consider that all sections of the TADDOL system can be readily modified, it is not surprising that more than 300 derivatives have been described in the literature (see supplementary material of [11]).

Scheme 3. *Products of Derivatization and of Mono- and Disubstitution of the OH Groups in the Parent TADDOL* ($R^1 = R^2$ = Me, X = O, Y = Z = OH, Aryl = Ph in *Scheme 1*). The derivatizations were achieved by treatment with halides such as $ClPR_2$, Cl_2PR, Cl_2SO, Cl_2SiR_2, Cl_3P, and a suitable base. The key-intermediates for substitutions are the mono- and the dichloride. In some of the compounds shown, the heteroatoms on the diphenylmethyl group are leaving groups, so that they may be lost, *i.e.*, replaced by OH, on contact with moisture in solution, and also during chromatography; thus, purification by recrystallization (from a non-nucleophilic solvent!) is recommended. The derivatives with *trans*-fused bicyclo[3.3.0]octane skeleton are surprisingly stable. The trityl derivative shown at the bottom right is formed in high yield from the dichloride and methyl(phenyl)amine.

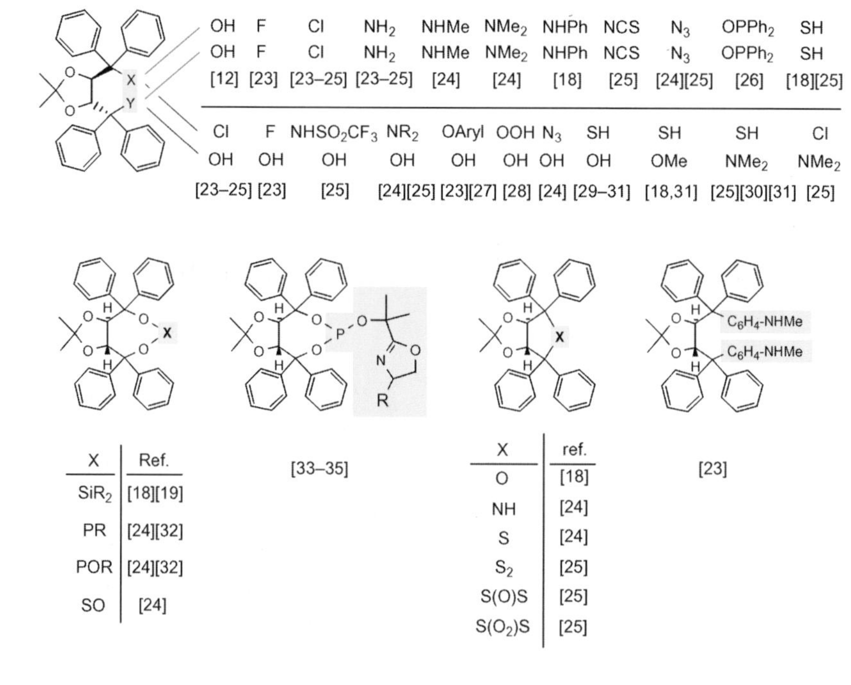

What is special about TADDOLs to make them so attractive to chemists that they have been prepared in such large numbers? TADDOLs have a predictable, rigid structure with the aryl groups arranged like on a propeller (see *Scheme 4*). The OH groups are pushed together and form an intramolecular H-bond (OH··· O), with a more acidic OH group ready for intermolecular H-bonding. The conformational lock provided by the *geminal* Ph groups also holds the attached heteroatoms in place for chelation of metal centers (*cf.* the *Thorpe-Ingold* effect). The aryl groups occupy *quasi*-axial and *quasi*-equatorial positions and mold a chiral environment around the center in between the two heteroatoms in such a way that there is room in two C_2-symmetrically related areas (upper left and lower right-hand side, with (*R,R*)-TADDOL derivatives) and steric hindrance in two others (upper right- and lower left-hand side; see *Scheme 4*). The *quasi*-axial Ph groups have a slight preference for a so-called 'edge-on', the *quasi*-equatorial ones for a 'face-on' conformation,

Scheme 4. *Structures of TADDOL Derivatives*. With very few exceptions, TADDOLs and their analogs have structures (in solution [31][36], in the solid-state [14][16][23][25][37], and according to calculations [14][38]) with juxtaposition of the heteroatoms on the two diarylmethyl groups. The three views from orthogonal directions are superpositions of 20 (*R,R*)-TADDOL crystal structures, with the substituents at C(2) of the dioxolane omitted for clarity (structures of (*S,S*)-TADDOLs have been inverted for inclusion in the superposition). The *quasi*-equatorial Ph groups may be replaced by Me groups, with little effect on the stereoselectivity of the Et_2Zn addition to PhCHO, replacement of the axial Ph by Me or of all four Ph by Me leads to total loss of stereoselectivity [14]. The insert shows the crystal structure of the $PdCl_2$ complex of TADDOP, with a nine-membered chelate ring in which the chiral ligand sphere around the metal is dictated by the two far remote stereogenic centers of the TADDOL moiety.

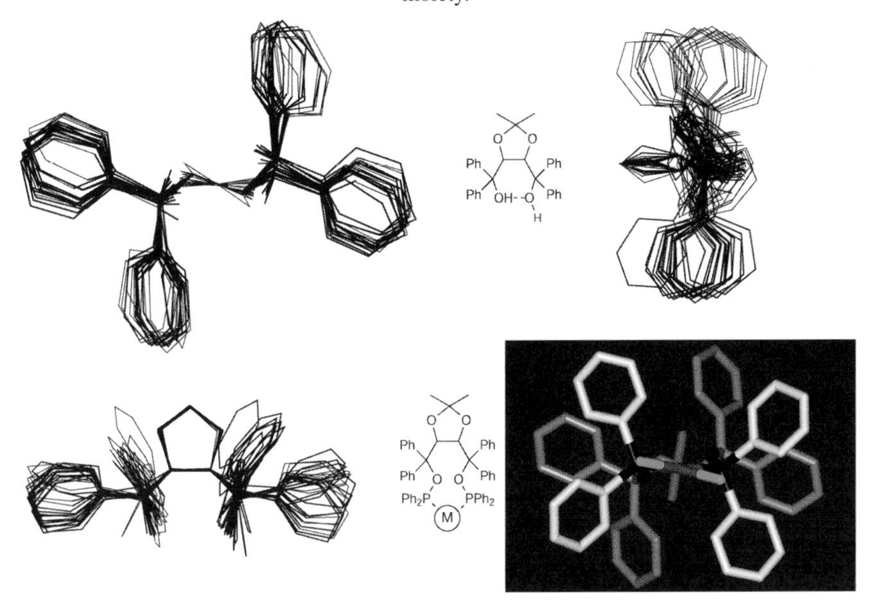

as viewed from the direction where a complexed metal ion is expected to reside (see the structural superpositions in *Scheme 4*). From numerous X-ray crystal structures (*ca.* 90 are in the *Cambridge Crystallographic Data Centre*), it is evident that substituents in the 2-position of the dioxolane ring may influence, in a subtle way, the conformation around the axial C–Aryl bonds, and thus be important for fine-tuning reactivity in the ligand sphere of a chelated metal center. The TADDOL moiety may have far-reaching conformational effects.

What have TADDOL derivatives been used for? They can be called chiral auxiliaries in the broadest sense of the term. A list of applications is presented in *Scheme 5*, and selected topics out of this list are discussed in the following sections. TADDOL itself may be considered as a chiral, propeller-like, rather rigid H-bond-donor molecule with a great tendency to crystallize by inclusion of smaller molecules (the combination of the bulky diarylmethyl groups with the tiny dioxolane ring in the same molecule seems to create packing problems, so that host molecules (preferably H-bond acceptors) are included in crystals, of course enantioselectively so!). The TADDOLates, on

Scheme 5. *The Width and Wealth of Applications of the TADDOL System.* Some of the topics listed will be the subject of the following *Schemes* (cholesteric phases; resolutions; solid-state reactions; conversions involving Mg, Al, Cu, Zn, Ti, Mo, Rh, Pd; polymer-bound TADDOLs), therefore references are given herein only for those applications which will not be further mentioned.

Applications of TADDOL Derivatives?

- chiral additives in liquid crystals (nematic ⟶ cholesteric phases)
- enantiomer separation (hydrogen bonding, clathrate formation) crystallizing resolutions; NMR shift reagents [39]
- solid-state reactions, including photochemistry
- ligands and ligand components on metal centers for stoichiometric and catalytic (homogeneous and heterogeneous) stereoselective reactions

Li	Mg	B	Al	Si	Cu	Zn	Ce	Ti	Zr	Mo	Rh	Ir	Pd
[40]		[41]					[42]		[37]			[34]	

• nucleophilic 1,2- and 1,4-additions • allylations

• [2+1], [2+2], [3+2], [4+2] cycloadditions • iodolactonizations

• metatheses • ene reactions

- stoichiometric chiral reagents (*cf.* protonations, oxidations)

the other hand, can act as bidentate chiral ligands on the polar main-group and early-transition-group metal centers, and their nitrogen, phosphorous, and sulfur derivatives on the late-transition-group metal centers.

As first applications we should like to mention examples for the use of TADDOLs, which do not involve chemical reactions. In the field of liquid crystals (material science), it is useful to be able to convert an achiral (nematic) phase into a chiral (cholesteric) one, and this can be achieved by adding small amounts of chiral additives (*dopants*, compare with *catalysts*). It turns out that TADDOLs are the most efficient dopants known (see *Scheme 6*). The above-mentioned ability of TADDOLs to form inclusion or *guest-host* compounds has been recognized by the *Toda* group [44] to provide a third, fundamental method of crystallizing resolution, besides the use of chiral acids or bases for resolving *rac*-bases and *rac*-acids, respectively. Thus, *rac*-compounds containing O or N H-bond acceptors and TADDOLs form inclusion compounds with enantiomer differentiation. Indeed, a large fraction of papers on TADDOLs describe examples of this type, including the corresponding X-ray crystal structures. A single crystallization is often sufficient to obtain almost the theoretical amount of 50% of one enantiomer in the inclusion compound. Of the dozens of examples known, we show a few in *Scheme 7*. The versatility of the method is impressive, and it is fair to say that TADDOL resolutions can be compared with the classical acid-base resolutions, also in view of the fact that there is a large *family* [49] of TADDOLs available.

Another use of TADDOL inclusion compounds is to carry out a solid-state reaction with a TADDOL crystal containing an *achiral* guest molecule, such that the latter is converted to a chiral product. There are again numerous highly successful applications of this procedure by the *Toda* group [50]. Especially intriguing are enantioselective photoreactions with inter- and intramolecular C,C-bond formation, for which three examples are shown in *Scheme 8*.

TADDOL was originally designed as a bidentate alkoxide ligand for titanium, and the majority of applications involve Ti-TADDOLates that are used to generate stoichiometrically applied chiral reagents, or to mediate or catalyze reactions depending on *Lewis*-acid activation. In several cases, it has been proven that Ti-TADDOLates accelerate reactions more strongly than comparable simple Ti-alkoxides, such as $(i\text{-PrO})_4Ti$ [14][38][56], an effect which has been referred to as 'ligand acceleration' [57]. In *Scheme 9*, we show a few examples of enantioselective nucleophilic additions to aldehydes, enones, nitro-olefins, and anhydrides, in which Ti-TADDOLates are employed as additives, or in which reagents of the type TADDOLate·Ti(X)Nu are prepared and subsequently combined with the corresponding electrophile (Nu may be an alkyl, vinyl, aryl, allyl, enolate, alkoxide, or CN group).

Scheme 6. *Helical-Twisting Power (HTP) of TADDOLs* (exceeds the value 400 μm^{-1} [19]). HTP is a measure for the efficiency with which a small amount of a chiral additive (*dopant*) causes an achiral liquid crystal to become cholesteric (a highly valued property of liquid crystals [43]). Prior to the tests with TADDOLs, a HTP of 100 μm^{-1} was considered high. Analysis of UV and CD spectra, and of the HTP measurements provides information about the specific conformation of the TADDOL molecules in the liquid-crystalline medium.

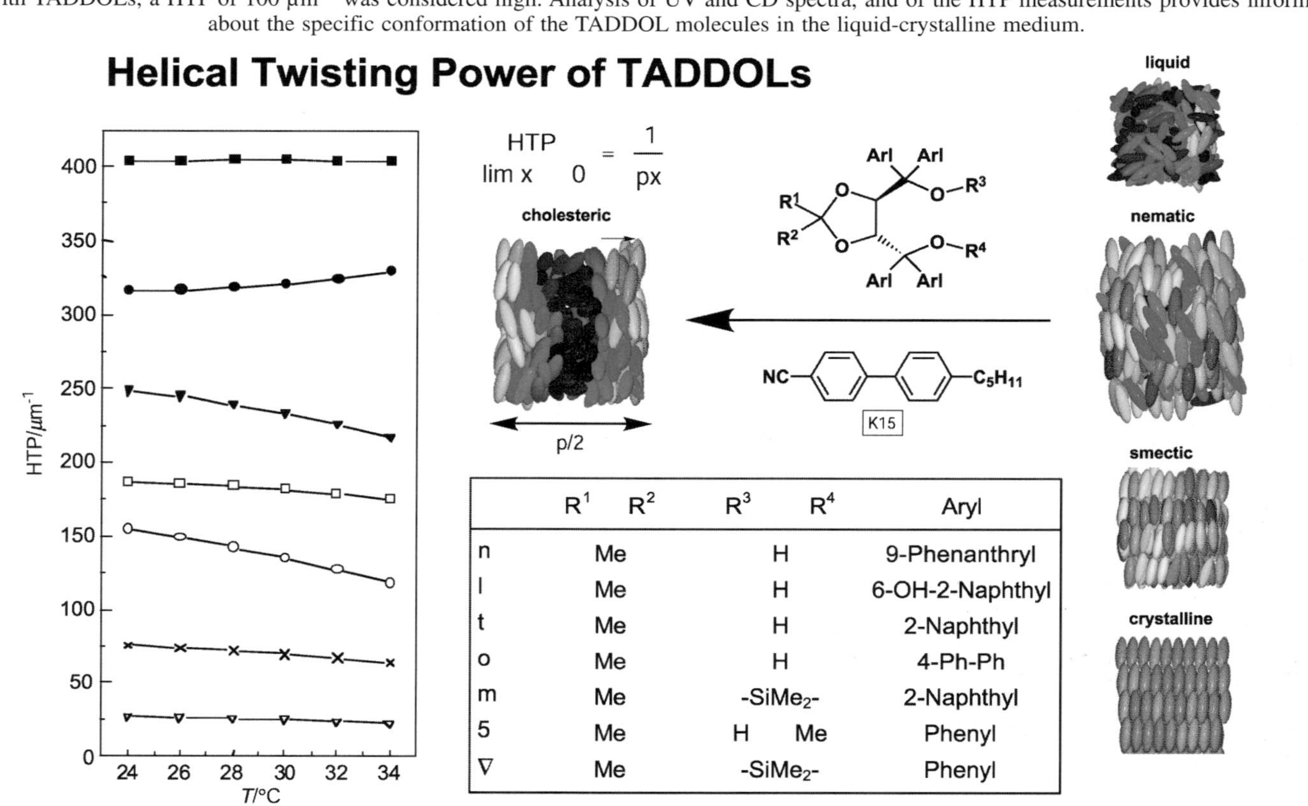

	R^1 R^2	R^3 R^4	Aryl
n	Me	H	9-Phenanthryl
l	Me	H	6-OH-2-Naphthyl
t	Me	H	2-Naphthyl
o	Me	H	4-Ph-Ph
m	Me	-SiMe$_2$-	2-Naphthyl
5	Me	H Me	Phenyl
∇	Me	-SiMe$_2$-	Phenyl

Scheme 7. *Examples for Enantiomer Separations by Crystallization with TADDOLs*. Besides the original TADDOL (from tartrate acetonide and PhMgX), *Toda et al.* [44] have often used the cyclopentanone- and cyclohexanone-derived analogs. The dynamic resolution (resolution with *in-situ* recycling) of 2-(2-methoxyethyl)cyclohexanone was reported by *Tsunoda et al.* The resolved compounds shown here are only a small selection from a large number of successful resolutions, which include alcohols, ethers, oxiranes, ketones, esters, lactones, anhydrides, imides, amines, aziridines, cyanohydrins, and sulfoxides. The yields given refer to the amount of guest compound isolated in the procedure given. Since we are not dealing with reactions (for which we use % es to indicate enantioselectivity with which the major enantiomer is formed), we use % ep (enantiomeric purity of the enantiomer isolated from the inclusion compound) in this *Scheme*.

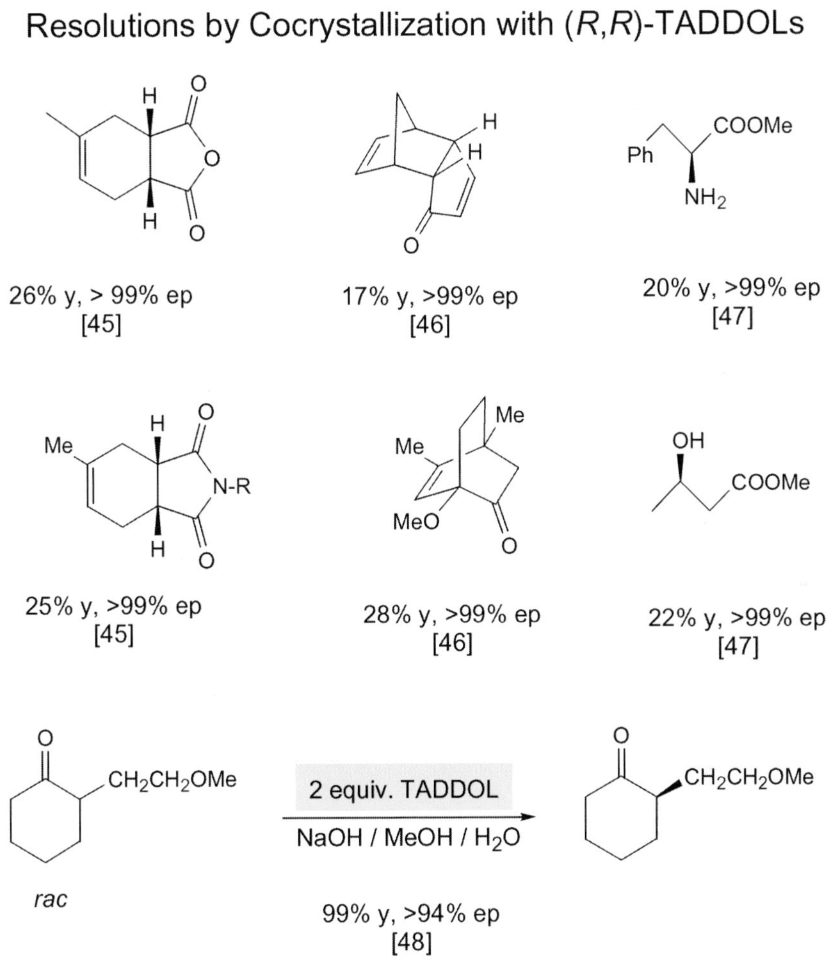

Resolutions by Cocrystallization with (*R*,*R*)-TADDOLs

26% y, > 99% ep [45]

17% y, >99% ep [46]

20% y, >99% ep [47]

25% y, >99% ep [45]

28% y, >99% ep [46]

22% y, >99% ep [47]

rac

2 equiv. TADDOL

NaOH / MeOH / H_2O

99% y, >94% ep [48]

Scheme 8. *Enantioselective Photoreactions in TADDOL Inclusion Compounds with a Coumarin, a Methacryl Anilide, and an Oxocyclohexenyl-carboxamide*. In the first case, the packing of the coumarin molecules in the mixed crystal is such that the double bonds are predisposed for the (2+2) cycloaddition. In the second example, a photochemical electrocyclic reaction is followed by a sigmatropic H shift. The third reaction is an intramolecular (2+2) cycloaddition with dia- and enantioselective formation of three new stereogenic centers. There are several more reactions of this type, described in the literature [54], and the *Toda* group has determined the crystal structures of a number of inclusion compounds to show the correlation between the crystal packing and the configuration of the photoproducts. Diastereoselective solid-phase reactions of chiral guests in TADDOL-host lattices have also been described by the *Toda* group [55].

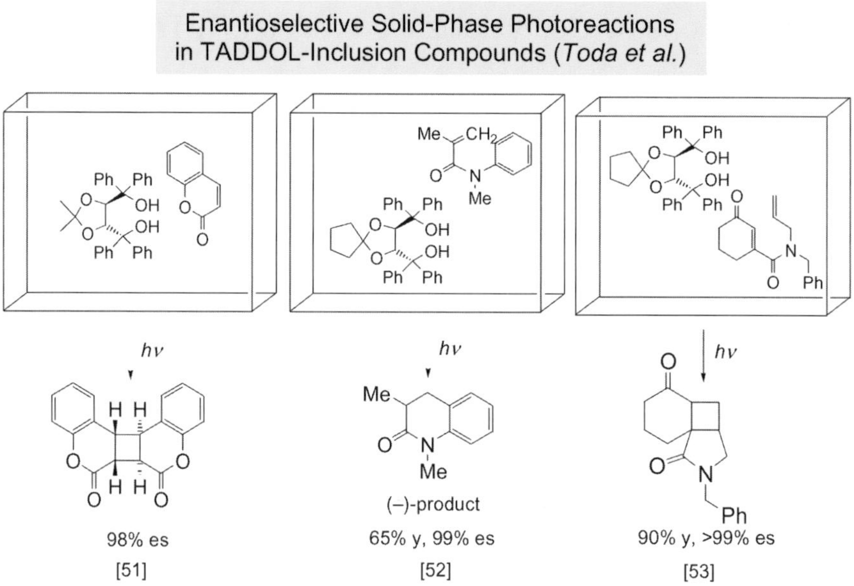

Enantioselective Solid-Phase Photoreactions
in TADDOL-Inclusion Compounds (*Toda et al.*)

98% es
[51]

(–)-product
65% y, 99% es
[52]

90% y, >99% es
[53]

In *Scheme 10*, examples of enantioselective cycloadditions with formation of three-, four-, five-, and six-membered carbocyclic or heterocyclic rings are presented, to demonstrate the use of Ti-TADDOLates as *Lewis* acids. The *Diels-Alder* reaction has, so far, been studied most extensively, and many protocols have been proposed for the preparation of the Ti catalyst, of which as little as 5 mol-% and as much as several equivalents were employed (more often than not, *Lewis* acids bind strongly to the products of reactions they have mediated; *cf.* the classical *Friedel-Crafts* acylation with $AlCl_3$!).

Other polar metal centers (Li, B, Mg, Al, Zr, Ta, Mo) have also been complexed with TADDOL, and its derivatives. Thus, a chiral lithium aluminium hydride can be generated for enantioselective ketone reductions; smaller excesses are required for comparable high selectivities, as compared to other chiral complex hydrides, and there is an additional benefit by enantiomer

Scheme 9. *Enantioselective Nucleophilic Additions Mediated by Ti-TADDOLates*. In the cyanohydrin reaction, in the aldol addition/allylation, and in the ring opening of *meso*-anhydrides, TADDOLates bearing the nucleophilic groups (NC–, R^2–CH=C(O^{t}Bu)–, R^2–CH=CH–CH_2– and i-PrO–) are prepared and used stoichiometrically. In the additions of alkyl and aryl Zn and Ti derivatives to aldehydes, in the ene reaction, and in the conjugate additions, substoichiometric amounts of a Ti-TADDOLate are present in the reaction mixtures. The most intensively studied and best understood reaction is that of R_2Zn or $(i\text{-PrO})_3TiR^2$ with R^1CHO; 5–20 mol-% of $(i\text{-PrO})_2Ti \cdot$ TADDOLate and an excess of $(i\text{-PrO})_4Ti$ are employed; with the 2-naphthyl-TADDOL, highest enantioselectivities are obtained with all types of R^1 and R^2 groups (as long as the reactands do not contain heteroatoms in *chelating* positions!).

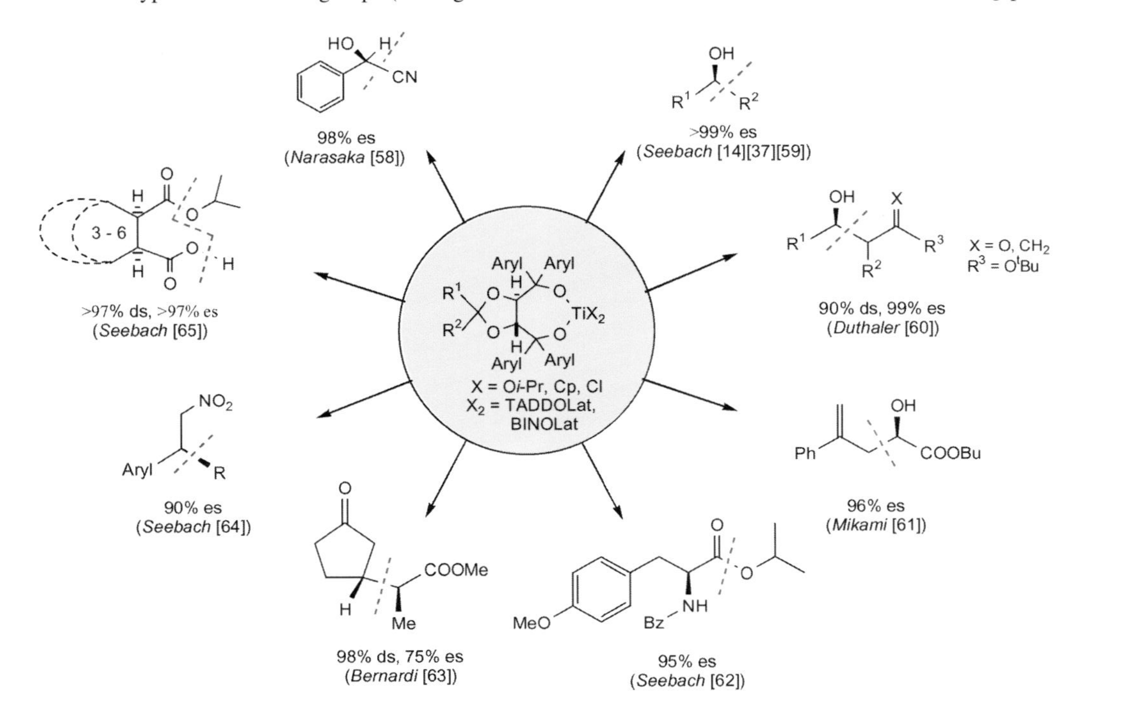

Scheme 10. *Cycloadducts Prepared Enantioselectively in the Presence of Various Ti-TADDOLates*. The *Lewis* acidity of the titanates can be modified by using X groups of various electronegativity. Besides the original TADDOL ($R^1 = R^2$ = Me, Aryl = Ph), the diethyl and the hexaphenyl ($R^1 = R^2$ = Et or Ph), and the C_1-symmetrical methyl-pentaphenyl (R^1 = Me, R^2 = Ph) derivatives have been used in these reactions. Although careful optimization and comparison have not always been carried out, it is obvious that structural variation of the chiral ligand can serve as fine-tuning for a given pair of reactants. The reaction leading to the cyclopropanol (from AcOEt and (2-phenylethyl)magnesium bromide) is the first example of the use of a low-valent Ti-TADDOLate. In the 1,3-dipolar cycloadduct of diphenylnitrone and in the *Diels-Alder* adduct to cyclopentadiene, the X-group is 2-oxo-1,3-oxazolidin-3-yl.

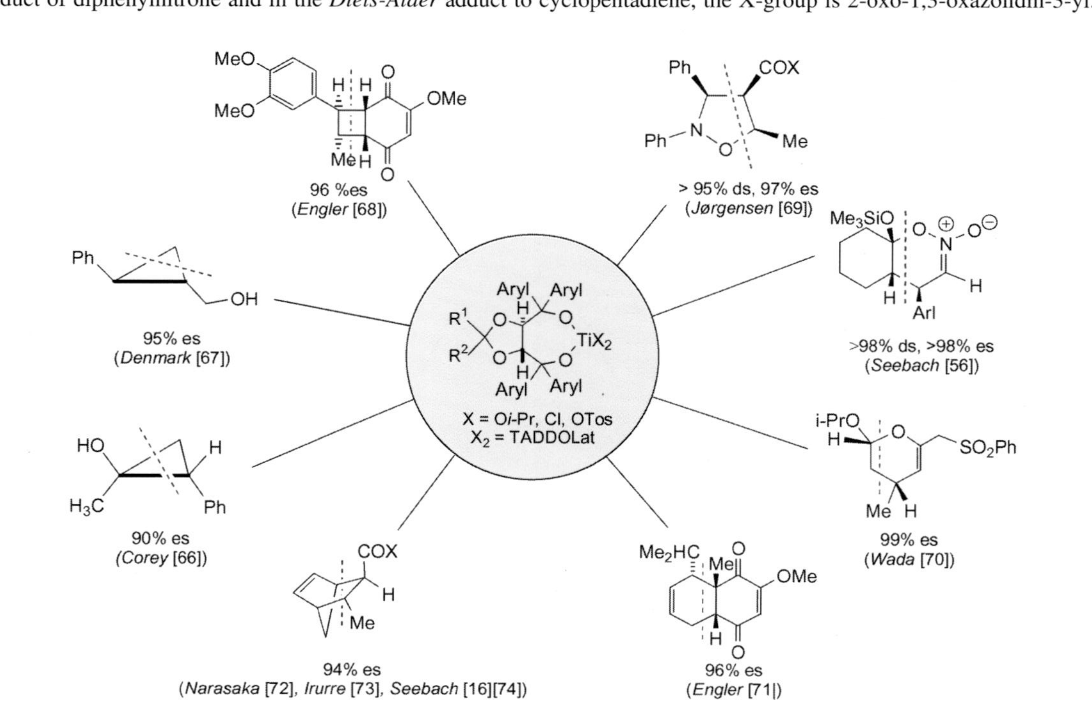

enrichment during workup (due to selective formation of TADDOL-inclusion compounds); the Mg-TADDOLate mediates highly enantioselective additions of alkyl *Grignard* reagents to ketones, with formation of tertiary alcohols; an imino-TADDOLate-carbene complex of Mo has been used for ring-opening metathesis polymerization with excellent tacticity and good monodispersity of the polymer formed, (see *Scheme 11*, top). More importantly, access to N, P, and S derivatives of TADDOLs paved the way to the late-transition metals (Rh, Ir, Pd, Cu), and first applications in enantioselective syntheses are included in *Scheme 11* (*bottom*). Since the corresponding deriva-

Scheme 11. *TADDOL Complexes of Metals Other Than Ti for Enantioselective Reactions*. The $LiAlH_4$ derivative and the Mg-TADDOLate are generated *in situ* from TADDOL, $LiAlH_4$, and EtOH, or from TADDOL and a *Grignard* reagent (MeMgBr). *Schrock*'s carbenemolybdenum complex is used for diastereoselective ring-opening metathesis polymerization (ROMP): conjugate addition of RMgX to cyclic enones is catalyzed by 2 mol-% of the R*SCu complex shown. Typical Rh- and Pd-catalyzed reactions, such as the hydrosilylation of ketones or the allylation of nucleophiles (like malonates), are enantioselective in the presence of TADDOL phosphinite, phosphonite, and phosphite esters; the ligand specified by R = oxazolin is a phosphite carrying a second chiral group (1-(4,5-dihydro-4-isopropyloxazol-2-yl)-1-methylethoxy) [33][34]. The crystal structure of the bis(diphenylphosphinite)palladium complex is shown in *Scheme 4*.

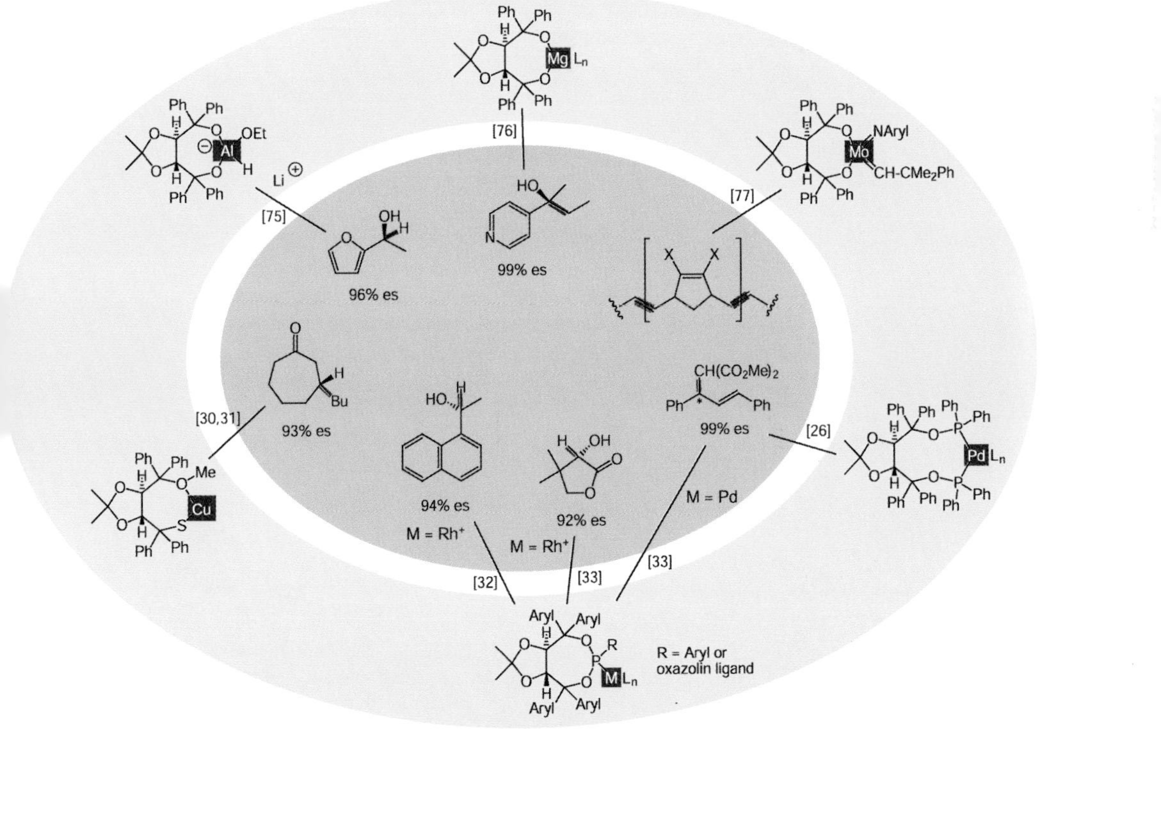

tives are readily available in one, two, or three simple steps from commercial TADDOL(s), their use will undoubtedly become common.

TADDOLs have a large tendency to crystallize, they are soluble in nonpolar media, they have small R_f-values, they are not volatile at all, and they have high molecular weights (for the original TADDOL: m.w. 466.6, m.p. 194-195°), so that they are easily recovered from reaction mixtures and separated from the products (for instance by bulb-to-bulb distillation of the latter). Especially when TADDOLs are used in stoichiometric amounts, how-

Scheme 12. *TADDOL Precursors for Grafting* (on *Merrifield* resin or on controlled-pore glass) *and for Cross-Linking Suspension Copolymerization with Styrene.* These TADDOL derivatives are prepared in the usual way from (suitably protected) components [21][78–81]. The zero- and first-generation *Fréchet*-type [82] branches were attached by reaction of branch benzyl bromides with a TADDOL bearing four phenolic OH groups [22][78][79]. The cross-linked polymer beads we used had a diameter of *ca.* 400 μm. To keep the styryl derivatives from uncontrolled polymerization, they should be kept in the dark and in the cold.

TADDOLs for Grafting and for Cross-Linking Copolymerizations

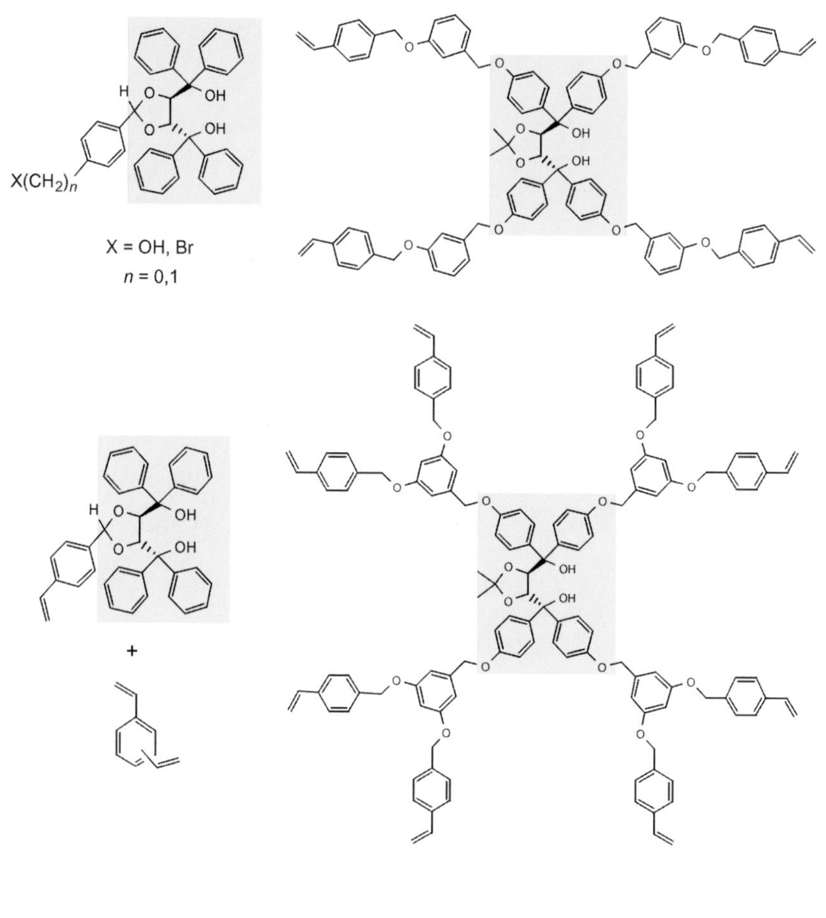

$X(CH_2)_n$

X = OH, Br

n = 0,1

+

ever, a crude product mixture may largely consist of the auxiliary. We have, therefore, started a program to prepare and use solid-phase-bound TADDOLs. Four different approaches were chosen: *i*) grafting on a given polymer, such as the *Merrifield* resin; *ii*) grafting on an inorganic support, such as porous silica gel; *iii*) cross-linking suspension copolymerization, for instance of a styryl TADDOL with divinyl benzene and styrene; and *iv*) suspension polymerization of styrene with a cross-linking, dendritically enlarged TADDOL carrying peripheral styryl groups.

Scheme 13. *Comparison of Two Polymer-Bound Ti-TADDOLates Generated after Cross-Linking Polymerization of Styrene with Styryl-Substituted TADDOL Monomers* (see *Scheme 12*). Best reproducibility is observed with materials of low loading degree (0.10 mmol/g). According to elemental analysis (and rate measurements), *ca.* 90% of the TADDOL moieties introduced into the polymer (with the monomeric, cross-linking styryl-TADDOLs) carry a Ti-atom [78] [79] and are thus not buried inaccessibly in the cross-linked polymer! The dendritic polymer performs better, as far as enantioselectivity of the Et_2Zn addition to PhCHO is concerned: there is no erratic up and down, and the value obtained in the 20th run is identical to that of the first run (within experimental error) [79].

Polystyrene Cross-Linking Ti-TADDOLates

Multiple additions of Et_2Zn to PhCHO using these polymer-bound Ti-TADDOLates

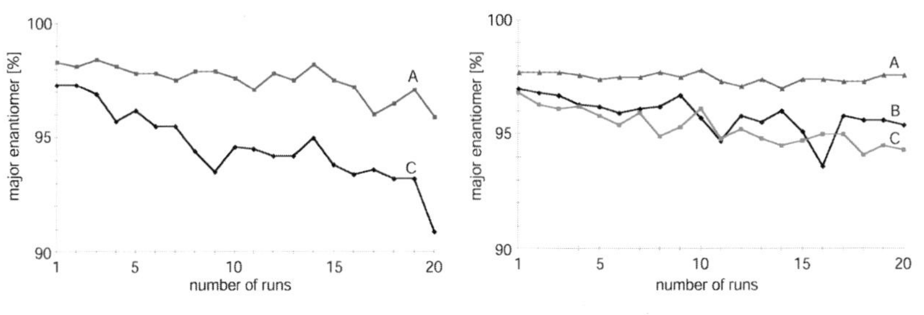

Loading : A : 0.10 mmol/g; B : 0.14 mmol/g; C : 0.25 mmol/g

Again, the variability of TADDOL synthesis was instrumental for preparing the necessary precursors (*Scheme 12*). It turned out that it was worthwhile to learn the technique of suspension polymerization [21] for the preparation of self-made beads loaded with TADDOLs: multiple applications [83] with dendritically embedded TADDOL disclosed unique material properties of this new type of cross-linked polystyrene (*Scheme 13*). The enantioselectivity, the rate of reactions (*e.g.*, in the Ti-TADDOLate-catalyzed addition of Et_2Zn to PhCHO), and the swelling properties of the material remained constant after 20 and more runs (with the same batch of beads) (see *Scheme 14*).

Scheme 14. *Rates of Reaction and Swelling Performances of Polymer-Bound Ti-TADDOLate in Multiple Applications* (with the standard-test reaction shown) [78][79]. *Left:* the monomeric Ti-TADDOLate in homogeneous solution gives a higher rate than the polymeric one made from it. *Right:* the polymer-bound dendritic Ti-TADDOLate gives a higher rate than its monomeric precursor; the dendritic material gives the same rate with and without stirring of the reaction mixture; there is no change of the swelling factor after 20 runs with the dendritically embedded Ti-TADDOLate, while the 'simple' derivative has lost some of its swelling ability after 20 runs. Both types of dendritic Ti-TADDOLates are slower than the non-dendritic ones.

Comparison of Rates of Homogeneous and Heterogeneous Reactions

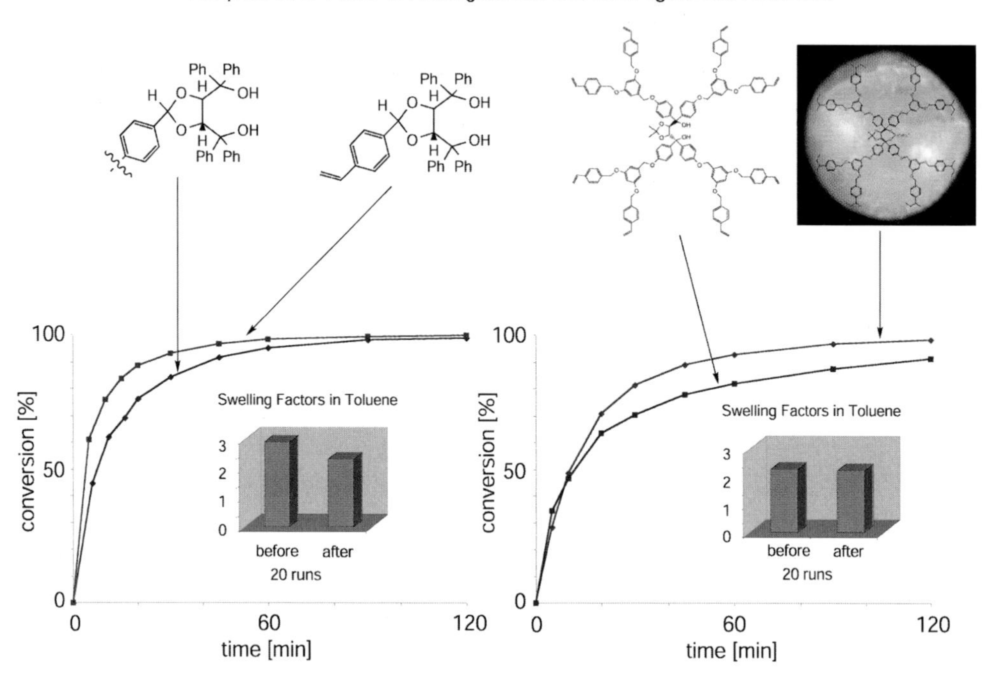

Loading of Polymers : 0.10 mmol TADDOL/g Polymer

$$Et_2Zn + PhCHO \xrightarrow[\text{1.2 equiv. Ti(OiPr)}_4\text{, toluene, -20 °C}]{\text{0.2 equiv. polym. Ti-TADDOLate}} \text{PhCH(OH)CH}_2\text{CH}_3$$

Also, while the dendritic TADDOL polymer gave a somewhat slower titanate catalyst for the above mentioned reaction than the simple copolymerized TADDOL, a unique feature emerged with the former (*Scheme 14*): the monomeric, dendritically substituted Ti-TADDOLate was found to be slower than the polymeric one (under diffusion-controlled conditions)! Thus, our investigations on chiral dendrimers (review: [84]) have fertilized the TADDOL work, and they may lead to a new type of high-performance polymer-bound reagents, beyond the TADDOLs.

The newest development of TADDOL research in our group is the grafting on controlled-pore glass (CPG) (*Scheme 15*). The advantage is that the inorganic support does not depend on swelling properties, *i.e.*, on the flexibility of a polymer backbone, which can be subject to aging effects. Still, there are CPG qualities with extremely high surface and, thus, loading capacity. We have tested Ti-TADDOLates grafted on CPG (with Me_3Si hydrophobization) in two different types of reactions to find that the enantioselectivities

Scheme 15. *SiO_2-Bound Ti-TADDOLates Tested in Two Reactions* [85]. The controlled-pore glass (CPG, supplied by the *Grace Company*) has an enormously large surface/g. Up to 0.3 mmol/g of TADDOL have been loaded by first attaching Si-$(CH_2)_3$S-Trityl and then the TADDOL moiety (through *S*-alkylation with the corresponding benzylic bromide); unreacted SiOH groups on the surface are trimethylsilylated. The rates and stereoselectivities of the Et_2Zn-to-PhCHO addition and of the diphenylnitrone 1,3-dipolar addition shown are comparable to those observed under homogeneous conditions. The material can be washed with aqueous HCl to restore, after reloading with titanate, whatever activity of the solid-state-bound catalyst had been lost after several catalytic cycles [85].

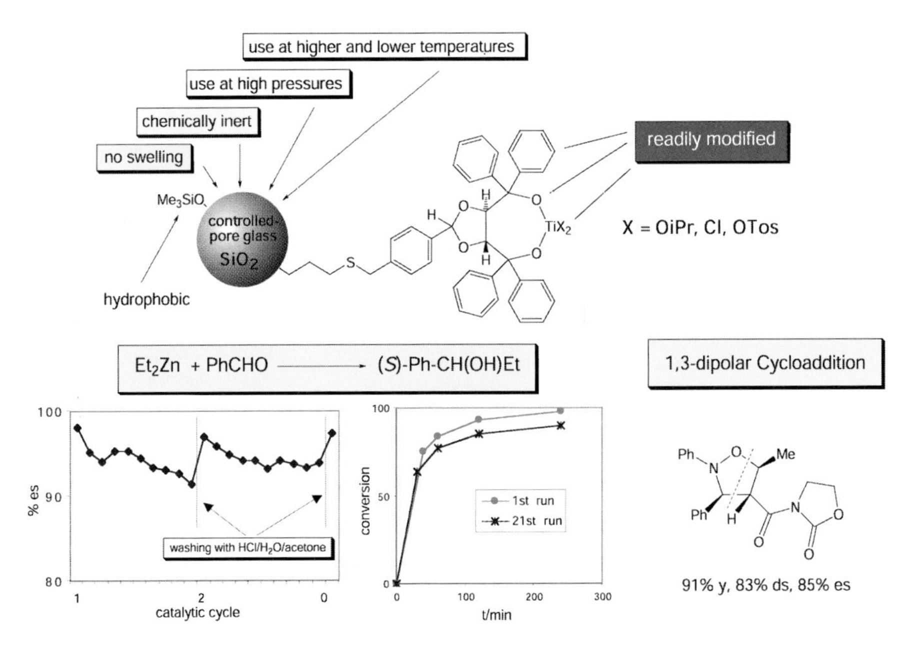

are essentially identical to those in homogeneous solution. Again, an effect was discovered which might be of importance far beyond the TADDOL field: a decrease of enantioselectivity after 10 runs with the same CPG particles (from 98 to 92% es) could be corrected by washing with $HCl/H_2O/Me_2CO$. This procedure did not remove the Me_3Si groups from the CPG surface and fully restored the enantioselectivity (*Scheme 15*).

The short preparative review presented here – without any consideration of mechanistic aspects – has hopefully demonstrated why we consider the 'TADDOL system' a candidate for a universally applicable chiral auxiliary. It has all the features, first of all the synthetic availability and almost unlimited flexibility to be used not only for all types of enantioselective syntheses and resolutions (to prepare enantiomerically pure compounds, EPC [86][87]),

Scheme 16. *Comparison of TADDOL with BINOL and Future Goals.* Derivatives of (*R,R*)-TADDOLs and of (*P*)- or (*S*)-BINOL, when employed in the same reaction, often give the same product enantiomer preferentially (*cf.* the C_2-symmetrical structural similarity). While TADDOLs are *much* easier to prepare and to modify, they are many orders of magnitude less acidic than BINOLs (TADDOLates are less polar ligands to a metal). Some TADDOL derivatives contain CPh_2X groups which are labile to (undesired) nucleophilic substitution; the dioxolane group of TADDOLs, on the other hand, is surprisingly stable to hydrolysis. BINOL (a naphthol derivative!) can be very sensitive to oxidation and (undesired) electrophilic aromatic substitution, and there are conditions under which it may racemize. Some goals to increase the usefulness of the TADDOL system are shown at the bottom of the *Scheme*: other P derivatives, more acidic derivatives, reagents for enantioselective protonation, deprotonation, oxidation, and methylenation.

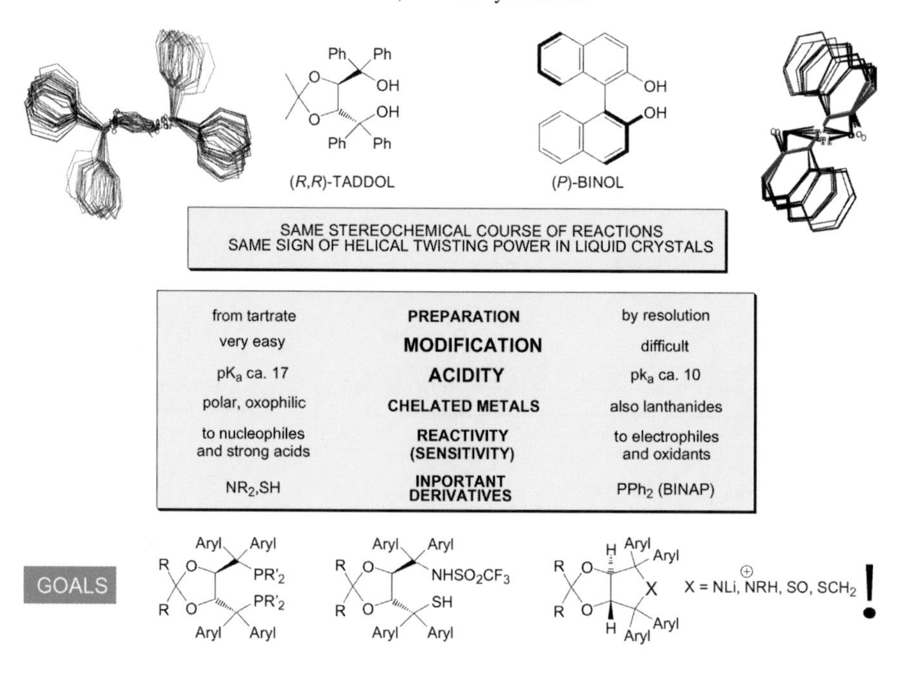

but also for material science and solid-state reactions and immobilization of catalysts.

There is another system, that of BINOL/BINAP, the Japanese dream of a universal chiral auxiliary [88], and there are striking similarities, but also fundamental differences, when compared with the TADDOLs (see *Scheme 16*). There is still a lot to do, and there are other competitors out there, sometimes referred to as *preferred ligands* (such as the semicorrine and oxazoline [89], the metallocene [90], the diaminocyclohexane [91], and the quinine/cinchonine [92] systems). Everybody in the field of EPC synthesis dreams of hav-

Scheme 17. *Diphenylmethanol Derivatives Other Than of the TADDOL Type, Readily Prepared from the Corresponding Esters and Phenyl* Grignard *Reagents.* The two geminal Ph groups favor the formation of rings (probably also in the alkoxide-enolate derived from mandelic acid). Furthermore, the two diastereotopic phenyl groups in these derivatives are locked in conformations, such that the stereochemical bias between diastereotopic faces and groups is increased. As has been shown for the oxazolidinones, the two Ph groups may sterically protect the carbonyl group in the ring from undesired nucleophilic attack. Finally, the two Ph (or other aryl) groups increase crystallinity and thus are useful for diastereoisomer enrichment and for avoiding chromatographic purification steps with intermediates.

Other Diaryl Methanol Derivatives

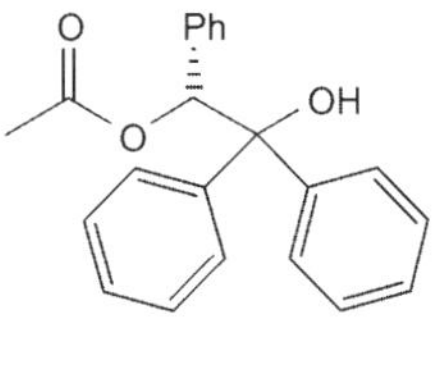

(*from* (R)*-mandelic acid*) for preparation of β-hydroxy-acids [94]

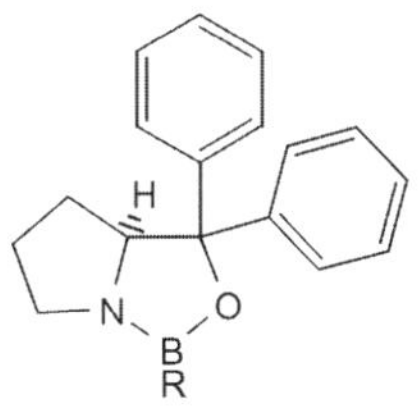

(*from* (S)*-proline*) for *Corey-Itsuno* reduction of ketones [95–97]

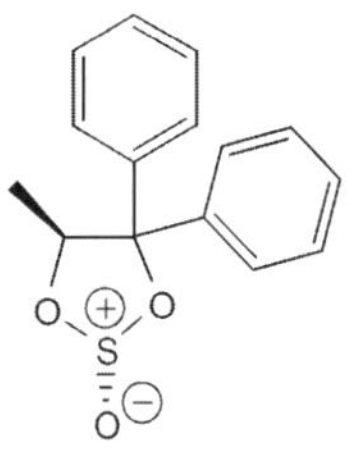

(*from* (S)*-lactic acid*) for preparation of sulfoxides R^1R^2SO [98]

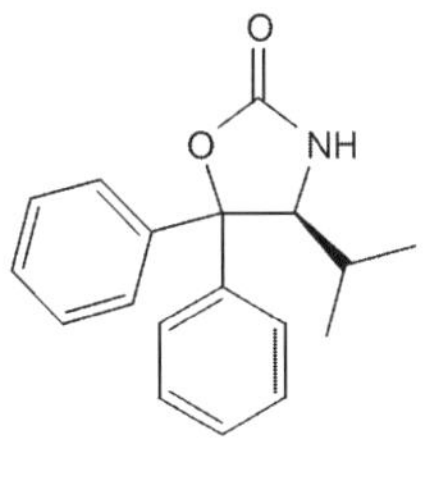

(*from* (S)*-valine*) improved *Evans* auxiliary [99–101]

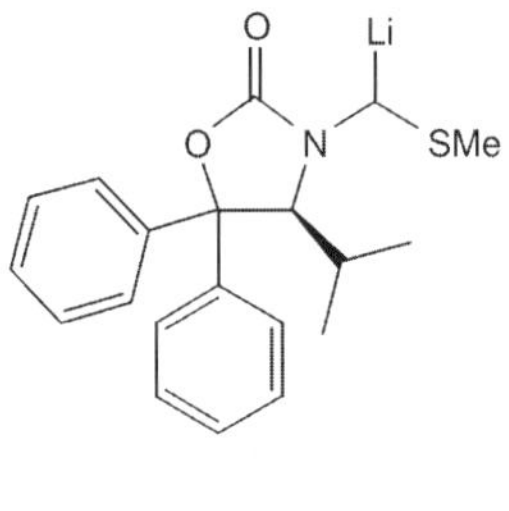

(*from* (S)*-valine*) for preparation of α-hydroxy acid derivatives [102]

ing discovered or of discovering *the* universal system, which might not even exist (remember *Emil Fischer*'s lock-and-key allegory! [93]).

One of the governing principles in the TADDOL system is the conformational lock and steric bias provided by the *geminal* diarylmethyl group. The question arises whether there are other chiral auxiliaries exploiting this geminal-*diaryl effect*. Indeed there are; some are shown in *Scheme 17*. We expect to see more examples for the preparation and use of diarylmethanol derivatives in EPC synthesis and other chirality-creating processes.

REFERENCES

[1] D. Seebach, *Synthesis* **1969**, 17 (*Nucleophile Acylierung mit 2-Lithium-1,3-dithianen bzw. 1,3,5-trithianen*); B. T. Gröbel, D. Seebach, *Synthesis*, **1977**, 357 (*Umpolung of the Reactivity of Carbonyl Compounds Through Sulfur-Containing Reagents*).

[2] D. Seebach, D. Enders, *Angew. Chem.* **1975**, *87*, 1; *Angew. Chem., Int. Ed.* **1975**, *14*, 15 (*Umpolung of the Reactivity. Nucleophilic α-(Secondary Amino)-alkylation via Metalated Nitrosamines*).

[3] D. Seebach, E. W. Colvin, F. Lehr, T. Weller, *Chimia* **1979**, *33*, 1 (*Nitroaliphatic Compounds – Ideal Intermediates in Organic Synthesis?*).

[4] D. Seebach in 'Proceedings of The Robert A. Welch Foundation. Conferences on Chemical Research. XXVII. Stereospecificity in Chemistry and Biochemistry, Nov. 7.–9. 1983', Houston, Texas, 1984, pp. 93–145 (*Crystal Structures and Stereoselective Reactions of Organic Lithium Derivatives*).

[5] D. Seebach, *Angew. Chem.* **1988**, *100*, 1685; *Angew. Chem., Int. Ed.* **1988**, *27*, 1624 (*Structure and Reactivity of Lithium Enolates. From Pinacolone to Selective* C-*Alkylations of Peptides. Difficulties and Opportunities Afforded by Complex Structures*).

[6] D. Seebach, A. K. Beck, M. Schiess, L. Widler, A. Wonnacott, *Pure Appl. Chem.* **1983**, *55*, 1807 (*Some Recent Advances in the Use of Titanium Reagents for Organic Synthesis*); The original TADDOL has first been prepared by *A. K. Beck* in November 1982, ETH-Zürich.

[7] B. Weidmann, D. Seebach, *Angew. Chem.* **1983**, *95*, 12; *Angew. Chem., Int. Ed.* **1983**, *22*, 31 (*Organometallic Compounds of Titanium and Zirconium as Selective Nucleophilic Reagents in Organic Synthesis*).

[8] D. Seebach, B. Weidmann, L. Widler, in 'Modern Synthetic Methods 1983', Ed. R. Scheffold, Salle + Sauerländer, Aarau, 1983, Vol. 3, p. 217–353 (*Titanium and Zirconium Derivatives in Organic Synthesis. A Review with Procedures*).

[9] C. Betschart, D. Seebach, *Chimia* **1989**, *43*, 39 (*Anwendungen niedervalenter Titan-Reagentien in der Organischen Synthese*).

[10] For an extensive review of independent work along these lines, see: M. T. Reetz, 'Organotitanium Reagents in Organic Synthesis', Springer Verlag, Berlin, 1986.

[11] D. Seebach, A. K. Beck, A. Heckel, *Angew. Chem.* **2001**, *113*, 96; *Angew. Chem., Int. Ed.* **2001**, *40*, 92 (*TADDOLs, Their Derivatives, and TADDOL Analogous: Versatile Chiral Auxiliaries*).

[12] A. K. Beck, B. Bastani, D. A. Plattner, W. Petter, D. Seebach, H. Braunschweiger, P. Gysi, L. La Vecchia, *Chimia* **1991**, *45*, 238; A. K. Beck, P. Gysi, L. La Vecchia, D. Seebach, *Org. Synth.* **1999**, *76*, 12.

[13] The TADDOL analog from (*S,S*)-cyclohexane-1,2-dicarboxylate and phenyl *Grignard* reagent was prepared as early as 1934: G. Wittig, G. Waltnitzki, *Ber. dtsch. chem. Ges. A*, **1934**, *67*, 667.

[14] Y. N. Ito, X. Ariza, A. K. Beck, A. Bohác, C. Ganter, R. E. Gawley, F. N. M. Kühnle, J. Tuleja, Y. M. Wang, D. Seebach, *Helv. Chim. Acta* **1994**, *77*, 2071.

[15] R. O. Duthaler, Central Research, Ciba Geigy, Basel, 1995.
[16] D. Seebach, R. Dahinden, R. E. Marti, A. K. Beck, D. A. Plattner, F. N. M. Kühnle, *J. Org. Chem.* **1995**, *60*, 1788, 5364.
[17] K. Narasaka, M. Inoue, N. Okada, *Chem. Lett.* **1986**, 1109.
[18] A. K. Beck, C. Müller, P. Monard, P. B. Rheiner, A. Pichota, unpublished results, ETH-Zürich 1997/1998.
[19] H. G. Kuball, B. Weiss, A. K. Beck, D. Seebach, *Helv. Chim. Acta* **1997**, *80*, 2507; B. Weiss, Ph. D. Thesis, Universität Kaiserslautern 1999 (*TADDOLe als chirale Dotierstoffe in nematischen Phasen*).
[20] D. Seebach, P. B. Rheiner, A. K. Beck, F. N. M. Kühnle, B. Jaun, *Pol. J. Chem.* **1994**, *68*, 2397.
[21] D. Seebach, R. E. Marti, T. Hintermann, *Helv. Chim. Acta* **1996**, *79*, 1710.
[22] P. B. Rheiner, D. Seebach, *Chem. Eur. J.* **1999**, *5*, 3221.
[23] D. Seebach, A. Pichota, A. K. Beck, A. B. Pinkerton, T. Litz, J. Karjalainen, V. Gramlich, *Org. Lett.* **1999**, *1*, 55.
[24] D. Seebach, M. Hayakawa, J. Sakaki, W. B. Schweizer, *Tetrahedron* **1993**, *49*, 1711.
[25] D. Seebach, A. K. Beck, M. Hayakawa, G. Jaeschke, F. N. M. Kühnle, I. Nägeli, A. B. Pinkerton, P. B. Rheiner, R. O. Duthaler, P. M. Rothe, W. Weigand, R. Wünsch, S. Dick, R. Nesper, M. Wörle, V. Gramlich, *Bull. Soc. Chim. Fr.* **1997**, *134*, 315.
[26] D. Seebach, E. Devaquet, A. Ernst, M. Hayakawa, F. N. M. Kühnle, W. B. Schweizer, B. Weber, *Helv. Chim. Acta* **1995**, *78*, 1636.
[27] The bis[4-(*tert*-butyl)phenoxy] derivative has also been prepared: A. Pichota, unpublished results, ETH-Zürich 1999.
[28] M. Aoki, D. Seebach, *Helv. Chim. Acta* **2001**, *84*, 187.
[29] O. De Lucchi, P. Maglioli, G. Delogu, G. Valle, *Synlett* **1991**, 841.
[30] D. Seebach, G. Jaeschke, A. Pichota, L. Audergon, *Helv. Chim. Acta* **1997**, *80*, 2515.
[31] A. Pichota, P. S. Pregosin, M. Valentini, M. Wörle, D. Seebach, *Angew. Chem.* **2000**, *112*, 157; *Angew. Chem., Int. Ed.* **2000**, *39*, 153.
[32] J. Sakaki, W. B. Schweizer, D. Seebach, *Helv. Chim. Acta* **1993**, *76*, 2654.
[33] D. K. Heldmann, D. Seebach, *Helv. Chim. Acta* **1999**, *82*, 1096; and unpublished results, ETH-Zürich 1998/1999.
[34] R. Hilgraf, A. Pfaltz, *Synlett* **1999**, 1814.
[35] For phosphites of this type with other chiral OR groups, see: A. Alexakis, J. Vastra, J. Burton, C. Benhaim, P. Mangeney, *Tetrahedron Lett.* **1998**, *39*, 7869; A. Alexakis, C. Benhaim, X. Fournioux, A. van den Heuvel, J. Levêque, S. March, S. Rosset, *Synlett* **1999**, 1811.
[36] C. Haase, C. R. Sarko, M. DiMare, *J. Org. Chem.* **1995**, *60*, 1777.
[37] D. Seebach, D. A. Plattner, A. K. Beck, Y. M. Wang, D. Hunziker, *Helv. Chim. Acta* **1992**, *75*, 2171.
[38] A. K. Beck, M. Dobler, D. A. Plattner, *Helv. Chim. Acta* **1997**, *80*, 2073.
[39] C. von dem Bussche-Hünnefeld, A. K. Beck, U. Lengweiler, D. Seebach, *Helv. Chim. Acta* **1992**, *75*, 438; K. Tanaka, M. Ootani, F. Toda, *Tetrahedron: Asymmetry* **1992**, *3*, 709.
[40] E. Juaristi, A. K. Beck, J. Hansen, T. Matt, T. Mukhopadhyay, M. Simson, D. Seebach, *Synthesis* **1993**, 1271.
[41] G. Giffels, C. Dreisbach, U. Kragl, M. Weigerding, H. Waldmann, C. Wandrey, *Angew. Chem.* **1995**, *107*, 2165; *Angew. Chem., Int. Ed.* **1995**, *34*, 2005.
[42] N. Greeves, J. E. Pease, *Tetrahedron Lett.* **1996**, *37*, 5821; N. Greeves, J. E. Pease, M. C. Bowden, S. M. Brown, *Tetrahedron Lett.* **1996**, *37*, 2675.
[43] F. Meyer, K. Siemensmeyer, H. G. Kuball, B. Weiss, D. Seebach, to *BASF AG*, Germany, PCT Int. Appl. 97 34,886; bistable LC displays can be prepared with TADDOLs as chiral dopants: F. Kuschel, *MLS GmbH*, D-Leuna, private communication.
[44] For reviews see: F. Toda, *Supramol. Sci.* **1996**, *3*, 139; F. Toda, in 'Solid-State Supramolecular Chemistry: Crystal Engineering', Eds. D. D. MacNicol and F. Toda, Elsevier, Oxford, 1996, Comprehensive Supramolecular Chemistry, Vol. 6, pp. 465–516.
[45] F. Toda, H. Miyamoto, H. Ohta, *J. Chem. Soc., Perkin Trans. 1* **1994**, 1601.

[46] F. Toda, K. Tanaka, D. Marks, I. Goldberg, *J. Org. Chem.* **1991**, *56*, 7332.
[47] F. Toda, A. Sato, L. R. Nassimbeni, M. L. Niven, *J. Chem. Soc., Perkin Trans. 2* **1991**, 1971.
[48] T. Tsunoda, H. Kaku, M. Nagaku, E. Okuyama, *Tetrahedron Lett.* **1997**, *38*, 7759.
[49] T. Vries, H. Wynberg, E. van Echten, J. Koek, W. ten Hoeve, R. M. Kellogg, Q. B. Broxterman, A. Minnaard, B. Kaptein, S. van der Sluis, L. Hulshof, J. Kooistra, *Angew. Chem.* **1998**, *110*, 2491; *Angew. Chem., Int. Ed.* **1998**, *37*, 2349.
[50] F. Toda, *Synlett* **1993**, 303 (*Solid State Organic Reactions*); F. Toda, *Acc. Chem. Res.* **1995**, *28*, 480 (*Solid State Organic Chemistry: Efficient Reactions, Remarkable Yields, and Stereoselectivity*).
[51] K. Tanaka, F. Toda, *J. Chem. Soc., Perkin Trans. 1* **1992**, 943.
[52] K. Tanaka, O. Kakinoki, F. Toda, *J. Chem. Soc., Chem. Commun.* **1992**, 1053.
[53] F. Toda, H. Miyamoto, S. Kikuchi, *J. Chem. Soc., Chem. Commun.* **1995**, 621.
[54] See also the Type B enone rearrangement in TADDOL inclusion compounds: H. E. Zimmerman, I. V. Alabugin, V. N. Smolenskaya, *Tetrahedron* **2000**, *56*, 6821.
[55] F. Toda, K. Kiyoshige, M. Yagi, *Angew. Chem.* **1989**, *101*, 329 ; *Angew. Chem., Int. Ed.* **1989**, *28*, 320.
[56] D. Seebach, I. M. Lyapkalo, R. Dahinden, *Helv. Chim. Acta* **1999**, *82*, 1829.
[57] D. J. Berrisford, C. Bolm, K. B. Sharpless, *Angew. Chem.* **1995**, *107*, 1159; *Angew. Chem., Int. Ed.* **1995**, *34*, 1059 (*Ligand-accelerated catalysis*).
[58] H. Minamikawa, S. Hayakawa, T. Yamada, N. Iwasawa, K. Narasaka, *Bull. Chem. Soc. Jpn.* **1988**, *61*, 4379.
[59] J. L. von dem Bussche-Hünnefeld, D. Seebach, *Tetrahedron* **1992**, *48*, 5719; L. Behrendt, D. Seebach, in 'Synthetic Methods of Organometallic and Inorganic Chemistry', Eds. W. A. Herrmann and A. Salzer, Thieme Verlag, Stuttgart, 1996, Vol. 1, pp. 103–104; B. Weber, D. Seebach, *Tetrahedron* **1994**, *50*, 7473; D. Seebach, A. K. Beck, B. Schmidt, Y. M. Wang, *Tetrahedron* **1994**, *50*, 4363.
[60] R. O. Duthaler, A. Hafner, M. Riediker, *Pure Appl. Chem.* **1990**, *62*, 631 (*Asymmetric C–C-Bond Formation With Titanium Carbohydrate Complexes*); R. O. Duthaler, A. Hafner, *Chem. Rev.* **1992**, *92*, 807 (*Chiral Titanium Complexes for Enantioselective Addition of Nucleophiles to Carbonyl Groups*).
[61] K. Mikami, S. Matsukawa, T. Volk, M. Terada, *Angew. Chem.* **1997**, *109*, 2936; *Angew. Chem., Int. Ed.* **1997**, *36*, 2768.
[62] K. Gottwald, D. Seebach, *Tetrahedron* **1999**, *55*, 723.
[63] A. Bernardi, K. Karamfilova, S. Sanguinetti, C. Scolastico, *Tetrahedron* **1997**, *53*, 13009.
[64] H. Schäfer, D. Seebach, *Tetrahedron* **1995**, *51*, 2305.
[65] D. Seebach, G. Jaeschke, K. Gottwald, K. Matsuda, R. Formisano, D. A. Chaplin, *Tetrahedron* **1997**, *53*, 7539; G. Jaeschke, D. Seebach, *J. Org. Chem.* **1998**, *63*, 1190.
[66] E. J. Corey, S. A. Rao, M. C. Noe, *J. Am. Chem. Soc.* **1994**, *116*, 9345.
[67] S. E. Denmark, S. P. O'Connor, *J. Org. Chem.* **1997**, *62*, 584.
[68] T. A. Engler, M. A. Letavic, J. P. Reddy, *J. Am. Chem. Soc.* **1991**, *113*, 5068; T. A. Engler, M. A. Letavic, R. Iyengar, K. O. LaTessa, J. P. Reddy, *J. Org. Chem.* **1999**, *64*, 2391.
[69] K. V. Gothelf, I. Thomsen, K. A. Jørgensen, *J. Am. Chem. Soc.* **1996**, *118*, 59.
[70] E. Wada, H. Yasuoka, S. Kanemasa, *Chem. Lett.* **1994**, 1637.
[71] T. A. Engler, M. A. Letavic, F. Takusagawa, *Tetrahedron Lett.* **1992**, *33*, 6731; T. A. Engler, M. A. Letavic, K. O. Lynch, F. Takusagawa, *J. Org. Chem.* **1994**, *59*, 1179.
[72] K. Narasaka, N. Iwasawa, M. Inoue, T. Yamada, M. Nakashima, J. Sugimori, *J. Am. Chem. Soc.* **1989**, *111*, 5340. The *Narasaka* group has published numerous other inter- and intramolecular [4+2] and [2+2] cycloadditions with a Ti-TADDOLate as a *Lewis* acid. For a most sophisticated study of Ti-TADDOLate mediated *Diels-Alder* reactions for the construction of steroids see: G. Quinkert, M. Del Grosso, in 'Stereoselective Synthesis', Eds. E. Ottow, K. Schöllkopf, and B. G. Schulz, Springer Verlag, Berlin, 1993, pp. 109–134; G. Quinkert, M. Del Grosso, A. Döring, W. Döring, R. I. Schenkel, M. Bauch, G. T. Dambacher, J. W. Bats, G. Zimmermann, G. Dürner, *Helv. Chim. Acta* **1995**, *78*, 1345.
[73] J. Irurre, X. Tomas, C. Alonso-Alija, M. D. Carnicero, *Afinidad* **1993**, *50*, 361; J. Irurre, C. Alonso-Alija, A. Fernandez-Serrat, *Afinidad* **1994**, *51*, 413.

[74] D. Seebach, A. K. Beck, R. Imwinkelried, S. Roggo, A. Wonnacott, *Helv. Chim. Acta* **1987**, *70*, 954.

[75] D. Seebach, A. K. Beck, R. Dahinden, M. Hoffmann, F. N. M. Kühnle, *Croat. Chem. Acta* **1996**, *69*, 459; A. K. Beck, R. Dahinden, F. N. M. Kühnle, ACS Symp. Ser. 1996, 641, 52 (*Tartrate-Derived Ligands for the Enantioselective $LiAlH_4$ Reduction of Ketones. A Comparison of TADDOLates and BINOLates*).

[76] B. Weber, D. Seebach, *Tetrahedron* **1994**, *50*, 6117.

[77] D. H. McConville, J. R. Wolf, R. R. Schrock, *J. Am. Chem. Soc.* **1993**, *115*, 4413.

[78] P. B. Rheiner, H. Sellner, D. Seebach, *Helv. Chim. Acta* **1997**, *80*, 2027.

[79] H. Sellner, D. Seebach, *Angew. Chem.* **1999**, *111*, 2039; *Angew. Chem., Int. Ed.* **1999**, *38*, 1918.

[80] J. Irurre, A. Fernandez-Serrat, M. Altayo, M. Riera, *Enantiomer* **1998**, *3*, 103.

[81] B. Altava, M. I. Burguete, B. Escuder, S. V. Luis, R. V. Salvador, J. M. Fraile, J. A. Mayoral, A. J. Royo, *J. Org. Chem.* **1997**, *62*, 3126.

[82] C. J. Hawker, J. M. J. Fréchet, *J. Am. Chem. Soc.* **1990**, *112*, 7638.

[83] P. J. Comina, A. K. Beck, D. Seebach, *Org. Process Res. Dev.* **1998**, *2*, 18.

[84] D. Seebach, P. B. Rheiner, G. Greiveldinger, T. Butz, H. Sellner, in 'Topics in Current Chemistry: Dendrimers', Volume-Ed. F. Vögtle, Springer Verlag, Heidelberg, 1998, Vol. 197, pp. 125–164 (*Chiral Dendrimers*).

[85] A. Heckel, D. Seebach, *Angew. Chem.* **2000**, *112*, 165; *Angew. Chem., Int. Ed.* **2000**, *39*, 163.

[86] D. Seebach, E. Hungerbühler, in 'Modern Synthetic Methods 1980', Ed. R. Scheffold, Salle + Sauerländer, Frankfurt/Aarau, 1980, Vol. 2, pp 91–173 (*Syntheses of Enantiomerically Pure Compound (EPC – Syntheses). Tartaric Acid, an Ideal Source of Chiral Building Blocks for Syntheses?*).

[87] For definitions in stereochemistry and 'problematic terms' see: G. Helmchen, in 'Stereoselective Synthesis', Eds. G. Helmchen, R. W. Hoffmann, J. Mulzer, and E. Schaumann, Georg Thieme Verlag, Stuttgart, 1995, Houben-Weyl/Methods of Organic Chemistry, Vol. E 21a, pp. 1–74.

[88] *Books*: I. Ojima, 'Catalytic Asymmetric Synthesis', VCH, New York, 1993; R. Noyori, 'Asymmetric Catalysis in Organic Synthesis', John Wiley & Sons, New York, 1994; T. Hayashi, K. Tomioka, O. Yonemitsu, 'Asymmetric Synthesis', Kodansha/Gordon and Breach Science Publishers, Tokyo, 1998; J. Seyden-Penne, 'Chiral Auxiliaries and Ligands in Asymmetric Synthesis', John Wiley & Sons, New York, 1995; *Recent Reviews:* M. Shibasaki, H. Sasai, T. Arai, *Angew. Chem.* **1997**, *109*, 1290; *Angew. Chem., Int. Ed.* **1997**, *36*, 1237 (*Asymmetric Catalysis with Heterobimetallic Compounds*); S. Kobayashi, *Pure Appl Chem.* **1998**, *70*, 1019 (*New Types of Lewis Acids Used in Organic Synthesis*); K. Mikami, *Pure Appl. Chem.* **1996**, *68*, 639 (*Asymmetric Catalysis of Carbonyl-Ene Reactions and Related Carbon-Carbon Bond Forming Reactions*).

[89] A. Pfaltz, in 'Modern Synthetic Methods 1989', Ed. R. Scheffold, Springer Verlag, Berlin, 1989, Vol. 5, pp. 199–248 (*Enantioselective Catalysis with Chiral Cobalt and Copper Complexes*); A. Pfaltz, in 'Asymmetric Chemical Transformations', Ed. P. M. Doyle, JAI Press Inc., Greenwich, USA, 1995, Advances in Catalytic Processes, Vol. 1, pp 61–94 (*Chiral Semicorrin and Bis-Oxazoline Ligands for Asymmetric Catalysts*); D. A. Evans, S. J. Miller, T. Lectka, P. von Matt, *J. Am. Chem. Soc.* **1999**, *121*, 7559; D. A. Evans, D. M. Barnes, J. S. Johnson, T. Lectka, P. von Matt, S. J. Miller, J. A. Murry, R. D. Norcross, E. A. Shaughnessy, K. R. Campos, *J. Am. Chem. Soc.* **1999**, *121*, 7582.

[90] A. Togni, T. Hayashi, 'Ferrocenes', VCH, Weinheim, 1995; F. A. Hicks, S. L. Buchwald, *J. Am. Chem. Soc.* **1999**, *121*, 7026.

[91] E. N. Jacobsen, in 'Comprehensive Organometallic Chemistry II: A Review of the Literature 1982–1994', Eds. E. W. Abel, F. G. A. Stone, G. Wilkinson, Pergamon Press, Oxford, U.K., 1995, Vol. 12, 1097–1136 (*Transition Metal-Catalyzed Oxidations: Asymmetric Epoxidation*); T. Katsuki, *Coord. Chem. Rev.* **1995**, *140*, 189–214 (*Catalytic Asymmetric Oxidations Using Optically Active (Salen)Manganese(III) Complexes as Catalysts*); B. M. Trost, C. Heinemann, X. Ariza, S. Weigand, *J. Am. Chem. Soc.* **1999**, *121*, 8667.

[92] H. C. Kolb, M. S. Vannieuwenhze, K. B. Sharpless, *Chem. Rev.* **1994**, *94*, 2843 (*Catalytic Asymmetric Dihydroxylation*).
[93] E. Fischer, *Ber. dtsch. chem. Ges.* **1894**, *27*, 2985; F. W. Lichtenthaler, *Angew. Chem.* **1994**, *106*, 2456; *Angew. Chem., Int. Ed.* **1994**, *33*, 2364 (*100 Years Schlüssel-Schloss-Prinzip – What Made Emil Fischer Use This Analogy?*).
[94] M. Braun, R. Devant, *Tetrahedron Lett.* **1984**, *25*, 5031; M. Braun, *Angew. Chem.* **1996**, *108*, 565; *Angew. Chem., Int. Ed.* **1996**, *35*, 519 (*The "Magic" Diarylhydroxymethyl Group*).
[95] S. Itsuno, M. Nakano, K. Miyazaki, J. Masuda, K. Ito, *J. Chem. Soc, Perkin Trans. 1* **1985**, 2039.
[96] E. J. Corey, R. K. Bakshi, S. Shibata, *J. Am. Chem. Soc.* **1987**, *109*, 5551.
[97] E. J. Corey, C. J. Helal, *Angew. Chem.* **1998**, *110*, 2092; *Angew. Chem. Int. Ed.* **1998**, *37*, 1986 (*Reduction of Carbonyl Compounds with Chiral Oxazaborolidine Catalysts: A New Paradigm for Enantioselective Catalysis and a Powerful New Synthetic Method.*); S. Itsuno, in 'Organic Reactions', Ed. L. A. Paquette, John Wiley & Sons, New York, 1998, Vol. 52, pp. 395–576 (*Enantioselective Reduction of Ketones*).
[98] F. Rebiere, H. B. Kagan, *Tetrahedron Lett.* **1989**, *30*, 3659.
[99] T. Hintermann, D. Seebach, *Helv. Chim. Acta* **1998**, *81*, 2093; M. Brenner, D. Seebach, *Helv. Chim. Acta* **1999**, *82*, 2365.
[100] C. L. Gibson, K. Gillon, S. Cook, *Tetrahedron Lett.* **1998**, *39*, 6733.
[101] S. D. Bull, S. G. Davies, S. Jones, H. J. Sanganee, *J. Chem. Soc., Perkin Trans. 1* **1999**, 387.
[102] C. Gaul, D. Seebach, *Org. Lett.* **2000**, *2*, 1501.

Dynamic Combinatorial Chemistry and Virtual Combinatorial Libraries[1])

by **Jean-Marie Lehn**

ISIS, Université Louis Pasteur, 4, rue Blaise Pascal, F-67000 Strasbourg

Introduction

The discovery of biologically active substances and, in particular, drug discovery require finding molecules that interact selectively with given biological targets. In recent years, a combinatorial chemistry (CC) approach has been very actively pursued, fueled especially by the hope to gain quick access to novel pharmaceuticals through rapid generation and screening of vast collections of molecules.

The corresponding combinatorial libraries (CLs) consist in large, static populations of different, discrete molecules prepared by the methodologies of molecular chemistry and derived from a set of units connected in various sequences by the repetitive application of specific chemical reactions, with the aim of producing as high a structural diversity as possible [1–6]. This procedure can, of course, be extended to other areas, such as the combinatorial preparation of multicomponent materials and the rapid screening for their physical properties or the discovery of novel catalyst for specific reactions.

In the present essay, we wish to outline the conceptual framework of another approach to this new area of chemistry, to provide some illustration from our own work and related studies, and to point to possible extensions and perspectives.

---

[1]) Adapted from : J.-M. Lehn, *Chem. Eur. J.* **1999**, *5*, 2455.

Discussion

Basic Concepts: from Static to Dynamic, from Real to Virtual, and from Prefabricated to Adaptive

This conceptually different approach resides in *dynamic combinatorial chemistry* (DCC). It is based on dynamically generated combinatorial libraries (DCLs), which are *virtual* in their generality, the actual constituents present at any moment being just the *real* sub-set of all those that are potentially accessible[2]). Its main features are summarized in *Fig. 1*, in comparison to those of CLs themselves.

DCC relies on reversible connection processes for the spontaneous and continuous generation of the *constituents* of the DCL, *i.e.*, of all possible combinations in nature, number, and arrangement of a set of basic *components*. Since such a multicomponent self-assembly makes virtually available all structural and interactional features that these combinations may display, it amounts in effect to the presentation of a *virtual combinatorial library* (VCL) that is dynamic, polymorphic, and multipotent. This DCC/VCL approach is *target-driven*: it rests on the conjecture that, in presence of the target, that member of the VCL possessing the features best suited for binding most strongly to the target site, *i.e.*, the constituent presenting highest *molecular recognition*, will be selectively expressed by the recruiting of the correct components from the set of those available. The VCL will narrow down to the thermodynamically driven, preferential expression of a real constituent through receptor-substrate molecular recognition.

---

[2]) The use of the term *'virtual'* to designate the present combinatorial libraries made accessible by DCC, deserves some comments. The designation as dynamic combinatorial libraries is correct but incomplete, highlighting only one feature, *i.e.*, the ability of their constituents to reversibly interconvert. However, they contain all the *potentially* possible combinations of the components undergoing dynamic random connection, whether these combinations are or are not actually present in the conditions used. Thus, the *'virtuality'* expresses *both* the potential combinatorial space available to the system and the dynamic accessibility of these combinations. The actual *'reification'*, *i.e.*, the generation of a real entity from the virtual set, occurs in presence of the target whose preferential interaction with a given member of the set leads to its expression (or to its amplification within an equilibrating mixture of preformed constituents).

The *Unabridged Webster* Dictionary understands virtual as being a hypothetical particle whose existence is inferred from indirect evidence, while the *Oxford Dictionary of New Words*, 1997, points out that the meaning of *'virtual'* is more and more shifted to *'not physically existing but made to appear so from the point of view of the user'*. We prefer to return to the philosophical/literary meaning of *'virtual'*, *'which is in the state of simple possibility in a real being'* or, more commonly, *'which contains in itself all the essential conditions to its own realization'* (*Dictionnaire de la langue française, P. Robert*). The latter corresponds to the meaning we wish to convey in designating the present combinatorial libraries as *'virtual'*.

VIRTUAL COMBINATORIAL LIBRARIES through DYNAMIC COMBINATORIAL CHEMISTRY

⇨ Dynamic Generation of MOLECULAR and SUPRAMOLECULAR DIVERSITY by TARGET-RECOGNITION-DIRECTED SELF-ASSEMBLY

⇨Combinatorial Library CL	Virtual Combinatorial Library VCL
• molecular constituents	• molecular or supramolecular constituents
• real set	• virtual set
• collection of molecules	• collection of components
• covalent	• covalent or non-covalent
• non reversible	• reversible
• neutral, uninformed	• instructed ➢internally (self-recognition) ➢externally (species binding) ⇨adaptive
• systematic	• recognition-directed
• preformed by synthesis	• self-assembled
• in absence of the target	• in presence of the target

Fig. 1. *Static and dynamic combinatorial chemistry: comparative basic features of real and dynamic virtual combinatorial libraries*

DCC is thus rooted in supramolecular chemistry [7], being based on two of its main themes, *self-assembly* in the generation of the library constituents and *molecular recognition* in their interaction with the target entity. Self-assembly in a multicomponent system is a combinatorial process with a search-procedure directed by the kinetic and thermodynamic parameters imposed by the nature of the components and their connections.

1) *Reversibility* is an essential feature of DCC. It gives access to VCLs by the continuous recombination of the components to make available at any instant all possible constituents of the library. Whereas combinatorial chemistry is static, based on libraries of stable non-interconverting molecules, the present approach may be termed *dynamic* since it consists in the reversible combination of components.

2) CLs are molecular, real, and their constituents are preformed by step-wise synthesis through covalent, non-reversible linkages. VCLs are molecular or supramolecular; they may be termed *virtual* because their constituents may not, and need not, exist in significant amount in absence of the assembling target[2]); they are generated spontaneously through a reversible assembly process.

The formation of the constituents of a VCL results from the combination of its components either by covalent assembly through a reversible chemical reaction (*molecular VCL*) or by self-assembly through reversible non-covalent binding interactions (*supramolecular VCL*). VCLs, thus, rely on the spontaneous generation of *dynamic molecular or supramolecular diversity* for the efficient and economical exploration of the structure/energy hypersurface through reversible recombination of a set of components. They display a sort of molecular and supramolecular polymorphism.

The degree of completeness of the set of components depends on the extent to which their possible combinations cover the geometrical and interactional spaces of the target site. The number of constituents that may be generated amounts to $[n\,(n+1)\,(n+2)\cdots(n+p-1)]/p!$ for the combination of n components p to p (without order) or to n^p arrangements when the combinations are ordered. The dynamic library of real constituents that actually exist at a given moment is a sub-set of the VCL of all possible constituents.

In a CL, there is a 1/1 correspondence between a member of the library and the target, whilst a VCL is a multiplexing system that establishes a n/1 *adaptive* correspondence between a set of components and the target site.

3) The expression of a given member of a VCL results from the fact that the dynamic process is conducted *in presence of the target*, whereas the members of a CL are fabricated independently, in absence of the target. The entity expressed from the virtual set of constituents of a VCL is that forming the optimal supramolecular entity with the target site. Like protein folding cannot occur through the successive exploration of all possible conformations at all positions of the chain, combinatorial chemistry cannot conduct a full exploration of all possible combinations and arrangements of molecular fragments into library molecules. DCC makes use of a search procedure that is thermodynamically driven and bypasses the need to generate and screen all possible molecules by letting the target direct the evolution of the system towards the product that presents the strongest interaction.

4) To this end, it is desirable to achieve an unbiased, *isoenergetic DCL*, whose equilibrating constituents be of similar free energy, so as to generate a *Boltzmann* distribution displaying comparable population for its different states/constituents. With a biased library, where one or a few constituents would be highly favored, the preferred interaction of a minor constituent with the target may not be strong enough to overturn the equilibrium situation.

5) Dynamic combinatorial diversity may result in the generation of a given type of species either by virtue of internal properties of the product(s) (self-selection) [7b] [8] (for the possibilities that may be offered by mixed ligands/metal ions combinations, see [8b]) or through interaction with external entities. The latter process amounts to the *generation of the fittest* and presents *adaptation* and *evolution* by spontaneous recombination under the selection pressure exerted by changes in the partner(s) or in the environmental conditions. It thus embodies a sort of *(supra)molecular Darwinism*!

Components and Processes

DCC presents several basic requirements: in addition to the selection of a satisfactory set of *components*, *i*) a major task in the development of VCLs is the search for suitable *reversible processes* to connect them; *ii*) it is desirable to devise procedures for quenching these processes so as to *lock-in* irreversibly the constituent(s) expressed; *iii*) library *deconvolution* methods for the rapid identification of the components contained in the active constituent(s) are needed; finally, practical, but not minor, questions concern the *characterization* of the constituent of the VCL that is being expressed.

1) The *basic components* of DCL constituents may be either interactional, bearing the sites providing interaction with the target, or organizational, serving as framework on which to assemble and arrange the first ones, or they combine both features.

The choice of these components must take into consideration two main characteristics: the structural and interactional features for binding to the target, and the functional group(s) for reversible connection. Thus, the basic set (alphabet) comprises a variety of fragments (letters) of different shapes, sizes, and constitutions: aliphatic or aromatic, carbocyclic or heterocyclic, positively or negatively charged, polar or non-polar, electron donor or acceptor, hydrophilic or lipophilic, *etc.*

2) The *reversible connections* between the components concern two levels.

At the *molecular level*, various reactions producing covalent combinations between reagents may be considered (*Fig. 2*). Functional groups involving a carbonyl unit (imines, esters, amides) are of special interest since they may undergo disconnection/reconnection cycles (trans-imination, trans-esterification, trans-amidation). In particular, the (amine + carbonyl) condensation into imine-type compounds (such as imines, oximes, hydrazones) takes place under mild conditions, and its products can be trapped irreversibly by reduction to amines. Reactions such as thiol exchange in disulfides or alcohol exchange in borate esters *etc.* are further candidates, as well as reversible *Diels-Alder* and *Michael* condensations, or olefin metathesis using catalysts that may be water-soluble. Photoinduced interconversions represent another possibility leading to *photodynamic* combinatorial processes.

At the *supramolecular level*, the connections between the components may involve organic (H-bonding, donor-acceptor, *Van der Waals*, ...) or inorganic (metal-ion binding) interactions. The latter are particularly attractive in view of the variety of arrangement geometries, binding energies, and formation kinetics provided by metal-ion coordination.

In addition, molecular diversity can also be generated through modifications in shape and spatial disposition provided by *intramolecular dynamic*

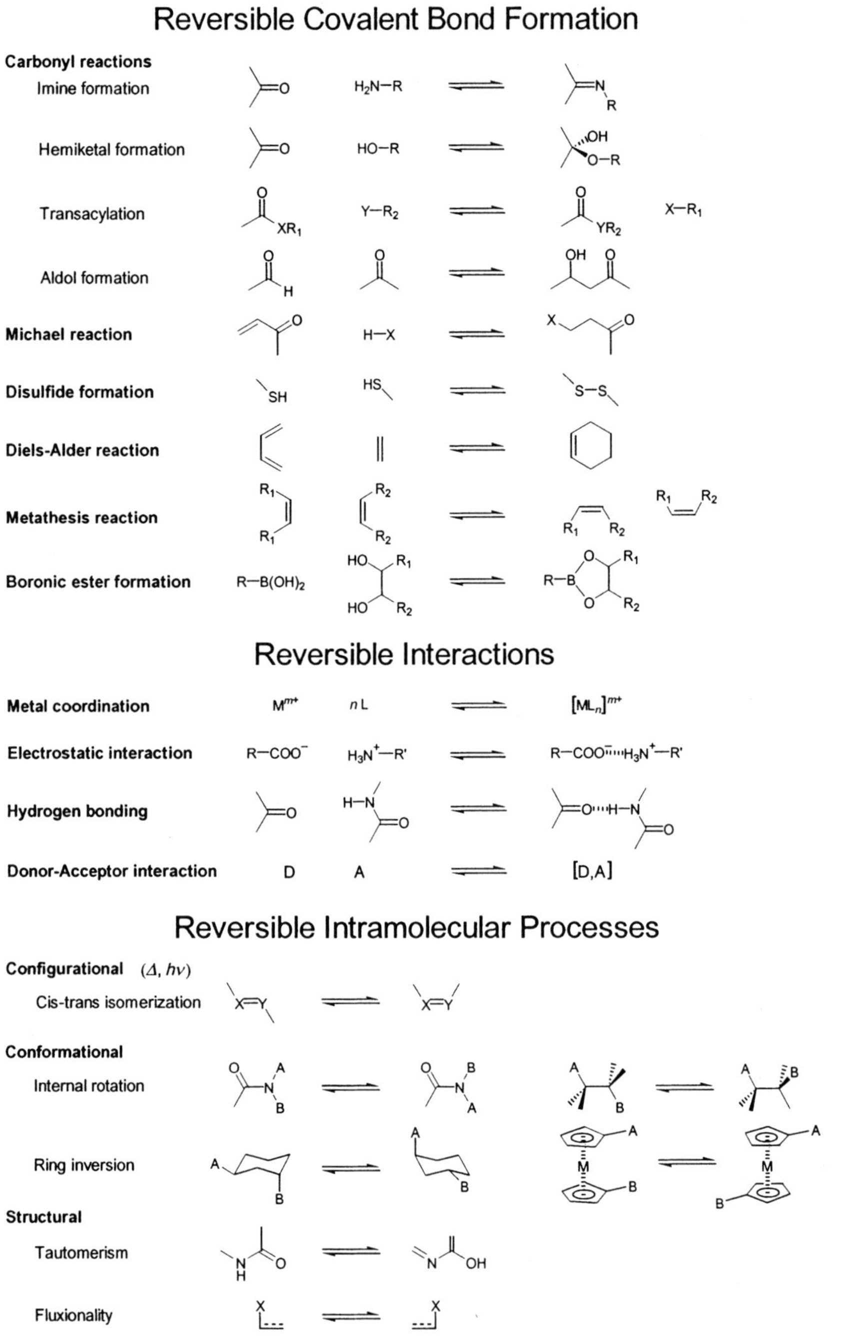

Reversible Covalent Bond Formation

Carbonyl reactions
Imine formation
Hemiketal formation
Transacylation
Aldol formation
Michael reaction
Disulfide formation
Diels-Alder reaction
Metathesis reaction
Boronic ester formation

Reversible Interactions

Metal coordination M^{m+} nL ⇌ $[ML_n]^{m+}$
Electrostatic interaction R–COO^- H_3N^+–R' ⇌ R–COO^-···H_3N^+–R'
Hydrogen bonding
Donor-Acceptor interaction D A ⇌ [D,A]

Reversible Intramolecular Processes

Configurational (Δ, $h\nu$)
Cis-trans isomerization
Conformational
Internal rotation
Ring inversion
Structural
Tautomerism
Fluxionality

Fig. 2. *A selection of potentially reversible reactions, of interactions, and of intramolecular processes for the dynamic generation of virtual molecular and supramolecular combinatorial libraries/diversity*

processes, conformational (internal rotation, ring or site inversion, rotation around metallic rotules as in sandwich complexes, *e.g.*, ferrocene derivatives, *etc.*) and configurational (*cis* ⇌ *trans* interconversion) changes or internal reversible rearrangements (including fluxional changes in metallo-organic species) between multiple states of not too different energies so as to be populated to some extent. Tautomerism, in particular in heterocyclic compounds, gives rise to dynamic diversity, allowing a shift towards one of the tautomeric states in response to interaction with the target or to environmental factors. Protein folding makes, in principle, accessible a very large variety of virtual tertiary structures that may be expressed according to circumstances. An intriguing case of dynamic presentation of different arrangements of functional groups would be provided by multiply substituted derivatives of the fluxional bullvalene molecule.

Much higher diversity may be generated through multiple combinations, by using *polyvalent frameworks* of various types and shapes (linear, branched, globular *etc.*, including polymeric and dendrimeric structures, as well as functionalized inorganic architectures assembled around metal ions), bearing several functional groups and/or interaction sites for reacting with / binding and collecting several different components.

Multiple dynamic systems, providing vastly increased diversity, may be set up by putting into action two or more reversible processes (molecular or supramolecular, intra- or intermolecular). They may be *orthogonal*, if the processes operate independently without interference (*e.g.*, imine formation and metal-ion coordination, or imine formation and disulfide exchange) or *non-orthogonal*, if cross-combination can occur (*e.g.*, imine formation and *Michael* addition).

To achieve an isoenergetic DCL (see above), it may be desirable to introduce a *spacer* between the reactive functional groups and the interaction/recognition subunits so that all connections between different components have similar strength.

Thus, the diversity of the VCLs made accessible by DCC may originate in the structural plasticity resulting from *a*) recombination of two or more components in a molecular or *b*) in a supramolecular fashion, through (multiple) reversible reactions or through organic/inorganic interactions, respectively, and *c*) intramolecular interconversions in entities subject to conformational, configurational, or fluxional changes, or to thermal or photoinduced rearrangements.

3) The *lock-in* of the self-assembled structures may be achieved by performing a chemical reaction that irreversibly links together the components of the entity represented by the optimal combination. This is the case, for instance, in the reduction of imines to amines. On the other hand, a simple change in conditions, such as temperature or pH, may slow down the connective process so as to hinder the reversibility and stabilize the product(s) sufficiently for practical purposes of retrieval.

4) In addition to their difference in constitution, CLs and VCLs also differ in the requirements for *screening* of the library and *retrieval* of a given entity. Whereas the complex mixture represented by a CL requires encoding procedures (such as nucleotide sequences, chemical barcodes, radio frequency microchips, *etc.*) [9] and the development of high through-put screening (HTS) technologies, a VCL may, in principle, be reduced to a few constituents or even to a single one, thus bypassing the need to screen all constituents and greatly facilitating detection and identification of the 'active' substance produced. Thus, DCC possesses *self-screening* capacity.

5) To efficiently determine the composition of the active compound(s), one may proceed by sequentially omitting one of the components in the full library. This procedure amounts to a *'dynamic deconvolution'*, taking advantage of the dynamic features of the library, since, by removal of a given building block, the remaining components will redistribute, and all constituents that contain this unit will automatically be deleted from the equilibrating library. A decrease in the activity of the library will indicate that the removed component is an important element in the generation of an active compound in the dynamic mixture.

Such a procedure povides an efficient way to converge on the active constituent(s) of a DCL. In more complex cases with many more library members, it may go through successive steps, the experiments with removal of a single component being followed by tests involving the removal of two or more components, thus rendering the procedure convergent. Simultaneous removal of two or more components makes possible the identification of components that may contribute to activity although less than the optimal one(s).

6) The *characterization* of the expressed constituent of a VCL is greatly facilitated by the availability of efficient analytical methods such as mass and NMR spectrometries, capillary electrophoresis, micro-HPLC, *etc.* These can be applied either to the isolated constituent or even directly to its complex with the target (for the use of mass spectrometry, see [10]). It is clear that the development of micro-methods and laboratory-on-a-chip [11] procedures will have a strong impact on the implementation of combinatorial chemistry approaches.

7) Finally, it may not be possible to conduct the dynamic equilibration in the presence of the biological target, for instance, if the conditions for chemical reversibility are incompatible with target stability. This leads to the consideration of two procedures in the implementation of the DCC approach depending on whether library generation and screening is performed in a single step or in two steps, thus defining two types of dynamic libraries:

a) *Adaptive Combinatorial Libraries:* the generation of the library constituents is conducted in a single step in presence of the target, so that the library composition may adjust, leading to selection and amplification of the preferred substrate(s); screening by the target occurs in parallel with the reversible generation of the library constituents; this is the approach where the dynamic features are operative over the whole process, *i.e.*, the fully dynamic procedure as described above.

b) *Pre-equilibrated Dynamic Combinatorial Libraries* (pDCL): the constituents of the library are generated by equilibration in absence of the target, which is introduced in a second step under conditions where inerconversion may be very slow or stopped; this has the advantage that one may use reversible reactions that are not compatible with the presence of the target, but the process is not adaptive, and no amplification of the preferred substrate can result; it is, however, sufficient for lead generation, *i.e.*, the discovery of species presenting the activity sought for; in its second phase it amounts to the usual, static combinatorial chemistry approach where an actual, real library is screened by the target; the resulting library may be termed pre-equilibrated or post-dynamic combinatorial library (pDCL).

Complementary Morphogenesis: Casting and Molding

The generation of CLs or of VCLs may be applied either to the discovery of a substrate for a given receptor or to the construction of a receptor for a given substrate.

In the spirit of *Emil Fischer*'s *'Lock and Key'* metaphor, the constitution of a CL of substrates amounts to the fabrication of a large collection of keys, with the goal (hope?) that one of them would fit the target lock/receptor and be retrievable from the mixture. On the other hand, a substrate VCL derives from a set of parts that may spontaneously and reversibly assemble so as to generate potentially a large set of different keys, one of which could fit the lock/receptor, the degree of complementarity depending on the way in which the features of the available parts are able to cover the space of all possible shapes. The same image can be applied to CLs and VCLs of receptors for a given substrate.

Thus, in the framework of VCLs two processes may be considered, depending on whether a receptor or a substrate acts as target-template for the assembly of the other partner: *casting* consists in the receptor-induced assembly of a substrate that fits the receptor; conversely, *molding* consists in the substrate-induced assembly of a receptor that optimally binds/fits the substrate (*Fig. 3*). Both processes involve *1*) a set of components, *2*) their

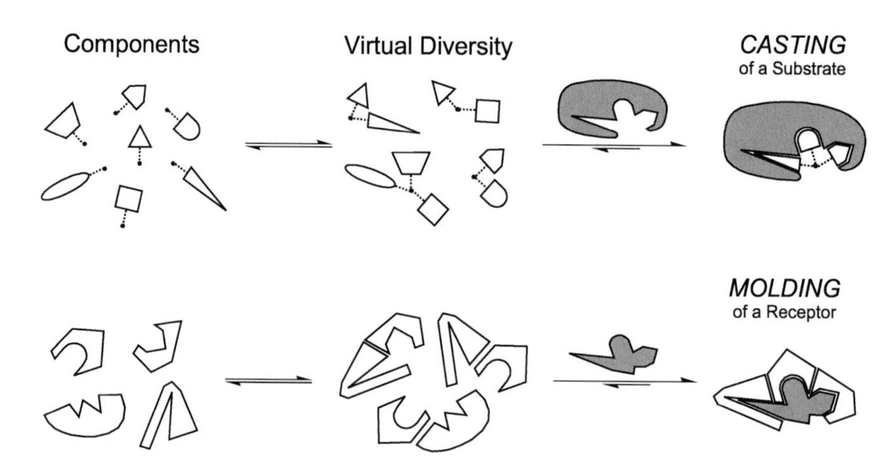

Fig. 3. *Dynamic generation of virtual combinatorial libraries. Top: Casting process:* receptor-induced self-assembly of the complementary substrate from a collection of components serving as building blocks; it amounts to the selection of the optimal substrate from a virtual substrate library. *Bottom: Molding process:* substrate-induced self-assembly of the complementary receptor from a collection of structural components; it amounts to the selection of the optimal receptor from a virtual receptor library. The diverse potential constituents of the libraries (*center, top* and *bottom*) are either covalently linked or noncovalently bound, reversibly generated species that may or may not exist in significant amount(s) in the free state, in absence of the partner. The components may either be directly connected or assemble reversibly on polyfunctional supporting frameworks of various structural types.

reversible combination for spontaneous diversity generation, *3*) the recognition-directed selection of one partner by the other one (in fact, both partners could, in principle, be self-assembled species). In these processes, the role of the target is related to the classical template effect of coordination chemistry.

Implementation of the Dynamic Combinatorial Approach

Although its basic concepts have been formulated only recently, the DCC/VCL approach has already been implemented in inorganic, organic, and bioorganic processes. It may operate in solution and also in organized phases, on a surface (onto which one of the components could self-assemble or be attached) or in molecular layers or membranes (containing freely diffusing components).

Coordination chemistry offers, by essence, the possibility to generate chemical diversity. Mixtures of ligands and metal ions in exchange define a *dynamic combinatorial coordination chemistry*. It is in the context of the reversible connection of ligand components by means of metal-ion coordination that we were first led to consider such a dynamic combinatorial process. The concept of the dynamic generation of molecular and supramolecular diversity from the reversible recombination of building blocks emerged initially from our studies on the self-selection of ligands occuring in the formation of double-helical metal complexes (helicates) where only the correctly paired double helicates were produced from a mixture of ligands and metal ions in dynamic coordination equilibrium [8]. Indeed, such self-recognition already displays basic features of DCC and VCLs: *i)* the *self-assembly* of a supramolecular structure, *ii)* by a *reversible* process with *iii)* *selection* of the correct partner.

The VCL concept was then introduced and exemplified in the dynamic, multicomponent self-assembly of circular double helicates [12]. The coordination of several tris(2,2′-bipyridine) strands with hexacoordinated metal ions may, in principle, generate circular helicates of any size, giving thus reversibly access to a VCL of oligomeric circular helicates, as schematically represented in *Fig. 4*. The actual complex obtained depends on the conditions. Thus, the pentameric entity **1** is expressed quantitatively in presence of chloride anions, due to the strong binding of a single chloride ion in the central cavity, a process that amounts to a molding event. The corresponding hexameric species **2** is formed when the larger tetrafluoroborate or sulfate anions are used. One may point out that, in analogy to biological processes, the chloride anion plays here the role of a *chaperone* species, directing the assembly towards a given architecture, that would not be formed in its absence and that is conserved when the anion is removed (under non-equilibrating conditions) after the structure has been formed.

The DCC/VCL concept was presented in detail and implemented through a casting process involving the recognition-induced assembly of inhibitors of the enzyme carbonic anhydrase [13]. In this study, it was found that the reversible recombination of structural fragments bearing aldehyde and amine groups into a library of imines (*Fig. 5*) led to a shift of the equilibrium population towards that imine product that was closest in structure to a known

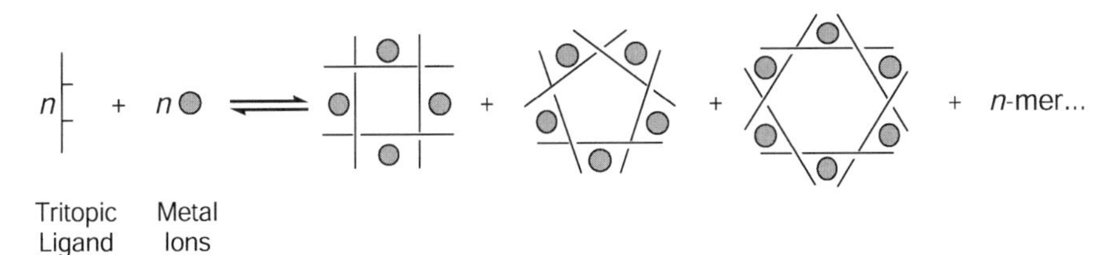

Fig. 4. *A virtual dynamic combinatorial library of oligomeric circular helicates generated from a tritopic tris(2,2′-bipyridine) ligand and metal ions of octahedral coordination.* All constituents of the dynamic library are potentially accessible at any time by reversible interconversion.

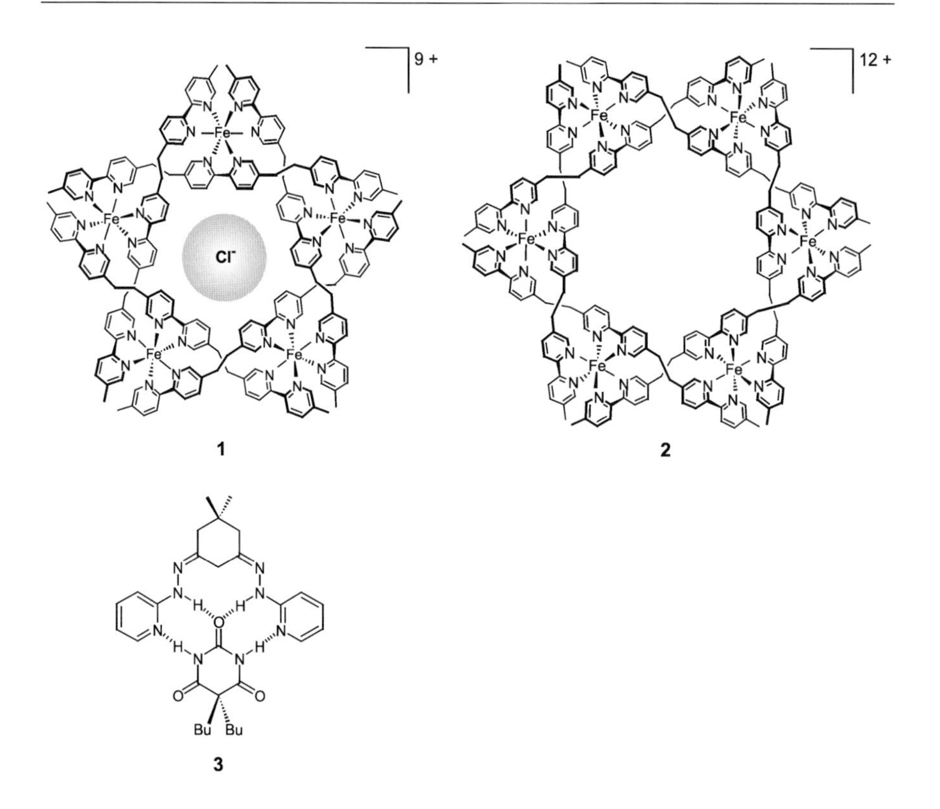

strong inhibitor of the enzyme. A relevant study described the template-directed imine formation between two trinucleotides in presence of the complementary hexamer [14].

One may note that DCC should be well-suited for the exploration of protein surfaces and the inhibition of protein-protein interactions, in particular towards the discovery of allosteric effectors binding to locations not put to use in the natural processes.

A molding process operates in the induced fit selection of a receptor for dibutyl barbiturate **3** from a dynamic library of interconverting structural and configurational/conformational isomers [15a]. Along the same lines, barbiturate binding induces the deconvolution of a large library of folding isomers of a linear molecule [15b]. This process is of special significance for biopolymers where different foldamers may present specific substrate binding and allosteric features. *N*-Alkylated oligopeptides give rise to conformational diversity [15c]. Other illustrations from work in our laboratory concern the formation in solution, of an equilibrating collection of copper(I) coordination architectures which narrows down to the 'expression' of a single constituent in the solid state [16], the directed assembly of receptors based on 2,2′-bipyridine metal complexes by substrate H-bonding [17], the selective binding by

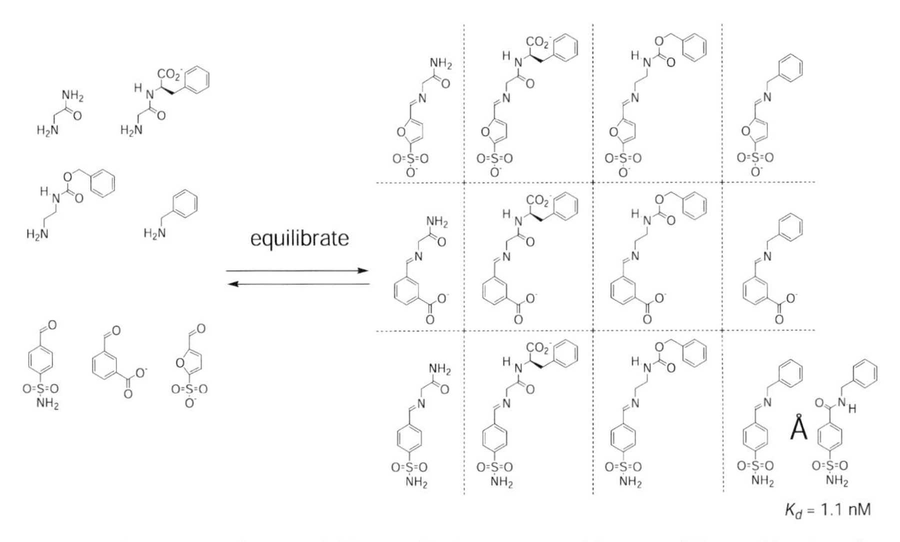

Fig. 5. *A dynamic combinatorial library of imines generated by reversible combination of a set of aldehydes and of amines.* One member of the library (*bottom right corner*) has features close to those of a strong inhibitor of carbonic anhydrase.

the lectin concanavalin A of mannose-containing constituents of either an adaptative or a pre-equilibrated DCL of carbohydrate dimers [18], the identification of acetylcholine esterase inhibitors by dynamic deconvolution of a pre-equilibrated DCL [19]. Double-level orthogonal DCLs have been described, based on two dynamic processes: metal-ion coordination by a functionalized ligand and imine formation with the aldehyde groups on the ligand [20].

Numerous other applications may be imagined and a great variety of extensions may be envisaged, also into non-biological areas.

The DCC/VCL approach bears relation to a number of dynamic processes scattered through the chemical and biochemical literature (see refs. 6–19 in [13]), and some of its characteristic features may appear *a posteriori* to be implicitly present in earlier studies (see, for example, biochemical [21a, b] and inorganic [21c] cases). Its timeliness is indicated by recent reports describing, for instance, the interconversion of macrocyclic structures [22], the dynamic screening of a peptide library by binding to a molecular trap [21b], the generation of peptide recognition sites in mixed peptide monolayers [23], and of DNA-binding compounds from an equilibrating mixture of zinc complexes [24], the assembly of inorganic architectures depending on substrate binding [25], amplification in libraries of self-assembled capsules [26a] and of covalent pseudo-peptide macrocycles [26b], the selection of a bis-macrocyclic receptor for a tripeptide [27], the generation of a library of calixarene-derived hydrogen-bonded assemblies under thermodynamically controlled conditions [28a], with receptor amplification [28b]. A case of fluxional-configurational diversity is represented by the dynamic molecular recognition of a metal complex bearing carbohydrate residues by lectins [29]. An olefin library for potential use in receptor-assisted combinatorial synthesis (RACS) has been produced by the metathesis reaction [30]. Molecular imprinting and template processes [31] display analogies through the adaptation of the matrix to the substrate (for an adjustment of inorganic cluster cages to included ions, see, *e.g.*, [32a]; for an approach based on diversity in random copolymers, see [32b]). Of particular interest is the generation of oligopeptide libraries through reversible pairing of oligonucleotides bearing peptide units [33]. A recent report describes diversity in macrocyclic structures interconverting through tautomerism [34]. Addition of the D-Ala-D-Ala target leads to amplification and rate acceleration of the formation of specific constituents of a DCL of vancomycin dimers [35].

The relative degree of amplification depends, of course, on the difference in target affinity among the library components. A theoretical analysis of a dynamic library of polymeric chains with a given continuous distribution of target affinity indicated that the amplification in such a system is limited to *ca.* two orders of magnitude and decreases with increasing target saturation [36].

Diversity generation that is neither due to multicomponent assembly nor thermally spontaneous, but is based on photoinduced changes within a single molecule, is found in the selection of a receptor molecule for arginine [37]; in this instance, the structural changes are produced by *cis-trans* isomerization around double bonds requiring light irradiation. They represent examples of photodynamic combinatorial processes.

It should be noted that, among the various studies reported, some are limited to dynamic library generation in absence of target/template, while the others concern full DCC systems involving the presence of the target (receptor or substrate) and amplification through target binding.

Recent reports have already reviewed and commented upon the early stages of development of this highly promising DCC approach [38–40].

Dynamic Combinatorial Materials

The combinatorial approach has also been applied to the fabrication of new materials and the discovery of novel properties [41]. This is the case, for instance, in the exploration of the electrical or optical properties of solids. A significant further step is brought about by the introduction of reversibility, which confers the dimension of dynamic diversity to the generation of novel materials.

We define *dynamic materials* as materials whose constituents are linked through *reversible* connections and undergo spontaneous and continuous assembly/deassembly processes in a given set of conditions. They are, in fact, either of molecular or of supramolecular nature depending on whether the links between the components are reversible covalent connections or non-covalent ones.

Because of their intrinsic ability to exchange their constituents, dynamic materials also offer combinatorial capability thus giving access to *Dynamic Combinatorial Materials* (DCMs), whose composition, and, therefore, also properties, may change by the reversible incorporation of different components in response to internal or external factors (*Fig. 6*) such as structural compatibility, an electric or magnetic field, temperature, pressure, medium, *etc.* The selection of a given constitution occurs here on the basis of a given property rather than of a binding/recognition process. This applies of course as well to biomaterials based on derivatives of biomolecules such as amino acids, nucleosides, or saccharides.

There is no doubt that the merging of dynamic with combinatorial features offered by the extension of the DCC/VCL concept to material science provides a range of novel perspectives and may be expected to rapidly become an area of active investigation and of great potential for application.

Supramolecular materials are by nature dynamic; they rely on the explicit manipulation of the non-covalent forces that hold the components togeth-

DYNAMIC COMBINATORIAL MATERIALS

★ DYNAMIC — reversible incorporation of components
— responsive to environmental factors

★ COMBINATORIAL — property-driven selection of incorporated components and amplification of preferred combination

★ FUNCTIONAL — functional components

⇨ ADAPTIVE FUNCTIONAL MATERIALS

Fig. 6. *Basic features of dynamic combinatorial materials*

er, and of the recognition processes that they underlie, for their controlled and reversible buildup from suitable units by self-assembly. Supramolecular materials thus are *instructed*, *dynamic*, and *combinatorial*; they may, in principle, select their components in response to external stimuli or environmental factors and thus behave as *adaptive materials* (*Fig. 6*).

Supramolecular Polymer Chemistry and Dynamic Combinatorial Polymers

The combination of polymer chemistry with supramolecular chemistry defines a supramolecular polymer chemistry [7a] [42] [43]. It involves the designed manipulation of molecular interactions (hydrogen bonding, donor-acceptor effects, *etc.*) and recognition processes to generate main-chain (or side-chain) supramolecular polymers by the self-assembly of complementary monomeric components (or by binding to lateral groups). In view of the lability of these associations, such entities present features of 'living' polymers capable of growing or shortening, of rearranging their interaction patterns, of exchanging components, of undergoing annealing, healing, and adaptation processes (*Fig. 7*).

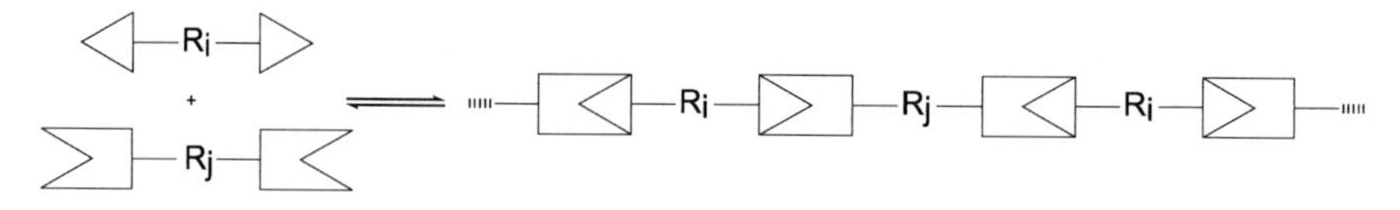

Fig. 7. *Formation of main-chain supramolecular polymers by polyassociation of complementary components.* R_i and R_j represent different subunits fitted with recognition groups; a variety of such groups may be used. Cross-linking components may also be introduced. The process is instructed, dynamic, and combinatorial.

Supramolecular polymer chemistry is thus both dynamic and combinatorial, and supramolecular polymers may be considered as DCMs.

Similar considerations apply to the generation and behavior of *supramolecular liquid crystals* [42], and extensions to other properties such as optical ones (for a case of combinatorial color generation where a dynamic aspect is introduced by time-dependence of coloration, see [44]) may be envisaged.

Nanochemistry – Nanomaterials

Nanoscience and nanotechnology have become very active areas of investigation, in view of both their basic interest and their potential applications. The spontaneous but controlled generation of well-defined, functional molecular or supramolecular architectures of nanometric size through the programmed self-assembly from mixtures of instructed components offers a very powerful alternative to nanofabrication and to nanomanipulation [7]. It provides a chemical approach to nanoscience and technology, which does not have to resort to stepwise construction or to top-down prefabrication of specific nanostructures. Self-assembling nanostructures possess dynamic and combinatorial features that confer to them the potential to undergo healing and adaptation, as required for the development of *'smart' nanomaterials*.

Dynamic Combinatorial, What Else?

In addition, to the extension to materials science, one may envisage the implementation of dynamic combinatorial approaches in other areas.

Of major interest is the development of combinatorial methodologies for the discovery of novel *catalysts* [45]. The DCC/VCL scheme may be applied towards the same goal by grafting reactive groups to the basic components so as to generate DCLs of potentially catalytic constituents. The self-assembly of an inorganic catalyst represents a case in point [46].

Transport processes and *signal transduction* may also take advantage of DCC. For instance, the dynamic assembly of several different peptide chains in a membrane offers the possibility to generate combinatorial multisubunit ion channels (for an analogy, see the oligomeric alamethicin ion pore [47]; for the cation-dependent self-assembly of ionophores, see, *e.g.*, [48]; for a combinatorial approach towards gene-delivery agents, see [49]). Dynamic combinatorial processes may be envisaged in and between molecular assembles such as monolayers, membranes, vesicles containing suitably designed and functionalized lipids. Information and signal transduction through a membrane may make use of the dynamic combinatorial positioning of several components in the plane and out of the plane of a membrane to provide a very large number of transmembrane patterns (see Fig. 28, p. 126 in [7a]).

Thus, the selection leading to the expression of a specific constituent of a VCL is, in its generality, *function-driven*; it concerns not only target-binding/recognition processes or physical properties, but also chemical transformation and catalysis, as well as translocation and transduction events.

The DCC/VCL approach bears conceptual or formal analogies to processes belonging to other areas of science, in particular to biology. Such is for instance the case for the operation of the immune system that leads to the amplification of the 'active' component of the full antibody library. Combinatorial aggregates of regulatory proteins allow the buildup of a large number of regulatory circuits in cell function and in developmental biology. Also in the assembly-forming-connection scheme of brain function [50], individual brain cells rapidly change the partners with which they synchronize their responses, so that the same cells are used in different constellations, in a sort of VCL of neurons.

Conclusion

The DCC/VCL concept provides a unifying framework within which the various entities and processes considered above can be brought together in a coherent fashion. It emphasizes that *informed diversity* is the goal, not diversity by sheer number. It opens the way to the development of instructed, target- or property-directed combinatorial chemistry, *i.e.*, of 'smart' combinatorial chemistry, where the sought-after property does the driving!

It is clear that we are just at the start of the exploration of the potential of dynamic combinatorial systems and of virtual diversity presentation. Rapid and vigorous development may be expected along a variety of directions.

In a quite different vein, combinatorial procedures with dynamic features have also been implemented in the letters and the arts, as illustrated for instance in contemporary literature [51] and music [52].

Finally, the DCC/VCL approach contributes to the perception of chemistry as an information science, spanning the domains from biology to materials, and to its progression towards complexity [7] [53 a]. With respect to the general issue of selection *vs.* design, it represents a hybrid stage of 'informed selection' or 'self-design'. By its adjustability and evolutionary character, it participates in the emergence of an *adaptive chemistry* [53].

REFERENCES

[1] G. Lowe, *Chem. Soc. Rev.* **1995**, *24*, 309.
[2] N. K. Terrett, M. Gardner, D. W. Gordon, R. J. Kobylecki, J. Steele, *Tetrahedron* **1995**, *51*, 8135.
[3] *Acc. Chem. Res., Special Issue*, Eds., A. W. Czarnik, J. A. Ellman, **1996**, *29*, 112; *'Combinatorial Chemistry – Synthesis and Applications'*, Eds. S. R. Wilson, A. W. Czarnik, John Wiley & Sons, New York, 1997.
[4] F. Balkenhohl, C. von dem Bussche-Hünnefeld, A. Lansky, C. Zechel, *Angew. Chem., Int. Ed.* **1996**, *35*, 2289.
[5] *Curr. Opin. Chem. Biol.*, Special Issue, Eds. K. T. Chapman, G. F. Joyce, W. C. Still, **1997**, *1*, 1.
[6] *Chem. Rev.*, special thematic issue, Ed. J. W. Szostak, **1997**, *97*, 347–510.
[7] a) J.-M. Lehn, *Supramolecular Chemistry – Concepts and Perspectives*, VCH, Weinheim, 1995; b) *ibid.* see p.180; c) *ibid.* see Chapt. 10.
[8] a) R. Krämer, J.-M. Lehn, A. Marquis-Rigault, *Proc. Natl. Acad. Sci. U.S.A.* **1993**, *90*, 5394; b) V. S. Smith, J.-M. Lehn, *Chem. Commun.* **1996**, 2733.
[9] A. W. Czarnik, *Proc. Natl. Acad. Sci. U.S.A.* **1997**, *94*, 12738.
[10] B. Ganem, J. D. Henion, *Chemtracts* **1993**, *6*, 1; M. Przybylski, M. O. Glocker, *Angew. Chem., Int. Ed.* **1996**, *35*, 807.

[11] D. Craston, S. Cowen, *Science Progress* **1998**, *81*, 225.
[12] B. Hasenknopf, J.-M. Lehn, B. O. Kneisel, G. Baum, D. Fenske, *Angew. Chem., Int. Ed.* **1996**, *35*, 1838; B. Hasenknopf, J.-M. Lehn, N. Boumediene, A. Dupont-Gervais, A. Van Dorsselaer, B. Kneisel, D. Fenske, *J. Am. Chem. Soc.* **1997**, *119*, 10956.
[13] I. Huc, J.-M. Lehn, *Proc. Natl. Acad. Sci. U.S.A.* **1997**, *94*, 2106; *ibid.* **1997**, *94*, 8272.
[14] J. T. Goodwin, D. G. Lynn, *J. Amer. Chem. Soc.* **1992**, *114*, 9197.
[15] a) V. Berl, I. Huc, J.-M. Lehn, A. DeCian, J. Fischer, *Eur. J. Org. Chem.* **1999**, 3089; b) V. Berl, M. J. Krische, J.-M. Lehn, M. Schmutz, *Chem. Eur. J.* **2000,** *6*, 1938; c) T. Bunyapaiboonsri, J.-M. Lehn, unpublished work.
[16] P. N. W. Baxter, J.-M. Lehn, K. Rissanen, *Chem. Commun.* **1997**, 1323.
[17] I. Huc, M. J. Krische, D. P. Funeriu, J.-M. Lehn, *Eur. J. Inorg. Chem.* **1999**, 1415.
[18] O. Ramström, J.-M. Lehn, *ChemBioChem* **2000**, *1*, 41.
[19] T. Bunyapaiboonsri, O. Ramström, S. Lohmann, J.-M. Lehn, L. Peng, M. Goeldner, *ChemBioChem* **2001**, in press.
[20] V. Goral, M. I. Nelen, A. V. Eliseev, J.-M. Lehn, *Proc. Natl. Acad. Sci. U.S.A.* **2001**, *98*, 1347.
[21] a) B. A. Katz, J. Finer-Moore, R. Mortezai, D. H. Rich, R. M. Stroud, *Biochemistry* **1995**, *34*, 8364; b) P. G. Swann, R. A. Casanova, A. Desai, M. M. Frauenhoff, M. Urbancic, U. Slomczynska, A. J. Hopfinger, G. C. Le Breton, D. L. Venton, *Biopolymers* **1996**, *40*, 617; c) S. M. Nelson, C. V. Knox, M. McCann, M. G. B. Drew, *J. Chem. Soc., Dalton Trans.* **1981**, 1669; S. M. Nelson, *Inorg. Chim. Acta* **1982**, *62*, 39.
[22] P. A. Brady, J. K. M. Sanders, *J. Chem. Soc. Perkin Trans 1* **1997**, 3237.
[23] X. Cha, K. Ariga, T. Kunitake, *J. Am. Chem. Soc.* **1996**, *118*, 9545.
[24] B. Klekota, M. H. Hammond, B. L. Miller, *Tetrahedron Lett.* **1997**, *38*, 8639; B. Klekota, B. L. Miller, *Tetrahedron* **1999**, *55*, 11687.
[25] M. Fujita, S. Nagao, K. Ogura, *J. Am. Chem. Soc.* **1995**, *117*, 1649; S. B. Lee, S. Hwang, D. S. Chung, H. Yun, J.-I. Hong, *Tetrahedron Lett.* **1998**, *39*, 873; M. Albrecht, O. Blau, R. Frölich, *Chem. Eur. J.* **1999**, *5*, 48; S. Hiroaka, M. Fujita, *J. Am. Chem. Soc.* **1999**, *121*, 10239; S. Hiroaka, Y. Kubota, M. Fujita, *Chem. Commun.* **2000**, 1509.
[26] a) F. Hof, C. Nuckolls, J. Rebek, *J. Am. Chem. Soc.* **2000**, *122*, 4251; b) G. R. L. Cousins, R. L. E. Furlan, Y.-F. Ng, J. E. Redman, J. K. M. Sanders, *Angew. Chem., Int. Ed.* **2001**, *40*, 423.
[27] H. Hioki, W. C. Still, *J. Org. Chem.* **1998**, *63*, 904.
[28] a) M. C. Calama, R. Hulst, R. Fokkens, N. M. M. Nibbering, P. Timmerman, D. N. Reinhoudt, *Chem. Commun.* **1998**, 1021; b) M. C. Calama, P. Timmerman, D. N. Reinhoudt, *Angew. Chem., Int. Ed.* **2000**, *39*, 755.
[29] S. Sakai, Y. Shigemasa, T. Sasaki, *Tetrahedron Lett.* **1997**, *38*, 8145.
[30] T. Giger, M. Wigger, S. Audétat, S. A. Benner, *Synlett* **1998**, 688.
[31] K. Mosbach, *Trends Biochem. Sci.* **1994**, *2*, 166; G. Wulff, *Angew. Chem., Int. Ed.* **1995**, *34*, 1812.
[32] a) A. Müller, *J. Mol. Struct.* **1994**, *325*, 13; b) M. Jozefowicz, J. Jozefowicz, *Biomaterials* **1997**, *18*, 1633.
[33] C. Miculka, N. Windhab, G. Quinkert, A. Eschenmoser, German Patent, DE 196 19373 A1, 1997; *Chem. Abstr.* **1998**, *128*, 34984.
[34] A. Star, I. Goldberg, B. Fuchs, *Angew. Chem., Int. Ed.* **2000**, *39*, 2685.
[35] K. C. Nicolaou, R. Hughes, S. Y. Cho, N. Winsinger, C. Smethurst, H. Labischinski, R. Endermann, *Angew. Chem., Int. Ed.* **2000**, *39*, 3823.
[36] J. S. Moore, N. W. Zimmerman, *Org. Lett.* **2000**, *2*, 915.
[37] A. V. Eliseev, M. I. Nelen, *J. Am. Chem. Soc.* **1997**, *119*, 1147; A. V. Eliseev, M. I. Nelen, *Chem. Eur. J.* **1998**, *4*, 825.
[38] P. A. Brady, J. K. M. Sanders, *Chem. Soc. Rev.* **1997**, *26*, 327; G. R. L. Cousins, S.-A. Poulsen, J. K. M. Sanders, *Curr. Opin. Chem. Biol.* **2000**, *4*, 270.
[39] A. Ganesan, *Angew. Chem., Int. Ed.* **1998**, *37*, 2828; B. Klekota, B. L. Miller, *TIBTECH* **1999**, *17*, 205; P. Timmerman, D. N. Reinhoudt, *Adv. Mater.* **1999**, *11*, 71.
[40] A. V. Eliseev, *Curr. Opin. Drug Discov. Develop.* **1998**, *1*, 106.; A. V. Eliseev, J.-M. Lehn, *Curr. Top. Microbiol. Immunol.* **1999**, *243*, 159.

[41] a) D. R. Liu, P. G. Schultz, *Angew. Chem., Int. Ed.* **1999**, *38*, 36; b) T. Bein, *Angew. Chem., Int. Ed.* **1999**, *38*, 323; c) W. F. Maier, *Angew. Chem., Int. Ed.* **1999**, *38*, 1216.

[42] J.-M. Lehn, *Makromol. Chem., Macromol. Symp.* **1993**, *69*, 1; J.-M. Lehn, in *'Supramolecular Polymers'*, Ed. A. Ciferri, Marcel Dekker, New York, 2000, pp. 615–641.

[43] M. Antonietti, S. Heinz, *Nachr. Chem. Techn. Lab.* **1992**, *40*, 308; C. M. Paleos, D. Tsiourvas, *Angew. Chem., Int. Ed.* **1995**, *34*, 1696.

[44] A. Fernandez-Acebes, J.-M. Lehn, *Adv. Mater.* **1999**, *11*, 910.

[45] A. Hoveyda, *Chem. Biol.* **1998**, *5*, R187; S. M. Şenkan, S. Öztürk, *Angew. Chem., Int. Ed.* **1999**, *38*, 791.

[46] C. L. Hill, X. Zhang, *Nature* **1995**, *373*, 324.

[47] R. Nagaraj, P. Balaram, *Acc. Chem. Res.* **1981**, *14*, 356; R. O. Fox, Jr., F. M. Richards, *Nature* **1982**, *300*, 325.

[48] J. T. Davis, S. Tirumala, J. R. Jensen, E. Radler and D. Fabris, *J. Org. Chem.*, **1995**, *60*, 4167.

[49] J. E. Murphy, T. Uno, J. D. Hamer, F. E. Cohen, V. Dwarki, R. N. Zuckermann, *Proc. Natl. Acad. Sci. U.S.A.* **1998**, *95*, 1517.

[50] W. Singer, *Science* **1995**, *27*, 758.

[51] See the 'OULIPO' (Ouvroir de littérature potentielle, Workshop of potential literature); one may cite *'Cent Mille Milliards de Poèmes'* by *R. Queneau* (the verses of four sonnets are printed on superposed separate paper strips so that novel poems can be generated at will by turning the strips), and *'Composition N° 1'* by *M. Saporta* (a novel printed on free pages that may be redistributed like a stack of game cards, producing a new arrangement/novel each time); in both cases, a specific arrangement produces one specific expression from the very large set of all virtual combinations possible.

[52] Consider for instance: P. Boulez, *'... explosante-fixe ...'* based on dynamic combinations of basic musical cells and some aspects of I. Xenakis, *'Musiques formelles'*, la revue musicale, ed. Richard-Masse, Paris, 1963.

[53] a) J.-M. Lehn, *Chem. Eur. J.* **2000**, *6*, 2097; b) J.-M. Lehn, in 'Supramolecular Science: Where It Is and Where It Is Going', Eds. R. Ungaro, E. Dalcanale, Kluwer, Dordrecht, 1999, pp. 287–304.

The Importance of β-Alanine for Recognition of the Minor Groove of DNA

by **Peter B. Dervan*** and **Adam R. Urbach**

Division of Chemistry and Chemical Engineering,
California Institute of Technology, Pasadena, California, 91125, USA

1. Introduction

When compared with physics and biology, chemistry is a profoundly different field of study of the natural world. Chemists, like artists, are able to construct new three-dimensional objects (molecules/materials) that do not yet exist, but only in the mind of a person. An inspirational example is *Albert Eschenmoser*, who uses the power of synthetic organic chemistry to 'time travel' (admittedly, in a speculative 'what if' way) to ask about the 'properties' of early primitive molecules that might have existed billions of years ago that carried the genetic information of life on planet Earth. I became interested in the world of 'structure – function' shortly after arriving at Caltech in 1973. One cannot design without the brushes and paint of the craft. Indeed, organic chemists, standing on the shoulders of the pioneering synthetic achievements of *Woodward*, *Corey*, *Merrifield*, and others are able to construct novel molecular shapes different from those found in nature with new properties (*i.e.*, function). In early 1973, I was inspired by the work of *Lehn* and *Cram* in the field of molecular recognition where early studies were largely conducted in organic solvents (*e.g.*, cation-crown complexation). I decided that a pivotal step forward would be to understand in a 'predictive mechanistic sense' (reflecting my physical organic training) how to create ensembles of weak bonds between ligands and macromolecules in *water, the solvent of life*.

All living organisms on planet Earth from bacteria, yeast, flower, fruit fly, mouse to man store the genetic information in a common molecule, a two-stranded nucleic acid polymer wound in a double helix. In 1973, there were no crystal structures of DNA nor of protein-DNA complexes. We now know that the chemical principles by which nature's proteins 'read out' and control the genetic information are chemically complex. Nature had billions of years to solve this recognition problem and used selection over time from a vast library of protein surfaces. The question arises whether organic chemists could rationally create 'four chemical keys' that, when linked together, would distinguish each of the *Watson-Crick* A · T, T · A, G · C, and C · G base pairs and read directly any contiguous sequence of DNA. This structure-function program would be distinctly different from '*anti-sense*', wherein the *Watson-Crick* pairing rules uncovered in 1953 are used for 'sense-strand' recognition. Rather, this would be a design-synthesis exercise to read the 'edges' of intact *Watson-Crick* pairs in either the major or minor groove of the DNA double helix.

In this essay, I will not review that 25-year expedition. Rather, I will write a short essay on a specific topic that has been on my mind recently, which, I hope, my colleagues in the nucleic organic field will enjoy. This is an issue in the area of targeting the minor groove of DNA by synthetic polyamides. It is now well-recognized that there are two global motifs for targeting the minor groove of DNA based on analogs of the *N*-methylpyrrole-carboxamide ring of the natural product distamycin. There exist 1:1 and 2:1 stoichiometries with quite different rules of recognition and specificity [1–3]. This *ambiguity of sequence targeting depending on stoichiometry* raises a serious design issue for the DNA recognition field.

To date, 1:1 complexes have been mostly limited to oligopyrroles (Py) binding to pure A,T tracks [4]. The 2:1 motif appears more versatile and, importantly, more sequence-specific, wherein pairs of aromatic carboxamides comprising pyrrole (Py), imidazole (Im), and hydroxypyrrole (Hp) distinguish each of the four *Watson-Crick* base pairs (Im/Py = G · C; Py/Im = C · G; Hp/Py = T · A; Py/Hp = A · T) [5]. From high-resolution X-ray analysis, antiparallel dimers bind B-form DNA [6] [7]. However, 4–5 contiguous rings are overcurved with respect to the DNA helix, and it is not surprising that polyamides longer than 5 contiguous rings have decreased affinity for the minor groove in both the 1 : 1 motif [4] and 2 : 1 motif [8]. The introduction of β-alanine (β), which codes for A,T in both the 1:1 (β) and 2:1 motifs (β/β pairs), allows sequences larger than 6 bp to be bound with high affinity, presumably by relaxing the curvature of the crescent-shaped polyamide ligand [9–12].

After several years of experience targeting DNA with the 2:1 motif, we are identifying many DNA sequences that are resistant to the 'pairing rules' with regard to the criteria of high affinity ('nanobinders') and good sequence specificity. This is undoubtedly related to the *sequence-dependent microstructure of DNA*. Hairpin polyamides, antiparallel dimers connected by a γ-aminobutyric acid (γ) turn unit, are 'molecular calipers' for B-form DNA structures, and the 'low-affinity sequences' may reveal altered (non-optimal) minor groove width, depth, and curvature, which will require different ligand-design strategies. One clarifying example is the hairpin ImPyImPy-γ-ImPyImPy, which targets the sequence 5′-GCGC-3′, according to the pairing rules, but with low affinity [13]. By judicious replacement of Py with β-alanine (*i.e.*, Im-Py-Im→Im-β-Im), a second-generation polyamide hairpin Im-β-ImPy-γ-Im-β-ImPy restores the dissociation constants K_D to subnanomolar, affording a 'subset' of pairing rules wherein Im/β or Py/β pairs can substitute, in certain cases, for Im/Py and Py/Py pairs [13] (*Fig. 1*).

Recently, *Laemmli* and co-workers reported that certain β-linked Py/Im polyamides bind GAGAA tracks in a 1:1 stoichiometry in a single orientation [14]. The favorable Im-β-Im composition is present in their polyamide, leading us to ponder its importance and generality. On one hand, the 1:1 complex is an important observation by *Laemmli*, which could expand the sequence repertoire for DNA targeting. On the other hand, the fact that β-linked Py/Im polyamides can bind *both* 1:1 and 2:1 in the minor groove raises the issue that a single polyamide molecule may bind very different sequences depending on stoichiometry (*Fig. 2*).

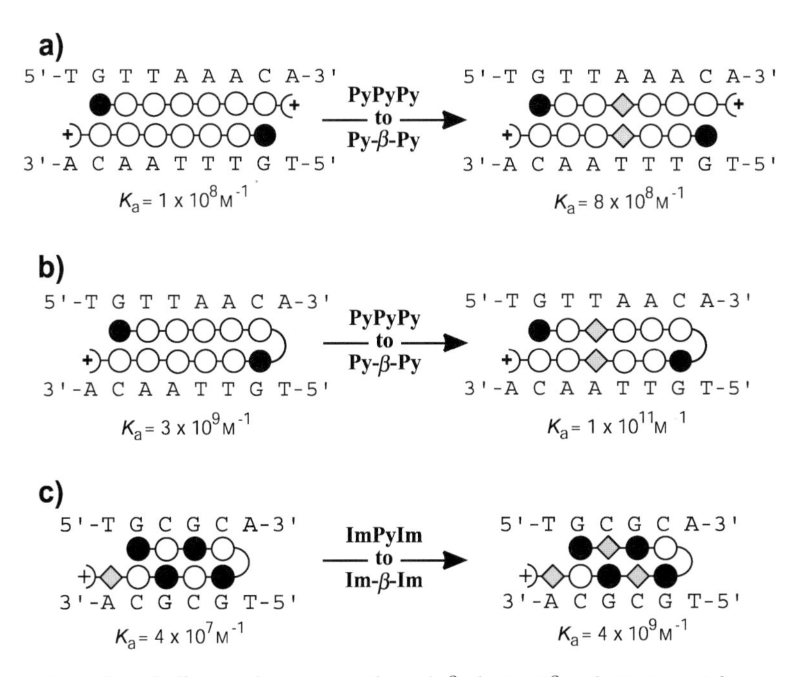

Fig. 1. *Examples of affinity enhancements through β-alanine (β) substitution:* a) *between two pyrroles in a 2:1 homodimeric complex* (β substitution increases affinity by 8-fold [10]); b) *between two pyrroles in a hairpin polyamide*, (β substitution enhances affinity by 33-fold [13]); *and* c) *between two imidazole residues* (the Im-β-Im unit increases affinity by 100-fold over the Im-Py-Im parent molecule [13]). Imidazole and pyrrole rings are represented as shaded and unshaded circles, respectively. The β-alanine residue is shown as a gray diamond. Measured equilibrium association constants, K_a, are listed below each complex.

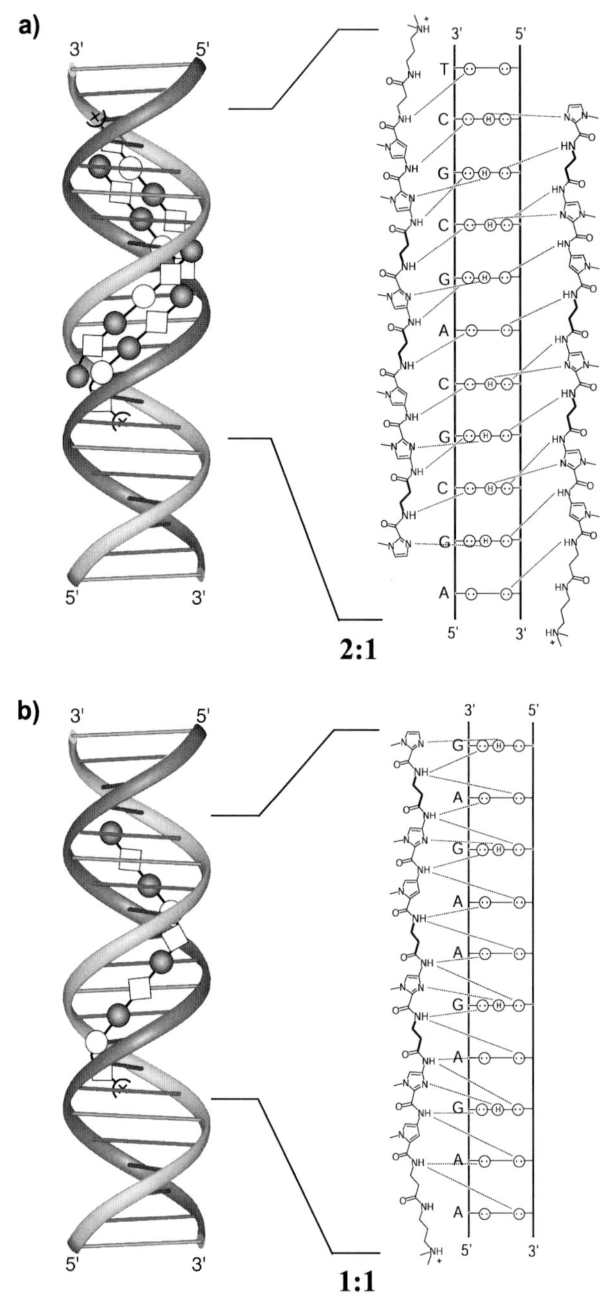

Fig. 2. *Hydrogen-bonding scheme showing observed binding in* a) *a 2:1 complex with the DNA sequence, 5'-AGCGCAGCGCT-3' and* b) *a 1:1 complex with 5'-AAGAGAAGAG-3'.* A circle with two dots represents the lone pair of N(3) of purines and O(2) of pyrimidines. Circles containing an H signify the exocyclic amino H-atom of guanine. Putative H-bonds are indicated by dotted lines.

To explore this issue further, we synthesized three polyamides: Im**Py**-ImPy-β-Im**Py**ImPy-β-Dp (**1**) and Im-β-ImPy-β-Im-β-ImPy-β-Dp (**2**), which vary in their Im-**X**-Im composition (X = Py or β), and Im-β-ImPy-**Py**Im-β-ImPy-β-Dp (**3**), wherein an internal β residue is changed to Py (*Fig. 3*).

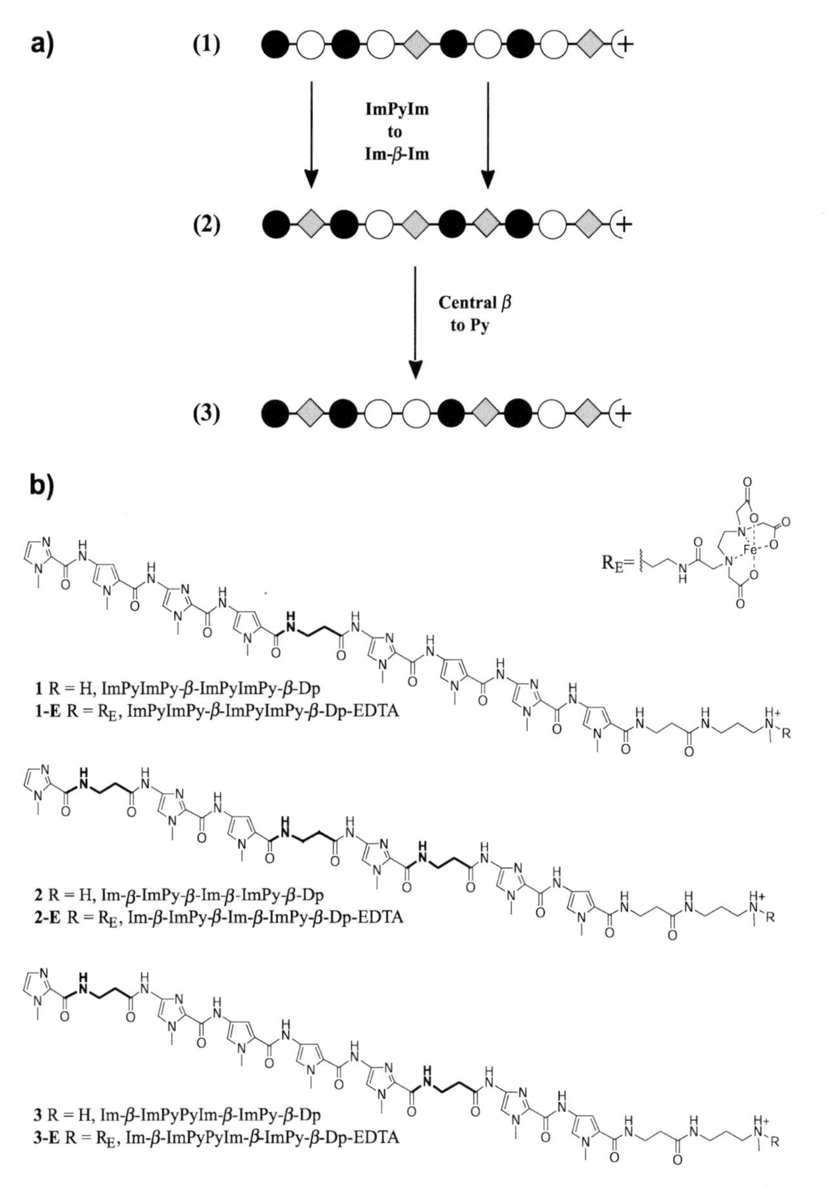

Fig. 3. a) *Scheme illustrating the selective Py → β substitutions for polyamides* **1**, **2**, *and* **3** *examined in this study.* b) *Chemical structures of these compounds and their EDTA conjugates,* **1-E**, **2-E**, *and* **3-E**. Polyamide sequences are indicated beneath each structure.

We find these polyamides bind two sites of similar size, but *quite different sequence composition, which is related to the stoichiometry of the complexation in the minor groove*. Footprinting and affinity cleavage experiments reveal 2 : 1 binding at 5′-AGCGCAGCGCT-3′ and 1 : 1 binding at 5′-AAGAGAAGAG-3′. Polyamide **2**, which contains two Im-β-Im units, has the highest affinity for both sites, with the 1 : 1 complex being energetically favored over 2:1.

2. Synthesis

Polyamides **1**–**3** were synthesized in 12 steps from Boc-β-Pam-resin (1.25 g resin, 0.26 mmol/g substitution) according to previously described solid-phase methods [15]. Non-terminal imidazole residues were introduced as dimers, Boc-Py-Im-COOH and Boc-β-Im-COOH. The polyamide was cleaved from the solid support by aminolysis of the resin-ester linkage with 3-(dimethylamino)propylamine (*N,N*-dimethylpropane-1,3-diamine; Dp) or 3,3′-diamino-*N*-methyldipropylamine (Dp-NH_2). Products were purified by reversed-phase preparatory HPLC to provide ImPyImPy-β-ImPyImPy-β-Dp (**1**), ImPyImPy-β-ImPyImPy-β-Dp-NH_2 (**1**-NH_2), Im-β-ImPy-β-Im-β-ImPy-β-Dp (**2**), Im-β-ImPy-β-Im-β-ImPy-β-Dp-NH_2 (**2**-NH_2), Im-β-ImPyPyIm-β-ImPy-β-Dp (**3**), and Im-β-ImPyPyIm-β-ImPy-β-Dp-NH_2 (**3**-NH_2). The purified polyamides with primary amines at the C-terminus were treated with an excess of ethylenediaminetetraacetic acid (EDTA) dianhydride (DMSO/*N*-methylphthalimide (NMP), DIEA, 55°, 15 min), and the remaining anhydride was hydrolyzed (0.1 N NaOH, 55°, 10 min). The EDTA conjugates were then purified by reversed-phase preparatory HPLC to yield **1-E**, **2-E**, and **3-E**.

3. MPE · Fe^{II} Footprinting and Affinity Cleaving

The plasmid pAU9 was designed with the two match sites, 5′-AGCGCAGCGCT-3′ and 5′-AAGAGAAGAG-3′, to examine 2 : 1 and 1 : 1 polyamide : DNA binding, respectively. MPE · Fe^{II} Footprinting [16] on the 3′- and 5′-^{32}P end-labelled 253 base pair *EcoRI*/*PvuII* restriction fragment (3′) or PCR product (5′) from plasmid pAU9 (28.6 mM [4-(2-hydroxyethyl)piperazin-l-yl]ethanesulfonic acid (HEPES), 285.7 mM NaCl, pH 7.0, 22°) reveals that polyamides **2** and **3**, each at 100 nM concentration, bind the designed sites 5′-AGCGCAGCGCT-3′ and 5′-AAGAGAAGAG-3′ (*Fig. 4,a*). No footprinting is observed for polyamide **1** at concentrations ≤1 μM.

Affinity-cleaving experiments [17] on the same DNA fragments (28.6 mM HEPES, 285.7 mM NaCl, pH 7.0, 22°) with the EDTA · Fe^{II} analogs **1-E**, **2-E**, and **3-E** reveals 3′-shifted cleavage patterns indicating minor groove binding for both polyamides **2-E** (at 10 nM) and **3-E** (at 100 nM) (*Fig. 4,a*). Polyamides **2-E** and **3-E** cleave both sides of site, 5′-AGCGCAGCGCT-3′, consistent with binding as an antiparallel dimer. However, **2-E** and **3-E** reveal cleavage only at the 5′-end of the site, 5′-AAGAGAAGAG-3′, suggesting one orientation and a 1:1 polyamide : DNA binding stoichiometry, similar to the finding of *Laemmli* and co-workers [14]. No cleavage is observed for polyamide **1-E** at concentrations ≤1 μM.

4. Quantitative DNase I Footprint Titrations

Quantitative DNase I footprint titration experiments [18] were performed to determine the apparent equilibrium dissociation constants (K_D = concentrations of polyamide bound at half-saturation) of **1**–**3** at each of the target sites (*Fig. 4,b*). The 5′-AGCGCAGCGCT-3′ site was bound by **2** and **3** with similar affinities, K_D = 0.63 nM and 1.9 nM, respectively, displaying cooperative binding isotherms ($n=2$) for both molecules. However, the 5′-AAGAGAAGAG-3′ site was bound 16-fold more tightly by polyamide **2** (K_D = 0.02 nM) than polyamide **3** (K_D = 0.32 nM), displaying non-cooperative binding isotherms ($n=1$) for both molecules, consistent with 1:1 stoichiometry. Polyamide **1** bound with poor affinity and at concentrations ≥100 nM in a non-specific manner (*Table*).

5. Discussion and Conclusion

Polyamides **2** and **3**, which preserve the Im-β-Im unit, bound DNA sites of similar size, 10- and 11-base pairs, but remarkably different sequence compositions. In the original 'lexitropsin' model based on 1 : 1 binding of polyamides to DNA, Im was proposed to bind GC/CG > AT/TA [1]. We now know this not to be the case. In a study to be published elsewhere, Im, in the sequence context of polyamide **2**, binds all four base pairs (within a factor of 2) [19]. From the crystal structure of the 1 : 1 netropsin : DNA complex, we understand the molecular mechanism by which Py specifies AT/TA > GC/CG [1]. However, this 1 : 1 recognition code of Py selecting AT/TA > GC/CG must now be modified to include the judicious placement of β for AT/TA recognition to reset the curvature in the 1 : 1 motif [9]. This is substantiated by our observation that Im-β-Im binds 5′-GAG-3′ with higher affinity than Im-Py-Im (*Table*).

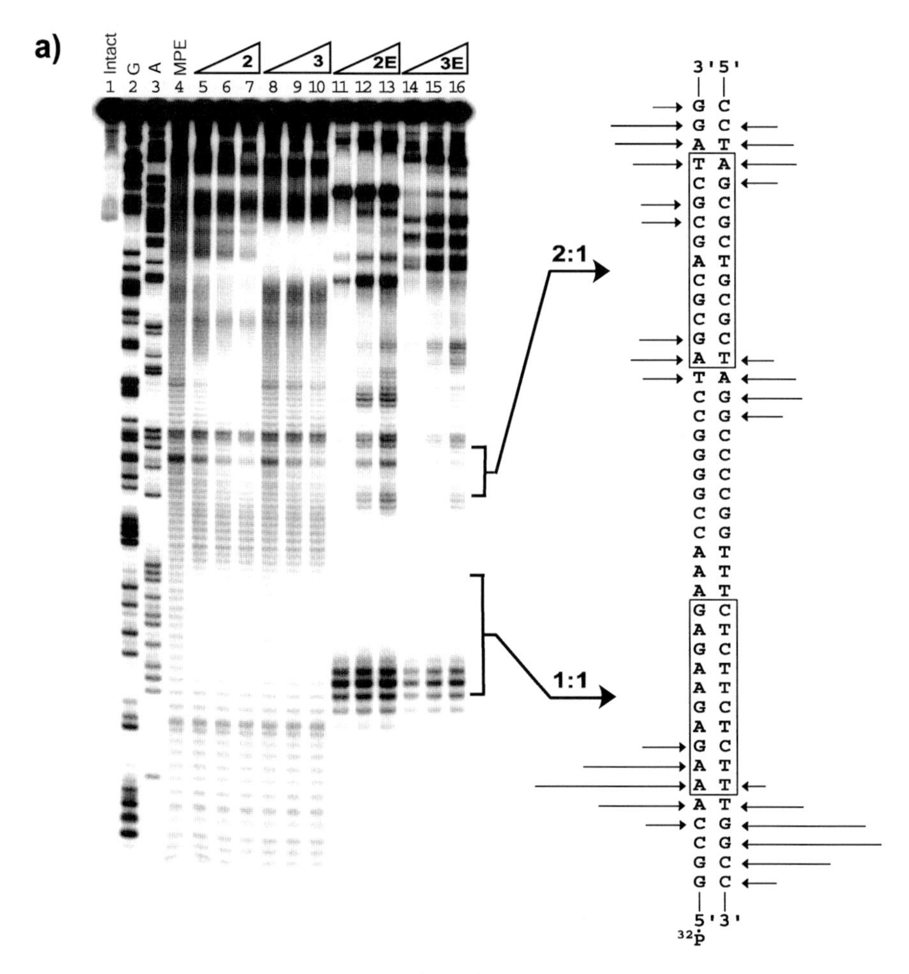

Fig. 4. a) *MPE Footprinting* (polyamides **2** *and* **3**) *and affinity-cleaving experiments* (**2-E** and **3-E**) *on the 5′-end-labelled 253 bp PCR product from pAU9. Lane 1*: intact DNA; *Lane 2*: G reaction; *Lane 3*: A reaction; *Lane 4*: MPE standard; *Lanes 5–7*: 10 nM, 30 nM, and 100 nM **2**, respectively: *Lanes 8–10*: 10 nM, 30 nM, and 100 nM **3**, respectively; *Lanes 11–13*: 1 nM, 3 nM, and 10 nM **2-E**, respectively, *Lanes 14–16*, 10 nM, 30 nM, and 100 nM **3-E**, respectively. Relative cleavage intensities and MPE footprints for each binding site are illustrated at right by arrows and boxes, respectively. b) Top: *Quantitative DNase I footprint titration experiments on the 5′-end-labelled 253 bp PCR product from pAU9* (left) *with Im-β-ImPy-β-Im-β-ImPy-β-Dp* (**2**) (*Lane 1*: intact DNA, *Lane 2*: G reaction, *Lane 3*: A reaction, *Lane 4*: DNase I standard, *Lanes 5–15*, 1 pM, 3 pM, 10 pM, 30 pM, 100 pM, 300 pM, 1 nM, 3 nM, 10 nM, 30 nM, and 100 nM **2**, respectively) *and* (right) *with Im-β-ImPyPyIm-β-ImPy-β-Dp,* (**3**): (*Lane 1*: intact DNA; *Lane 2*: G reaction; *Lane 3*: A reaction; *Lane 4*, DNase I standard, *Lanes 5–15*: 10 pM, 30 pM, 100 pM, 300 pM, 1 nM, 3 nM, 10 nM, 30 nM, 100 nM, 300 nM, and 1 μM **3**, respectively). 2:1 and 1:1 binding sites are shown boxed at the right of each gel. Bottom: *Binding isotherms for each DNase I footprint titration experiment directly above.* θ_{norm} *values were obtained according to published methods* [10–13]. The data points for the 5′-AGCGCAGCGCT-3′ (2:1) site are indicated by filled circles and the 5′-GAGAAGAGAA-3′ (1:1) site by open circles. The solid lines are best fit *Langmuir* binding titration isotherms obtained by a nonlinear least-squares algorithm.

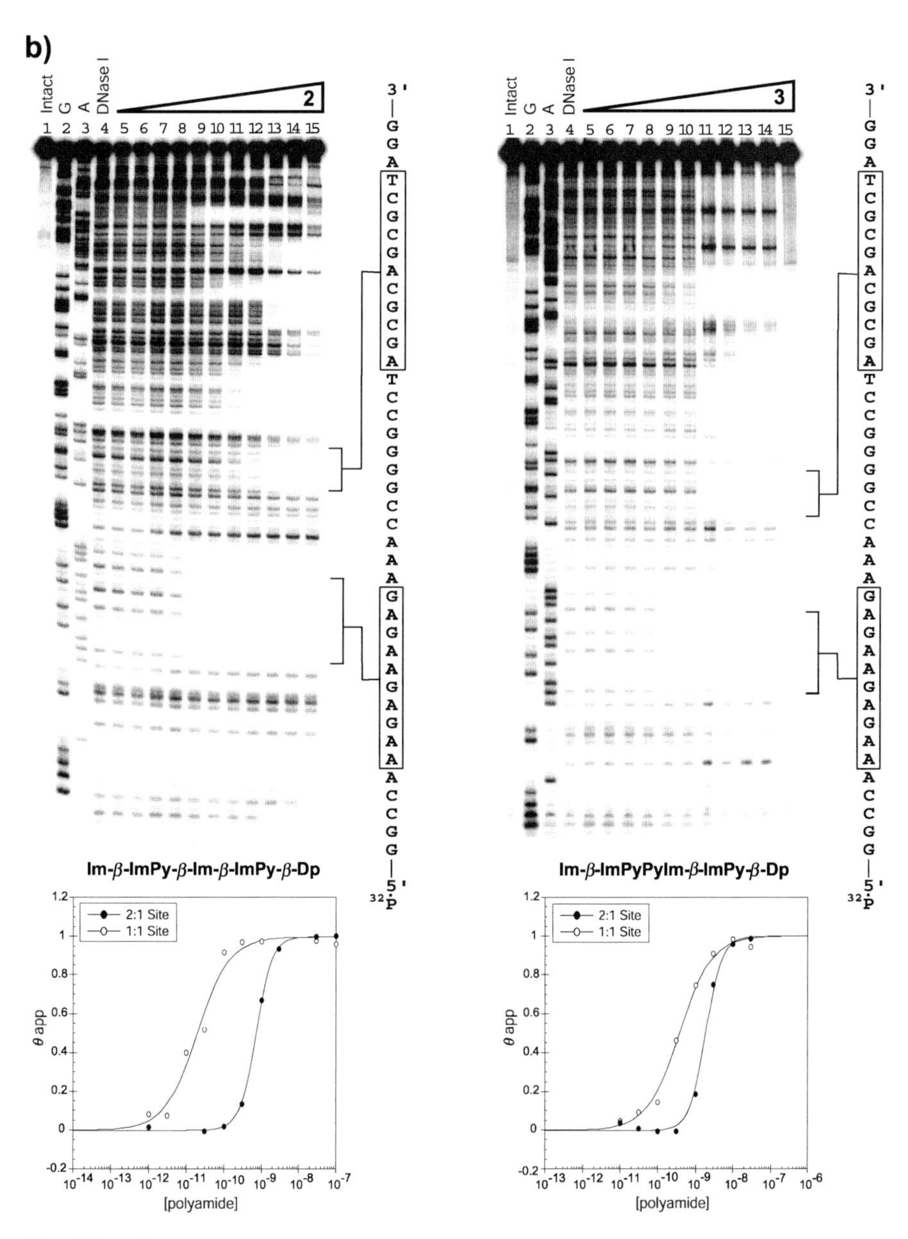
b)
Intact
G
A
DNase I
2
3
Im-β-ImPy-β-Im-β-ImPy-β-Dp
Im-β-ImPyPyIm-β-ImPy-β-Dp
2:1 Site
1:1 Site
θ app
[polyamide]
3'
5'
32P

Fig. 4 (cont.)

Table. *Apparent Equilibration Dissociation Constants* (K_D)

	Polyamide 1	Polyamide 2	Polyamide 3
2:1	5'-A G C G C A G C G C T-3' ●○●○◇●○●○◇(+ +)◇○●○●◇○●○● 3'-T C G C G T C G C G A-5' **³ 100 nM**	5'-A G C G C A G C G C T-3' ●◇●○◇●◇●○◇(+ +)◇○●◇●◇○●◇● 3'-T C G C G T C G C G A-5' **0.63 nM (±0.08)**	5'-A G C G C A G C G C T-3' ●◇●○○●◇●○◇(+ +)◇○●◇●○○●◇● 3'-T C G C G T C G C G A-5' **1.9 nM (±0.1)**
1:1	5'-A A G A G A A G A G-3' +)◇○●○●◇○●○● 3'-T T C T C T T C T C-5' **³ 100 nM**	5'-A A G A G A A G A G-3' +)◇○●◇●◇○●◇● 3'-T T C T C T T C T C-5' **0.020 nM (±0.002)**	5'-A A G A G A A G A G-3' +)◇○●◇●○○●◇● 3'-T T C T C T T C T C-5' **0.32 nM (±0.04)**

There are several implications for the addition of a 1:1 motif to current 2:1 targeting of mixed G,C/A,T sequences of DNA in the minor groove. Assuming cell permeability is related to size, one could argue that a 1:1 binding molecule will occupy larger binding-site sizes for the same molecular weight as a hairpin according to pairing rules. In fact, the difference of one CH_2 unit (β *vs.* γ) directs whether a polyamide binds extended as 1:1 complex or folds as a hairpin [20]. We anticipate less sequence discrimination in the 1:1 complex due to one ring per base pair *vs.* pairs of side-by-side rings reading each base pair. Therefore, one could target large binding-site sizes with smaller polyamides than hairpins, but with reduced specificity at a given base-pair position. We face the added complexity of the same molecule binding very different DNA sites depending on stoichiometry. If specificity is the goal, this raises the challenge for chemists to design third generation polyamides which enforce 2:1 *vs.* 1:1 binding in the minor groove. Recent examples are the 'cycle motif' or 'linked hairpins' [21][22].

We are grateful to the *National Institutes of Health* for support.

Experimental Part

1. Synthesis of Polyamides 1–3

Polyamides **1**–**3** were prepared by solid-phase methods and purified by reversed-phase HPLC as described in [17].

ImPyImPy-β-ImPyImPy-β-Dp (**1**): 70% yield. UV (H_2O): λ_{max} 244 nm (ε = 28500, measured), 308 nm (ε = 44300, measured). ^{1}H-NMR ((D_6)DMSO): 10.41 (*s*, 1 H); 10.33 (*s*, 1 H); 10.21 (*s*, 2 H); 10.01 (*s*, 1 H); 9.99 (*s*, 1 H); 9.91 (*s*, 1 H); 9.18 (br. *s*, 1 H); 8.06 (*m*, 3 H); 7.53 (*s*, 1 H); 7.52 (*s*, 1 H); 7.46 (*s*, 1 H); 7.38 (*m*, 2 H); 7.36 (*d*, *J* = 1.8, 1 H); 7.21 (*d*, *J* = 1.8, 1 H); 7.19 (*d*, *J* = 1.8, 1 H); 7.14 (*d*, *J* = 1.8, 1 H); 7.12 (*d*, *J* = 1.8, 1 H); 7.04 (*s*, 1 H); 6.94 (*m*, 2 H); 3.97 (*s*, 3 H); 3.95 (*s*, 6 H); 3.94 (*s*, 3 H); 3.84 (*s*, 3 H); 3.84 (*s*, 3 H); 3.80 (*s*, 3 H); 3.78 (*s*, 3 H); 3.42 (*m*, 2 H); 3.35 (*m*, 2 H); 3.09 (*q*, *J* = 6.3, 2 H); 2.98 (*m*, 2 H); 2.72 (*d*, *J* = 4.5, 6 H); 2.57 (*t*, *J* = 6.6, 2 H); 2.33 (*t*, *J* = 6.3, 2 H); 1.71 (*m*, 2 H). MALDI-TOF-MS (monoisotopic): 1210.6 ($C_{55}H_{68}N_{23}O_{10}$; calc. 1210.5).

Im-β-ImPy-β-Im-β-ImPy-β-Dp (**2**): 24% yield. UV (H_2O): λ_{max} 248 (ε = 27000, measured), 290 (ε = 28200, measured); ^{1}H-NMR ((D_6)DMSO): 10.36 (*s*, 1 H); 10.33 (*s*, 2 H); 9.32 (br. *s*, 1 H); 9.26 (*s*, 2 H); 8.40 (*t*, *J* = 6, 1 H); 8.05 (*m*, 3 H); 7.89 (*t*, *J* = 6, 1 H); 7.43 (*m*, 2 H); 7.39 (*s*, 1 H); 7.34 (*s*, 1 H); 7.19 (*d*, 2 H); 6.98 (*s*, 1 H); 6.92 (*s*, 1 H); 6.90 (*s*, 1 H); 3.91 (*m*, 9 H); 3.89 (*s*, 3 H); 3.78 (*s*, 6 H); 3.47 (*m*, 4 H); 3.36 (*m*, 4 H); 3.08 (*q*, *J* = 6, 2 H); 2.98 (*m*, 2 H); 2.72 (*d*, *J* = 4.5, 6 H); 2.58 (*t*, *J* = 6.6, 4 H); 2.53 (*t*, *J* = 7.2, 3 H); 2.32 (*t*, *J* = 7.2, 2 H); 1.71 (*m*, 2 H). MALDI-TOF-MS (monoisotopic): 1108.6 ($C_{49}H_{66}N_{21}O_{10}$; calc. 1108.5).

Im-β-ImPyPyIm-β-ImPy-β-Dp (**3**): 11% yield. UV (H_2O): λ_{max} 254 (ε = 32200, measured), 308 nm (ε = 43700, measured); ^{1}H-NMR ((D_6)DMSO): 10.34 (*s*, 1 H); 10.33 (*s*, 2 H); 10.00 (*s*, 1 H); 9.94 (*s*, 2 H); 9.22 (br. *s*, 1 H); 8.38 (*t*, *J* = 5.7, 1 H); 8.04 (*m*, 2 H); 7.91 (*t*, *J* = 6.0, 1 H); 7.50 (*d*, *J* = 1.2, 1 H); 7.45 (*d*, *J* = 1.2, 1 H); 7.41 (*d*, *J* = 1.2, 1 H); 7.33 (*s*, 1 H); 7.31 (*s*, 1 H); 7.25 (*s*, 1 H); 7.19 (*s*, 1 H); 7.13 (*s*, 1 H); 7.07 (*s*, 1 H); 6.96 (*t*, *J* = 1.2, 1 H); 6.91 (*s*, 1 H); 3.94 (*s*, 3 H); 3.93 (*m*, 6 H); 3.92 (*s*, 3 H); 3.82 (*s*, 6 H); 3.77 (*s*, 3 H); 3.49 (*m*, 4 H); 3.35 (*q*, *J* = 6.0, 2 H); 3.08 (*q*, *J* = 6.0, 2 H), 2.97 (*m*, 2 H); 2.71 (*d*, *J* = 3.9, 6 H); 2.59 (*m*, 4 H); 2.32 (*t*, *J* = 6.9, 2 H); 1.71 (*m*, 2 H). MALDI-TOF-MS (monoisotopic): 1159.6 ($C_{52}H_{67}N_{22}O_{10}$; calc. 1159.5).

2. Synthesis of EDTA Conjugates of Polyamides 1–3

A single-step aminolysis of the resin-ester linkage was accomplished using bis(3-aminopropyl)methylamine to afford polyamides with a primary amine at the C-terminus. After purification by HPLC, these were allowed to react with excess EDTA dianhydride, and the EDTA conjugates were purified by reversed phase HPLC to yield **1-E**, **2-E**, and **3-E**. MALDI-TOF-MS (monoisotopic): **1-E** (ImPyImPy-β-ImPyImPy-β-Dp-EDTA): 1527.7 ($C_{67}H_{87}N_{26}O_{17}$; calc. 1527.7); **2-E** (Im-β-ImPy-β-Im-β-ImPy-β-Dp-EDTA): 1425.8 ($C_{61}H_{85}N_{24}O_{17}$; calc. 1425.7); **3-E** (Im-β-ImPyPyIm-β-ImPy-β-Dp-EDTA): 1476.8 ($C_{64}H_{86}N_{25}O_{17}$; calc. 1476.7).

3. Construction of Plasmid DNA

Plasmid pAU9 was constructed by cloning the hybridized inserts, 5′-GATCCGGCCAAAGAGAAGAGAAACCGGGGCCTAGCGCAGCGCTAGGCCA-3′ · 5′-AGCTTGGCCTAGCGCTGCGCTAGGCCCCGGTTTCTCTTCTCTTTGGCCG-3′, into linear pUC19 *BamHI/HindIII* [23].

4. Preparation of 5′- and 3′-End-Labelled Restriction Fragments

For 3′-labelling, pUC19 was *Pvu*II/*Eco*RI-linearized and then 3′-filled with deoxyadenosine 5′-[α-^{32}P] and thymidine 5′-[α-^{32}P] triphosphates using *Klenow* polymerase. For 5′-labelling, two 20-base pair primer oligonucleotides, 5′-AATTCGAGCTCGGTACCCGG-3′ (forward) and 5′-CTGGCACGACAGGTTTCCCG-3′ (reverse), were constructed to complement the pUC19 *EcoRI* and *PvuII* sites, respectively, such that amplification by PCR would mimic the long, 3′-filled pUC19 *EcoRI/PvuII* restriction fragment. The forward primer was radiolabelled using γ-^{32}P-dATP and polynucleotide kinase. Labelled fragments were purified on a 7% non-denaturing preparatory polyacrylamide gel (5% cross-link) and isolated. Chemical-sequencing reactions were performed according to published protocols [24] [25].

5. MPE · Fe(II) Footprinting

All reactions were carried out in a volume of 400 µl. A polyamide stock soln. or H_2O (for reference lanes) was added to an assay buffer where the final concentrations were: 28.6 mM HEPES, 285.7 mM NaCl buffer (pH 7.0), and 20 kcpm 3′- or 5′-radiolabelled DNA. The solns. were allowed to equilibrate for 12 h. A fresh 5 μM MPE · Fe^{II} soln. was prepared from equal volumes of 10 μM MPE and 10 μM ferrous ammonium sulfate ($Fe(NH_4)_2(SO_4)_2 \cdot 6\,H_2O$) solns. MPE · Fe^{II} soln. (0.5 μM) was added to the equilibrated solns., and the reactions were allowed to proceed for 15 min. Cleavage was initiated by the addition of dithiothreitol (5 mM) and allowed to proceed for 30 min. Reactions were stopped by EtOH precipitation, and the pellets were washed in 75% EtOH, dried *in vacuo*, resuspended in 15 μl of H_2O, lyophilized to dryness, and resuspended in 100 mM *Tris*-borate-EDTA/80% formamide loading buffer. The solns. were then denatured at 90° for 9 min, and a 5 μl sample (*ca.* 8 kcpm) was loaded onto an 8% denaturing polyacrylamide gel (5% cross-link, 7M urea) and run at 2000 V.

6. Affinity Cleaving

All reactions were carried out in a volume of 400 μl. A polyamide stock soln. or H_2O (for reference lanes) was added to an assay buffer where the final concentrations were 28.6 mM HEPES/85.7 mM NaCl buffer (pH 7.0), 200 mM NaCl, and 20 kcpm 3′- or 5′-radiolabelled DNA. The solns. were allowed to equilibrate for 12 h. A fresh soln. of $Fe(NH_4)_2(SO_4)_2 \cdot 6\,H_2O$ (1 μM) was added to the equilibrated solns., and the reactions proceeded for 15 min. Cleavage was initiated by the addition of dithiothreitol (5 mM) and allowed to proceed for 11 min. Reactions were stopped by EtOH precipitation, and the pellets were washed in 75% EtOH, dried *in vacuo*, resuspended in 15 μl of H_2O, lyophilized to dryness, and resuspended in 100 mM *Tris*-borate-EDTA/80% formamide loading buffer. The solns. were then denatured at 90° for 9 min, and a 5-μl sample (*ca.* 8 kcpm) was loaded onto an 8% denaturing polyacrylamide gel (5% crosslink, 7M urea) and run at 2000 V.

7. Quantitative DNase I Footprint Titrations

All reactions were carried out in a volume of 400 μl. We note explicitly that no carrier DNA was used in these reactions until after DNase I cleavage. A polyamide stock soln. (or H_2O for reference and intact lanes) was added to an assay buffer where the final concentrations were 10 mM *Tris* · HCl buffer (pH 7.0), 10 mM KCl, 10 mM $MgCl_2$, 5 mM $CaCl_2$, and 20 kcpm 5′-radiolabelled DNA. The solns. were allowed to equilibrate for 18 h at 22°. Cleavage was initiated by the addition of 10 μl of a DNase I soln. (diluted with 1 mM DTT to 1.5 u/ml) and allowed to proceed for 7 min. at 22°. The reactions were stopped by adding 50 μl of a soln. containing 2.25M NaCl, 150 mM EDTA, 0.6 mg/ml glycogen, and 30 μM base-pair calf thymus DNA, and then EtOH-precipitated (2.1 volumes) at 14 krpm for 23 min. The pellets were washed with 75% EtOH, resuspended in 15 μl RNase-free H_2O, lyophilized to dryness, and then resuspended in 100 mM *Tris*-borate-EDTA/80% formamide loading buffer (with bromophenol blue as dye), denatured at 90° for 10 min, and loaded directly onto a pre-run 8% denaturing polyacrylamide gel (5% cross-link, 7M urea) at 2000 V for 1.2 h. The gels were dried *in vacuo* at 80° and then exposed to a storage phosphor screen (*Molecular Dynamics*). Equilibrium association constants were determined as previously described [10–13].

REFERENCES

[1] M. L. Kopka, C. Yoon, D. Goodsell, P. Pjura, R. E. Dickerson, *Proc. Natl. Acad. Sci. U.S.A.* **1985**, *82*, 1376.

[2] J. G. Pelton, D. E. Wemmer, *Proc. Natl. Acad. Sci. U.S.A.* **1989**, *86*, 5723.

[3] P. B. Dervan, R. W. Burli, *Curr. Opin. Chem. Biol.* **1999**, *3*, 688.

[4] R. S. Youngquist, P. B. Dervan, P. B. *Proc. Natl. Acad. Sci. U.S.A.* **1985**, *82*, 2565.

[5] S. White, J. W. Szewczyk, J. M. Turner, E. E. Baird, P. B. Dervan, *Nature* **1998**, *391*, 468.

[6] C. L. Kielkopf, E. E. Baird, P. B. Dervan, D. C. Rees, *Nat. Struct. Biol.* **1998**, *5*, 104.

[7] C. L. Kielkopf, S. White, J. W. Szewczyk, J. M. Turner, E. E. Baird, P. B. Dervan, *Science* **1998**, *282*, 111.
[8] J. J. Kelly, E. E. Baird, P. B. Dervan, *Proc. Natl. Acad. Sci. U.S.A.* **1996**, *93*, 6981.
[9] R. S. Youngquist, P. B. Dervan, *J. Am. Chem. Soc.* **1987**, *109*, 7564.
[10] J. W. Trauger, E. E. Baird, M. Mrksich, P. B. Dervan, *J. Am. Chem. Soc.* **1996**, *118*, 6160.
[11] S. E. Swalley, E. E. Baird, P. B. Dervan, *Chem. Eur. J.* **1997**, *3*, 1600.
[12] J. W. Trauger, E. E. Baird, P. B. Dervan, *J. Am. Chem. Soc.* **1998**, *120*, 3534.
[13] J. M. Turner, S. E. Swalley, E. E. Baird, P. B. Dervan, *J. Am. Chem. Soc.* **1998**, *120,* 6219.
[14] S. Janssen, T. Durussel, U. Laemmli, *Mol. Cells* **2000**, *6*, 999.
[15] E. E. Baird, P. B. Dervan, *J. Am. Chem. Soc.* **1996**, *118*, 6141.
[16] M. W. Van Dyke, P. B. Dervan, *Biochemistry* **1983**, *22*, 2373.
[17] P. G. Schultz, P. B. Dervan, *J. Biomol. Struct. Dyn.* **1984**, *1*, 1133.
[18] M. Brenowitz, D. F. Senear, M. A. Shea, G. K. Ackers, *Methods Enzymol.* **1986**, *130*, 132.
[19] A. R. Urbach, P. B. Dervan, *Proc. Natl. Acad. Sci. U.S.A.,* submitted.
[20] M. Mrksich, M. E. Parks, P. B. Dervan, *J. Am. Chem. Soc.* **1994**, *116*, 7983.
[21] D. M. Herman, J. M. Turner, E. E. Baird, P. B. Dervan, *J. Am. Chem. Soc.* **1999**, *121*, 1121.
[22] D. M. Herman, E. E. Baird, P. B. Dervan, *Chem. Eur. J.* **1999**, *5*, 975.
[23] J. Sambrook, E. F. Fritsch, T. Maniatis, *'Molecular Cloning'*, Cold Spring Harbor Laboratory, Cold Spring Harbor, NY, 1989.
[24] B. L. Iverson, P. B. Dervan, *Nucleic Acids Res.* **1987**, *15*, 7823.
[25] A. M. Maxam, W. S. Gilbert, *Methods Enzymol.* **1980**, *65*, 499.

Generating New Molecular Function: A Lesson from Nature[1])

by **David R. Liu** and **Peter G. Schultz**

Howard Hughes Medical Institute and Department of Chemistry,
University of California, Berkeley, Berkeley, CA 94720-1460, USA
(Fax: (+1) 510-643-6890, e-mail: pgschultz@lbl.gov)

1. Introduction

Our ability to create new molecular structures with novel physical, chemical, or biological properties has had a major impact on many fields of science including the biomedical sciences, chemistry, and solid-state physics. Examples range from the synthesis of antibiotics and genes to photoresists and high-temperature superconductors. As the structural and functional complexity of target molecules increases, our ability to define the precise structural requirements that result in a desired set of properties becomes increasingly limited. This in turn leads to a large increase in the number of molecules that must be iteratively synthesized and tested to identify those structures with a desired function – a time- and resource-intensive process. Among the strategies being developed to meet the challenges created by the ever growing need for molecules with new or enhanced properties are a number of biologically inspired approaches. The latter derive from our recognition that nature has produced an impressive array of complex molecular structures and assemblies with functions ranging from gene regulation and signal transduction to photosynthesis and protein biosynthesis. As we begin to understand the structures and molecular mechanisms involved in these processes, we can combine these biological insights, as well as the processes themselves, with the tools of the physical sciences to create new molecules with functions found neither in nature nor the laboratory.

One such example of this synergy between chemistry and biology is the development and application of combinatorial strategies. This approach, in which large, diverse collections, or 'libraries', of molecules are generated and subsequently screened or selected for novel functions, stems from the

[1]) This essay is based on a lecture delivered at the *36th IUPAC Congress* in Geneva, 1997.

combinatorial processes in nature. For example, both the humoral and cellular arms of the immune system have developed highly sophisticated combinatorial genetic mechanisms for generating molecular diversity and selecting receptors that can recognize foreign antigens with high affinity and selectivity. The mechanisms of these processes are relatively well understood [1–3]. The three-dimensional framework of the antibody molecule consists of two polypeptide chains (heavy (H) and light (L)) containing six loops on an eight-stranded β-sheet framework [4]. This structure is assembled from four different gene segments, variable (V), diversity (D), joining (J), and constant (C), each of which can be chosen from a number of distinct genes. The combinatorial association of V, D, and J gene segments with additional junctional diversity occurring at the V_L-J_L, V_H-D, and $D-J_H$ joining regions leads to a structurally diverse population of germ-line antibodies (*Fig. 1*). After a 'lead' antibody structure is selected from this pool based on its ability to bind a foreign substance, its affinity is increased as the immune response proceeds by somatic mutation. This process, which alters bases throughout the DNA sequences encoding the variable region, results in additional diversity. Thus, in its most basic form, the combinatorial process involves the synthesis of large numbers of distinct structures around a central framework from sets of building blocks. The choice of framework structure and building blocks is best made based on empirical and theoretical models at hand. Those structures with a desired function then are identified and further optimized by repeated rounds of screening or selection and mutation.

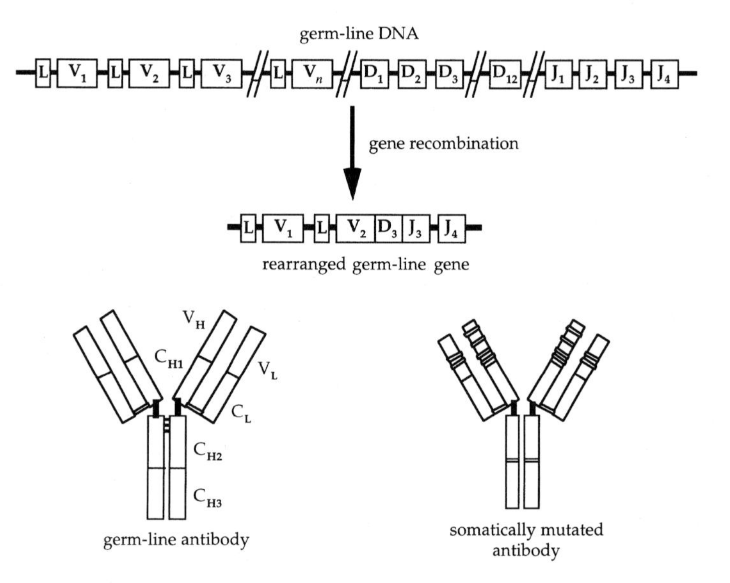

Fig. 1. *Combinatorial association of V, D, and J genes to form an antibody molecule*

This combinatorial approach is applicable to any molecular structure that can be assembled either in a stepwise or concerted fashion from a set of molecular precursors, and where a screen or selection for a desired function exists. This includes oligomeric molecules such as nucleic acids and peptides, nonpolymeric molecules such as natural products, or even solid-state materials such as superconductors or polymers. To illustrate the utility and scope of this new synthetic strategy, as well as the challenges that remain, a number of examples, both from our laboratory and many others, will be described. These applications cover a broad spectrum, ranging from biological catalysis and drug discovery to materials science.

2. Protein and Polypeptide Protein Diversity

2.1. Immunological Diversity and Catalytic Antibodies

The notion that natural immunological diversity can be used to generate novel chemical function was first illustrated with the generation of catalytic antibodies [5–7]. Immunological diversity is clonally selected on a time scale of weeks (in contrast to evolutionary diversity), and is thus 'programmable' in the laboratory setting. However, binding affinity rather than catalytic activity serves as the basis for antibody selection, and therefore mechanistic principles such as selective binding and stabilization of the transition state must be used to generate catalytic antibodies. For example, to generate an antibody that catalyzes the insertion of a metal ion into porphyrin (the last chemical step in heme biosynthesis), antibodies were generated against *N*-methylprotoporphyrin (**1**). This molecule mimics the putative transition state, a conformationally distorted porphyrin ring, of the metalation reaction. An antibody specific for **1** catalyzes the insertion of metal ion into mesoporphyrin with rates similar to that of the natural biosynthetic enzyme, ferrochelatase [8]. Resonance *Raman* studies revealed that both the enzyme and antibody facilitate the reaction by distorting the planar porphyrin substrate, which is consistent with the proposed reaction mechanism. This distortion involves doming of the porphyrin ring in the case of the enzyme, and for the antibody an up–down distortion mimicking that in *N*-methylmesoporphyrin [9] (*Fig. 2*). Crystallographic analysis of the antibody·*N*-methylprotoporphyrin complex has revealed the structural basis for this distortion, and shown that somatic mutations in response to **1** play an important mechanistic role [10].

A second example involves a transesterification reaction catalyzed by an antibody generated against a phosphonate diester transition state analogue [11]. The antibody catalyzes the corresponding acyl transfer reaction of thymidine and an alanyl ester with an effective molarity of 3×10^4 M

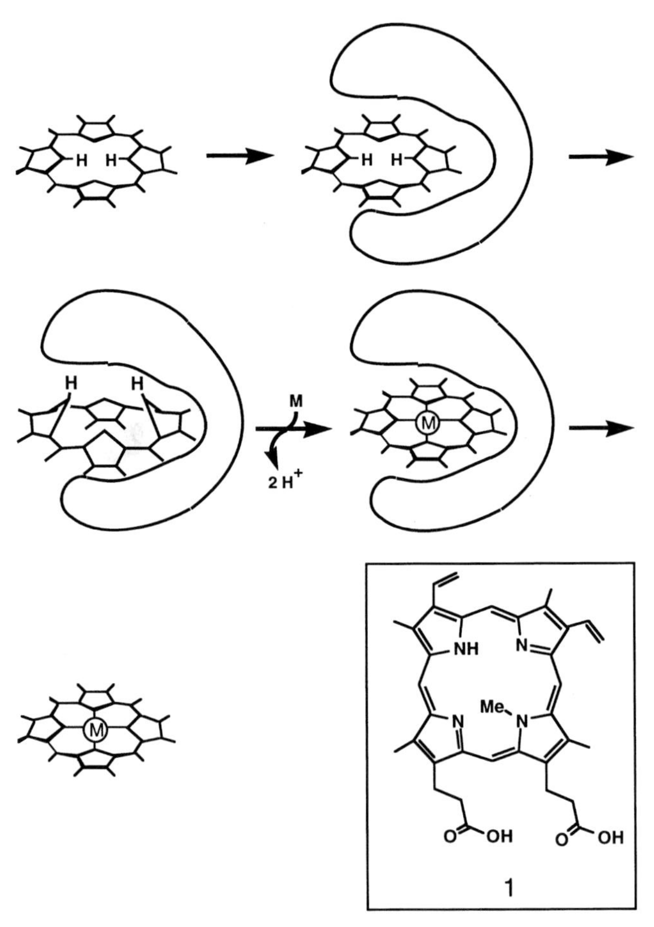

Fig. 2. *Representation of* N-*methylprotoporphyrin* (**1**) *distortion induced by an antibody with ferrochelatase activity* [10]

($(k_{cat}/K_m)/k_{uncat} \approx 10^8$). Moreover, the antibody does not catalyze acyl transfer to water. NMR Studies of the *Michaelis* complex suggest that the high catalytic efficiency and selectivity of this antibody result largely from an optimal orientation of the acyl donor and nucleophile in the antibody active site, reflecting the orientation of the phosphonate moiety in the hapten [12]. Thus, by selecting the natural diversity of the immune system on the basis of mechanistic criteria (for example, distortion of a planar porphyrin ring or proximity of reactive groups), it is possible both to generate antibodies with enzyme-like properties as well as to test basic principles of enzymic catalysis.

Other approaches have been developed to generate catalytic antibodies including covalent catalysis, proximity effects, and general acid–base catalysis [7]. In addition, strategies have been developed to directly select immunological diversity for catalytic function both *in vivo* and *in vitro*. Such approaches have been used to generate antibodies that catalyze a wide array of enzymic and nonenzymic reactions. These include stereoselective redox reactions, pericyclic reactions, rearrangement and acyl transfer reactions, as well as a number of 'difficult' chemical transformations including kinetically disfavored *exo-Diels–Alder* and *anti-Baldwin* cyclization reactions [13]. A recent example involved the generation of an aldolase antibody by means of a mechanism-based selection with a hapten capable of forming a covalent bond with active-site lysine residues. The catalytic efficiency and stereoselectivity of this antibody are remarkably similar to those of Class I aldolase enzymes [14]. Catalysis proceeds through imine formation between the ketone substrate and the ε-amino group of an active site-lysine residue, followed by enamine formation, condensation with an aldehyde, and subsequent hydrolysis to give the aldol product – a mechanism similar to that of the enzyme. At the same time the antibody tolerates a remarkably broad range

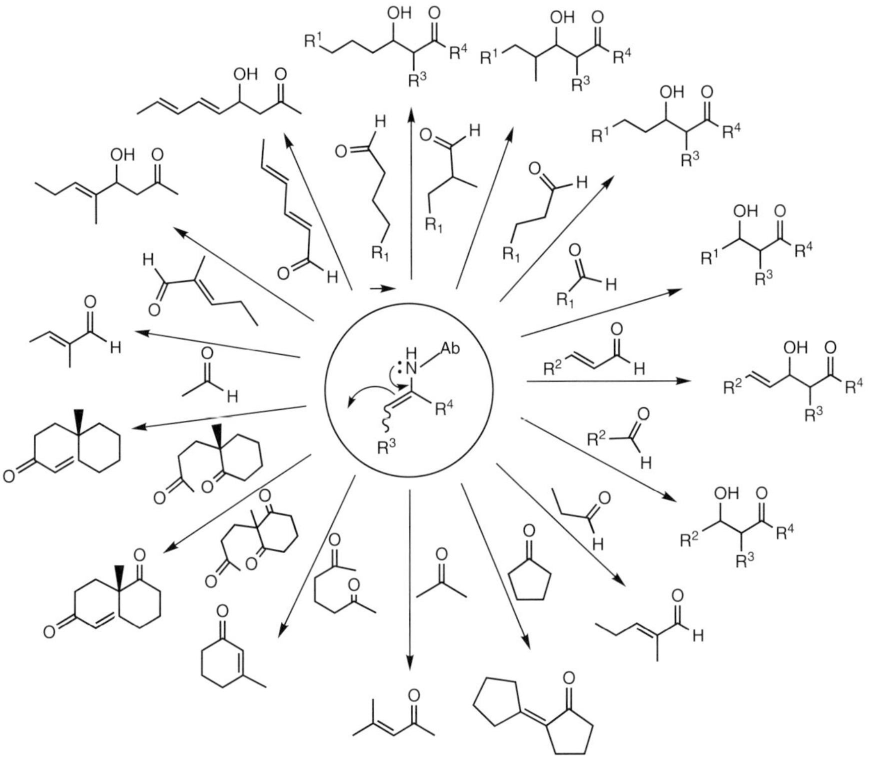

Fig. 3. *Schematic representation of an antibody (Ab) aldolase with broad substrate specificity* [14]. Shown in the sphere is the covalent enamine intermediate that is formed from the ketone substrate and the ε-amino group of an active-site lysine residue.

of substrate structures (*Fig. 3*) and thus represents one of the most general stereoselective catalysts to date for carbon–carbon bond formation [15]. This example beautifully illustrates the catalytic potential of immunological diversity that can be harnessed with proper chemical instruction.

The characterization of catalytic antibodies also provides important new insights into the evolution of binding and catalytic function, as well as the combinatorial processes of the immune system itself. For example, structural and mechanistic analyses of antibody-catalyzed oxy-*Cope* rearrangement and *Diels–Alder* reactions have illustrated a number of mechanisms whereby binding energy can be used affect the stereoelectronics of pericyclic reactions [16–18]. Detailed biophysical and structural studies have also been carried out to examine the affinity maturation of an esterolytic antibody generated against a nitrophenyl phosphonate transition state analogue. Nine somatic mutations resulted in a catalytic antibody that binds hapten with 35000-fold greater affinity than its germ-line precursor. In contrast to the affinity-matured antibody, which binds hapten in a 'lock-and-key' fashion, the germ-line antibody can adopt more than one combining site conformation Both antigen binding and somatic mutation stabilize the conformation with optimal hapten complementarity (*Fig. 4*). This result suggests that conformational diversity (a key element of *Pauling*'s chemical instruction model of antibody specificity) may play an important role in expanding the binding potential of the primary immune response [19–21]. This study further showed that the binding and catalytic properties of the antibody are greatly affected by mutations distant from the bound ligand, an important lesson for those involved in efforts to modify protein function. The study of catalytic antibodies has also provided insights into the structural basis for the polyspecificity of families of germ-line antibodies, another factor that likely contributes to the remarkable binding potential of the primary antibody repertoire [17]. Thus, by generating new functions from antibody diversity, new insights are being gained into the relationship between structure and function in the immune system itself.

Researchers have expanded upon the natural diversity generated by the immune system. In 1989 a novel vector system based on the bacteriophage lambda was used to express a synthetic combinatorial library of Fab fragments of the mouse antibody repertoire in *Escherichia coli* [22]. This system allowed rapid and easy identification of monoclonal Fab fragments that bind a given antigen, in a form suitable for genetic manipulation. For example,

Fig. 4. *Conformational diversity in an antibody esterase. Whether the amino acid belongs to a heavy or light chain is indicated with the letter H or L before the sequence number.* a) *Superposition of the germ-line Fab–hapten complex* (Fab=antigen-binding fragment of an antibody; purple) *and the affinity-matured Fab–hapten complex* (red). *Somatically mutated residues* (X→Y) *are shown in green.* b) *Superposition of the structures of the germ-line Fab domain*

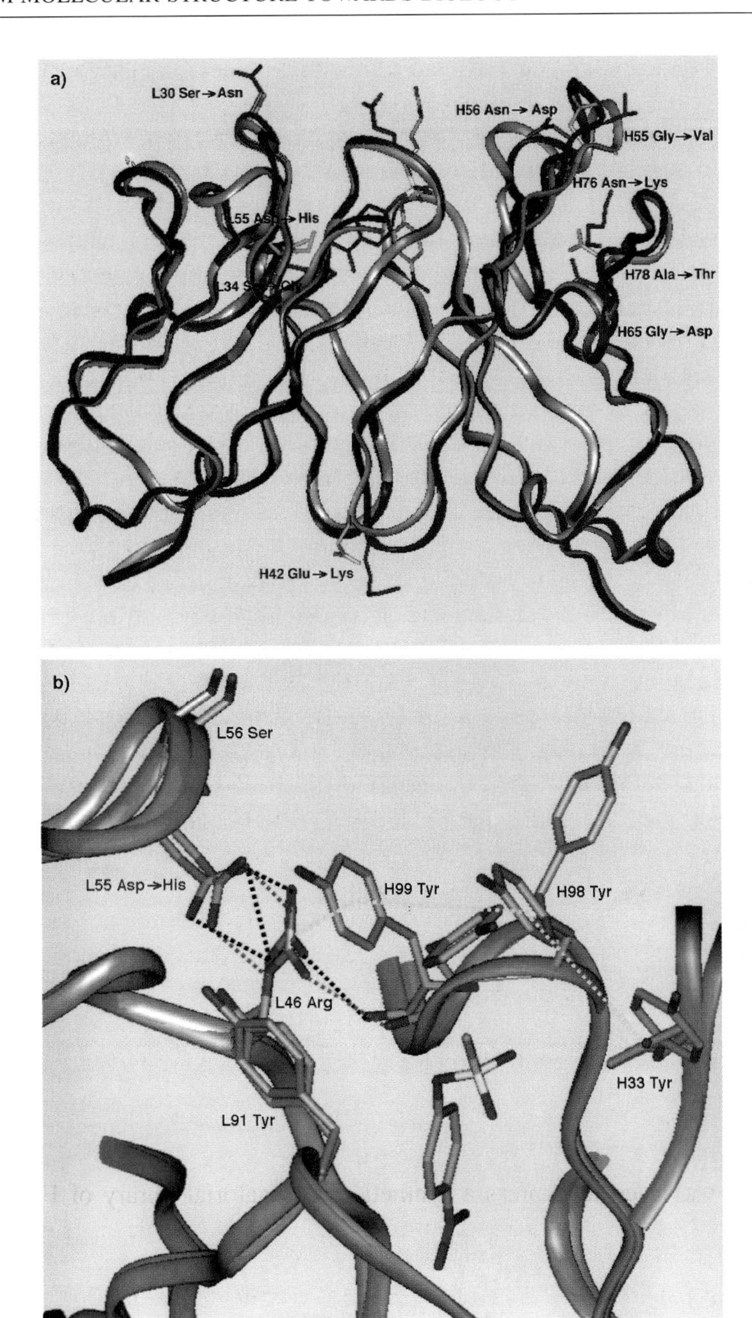

without hapten (blue) *and the germ-line Fab–hapten complex* (purple). Gray dotted lines denote hydrogen bonds in the structure of the germ-line Fab without hapten, while black dotted lines denote hydrogen bonds in the germ-line Fab–hapten complex [19]. Yellow dotted lines show the formation of a 'double T-stack' arrangement between the side chains of three tyrosine residues.

monoclonal Fab fragments against a transition state analogue hapten were generated using this technique.

Phage display [23–26], in which a peptide or protein is expressed on the surface of filamentous phage, as well as other schemes linking the polypeptide chain to the encoding DNA [27–29], have also proven to be very powerful techniques for generating diversity *in vitro*. By mutagenizing codons randomly or at specific sites in a gene with synthetic oligonucleotides, libraries of greater than 10^8 different polypeptide sequences can be generated and expressed on the phage surface. Subsequent screening by affinity-based techniques using immobilized ligands, followed by elution of bound phage, amplification, and additional rounds of screening can lead to the isolation of high-affinity receptors or ligands. The identity of a particular member of the library is determined by sequencing the phage DNA.

With phage display techniques, synthetic combinatorial libraries of antibody genes from multiple species (including human), or from large synthetic repertoires of antibody encoding genes, have been expressed on phage as single-chain Fv or Fab fragments (*Fig. 5*) [30] [31]. High-affinity antibodies have been isolated that bind a wide variety of small ligands and macromolecules including viral and tumor antigens, and that catalyze a number of chemical reactions [24] [32]. More recently, methods have been developed that allow the direct chemical selection of catalysts from phage displayed Fab libraries [33]. For example, an antibody glycosidase was isolated using a mechanism-based inhibitor that produced a reactive quinone methide, resulting in the covalent trapping of catalytic clones onto a solid support for fur-

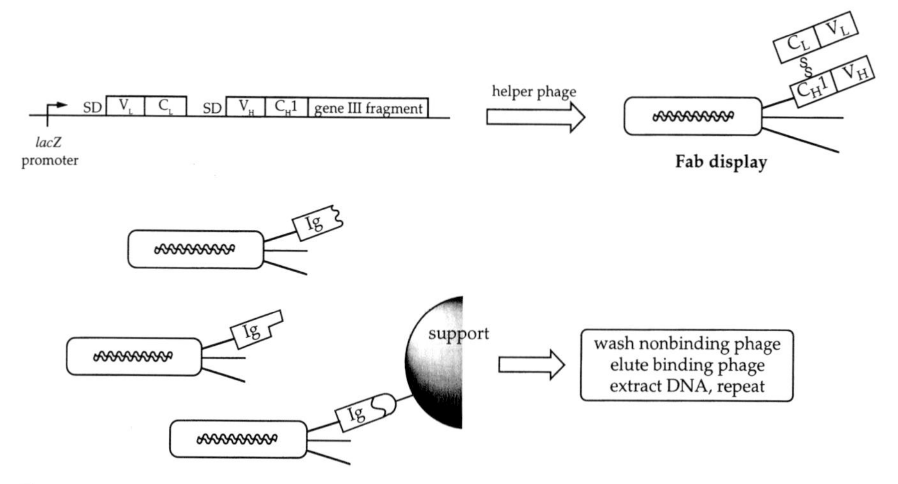

Fig. 5. *Phage display and screening of an antibody library.* SD = *Shine–Dalgarno* sequence (ribosome binding site on prokaryotic mRNA); Ig = immunoglobulin domain.

ther rounds of amplification and selection. *In vitro* methods for generating antibody diversity may become as important as hybridoma technology in the development of monoclonal antibodies for diagnostic, therapeutic, and chemical applications. Phage display methods are also being used to identify other protein frameworks, such as the loops of four-helix bundle proteins, which when randomized might function as miniantibodies [34].

Most recently, we have developed a general scheme for the *in vitro* evolution of protein catalysts that directly links substrate turnover to a selective advantage in a biologically amplifiable system [35]. Substrate is covalently and site-specifically attached by a flexible tether to the pIII coat protein of a filamentous phage that also displays the protein catalyst. Intramolecular conversion of substrate into product provides a basis for isolating active catalysts from a library of proteins, either by release or attachement of the phage to a solid support, or by the capture of the product by a product-specific antibody (*Fig. 6*). This methodology was developed using the enzyme staphylococcal nuclease (SNase) as a model. Phage displaying SNase can be enriched up to 1000-fold in a single step from a librarylike ensemble of phage displaying noncatalytic proteins. Additionally, this approach should allow one to functionally clone natural enzymes based on their ability to catalyze specific reactions (for example, glycosyl transfer, sequence-specific proteolysis or phosphorylation, polymerization), rather than their sequence or structural homology to known enzymes.

2.2. Peptide Diversity

Libraries of linear and cyclic peptides displayed on phage have been used to identify high-affinity, selective ligands which bind a range of biological receptors including antibodies, enzymes, lectins, cell surface receptors, signal transduction proteins, and even nucleic acids [24]. One of the most impressive examples was the isolation of a disulfide-bonded cyclic peptide containing 14 amino acids with the consensus sequence YXCXXGPXTWXCXP (where X can be several amino acids), that binds and activates the receptor for the cytokine erythropoietin (EPO) [36]. The successful isolation of this peptide required both variations in the stringency of the screening (by varying binding avidity and elution conditions) and the generation and screening of libraries derived from low-affinity peptides. The isolated peptide binds the extracellular domain of the EPO receptor (EPOR) with an IC_{50} of roughly 0.2 μM, and stimulates erythropoiesis in mice through a signaling pathway that appears to be identical to that induced by the natural ligand. The molecular basis for the agonist activity of this peptide was determined by analysis of the three-dimensional crystal structure of the complex of EPOR and a re-

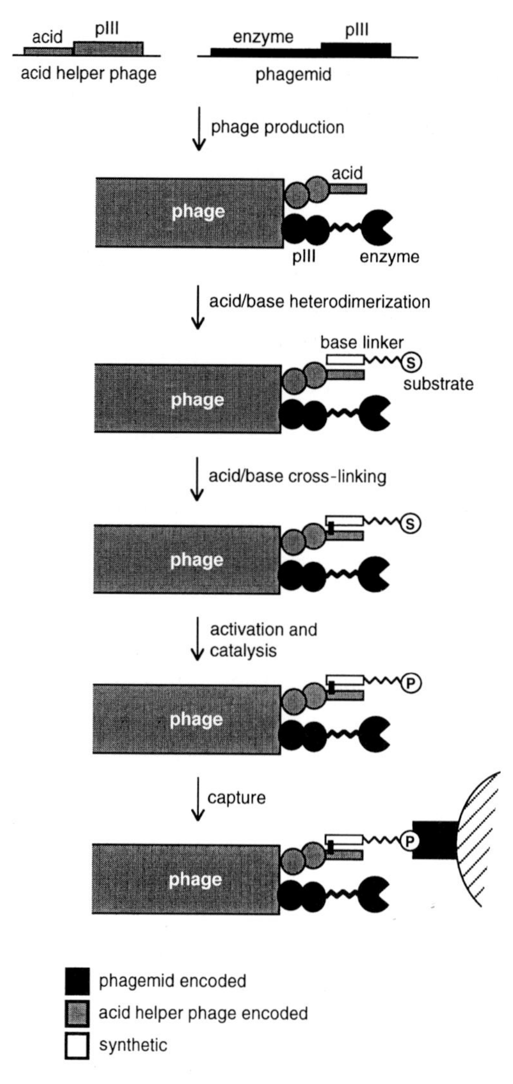

Fig. 6. *A general scheme for the* in vitro *evolution of protein catalysts*

lated 20 amino acid peptide (*Fig. 7*) [37]. The structure reveals that the peptide forms a dimer consisting of a four-stranded, anti-parallel β-pleated sheet and two type I β turns. The peptide induces an almost perfect twofold dimerization of the receptor through a combination of hydrophobic and hydrogen-bonding interactions with segments of four loop regions of EPOR. The successful isolation of this novel peptide, which may define a minimal epitope for activation of EPOR, suggests that peptides may be able to recapitulate many of the other functions of larger proteins.

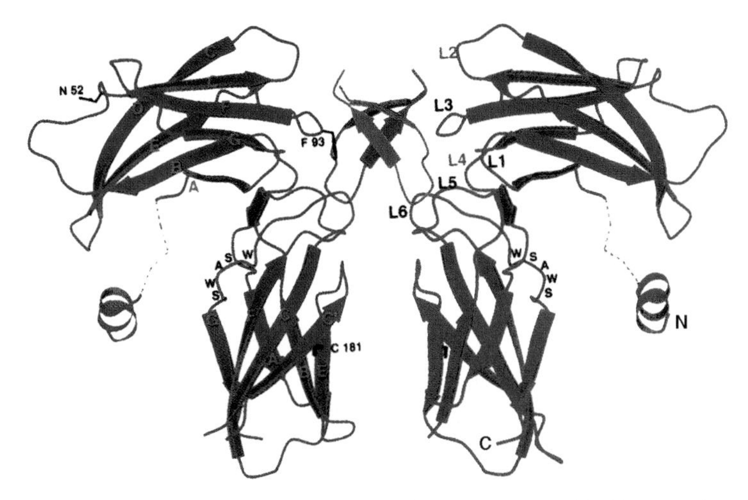

Fig. 7. *Complex of a 20-residue peptide isolated from a phage peptide library with the extracellular domain of the EPO receptor* [37]

Another interesting application of phage display techniques is the determination of substrate specificities of proteases by screening peptide libraries with an assay based on peptide hydrolysis [38]. In this case the amino terminal domain of the pIII protein of filamentous phage is fused to a tag by a randomized peptide linker. Phage are then bound to an affinity support specific for the tag. Upon protease treatment, phage displaying peptides that are good substrates are cleaved from the support. After several rounds of binding, proteolysis, and amplification, sensitive and resistant substrate sequences were identified for the proteases subtilisin BPN′ and factor X_a. Peptide libraries presented on phage have also been panned directly against cells to identify peptide sequences that bind and/or enter specific cell types. These peptides may ultimately provide the basis for intravenously delivered gene therapy vectors[39] or tissue-specific therapeutic agents [40].

2.3. Evolving Protein Function

The natural rate of genetic diversification by point mutation can be increased with UV, chemical, or enzymatic mutagenesis, or genetic methods such as the use of mutator strains. Proteins with random mutations generated by these methods can be subjected to an appropriate screen or selection to isolate mutants with altered properties. For example, these methods have been used to change the substrate specificity of β-galactosidase [41], ribitol dehydrogenase [42], and alkylamidases [43].

More recently, a number of highly efficient methods have been developed for generating larger and more diverse libraries of mutants including cassette mutagenesis [44] [45], error-prone polymerase chain reaction (PCR) [46], and the technique of DNA shuffling [47] [48]. These methods have accelerated our ability to carry out artificial evolution by introducing diversity at a higher rate, and have been used to modify the stability, catalytic activity, and specificity of proteins [49–54]. In particular, DNA shuffling mimics the combinatorial processes of the immune system and natural evolution by combining both random point mutagenesis of a gene with *in vitro* homologous recombination to generate libraries of structurally diverse mutants (*Fig. 8*). This approach allows one to more efficiently search large sequence spaces for mutations that lead to additive or even cooperative enhancements in binding or catalytic function. When directly compared with processes such as error-prone PCR that iteratively build up beneficial mutations without recombination, DNA shuffling has yielded evolved proteins with higher desired activities in fewer rounds of selection [47]. DNA Shuffling has been used to enhance the properties of a number of proteins including glycosidases, β-lactamases, antibodies, green fluorescent protein (GFP), and growth factors [48]. For example, three cycles of DNA shuffling and two cycles of backcrossing (recombination with the wild-type gene) with selection on increasing concentrations of the antibiotic cefotaxime resulted in a β-lactamase with six mutations which conferred to the cell a 32000-fold higher resistance to the antibiotic [47]. The mutations, like the somatic mutations in affinity-matured antibodies, are distributed throughout the protein, making the rational design of the mutant proteins difficult, if not impossible.

Recently, the technique of DNA shuffling has been applied to the evolution of an 'orthogonal' suppressor tRNA and aminoacyl-tRNA synthetase

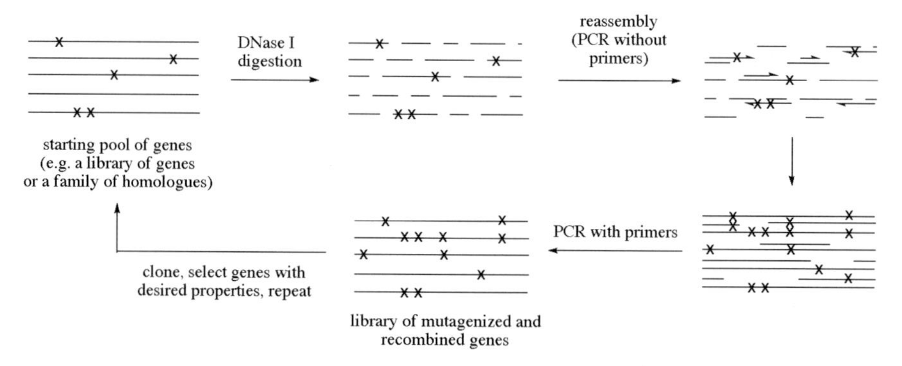

Fig. 8. *DNA Shuffling involves random fragmentation of a gene into discrete fragments, reassembly of these fragments by PCR without primers, final PCR amplification with primers, and cloning of the mutated and recombined genes* [47]

pair that may allow one to site-specifically incorporate into proteins in vivo amino acids with novel structural or electronic properties not specified in the genetic code [55]. The orthogonal suppressor tRNA was constructed by introducing eight mutations into $tRNA_2^{Gln}$ based on analyses of the X-ray crystal structure of the GlnRS–$tRNA_2^{Gln}$ complex (GlnRS = glutaminyl-tRNA aminoacyl synthetase) and previous biochemical data. The resulting tRNA satisfies the minimal requirements for the delivery of an unnatural amino acid: It is not acylated by any endogenous *E. coli* aminoacyl-tRNA synthetase, including GlnRS, and it functions efficiently in protein translation. Repeated rounds of DNA shuffling and oligonucleotide-directed mutagenesis were then used to generate a large library of directed and random mutants of GlnRS. This library was selected for mutant enzymes that efficiently acylate the engineered tRNA with glutamine *in vivo*, based on suppression of an amber codon (termination codon UAG) in the *lacZ* gene (*Fig. 9*). With this approach an 'orthogonal' mutant GlnRS/tRNA pair has been generated that functions *in vivo*. Currently, DNA shuffling and cassette mutagenesis are being used to select for synthetase mutants that acylate orthogonal tRNAs with novel amino acids [56]. Clearly, both structural information and diversity play a critical role in the design of these experiments.

A variety of other proteins with modified binding affinities and specificities, including growth hormones [57] and DNA-binding proteins [58–60], have been isolated from large libraries of mutants. For example, the zinc finger DNA-binding motif of Zif 268 protein has been expressed on filamentous phage, and libraries of mutants have been selected against duplex DNA sequences containing wild-type and mutant binding sites [61]. This approach

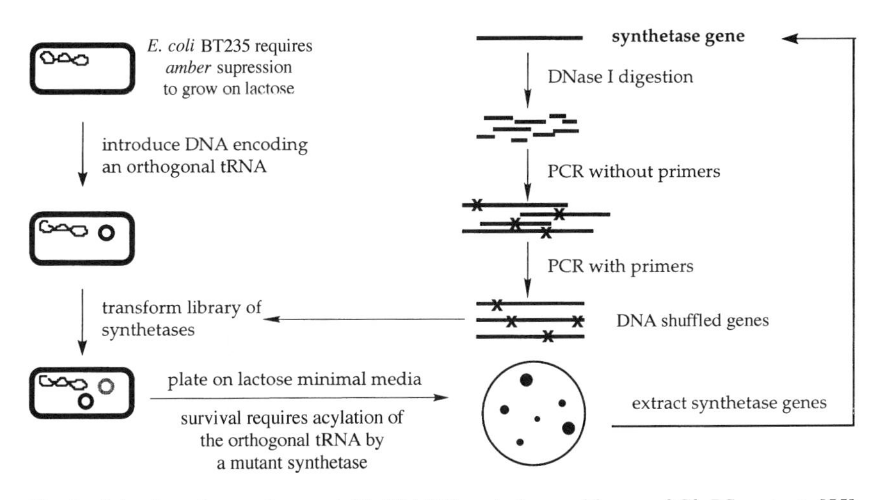

Fig. 9. *Selection of an orthogonal GlnRS/tRNA pair from a library of GlnRS mutants* [55]

has resulted in the generation of a DNA-binding peptide comprising three zinc fingers that binds to a unique nine base pair region of a BCR-ABL fusion oncogene with a dissociation constant K_d of 6.2×10^{-7} M. Its binding affinities for genomic BCR and C-ABL sequences (which differ by as little as one base pair) differ by an order of magnitude. Binding of the peptide to the target oncogene in transformed cells *in vitro* resulted in blockage of transcription. Techniques other than phage display can also be used to generate DNA-binding proteins with altered specificities. For example, helix-turn-helix and leucine zipper proteins with novel specificities have been generated from large libraries of mutants using selections based on interference with transcription of antibiotic resistance genes [62] [63]. Studies of this sort are also helping to define the recognition codes between the amino acid sequences of DNA-binding proteins and their cognate recognition sites.

Large libraries of mutant proteins have also been used to investigate the factors that influence the structure and stability of specific protein folds. For example, cassette mutagenesis has been used to randomly alter seven residues in the hydrophobic core of the N-terminal domain of phage λ repressor [64]. By selecting for functional repressors (phage-resistant transformants), it was shown that many different sequences can form a stable core. The main determinant of whether a particular sequence is compatible with the wild-type fold was hydrophobicity, although the van der Waals volume of the core and steric interactions between residues also limited the number of functional sequences. More recently, the sequence determinants that dictate the four helix bundle structure have been investigated [65]. A library of cytochrome b_{562} mutants was generated in which amino acids at defined positions were replaced with any of the hydrophobic amino acids Phe, Leu, Ile, Met, or Val, or any of the hydrophilic amino acids Glu, Asp, Lys, Asn, Gln, or His to generate a binary pattern of nonpolar and polar amino acids. Bacterial expression of soluble, protease-resistant protein was assayed spectrophotometrically by monitoring heme absorbance. It was found that 29 of 48 sequences examined expressed protein that was both soluble and resistant to intracellular degradation. Two proteins that were characterized by urea denaturation experiments were stabilized by 3.7 and 4.4 kcal mol^{-1} relative to the unfolded form. A similar strategy has been applied to a random sequence of 80 to 100 amino acids consisting of Gln, Leu, and Arg. Bacterial expression of these 'QLR proteins' revealed that 5% of the sequences were expressed at readily detectable levels. Several of the characterized proteins possessed features of native proteins such as α-helical content and highly cooperative folding, suggesting that folded proteins occur frequently in libraries of random amino acid sequences [66].

The relationship between protein sequence and structure has also been examined using libraries created neither *in vivo* nor *in vitro*, but rather by computational methods. Recently, a design algorithm based on chemical potential functions and stereochemical constraints was used to search 1.9×10^{27} amino acid sequences for those that adopt a β-α-β zinc finger motif. The resulting designed sequence (FSF-1) possesses little or no identity to any known protein sequence. The synthesis and subsequent NMR-spectroscopic investigation of FSF-1 in solution revealed a compact structure in excellent agreement with the design target. This experiment demonstrates the potential power of computational methods to screen combinatorial libraries ten or more orders of magnitude larger than any created in the laboratory [67].

3. Nucleic Acid Diversity

Combinatorial methods have played an important role in investigating the scope of nucleic acid function and its possible role in the prebiotic world. In these experiments libraries containing up to 10^{15} random single-stranded RNA or DNA sequences are generated using a combination of chemical synthesis, PCR, and runoff transcription (in the case of RNA) [68–71]. These libraries are then selected for individual molecules that either 1) bind specific ligands using affinity-based methods or 2) carry out specific chemical transformations using in vitro selections or screenings. Amplification of an enriched population of binders or catalysts is followed by subsequent rounds of selection. With these techniques, nucleic acids have been generated that selectively bind small molecules, proteins, and nucleic acids; and catalyze a number of reactions ranging from phosphoryl and acyl transfer reactions to metalation and pericyclic reactions.

3.1. Selecting for Binding Function

Single-stranded RNA and DNA oligomers (or aptamers) have been isolated from large, random oligonucleotide libraries that selectively bind a wide array of molecules including T4 DNA polymerase, human immunodeficiency virus (HIV), HIV Rev protein, thrombin, amino acids, organic dyes, and cofactors [68] [69] [72–78]. For example, by screening a library of 10^{13} transcripts with ATP immobilized on agarose, an RNA was isolated that binds adenosine with a dissociation constant K_d of 0.7 μM [74]. The solution structure of a 40 nucleotide RNA containing the consensus ATP-binding motif complexed with AMP (which binds with affinity similar to that of ATP) has been determined by NMR spectroscopy [79]. The AMP molecule is bound in a 'GNRA-like' hairpin fold [80] with the intercalated adenine involved in an A · G mismatch pair and the 2′- and 3′-hydroxy groups of the ribose sugar forming hydrogen bonds to the RNA. The binding of AMP was shown to induce a conformational transition in the ATP binding site on the RNA. RNA aptamers have also been isolated from oligonucleotide libraries that bind the purine theophylline with high affinity ($K_d = 0.1$ μM). These aptamers do not bind appreciably to caffeine ($K_d = 3500$ μM), which differs from theophylline by only a methyl group at nitrogen atom N(7) [81]. Experiments of this sort demonstrate that RNA, like proteins, can bind a wide array of small molecules with extremely high specificity, by a combination of van der Waals packing, hydrogen-bonding, and electrostatic interactions.

Aptamers have also been isolated that bind selectively to macromolecules, including proteins and nucleic acids. For example, single-stranded RNAs have been isolated that bind sequence-selectively to duplex DNA through the formation of triple-helical structures involving *Hoogsteen* base pairs [82]. Single-stranded DNAs have been isolated that bind to thrombin through the formation of a highly conserved guanine-rich structure of 14–17 nucleotides. In this case the binding affinity was 25–200 nM, and several aptamers inhibited thrombin-catalyzed fibrin clot formation in vitro at nanomolar concentrations [77].

3.2. RNA Catalysis

The earliest use of *in vitro* selection schemes to isolate RNAs with novel phenotypes from pools of mutants were those of Spiegelman et al. [83]. In these experiments Qβ replicase, a template-specific RNA-directed polymerase, was used in a *Darwinian* selection system to isolate a structured RNA that had reduced affinity for the intercalating dye ethidium bromide. The mutant arose by sequential mutation during the course of the experiments, and was not a preexisting variant of the original RNA pool.

The use of *in vitro* selections to isolate RNAs with novel catalytic activities has involved either libraries generated by mutagenesis of naturally occurring ribozymes, or synthetic libraries of randomized RNA sequences. For example, the *Tetrahymena* group I ribozyme has been used as a starting point to isolate a new ribozyme that cleaves DNA [70] [71]. The selection required the ribozyme to complete a phosphorylation reaction with an alternative DNA substrate that results in the addition of nine bases to the 3′-terminal hydroxyl group of the ribozyme (*Fig. 10*). This 3′-terminal extension permits successful annealing of an amplification primer necessary to complete the cycle of selection. With a much larger pool of mutants (10^{13}), variants were isolated that catalyze the reaction under physiological conditions at a rate up to 65 times faster than the wild-type ribozyme. Other selection schemes using naturally occurring ribozymes have yielded RNAs with self-copying functions [84] and altered metal requirements [85].

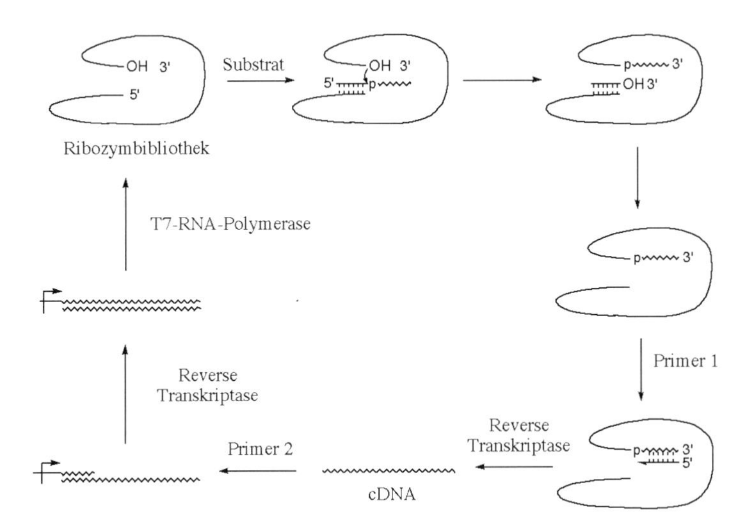

Fig. 10. *Selection of a* Tetrahymena *ribozyme mutant that uses an alternative DNA substrate* [70] [71]. Wavy lines represent DNA strands.

One of the most impressive examples of the selection of RNA catalysts from complex libraries of random sequences was the isolation of a RNA ligase that can act as a RNA polymerase [86] [87]. Beginning with a pool of approximately 10^{15} distinct transcripts, iterative rounds of amplification and selection were carried out based upon the ability of the ribozyme to ligate its own 5′-terminal triphosphate group to a substrate oligonucleotide containing a 5′ 'tag' (*Fig. 11*). A ribozyme that catalyzes 3′,5′-phosphodiester bond formation with k_{cat} greater than 1 s^{-1} was isolated [86] [87]; when fused to a template strand, this RNA ligase was shown to catalyze RNA primer extension by up to six nucleotides in a template-directed fashion [88]. Nucleotides complementary to the template are added up to 1000 times more efficiently than mismatched oligonucleotides. Although a number of obstacles remain to generate an RNA replicase (for example, strand separation after polymerization, nonspecific binding of template and primer, and sequence fidelity) these experiments illustrate the potential to evolve RNAs that catalyze reactions which may have been required in the prebiotic world.

Selections and affinity-based methods have also been used to expand the scope of RNA-catalyzed reactions to other classes of reactions including isomerization, carbon–carbon bond forming, metalation, and alkylation reactions. For example, by exploiting the same notions of transition state theory that were used in the initial generation of catalytic antibodies, an RNA was isolated that catalyzes the isomerization of a bridged biphenyl substrate to its diastereomer (*Fig. 12*) [89]. In this case a library of 10^{13} random RNA mole-

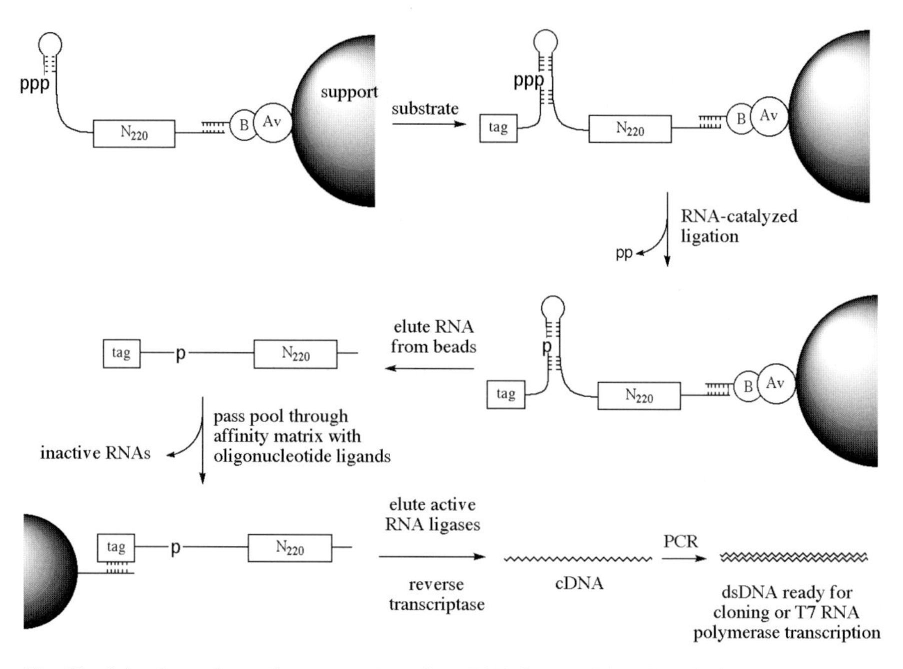

Fig. 11. *Selection scheme for generation of an RNA ligase* [86] [87]. Biotin and avidin are represented by B and Av, respectively.

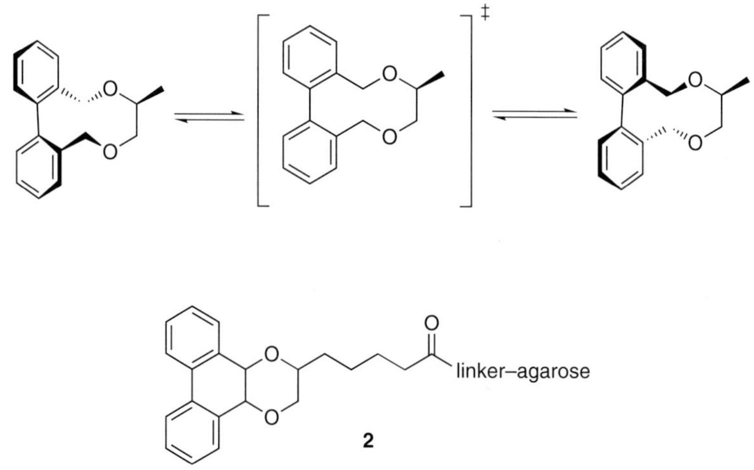

Fig. 12. *Atropisomerization between diastereoisomeric 1,1′-biphenyls.* Structure **2** is an analogue of the transition state for this reaction **2** [89].

cules was screened based on binding affinity to the planar biphenyl transition state analogue **2** immobilized on a solid support. A similar approach has been used to generate single-stranded RNAs and DNAs that catalyze porphyrin metalation [90] [91]. More recently, selections based on covalent linkage of a ribozyme to a solid support have been used to generate RNAs that catalyze a *Diels–Alder* reaction between a diene and dienophile [92].

3.3. Other Applications

Many other novel uses of nucleic acid libraries continue to appear including *in vivo* applications. One such example involves the use of genomic expression libraries for immunization, a technique that makes use of genetic immunization (immunization with the gene encoding the protein antigen rather than the protein itself) and the fact that all antigens of a pathogen are encoded in its DNA [93]. Indeed, it has been shown that even partial expression libraries made from the DNA of *Mycoplasma pulmonis*, a natural pathogen in rodents, provide protection against challenges from the pathogen. Libraries of nucleic acids have also been used *in vivo* to identify *cis*-acting sequences that lead to localization of RNAs in the nucleus of *Xenopus laevis* oocytes [94]. Clearly, libraries of nucleic acids, like those of peptides and proteins, will find many applications in the search for new biological and chemical functions, and in our efforts to better understand complex biological processes.

4. Diversity in Chemical Synthesis

The biosynthetic machinery used by living cells to generate diverse populations of molecules is limited to specific classes of chemical structures including oligomeric molecules such as polypeptides, oligonucleotides, and polysaccharides, and a variety of natural products such as the polyketides. Moreover, these structures are generated from restricted sets of building blocks. The development of methods for generating large, diverse populations of synthetic molecules has extended the scope of this approach to new classes of molecules with a broader range of chemical, biological, and physical properties.

4.1. Approaches to Synthetic Diversity

Virtually all of the efforts aimed at generating and screening large populations of synthetic diversity are based on the method of solid-phase synthesis developed by *Merrifield* [95]. The initial work focused on polypeptides, which can be rapidly assembled by iterative chemical synthesis from a diverse set of commonly occurring and unnatural amino acids. Methods were first developed that simply involved the simultaneous synthesis of multiple discrete peptide sequences using either polyacrylic acid grafted polyethylene pins arranged in microtiter format [96], or later resins sealed in porous polypropylene bags (the 'tea bag' technique) [97].

These experiments were followed by efforts aimed at generating more diverse populations of molecules by connecting sets of building blocks in all possible combinations, again mimicking the combinatorial strategies used by nature to efficiently generate molecular diversity. These methods involve either the generation of spatially separated libraries of molecules, or mixtures of resins each containing a discrete structure or small number of related structures. One example of the former approach involves the 'mimotope' strategy in which peptides are again synthesized on pins, but in this case by randomly incorporating residues from mixtures of activated amino acids to give all peptide sequences [98] [99]. Two positions in a peptide containing n residues are then iteratively defined by multipin synthesis of 400 discrete mixtures, each containing all possible $(n-2)$ peptide sequences with one of the 400 possible pairs of the common L-amino acids defined. This technique has been used in conjunction with ELISA (enzyme-linked immunosorbent assay) screening methods to identify peptides that selectively bind antibodies and other receptors.

A second method for the spatially addressable synthesis of combinatorial libraries makes use of photolithography together with solid-phase synthesis to generate high-density, combinatorial arrays of peptides on a glass substrate [100] [101]. By illuminating specific regions of the glass substrate containing growing peptide chains, in which the terminal amino group is protected with a photolabile group, the spatial distribution of individual amino acid coupling steps can be controlled (*Fig. 13*). With n binary masks in which half of the surface is photolyzed during each coupling step, 2^n compounds can be synthesized. The sequence of masks and coupling steps defines the identity of the peptide at each site. Binding of fluorescently labeled receptors to the surface is detected by fluorescence emission. Although this technique was developed for peptide synthesis, it has proven to be most useful for generating high-density arrays of oligonucleotides (see below).

Another approach that has been used to generate combinatorial libraries of peptides is the split synthesis method [102–104]. This technique involves dividing the resin support into n equal fractions, coupling each fraction with a single activated monomer (or in some cases, a small number of monomers), and then recombining the fractions (*Fig. 14*). Iteration for x cycles leads to a stochastic population of n^x peptides. This approach has been used in conjunc-

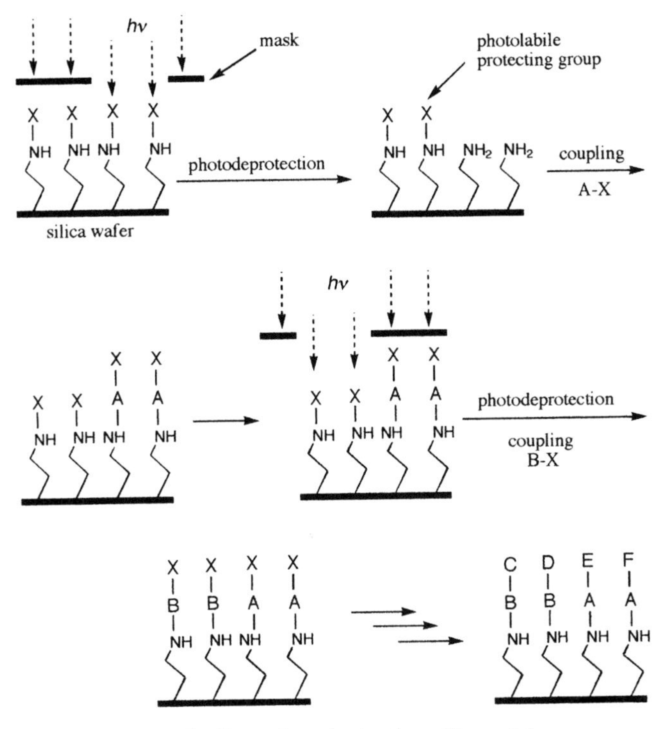

Fig. 13. *Light-directed synthesis of combinatorial arrays*

tion with the 'tea bag' technique and beaded resins equipped with noncleavable linkers to generate large combinatorial libraries of up to 50 million different sequences. However, since mixtures of resins are generated, this technique also requires methods to identify the sequence of a high-affinity ligand isolated from an affinity-based assay.

A number of strategies have been developed to determine the identity of a peptide ligand isolated from pools of random sequences. These include methods based on the iterative 'mimotope' deconvolution strategy described above [105], as well as direct *Edman* peptide microsequencing, which can be used with beads containing 80–100 pmol of peptide [106]. Mass spectrometry can also be used to identify the sequence of a peptide, if in each step of the synthesis a small fraction of the growing polypeptide chain is capped, resulting in a small amount of truncated chain at each step in the synthesis of the full length peptide [107].

More recently, a number of methods have been developed for generating libraries of peptides on resin beads in which each bead is attached to a molecular 'tag' that allows rapid and sensitive identification of the associated polypeptide [108–111]. The tags are attached to the bead coincident with

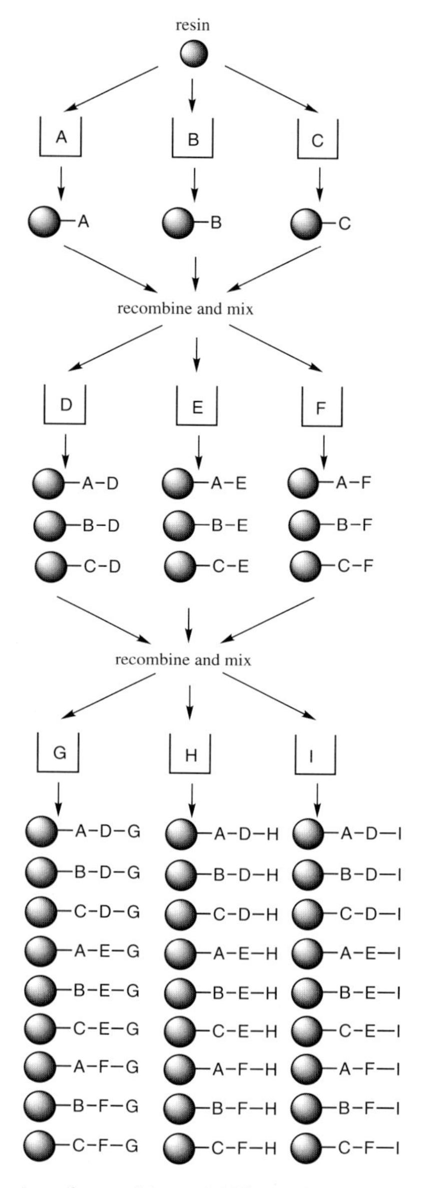

Fig. 14. *Preparation of a combinatorial library by the split synthesis method*

each monomer coupling step. Consequently, they must be stable under the conditions of polymer synthesis and deprotection, their attachment must not interfere with the synthesis of the polypeptide, and they must have high information content that can be detected with high sensitivity. A number of tagging techniques have been developed (*Fig. 15*), including the use of oligonu-

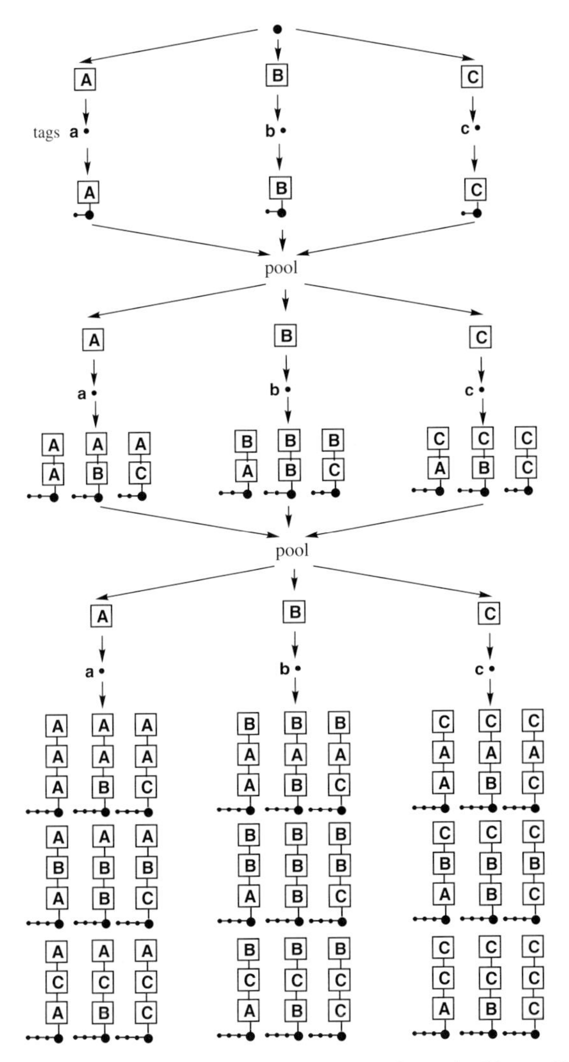

Fig. 15. *The use of molecular tags to identify members of a library* [108–111]

cleotide tags that can be detected by PCR amplification and DNA sequencing [109], and a set of related halocarbon tags that can be separated and detected be electron capture capillary gas chromatography [111]. The high detection sensitivity of both these tags should facilitate the synthesis of increasingly diverse chemical libraries.

The techniques described above have been used to identify natural and unnatural polypeptides that bind a host of molecules including antibodies, enzymes, and receptors [24]. For example, antimicrobial peptides with activities against both *Gram*-positive and *Gram*-negative bacteria have been generated [112]. The binding specificities of synthetic steroid- and macrocycle-based hosts for peptide ligands have also been determined [113–115]. In addition, these techniques have been used to analyze protein–protein interactions involved in cellular processes [116]. In one such study, two different SH3 domains conjugated to fluorescein were used to screen a peptide library for ligands that possessed affinity for SH3 domains. Analysis of the resulting sequences suggested a set of rules governing SH3–ligand interactions, and subsequent structural studies revealed a detailed mode of ligand binding by the SH3 domain [116].

4.2. Extension to Other Polymeric and Nonpolymeric Frameworks

Molecular diversity has also been used to explore the chemical and biological properties of heteropolymers composed of building blocks other than amino acids. These unnatural polymeric frameworks may have novel properties such as improved pharmacokinetic properties (including membrane permeability and biological stability) that increase bioavailability, or altered conformational or hydrogen-bonding properties that provide insights into biomolecular structure and folding. A number of novel 'unnatural biopolymers' have been generated including oligocarbamates [117], peptoids [118] [119], oligoureas [120–122], oligopyrrolinones [123], oligoazatides [124], oligosulfonamides [125], and β-peptides [126–128]. Because these oligomeric compounds can be assembled from a set of diverse building blocks, large molecular libraries can be rapidly generated and screened for desired properties. For example, oligocarbamates consisting of a chiral ethylene backbone with a variety of side chains linked through relatively rigid carbamate groups have been efficiently synthesized by the stepwise coupling of *N*-protected amino-*p*-nitrophenyl carbonate monomers. The monomers in turn are readily derived from the corresponding amino acids (*Fig. 16*) [117]. Libraries of these molecules have been generated, and high-affinity selective ligands have been identified that bind antibodies and integrin receptors [111] [129]. Libraries of polypeptoids [118] [119], a polypeptide backbone composed of *N*-substituted glycine units, have also been generated. Peptoids have been identified that selectively bind the α-adrenergic receptor and a number of enzymes [118]. Efforts are ongoing to determine the conformational and pharmacological properties of these molecules.

The application of combinatorial approaches to creating nonpolymeric small molecules came with the realization that many such molecules can also be synthesized by the iterative stepwise addition of building blocks on a solid support. The most notable early example was that of the combinatorial synthesis of a library of 1,4-benzodiazepines [130] [131], a pharmacophore

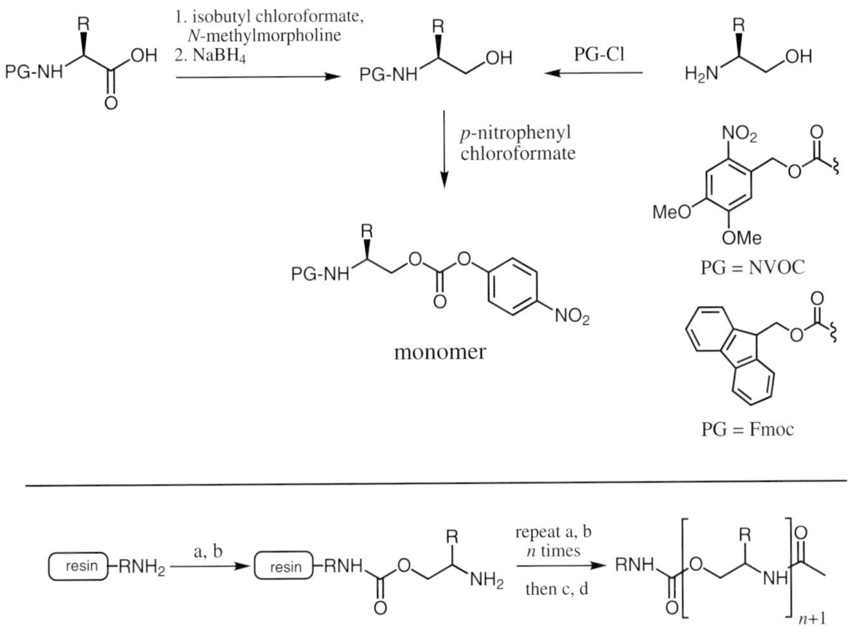

Fig. 16. *Synthesis of oligocarbamates from amino acid starting materials* [117]. a) Monomer, 1-hydroxy-1*H*-benzotriazole (HOBt), diisopropylethylamine, *N*-methylpyrrolidine (NMP); b) piperidine, NMP; c) acetic anhydride, pyridine; d) trifluoroacetic acid (TFA), CH_2Cl_2. PG=protecting group, NVOC=nitroveratrylmethoxycarbonyl, Fmoc=9-fluorenylmethoxycarbonyl.

common to a wide variety of clinically important therapeutic agents. A library of approximately 2000 structurally diverse 1,4-benzodiazepine derivatives containing a variety of chemical functionalities including amides, carboxylic acids, amines, phenols, and indoles was constructed by multipin synthesis from three components: 2-aminobenzophenones, amino acids, and alkylating agents (*Fig. 17*). Ligands that selectively bind p60src, a cholecytokinin A receptor, and DNA-binding antibodies have been isolated from this library.

Combinatorial libraries for a large number of small organic molecules have since been generated including β-turn mimetics [132], hydantoins [133], protease inhibitors [134], pyrrolidines [135], β-lactams [136], and tyrosine kinase inhibitors [137]. In addition, a library of 1300 di- and trisaccharides was recently synthesized on beads and screened against *Bauhinia purpurea* lectin conjugated to a reporter enzyme [138]. Two ligands were isolated that bound to the lectin more tightly than the natural ligand. Similar approaches provide promising routes to the identification of carbohydrate-based ligands for other receptors.

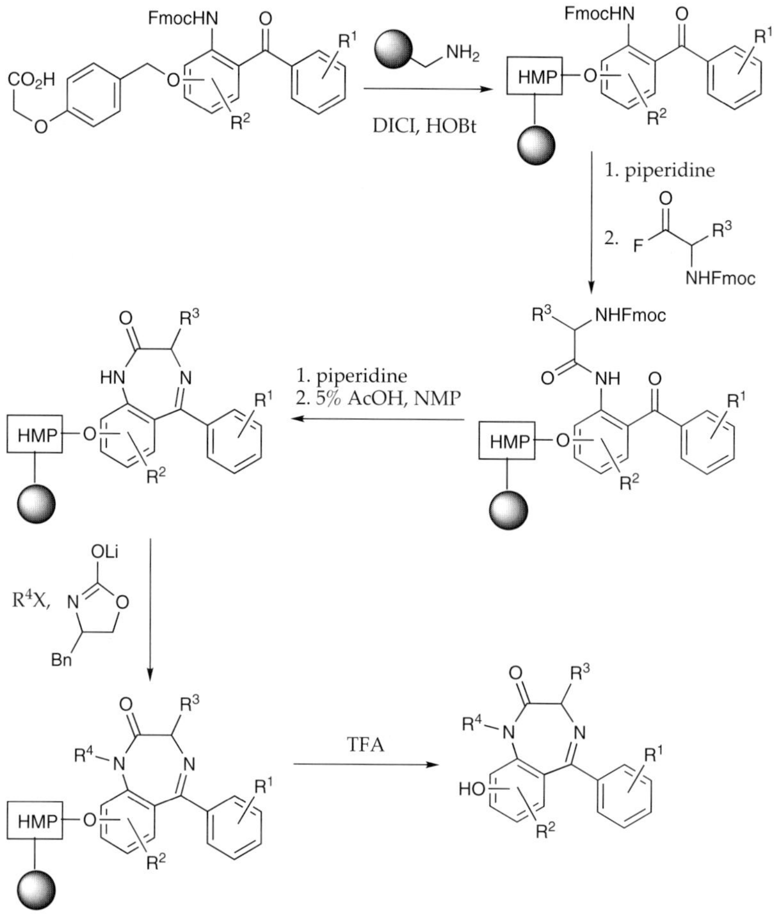

Fig. 17. *Synthesis of a library of 1,4-benzodiazepines on solid support* [130]. Bn=benzyl, DICI=diisopropylcarbodiimide, HMP=4-(hydroxymethyl)phenoxyacetic acid.

Finally, considerable effort is also being focused on the development of new solid-phase synthetic methodologies for generating small-molecule libraries [139] [140], as well as theoretical and empirical methods for library design. For example, the structure-based design of nonpeptide libraries containing the hydroxyethylamine pharmacophore of pepstatin yielded potent inhibitors (K_i = 9–15 nM) of cathepsin D [141]. Clearly, the use of libraries of synthetic oligomeric and nonpolymeric molecules will continue to have a significant impact on the pharmaceutical industry.

4.3. Natural Products

Not only can chemical synthesis be used to generate libraries of small organic molecules, recently it has been shown that the biosynthetic machinery of cells can be exploited to produce libraries of natural products such as the polyketides, which include the well-known antibiotics erythromycin, avermectin, and spiramycin [142]. These compounds are synthesized by modular polyketide synthases (PKSs), large multifunctional enzyme assemblies that catalyze the stepwise biosynthesis of these natural products [143] [144]. These modules encode sets of enzymes (ketosynthases, acyl transferases, ketoreductases, dehydratases, and enoyl reductases) that processively carry out rounds of polyketide chain elongation with concomitant alterations in oxidation level and stereochemistry after each elongation step. Deleting individual modules, altering the specificity of individual enzymes in a module, or adding new enzymatic activities results in the synthesis of a large number of different macrolides [142]. Most recently, it has been shown that exogenous addition of synthetic diketide mimics to deoxyerythronolide B synthase containing a genetic block in the KS-1 gene results in the synthesis of a novel polyketide containing the synthetic building block (*Fig. 18*) [145]. As more nontemplated biosynthetic pathways are isolated and characterized (*e.g.*, cyclic peptides, ene-diynes, and bleomycins) and our ability to manipulate their function increases, this approach is likely to provide an increasingly diverse collection of chemical structures to screen for novel biological properties.

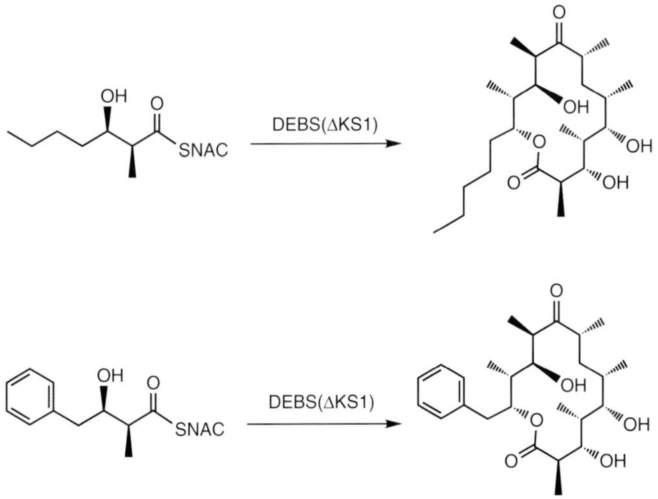

OH O SNAC DEBS(ΔKS1) O OH OH OH O O

OH O SNAC DEBS(ΔKS1) O OH OH OH O O

Fig. 18. *Precursor-directed biosynthesis of novel polyketides containing synthetic building blocks* [145]. NAC=*N*-acetylcysteamine, DEBS=deoxyerythronolide B synthase.

4.4. Genomics

Combinatorial synthetic methods have also found important applications in genomics. For example, the photolithographic chemical synthesis methods developed for peptides have made it possible to synthesize spatially addressable, high-density arrays of oligonucleotides of known sequence [146] [147]. This technique has been used to generate more than 260000 specifically chosen oligonucleotide probes (in an area of about 6.5 cm^2) that hybridize to all open reading frames (ORFs) in the yeast genome (twenty 25-mers complementary to segments within each ORF; *Fig. 19*) [148]. With this array one can monitor the expression levels of nearly all yeast genes in a quantitative, sensitive, and highly reproducible fashion due to the redundancy in the detection and analysis of the data. Arrays of this sort are likely to have a major impact on our ability to analyze the information content of genomic DNA, including mismatch scanning, homology comparisons, and the detection of genetic differences between strains. Other parallel, high-throughput methods have also been developed for generating oligonucleotide arrays [149–156], including spotting methods with presynthesized oligonucleotides [157].

A recent example demonstrates how combinatorial libraries of small molecules, structural information, and oligonucleotide arrays can be used synergistically to identify and characterize selective kinase inhibitors [158]. A combinatorial library of 2,6,9-trisubstituted purines was generated based on structural analysis of the novel binding mode of the purine olomoucine to the ATP binding pocket of the cell cycle kinase CDK2 [159] [160]. By iterating chemical library synthesis and biological screening, high-affinity, selec-

Fig. 19. *Fluorescence image of an array of oligonucleotide probes following hybridization of a fluorescently labeled RNA sample prepared from mRNA extracted from yeast cells*

tive inhibitors of human CDK2/cyclin A and yeast cdc28 kinases were identified. The CDK2/cyclin A kinase inhibitors include compounds that selectively inhibit the growth of colon cancer cell lines. The structural basis for the affinity and selectivity of these purine derivatives was determined by analysis of the three-dimensional crystal structure of their complex with CDK2, providing a basis for further enhancing the affinity and selectivity of the compounds. The cellular effects of this class of compounds were further examined and compared to those of the CDK2 inhibitor flavopiridol, as well as to those arising from mutations in cdc28 itself, by monitoring changes in mRNA expressions levels for all genes in treated cells of *Saccharomyces cerevisiae* using high-density oligonucleotide probe arrays [148] [161]. In particular, of the 105 transcripts that changed greater than threefold in response to both compounds (335 for the purine and 267 for flavopiridol) only seven were down regulated, all of which were associated with progression of the cell cycle. These experiments also began to delineate other common effects of these compounds on cellular metabolism, stress response, signaling, and growth, as well as reveal significant differences in the effects of flavopiridol and the purine, despite their apparent *in vitro* selectivity for CDKs. Information of this sort may, in general, be useful in comparing compounds prior to clinical studies or in identifying targets whose inhibition might potentiate the effects of a primary drug. Moreover, the use of gene expression profiles together with multiple copy overexpression libraries [162] may allow one to rapidly identify the targets of compounds with interesting biological activities isolated from phenotypic screening of whole cells. The combined use of chemical libraries, novel *in vitro* and cellular screens [163–165], and genomics is likely to provide important new insights into complex cellular processes.

5. Materials Science

The most recent application of combinatorial approaches has been to solid-state and materials science. The properties of many functional materials, such as high-temperature superconductors, heterogeneous catalysts, ferroelectric materials, magnets, and even structural materials such as alloys arise from complex interactions involving the host structure, dopants, defects, and interfaces, all of which are highly dependent on composition and processing. Unfortunately, the current level of theoretical and empirical understanding does not generally allow one to predict the structures and resulting properties of these materials [166]. The situation is further complicated by the complex compositions of many modern materials (four or more elements) and the fact that materials synthesis, unlike organic synthesis, is generally not a kinetically controlled process. Given approximately 60 elements in the periodic table that can be used to make compositions consisting of three, four, five, or even six elements, the universe of possible new compounds with interesting physical and chemical properties remains largely uncharted. Combinatorial methods should make it possible to rapidly synthesize, process, and analyze large libraries (hundreds to hundreds of thousands of members) of inorganic and organic materials and devices. This may lead to significant increases both in the efficiency of materials discovery and optimization, and in our information base relating materials structure and properties.

5.1. Synthesis and Screening

The first application of combinatorial methods to materials science involved the synthesis of libraries of thin-film copper oxides containing high-temperature superconductors [167]. Libraries designed to explore many different materials compositions were synthesized by sequentially sputtering precursors at different sites on a substrate using a series of precisely positioned shadow masks (*Fig. 20*). Low-temperature annealing followed by high-temperature processing resulted in formation of superconducting thin films. More recently, quaternary and shutter masking systems, together with photolithographic masking techniques and pulsed laser deposition, have made possible the synthesis of high-quality, diverse thin-film libraries of some 1000 to 10000 samples on an area of about 6.5 cm^2 [168–170]. Solution-based methods have also been applied to materials library synthesis. For example, scanning fluid delivery stems (inkjets) have been used to rapidly and accurately deliver nanoliter volumes of precursor solutions to generate libraries of metal oxides and organic polymers [171] [172]. Processing conditions play an important role in determining materials properties, and it is important to include these variables in the design of a materials library. Identical libraries synthesized simultaneously can be processed under different conditions to enhance diversity. Gradient temperature ovens have been used to identify optimal processing conditions for a library of metal oxides; similarly, it may be possible to vary oxygen partial pressure by introducing oxygen on one side of the chip [173].

Fig. 20. a) *In the quaternary masking scheme, deposition is carried out using a series of n different masks which subdivide the substrate into a series of nested quadrants.* Each mask is used in four sequential deposition steps with a 90° rotation of the mask in each step, generating up to 4^n compositions in $4n$ deposition steps. b) *Photograph of the 1024-membered library* (deposited on a Si substrate with an area of 2.54×2.54 cm^2) *under ambient light* (top) *and UV irradiation* (bottom).

a)

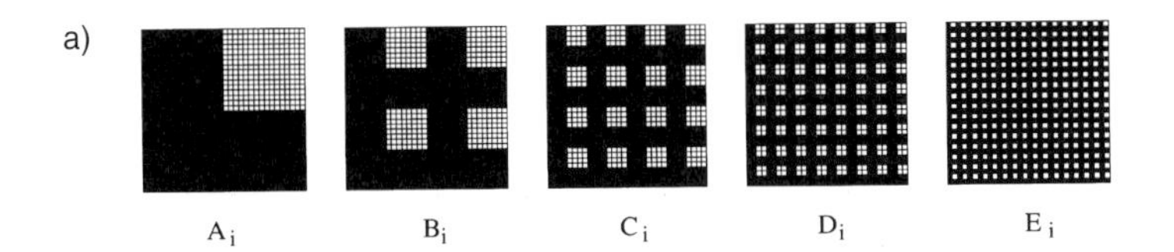
A_i B_i C_i D_i E_i

b)

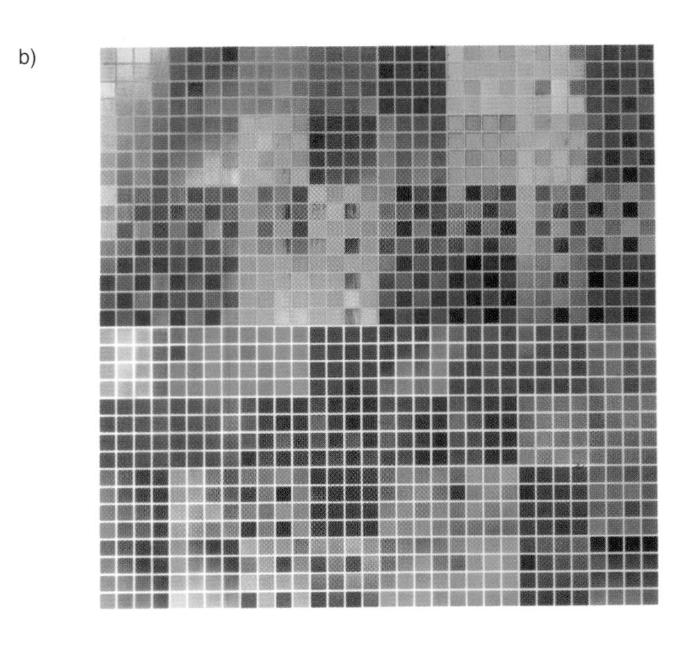

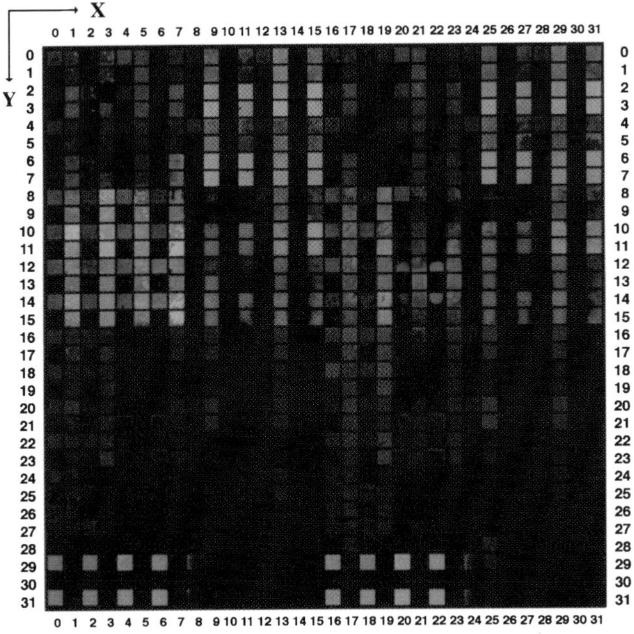
X
Y
0 1 2 3 4 5 6 7 8 9 10 11 12 13 14 15 16 17 18 19 20 21 22 23 24 25 26 27 28 29 30 31

A number of detection systems have been developed to date for screening materials libraries for optical, electronic, magnetic, or chemical properties of interest [173]. Optical imaging systems have been used to evaluate libraries of photoluminescent materials. For example, a scanning spectrophotometer and CCD array detectors (CCD=charge-coupled device) were used to evaluate the photon output and chromaticity of each member in phosphor libraries upon excitation with monochromatic UV light [168] [169]. A novel scanning-tip microwave near-field microscope (STMNM) has been developed to nondestructively measure dielectric constant and tangent loss of a library of ferroelectric materials with submicron spatial resolution and high sensitivity (*Fig. 21*) [170] [174]. More recently, X-ray microbeam techniques with a spot size of $3 \times 20\ \mu m^2$ have been used to characterize the composition and structure of samples in thin-film libraries [175]. Infrared imaging thermography has been used to qualitatively evaluate catalyst performance in a small member library of hydrogen oxidation catalysts [176], and mass spectrometry has been used to rapidly screen a library of oxidation catalysts with high sensitivity [177]. Magneto-optical detectors are being developed to image libraries of magnetic materials, and many other detector systems can be envisaged to measure a range of material properties (IR and Raman spectroscopy, surface plasma resonance, nanoindentation, light scattering, and polarized light microscopy).

5.2. Applications

Although there remains a great deal of work to be done in developing methods for the combinatorial synthesis, processing, and detection of materials libraries, a number of applications of this approach has already been reported. Combinatorial approaches have been used to identify a class of cobalt oxide magnetoresistive materials of the form $(La_{0.88}S_{0.12})CoO_3$ [178]. Magnetoresistance was found to increase as the size of the dopant ion increased, in contrast to Mn-containing compounds, in which the magnetoresistive effect increases as the size of the alkaline earth ion decreases. Combinatorial methods have also been applied to the optimization and identification of luminescent materials. Recently, two novel blue phosphorescent compounds, $SrCeO_4$ [179] and $Gd_3Ga_5O_{12}/SiO_x$ [168], were discovered in combinatorial libraries. Similar approaches have also been applied to ferroelectric

Fig. 21. a) *Schematic representation of a scanning tip microwave near-field microscope for analyzing ferroelectric and dielectric libraries.* b) *Composition of a region of a $BaTiO_3$ ferroelectric library with dopands indicated.* c) *Image of tangent loss* ($\tan \delta$) *from this library obtained with the near-field scanning microwave microscope* [174].

a)

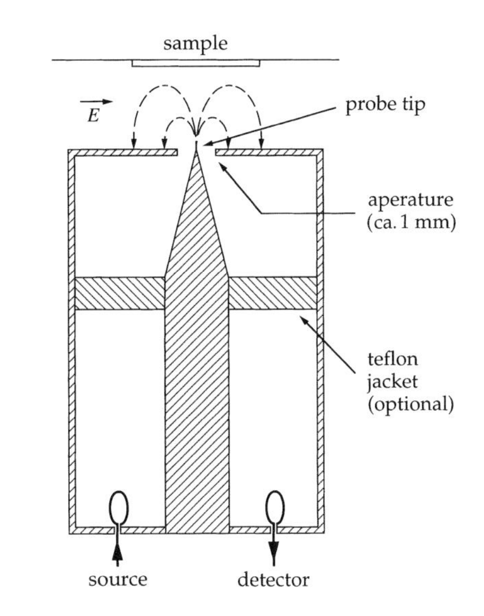

b)

Fe Mg Cr	Fe Y Cr	Fe Mg Mn	Fe Y Mn	W Mg Cr	W Y Cr	W Mg Mn	W Y Mn
Fe Cr	Fe La Cr	Fe Mn	Fe La Mn	W Cr	W La Cr	W Mn	W La Mn
Fe Mg	Fe Y	Fe Mg Ce	Fe Y Ce	W Mg	W Y	W Mg Ce	W Y Ce
Fe	Fe La	Fe Ce	Fe La Ce	W	W La	W Ce	W La Ce
Mg Cr	Y Cr	Mg Mn	Y Mn	Ca Mg Cr	Ca Y Cr	Ca Mg Mn	Ca Y Mn
Cr	La Cr	Mn	La Mn	Ca Cr	Ca La Cr	Ca Mn	Ca La Mn
Mg	Y	Mg Ce	Y Ce	Ca Mg	Ca Y	Ca Mg Ce	Ca Y Ce
	La	Ce	La Ce	Ca	Ca La	Ca Ce	Ca La Ce

c)

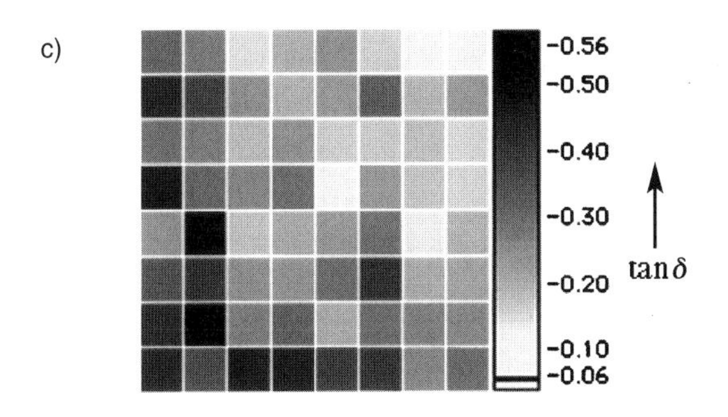

materials where the effects of transition metal dopants on the dielectric constant and tangent loss of a library of thin films of $(Ba_xSr_{1-x})TiO_3$ were determined [170]. Combinatorial methods are likely to have a significant impact on catalysis and polymer chemistry, and the first attempts to apply combinatorial methods in this area have already surfaced [173–176] [180–182]. It may also be possible to generate combinatorial libraries of entire devices such as capacitors or electroluminescent displays [183].

6. Conclusion

The development of strategies for generating large diverse libraries of biomolecules, small organic molecules, and solid-state materials, together with novel screens and selections for specific biological, chemical, or physical properties, is having an enormous impact on science. These approaches are significantly increasing the rate and efficiency with which the scientific method can be used both to identify molecules with novel properties and to understand the structural basis for their function. Clearly, many challenges lie ahead including 1) improved theoretical and computational methods for designing libraries and analyzing the output of library experiments; 2) the development of methods for increasing library size (especially for polypeptides, proteins, and synthetic molecules); 3) high-throughput screens for a broader range of biological (cellular), chemical, and physical properties; 4) extension of biological selections and amplification to synthetic diversity; and 5) further engineering improvements in miniaturization and robotics. Nonetheless, molecular diversity has already been exploited to produce a large number of molecules with functions that would have been difficult and, in many cases, impossible to realize using more traditional synthetic approaches.

A portion of this work was supported by the Department of Energy, the *National Institutes of Health*, and the *Office of Naval Research. D. R. L.* would like to acknowledge a predoctoral fellowship from the *Howard Hughes Medical Institute. P. G. S.* is a *Howard Hughes Medical Institute Investigator* and *W. M. Keck Foundation Investigator*.

REFERENCES

[1] F. M. Burnet, *The Clonal Selection Theory of Acquired Immunity*, Vanderbilt University Press, Nashville, TN, 1959, p. 53.
[2] D. W. Talmage, *Science* **1959**, *129*, 1649.
[3] S. Tonegawa, *Nature* **1983**, *302*, 57.
[4] D. R. Davies, S. Chacto, *Acc. Chem. Res.* **1993**, *26*, 421.
[5] S. J. Pollack, J. W. Jacobs, P. G. Schultz, *Science* **1986**, *234*, 1570.
[6] A. Tramontano, K. D. Janda, R. A. Lerner, *Science* **1986**, *234*, 1566.
[7] P. G. Schultz, R. A. Lerner, *Science* **1995**, *269*, 1835.

[8] A. G. Cochran, P. G. Schultz, *Science* **1990**, *249*, 781.
[9] M. E. Blackwood, Jr., T. S. Rush III, F. Romesberg, P. G. Schultz, T. G. Spiro, *Biochemistry* **1998**, *37*, 779.
[10] F. E. Romesberg, B. D. Santarsiero, B. Spiller, J. Yin, D. Barnes, P. G. Schultz, R. C. Stevens, *Biochemistry* **1998**, *37*, 14404.
[11] J. R. Jacobsen, J. R. Prudent, L. Kockersperger, S. Yonkovich, P. G. Schultz, *Science* **1992**, *256*, 365.
[12] E. Driggers, P. G. Schultz, unpublished results.
[13] P. G. Schultz, R. A. Lerner, *Acc. Chem. Res.* **1993**, *26*, 391–395.
[14] C. F. Barbas III, A. Heine, G. Zhong, T. Hoffmann, S. Gramatikova, R. Bjornestedt, B. List, J. Anderson, E. A. Stura, I. A. Wilson, *Science* **1997**, *278*, 2085–2092.
[15] G. F. Zhong, T. Hoffmann, R. A. Lerner, S. Danishefsky, C. F. Barbas III, *J. Am. Chem. Soc.* **1997**, *119*, 8131–8132.
[16] H. D. Ulrich, E. Mundorff, B. D. Santarsiero, E. M. Driggers, R. C. Stevens, P. G. Schultz, *Nature* **1997**, *389*, 271–275.
[17] F. E. Romesberg, B. Spiller, P. G. Schultz, R. C. Stevens, *Science* **1998**, *279*, 1929–1933.
[18] A. Heine, E. A. Stura, J. T. Yli-Kauhaluoma, C. Gao, Q. Deng, B. R. Beno, K. N. Houk, K. D. Janda, I. A. Wilson, *Science* **1998**, *279*, 1934.
[19] G. J. Wedemayer, P. A. Patten, L. H. Wang, P. G. Schultz, R. C. Stevens, *Science* **1997**, *276*, 1613–1756.
[20] P. A. Patten, N. S. Gray, P. L. Yang, C. B. Marks, G. J. Wedemayer, J. J. Boniface, R. C. Stevens, P. G. Schultz, *Science* **1996**, *271*, 1086.
[21] G. J. Wedemayer, L. H. Wang, P. A. Patten, P. G. Schultz, R. C. Stevens, *J. Mol. Biol.* **1997**, *268*, 390.
[22] W. D. Huse, L. Sastry, S. A. Iverson, A. S. Kang, M. Alting-Mees, D. R. Burton, S. J. Benkovic, R. A. Lerner, *Science* **1989**, *246*, 1275–1281.
[23] S. Cwirla, E. A. Peters, R. W. Barrett, W. J. Dower, *Proc. Natl. Acad. Sci. U.S.A.* **1990**, *87*, 6378–6382.
[24] M. A. Gallop, R. W. Barrett, W. J. Dower, S. P. A. Fodor, E. M. Gordon, *J. Med. Chem.* **1994**, *37*, 1233–1251.
[25] J. K. Scott, C. P. Smith, *Science* **1990**, *249*, 386–390.
[26] J. J. Delvin, L. C. Panganiban, P. E. Delvin, *Science* **1990**, *249*, 404–406.
[27] M. G. Cull, J. F. Miller, P. J. Schatz, *Proc. Natl. Acad. Sci. U.S.A.* **1992**, *89*, 1865–1869.
[28] L. C. Mattheakis, R. Bhatt, W. J. Dower, *Proc. Natl. Acad. Sci. U.S.A.* **1994**, *91*, 9022.
[29] R. W. Roberts, J. W. Szostak, *Proc. Natl. Acad. Sci. U.S.A.* **1997**, *94*, 12297–12302.
[30] A. S. Kang, C. F. Barbas, K. D. Janda, S. J. Benkovic, R. A. Lerner, *Proc. Natl. Acad. Sci. U.S.A.* **1991**, *88*, 4363–4366.
[31] J. McCafferty, A. D. Griffiths, G. Winter, D. J. Chiswell, *Nature* **1990**, *348*, 552–554.
[32] K. D. Janda, C. H. Lo, T. Li, C. F. Barbas III, P. Wirsching, R. A. Lerner, *Proc. Natl. Acad. Sci. U.S.A.* **1994**, *91*, 2532.
[33] K. D. Janda, L.-C. Lo, C.-H. Lo, M.-M. Sim, R. Wang, C.-H. Wong, R. A. Lerner, *Science* **1997**, *275*, 945–948.
[34] J. Ku, P. G. Schultz, *Proc. Natl. Acad. Sci. U.S.A.* **1995**, *92*, 6552.
[35] H. Pedersen, S. Holder, D. S. Sutherlin, U. Schwitter, D. S. King, P. G. Schultz, *Proc. Natl. Acad. Sci. U.S.A.* **1998**, *95*, 10523.
[36] N. C. Wrighton, F. X. Farrell, R. Chang, A. K. Kashyap, F. P. Barbone, L. S. Mulcahy, D. L. Johnson, R. W. Barrett, L. K. Jolliffe, W. J. Dower, *Science* **1996**, *273*, 458–463.
[37] O. Livnah, E. A. Stura, D. L. Johnson, S. A. Middleton, L. S. Mulcahy, N. C. Wrighton, W. J. Dower, L. K. Jolliffe, I. A. Wilson, *Science* **1996**, *273*, 464.
[38] D. J. Matthews, J. A. Wells, *Science* **1993**, *260*, 1113–1117.
[39] M. A. Barry, W. J. Dower, S. A. Johnston, *Nat. Med.* **1996**, *2*, 299–305.
[40] M. Barinaga, *Science* **1998**, *279*, 323–324.
[41] B. G. Hall, *Biochemistry* **1981**, *20*, 4042–4049.
[42] M. S. Neuberger, B. S. Hartley, *J. Gen. Microbiol.* **1981**, *122*, 181–191.
[43] A. Paterson, P. H. Clarke, *J. Gen. Microbiol.* **1979**, *114*, 75–85.

[44] J. Reidhaar-Olson, J. Bowie, R. M. Breyer, J. C. Hu, K. L. Knight, W. A. Lim, M. C. Mossing, D. A. Parsell, K. R. Shoemaker, R. T. Sauer, *Methods Enzymol.* **1991**, *208*, 564–586.
[45] J. A. Wells, M. Vasser, D. B. Powers, *Gene* **1985**, *34*, 315.
[46] R. C. Caldwell, G. F. Joyce, *PCR Methods Appl.* **1992**, *2*, 28–33.
[47] W. P. C. Stemmer, *Nature* **1994**, *370*, 389–391.
[48] P. A. Patten, R. J. Howard, W. P. C. Stemmer, *Curr. Opin. Biol.* **1997**, *8*, 724–733.
[49] C. Ho, M. Jasin, P. Schimmel, *Science* **1985**, *229*, 389.
[50] A. R. Oliphant, K. Struhl, *Proc. Natl. Acad. Sci. U.S.A.* **1989**, *86*, 9094.
[51] H. Liao, T. McKensie, R. Hageman, *Proc. Natl. Acad. Sci. U.S.A.* **1986**, *83*, 576.
[52] J. F. Reidhaar-Olson, R. T. Sauer, *Science* **1988**, *241*, 53.
[53] J. C. Moore, F. H. Arnold, *Nat. Biotechnol.* **1996**, *14*, 458.
[54] K. W. Munir, D. C. French, L. A. Loeb, *Proc. Natl. Acad. Sci. U.S.A.* **1993**, *90*, 4012.
[55] D. R. Liu, T. J. Magliery, M. Pastrnak, P. G. Schultz, *Proc. Natl. Acad. Sci. U.S.A.* **1997**, *94*, 10092–10097.
[56] D. R. Liu, M. Pastrnak, P. G. Schultz, unpublished results.
[57] S. Bass, R. Green, J. A. Wells, *Proteins: Struct. Funct. Genet.* **1990**, *8*, 309–314.
[58] Y. Choo, A. Klug, *Proc. Natl. Acad. Sci. U.S.A.* **1994**, *91*, 1163–1167.
[59] E. J. Rebar, C. O. Pabo, *Science* **1994**, *263*, 671–673.
[60] A. C. Jamieson, S.-H. Kim, J. A. Wells, *Biochemistry* **1994**, *33*, 5689–5695.
[61] Y. Choo, I. Sánchez-García, A. Klug, *Nature* **1994**, *372*, 642–645.
[62] L. Huang, T. Sera, P. G. Schultz, *Proc. Natl. Acad. Sci. U.S.A.* **1994**, *91*, 3969–3973.
[63] T. Sera, P. G. Schultz, *Proc. Natl. Acad. Sci. U.S.A.* **1996**, *93*, 2920–2925.
[64] W. A. Lim, R. T. Sauer, *Nature* **1989**, *339*, 31–36.
[65] S. Kamtekar, J. M. Schiffer, H. Xiong, J. M. Babik, M. H. Hecht, *Science* **1993**, *262*, 1680–1685.
[66] A. R. Davidson, R. T. Sauer, *Proc. Natl. Acad. Sci. U.S.A.* **1994**, *91*, 2146–2150.
[67] B. I. Dahiyat, S. L. Mayo, *Science* **1997**, *278*, 82–87.
[68] A. D. Ellington, J. W. Szostak, *Nature* **1990**, *346*, 818–822.
[69] C. Tuerk, L. Gold, *Science* **1990**, *249*, 505–510.
[70] D. L. Robertson, G. F. Joyce, *Nature* **1990**, *344*, 467.
[71] A. A. Beaudry, G. F. Joyce, *Science* **1992**, *257*, 635.
[72] C. Tuerk, S. MacDougal, L. Gold, *Proc. Natl. Acad. Sci. U.S.A.* **1992**, *89*, 6988.
[73] D. P. Bartel, M. L. Zapp, M. R. Green, J. Szostak, *Cell* **1991**, *67*, 529.
[74] M. Sassanfar, J. W. Szostak, *Nature* **1993**, *364*, 550–553.
[75] M. Famulok, J. W. Szostak, *J. Am. Chem. Soc.* **1992**, *114*, 3990.
[76] G. J. Connell, M. Illangesekare, M. Yarus, *Biochemistry* **1993**, *32*, 5497.
[77] L. Bock, L. Griffin, J. Latham, E. Vermaasm, J. Toole, *Nature* **1992**, *355*, 564–566.
[78] A. D. Ellington, J. W. Szostak, *Nature* **1992**, *355*, 850.
[79] F. Jiang, R. A. Kumar, R. A. Jones, D. J. Patel, *Nature* **1996**, *382*, 183–186.
[80] C. R. Woese, S. Winker, R. R. Gutell, *Proc. Natl. Acad. Sci. U.S.A.* **1990**, *87*, 8467–8471.
[81] R. D. Jenison, S. C. Gill, A. Pardi, B. Polisky, *Science* **1994**, *263*, 1425–1429.
[82] D. Pei, H. D. Ulrich, P. G. Schultz, *Science* **1991**, *253*, 1408–1411.
[83] I. Haruna, S. Spiegelman, *Science* **1965**, *150*, 886.
[84] R. Green, *Science* **1992**, *258*, 1910–1915.
[85] N. Lehman, G. F. Joyce, *Nature* **1993**, *361*, 182–185.
[86] D. P. Bartel, J. W. Szostak, *Science* **1993**, *261*, 1411–1418.
[87] E. H. Ekland, J. W. Szostak, D. P. Bartel, *Science* **1995**, *269*, 364–370.
[88] E. H. Ekland, D. P. Bartel, *Nature* **1996**, *382*, 373–376.
[89] J. R. Prudent, T. Uno, P. G. Schultz, *Science* **1994**, *264*, 1924–1927.
[90] M. M. Conn, J. R. Prudent, P. G. Schultz, *J. Am. Chem. Soc.* **1996**, *118*, 7012–7013.
[91] Y. Li, D. Sen, *Nat. Struct. Biol.* **1996**, *3*, 743–747.
[92] T. M. Tarasow, S. L. Tarasow, B. E. Eaton, *Nature* **1997**, *389*, 54–57.
[93] M. A. Barry, W. C. Lai, S. A. Johnston, *Nature* **1995**, *377*, 632–635.
[94] C. Grimm, E. Lund, J. E. Dahlberg, *EMBO J.* **1997**, *16*, 793–806.
[95] R. B. Merrifield, *J. Am. Chem. Soc.* **1963**, *85*, 2149–2154.

[96] H. M. Geysen, R. H. Meloen, S. J. Barteling, *Proc. Natl. Acad. Sci. U.S.A.* **1984**, *81*, 3998–4002.
[97] R. A. Houghten, *Proc. Natl. Acad. Sci. U.S.A.* **1985**, *82*, 5131–5135.
[98] H. M. Geysen, S. J. Rodda, T. J. Mason, G. Tribbick, P. G. Schoofs, *J. Immunol. Methods* **1987**, *102*, 259–274.
[99] H. M. Geysen, S. J. Rodda, T. J. Mason, *Mol. Immunol.* **1986**, *23*, 709–715.
[100] S. P. A. Fodor, J. L. Read, M. C. Pirrung, L. Stryer, A. T. Lu, D. Solas, *Science* **1991**, *251*, 767.
[101] J. W. Jacobs, S. P. A. Fodor, *Trends Biotechnol.* **1994**, *12*, 19–26.
[102] A. Furka, F. Sebestyen, M. Asgedom, G. Dibo, *Int. J. Pept. Protein Res.* **1991**, *37*, 487–493.
[103] K. S. Lam, S. E. Salmon, E. M. Hersh, V. J. Hruby, W. M. Kazmierski, R. J. Knapp, *Nature* **1991**, *354*, 82–84.
[104] R. A. Houghten, C. Pinilla, S. E. Blondelle, J. R. Appel, C. T. Dooley, J. H. Cuervo, *Nature* **1991**, *354*, 84–86.
[105] C. Pinilla, J. R. Appel, P. Blanc, R. A. Houghten, *Biotechniques* **1992**, *13*, 901–905.
[106] K. S. Lam, V. J. Hruby, M. Lebl, R. J. Knapp, W. M. Kazmierski, E. M. Hersh, S. E. Salmon, *Bioorg. Med. Chem. Lett.* **1993**, *3*, 419–424.
[107] G. P. Smith, D. A. Schultz, J. E. Ladbury, *Gene* **1993**, *128*, 37–42.
[108] S. Brenner, R. A. Lerner, *Proc. Natl. Acad. Sci. U.S.A.* **1992**, *89*, 5181–5183.
[109] M. N. Needels, D. G. Jones, E. H. Tate, G. L. Heinkel, L. M. Kochersperger, W. J. Dower, R. W. Barrett, M. A. Gallop, *Proc. Natl. Acad. Sci. U.S.A.* **1993**, *90*, 10700–10704.
[110] V. Nikolaiev, A. Stierandova, V. Krchnak, B. Sekigmann, K. S. Lam, S. E. Salmon, M. Lebl, *Pept. Res.* **1993**, *6*, 161–170.
[111] M. H. J. Ohlmeyer, R. N. Swanson, L. W. Dillard, J. C. Reader, G. Asouline, R. Kobayashi, M. Wigler, W. C. Still, *Proc. Natl. Acad. Sci. U.S.A.* **1993**, *90*, 10922–10926.
[112] R. A. Houghten, J. R. Appel, S. E. Blondelle, J. H. Cuervo, C. T. Dooley, C. Pinilla, *Biotechniques* **1992**, *13*, 412–421.
[113] M. Torneiro, W. C. Still, *J. Am. Chem. Soc.* **1995**, *117*, 5887–5888.
[114] S. S. Yoon, W. C. Still, *Tetrahedron* **1995**, *51*, 567–578.
[115] A. Borchardt, W. C. Still, *J. Am. Chem. Soc.* **1994**, *116*, 373–374.
[116] J. K. Chen, S. L. Schreiber, *Angew. Chem.* **1995**, *107*, 1041; *Angew. Chem. Int. Ed.* **1995**, *34*, 953–969.
[117] C. Y. Cho, E. J. Moran, S. R. Cherry, J. C. Stephans, S. P. A. Fodor, C. L. Adams, A. Sundaram, J. W. Jacobs, P. G. Schultz, *Science* **1993**, *261*, 1303–1305.
[118] R. J. Simon, R. S. Kania, R. N. Zuckermann, V. D. Huebner, D. A. Jewell, S. Banville, S. Ng, L. Wang, S. Rosenberg, C. K. Marlowe, *Proc. Natl. Acad. Sci. U.S.A.* **1992**, *89*, 9367–9371.
[119] R. N. Zuckermann, E. J. Martin, D. C. Spellmeyer, G. B. Stauber, K. R. Shoemaker, J. M. Kerr, G. M. Figliozzi, D. A. Goff, M. A. Siana, R. J. Simon, *J. Med. Chem.* **1994**, *37*, 2678–2685.
[120] J. S. Nowick, N. A. Powell, E. J. Martinez, E. M. Smith, G. Noronha, *J. Org. Chem.* **1992**, *57*, 3763–3765.
[121] K. Burgess, J. Ibarzo, D. S. Linthicum, D. H. Russell, H. Shin, A. Shitangkoon, R. Totani, A. J. Zhang, *J. Am. Chem. Soc.* **1997**, *119*, 1556–1564.
[122] J.-M. Kim, T. E. Wilson, T. C. Norman, P. G. Schultz, *Tetrahedron Lett.* **1996**, *37*, 5305–5308.
[123] A. B. Smith, T. P. Keenan, R. C. Holcomb, P. A. Sprengeler, M. C. Guzman, J. L. Wood, P. J. Carroll, R. Hirschmann, *J. Am. Chem. Soc.* **1992**, *114*, 10672–10674.
[124] H. Han, K. D. Janda, *J. Am. Chem. Soc.* **1996**, *118*, 2539–2544.
[125] C. Gennari, B. Salom, D. Potenza, A. Williams, *Angew. Chem.* **1994**, *106*, 2181; *Angew. Chem. Int. Ed.* **1994**, *33*, 2067–2069.
[126] M. Hagihara, N. J. Anthony, T. J. Stout, J. Clardy, S. L. Schreiber, *J. Am. Chem. Soc.* **1992**, *114*, 6568–6570.
[127] D. H. Appella, L. A. Christianson, D. A. Klein, D. R. Powell, X. Huang, J. J. Barchi, S. H. Gellman, *Nature* **1997**, *387*, 381–384.

[128] J. L. Matthews, M. Overhand, F. N. M. Kühnle, P. E. Ciceri, D. Seebach, *Liebigs Ann. Chem.* **1997**, 1371–1379.
[129] C. Y. Cho, R. S. Youngquist, S. J. Paikoff, M. H. Beresini, A. R. Hebert, C. W. Liu, D. E. Wemmer, T. Keough, P. G. Schultz, *J. Am. Chem. Soc.* **1998**, in press.
[130] B. A. Bunin, J. A. Ellman, *J. Am. Chem. Soc.* **1992**, *114*, 10997–10998.
[131] B. A. Bunin, M. J. Plunkett, J. A. Ellman, *Proc. Natl. Acad. Sci. U.S.A.* **1994**, *91*, 4708–4712.
[132] A. A. Virgilio, A. A. Bray, W. Zhang, L. Trinh, M. Snyder, M. M. Morrissey, J. A. Ellman, *Tetrahedron* **1997**, *53*, 6635–6644.
[133] S. H. De Witt, J. S. Kiely, C. J. Stankovic, M. C. Schroeder, D. M. Cody, M. R. Pavia, *Proc. Natl. Acad. Sci. U.S.A.* **1993**, *90*, 6909–6913.
[134] D. A. Campbell, J. C. Bermak, T. S. Burkoth, D. V. Patel, *J. Am. Chem. Soc.* **1995**, *117*, 5381–5382.
[135] M. M. Murphy, J. R. Schullek, E. M. Gordon, M. A. Gallop, *J. Am. Chem. Soc.* **1995**, *117*, 7029–7030.
[136] B. Ruhland, A. Bhandari, E. M. Gordon, M. A. Gallop, *J. Am. Chem. Soc.* **1996**, *118*, 253–254.
[137] J. Green, *J. Org. Chem.* **1995**, *60*, 4287–4290.
[138] R. Liang, L. Yan, J. Loebach, M. Ge, Y. Uozumi, K. Sekanina, N. Horan, J. Gildersleeve, C. Thompson, A. Smith, *Science* **1996**, *274*, 1520–1522.
[139] P. H. H. Hermkens, H. C. J. Ottenheijm, D. Rees, *Tetrahedron* **1996**, *52*, 4527–4554.
[140] P. H. H. Hermkens, H. C. J. Ottenheijm, D. Rees, *Tetrahedron* **1997**, *53*, 5643–5678.
[141] E. K. Kick, D. C. Roe, A. G. Skillman, G. Liu, T. J. A. Ewing, Y. Sun, I. D. Kuntz, J. A. Ellman, *Chem. Biol.* **1997**, *4*, 297–307.
[142] C. Khosla, *Chem. Rev.* **1997**, *97*, 2577–2590.
[143] J. Cortes, S. F. Haydock, G. A. Roberts, D. J. Bevitt, P. F. Leadlay, *Nature* **1990**, *348*, 176–178.
[144] S. Donadio, M. J. Staver, J. B. McAlpine, S. J. Swanson, L. Katz, *Science* **1991**, *252*, 675–679.
[145] J. R. Jacobsen, C. R. Hutchinson, D. E. Cane, C. Khosla, *Science* **1997**, *277*, 367–369.
[146] A. C. Pease, D. Solas, E. J. Sullivan, M. T. Cronin, C. P. Holmes, S. P. A. Fodor, *Proc. Natl. Acad. Sci. U.S.A.* **1994**, *91*, 5022–5026.
[147] M. S. Chee, X. Huang, R. Yang, E. Hubbell, A. Berno, D. Stern, J. Winkler, D. J. Lockhart, M. S. Morris, S. P. A. Fodor, *Science* **1996**, *274*, 610–614.
[148] L. Wodicka, H. Dong, M. Mittman, M.-H. Ho, D. J. Lockhart, *Nat. Biotechnol.* **1997**, *15*, 1359–1367.
[149] M. D. Adams, J. M. Kelley, J. D. Gocayne, M. Dubnick, M. H. Polymeropoulos, H. Xiao, C. R. Merril, A. Wu, B. Olde, R. F. Moreno, A. R. Kerlavage, W. R. McCombie, J. C. Venter, *Science* **1991**, *252*, 1651–1656.
[150] G. G. Lennon, H. Lehrach, *Trends Genet.* **1991**, *7*, 314–317.
[151] P. Liang, A. B. Pardee, *Science* **1992**, *257*, 967–971.
[152] K. Okubo, N. Hori, R. Matoba, T. Niiyama, A. Fukushima, Y. Kojima, K. Matsubara, *Nat. Genet.* **1992**, *2*, 173–179.
[153] S. Meier-Ewert, E. Maier, A. Ahmadi, J. Curtis, H. Lehrach, *Nature* **1993**, *361*, 375–376.
[154] V. E. Velculescu, L. Zhang, B. Vogelstein, K. W. Kinzler, *Science* **1995**, *270*, 484–487.
[155] N. Zhao, H. Hashida, N. Takahashi, Y. Misumi, Y. Sakaki, *Gene* **1995**, *156*, 207–213.
[156] C. Nguyen, D. Rocha, S. Granjeaud, M. Baldit, K. Bernard, P. Naquet, B. R. Jordan, *Genomics* **1995**, *29*, 207–216.
[157] M. Schena, D. Shalon, R. W. Davis, P. O. Brown, *Science* **1995**, *270*, 467–470.
[158] N. S. Gray, L. Wodicka, A.-M. Thunnissen, T. C. Norman, S. Kwon, F. H. Espinoza, D. O. Morgan, G. Barnes, S. LeClerc, L. Meijer, S.-H. Kim, D. J. Lockhart, P. G. Schultz, *Science* **1998**, *281*, 533–538.
[159] N. S. Gray, S. Kwon, P. G. Schultz, *Tetrahedron Lett.* **1997**, *38*, 1161–1164.
[160] T. C. Norman, N. S. Gray, J. T. Koh, *J. Am. Chem. Soc.* **1996**, *118*, 7430–7431.
[161] J. L. DeRisi, V. R. Iyer, P. O. Brown, *Science* **1997**, *278*, 680–686.
[162] D. Hirata, K. Yano, T. Miyakawa, *Mol. Gen. Genet.* **1994**, *242*, 250.

[163] G. Zlokarnik, P. A. Negulescu, T. E. Knapp, L. Mere, N. Burres, L. Feng, M. Whitney, K. Roemer, R. Y. Tsien, *Science* **1998**, *279*, 84–88.
[164] S. D. Liberles, S. T. Diver, D. J. Austin, S. L. Schreiber, *Proc. Natl. Acad. Sci. U.S.A.* **1997**, *94*, 7825–7830.
[165] E. J. Licitra, J. O. Liu, *Proc. Natl. Acad. Sci. U.S.A.* **1996**, *93*, 12817–12821.
[166] F. J. DiSalvo, *Science* **1990**, *247*, 649.
[167] X.-D. Xiang, X. Sun, G. Briceño, Y. Lou, K.-A. Wang, H. Chang, W. G. Wallace-Freedman, S.-W. Chen, P. G. Schultz, *Science* **1995**, *268*, 1738–1740.
[168] J. Wang, Y. Yao, C. Gao, I. Takeuchi, X. Sun, H. Chang, X. D. Xiang, P. G. Schultz, *Science* **1998**, *279*, 1712–1714.
[169] E. Danielson, J. H. Golden, E. W. McFarland, C. M. Reaves, W. H. Weinberg, X. D. Wu, *Nature* **1997**, *389*, 944.
[170] H. Chang, C. Gao, I. Takeuchi, Y. Yoo, J. Wang, R. P. Sharma, M. Downes, T. Venkatesan, P. G. Schultz, X.-D. Xiang, *Appl. Phys. Lett.* **1998**, *72*, 2185–2187.
[171] X. Sun, K.-A. Wang, Y. Yoo, W. G. Wallace-Freedman, C. Gao, X.-D. Xiang, P. G. Schultz, *Adv. Mater.* **1998**, *9*, 1046.
[172] R. Nielsen, personal communication, **1997**.
[173] P. G. Schultz, X.-D. Xiang, *Curr. Opin. Solid State Mater. Sci.* **1998**, *3*, 153–158.
[174] Y. Lu, T. Wei, F. Duewer, Y. Lu, N. B. Min, P. G. Schultz, X.-D. Xiang, *Science* **1997**, *276*, 2004.
[175] E. D. Isaacs, M. Kao, G. Aeppli, X.-D. Xiang, X. Sun, P. G. Schultz, M. A. Marcus, G. S. I. Cargill, R. Haushalter, *Appl. Phys. Lett.*, *73*, 1820–1822.
[176] F. C. Moates, M. Somani, J. Annamalai, J. T. Richardson. *Ind. Eng. Chem. Res.* **1996**, *35*, 4801.
[177] W. H. Weinberg, personal communication, **1998**.
[178] G. Briceño, H. Chang, P. G. Schultz, X.-D. Xiang, *Science* **1995**, *270*, 273–275.
[179] E. Danielson, M. Devenney, D. M. Giaquinta, J. H. Golden, R. C. Haushalter, E. W. McFarland, D. M. Poojary, C. M. Reaves, W. H. Weinberg, X. D. Wu, *Science* **1998**, *279*, 837–839.
[180] C. L. Hill, R. D. Gall, *J. Mol. Catal. A* **1996**, *114*, 103.
[181] S. Brocchini, K. James, V. Tangpasuthadol, J. Kohn, *J. Am. Chem. Soc.* **1997**, *119*, 4553.
[182] T. A. Dickinson, D. R. Walt, J. While, J. S. Kauer, *Anal. Chem.* **1997**, *69*, 3413.
[183] I. Takeuchi, W. H. Weinberg, unpublished results.

Ethical Limits to 'Molecular Medicine'

by **Ernst-Ludwig Winnacker**

Deutsche Forschungsgemeinschaft, D-53175 Bonn
(Fax: 0049 228 885 2770)

The *Nature* issue of September 2nd, 1999, contained an article entitled *'Young receptors make smart mice'* [1]. In this paper, the authors described a mouse transgenic for a membrane protein called the NMDA receptor, in which associative learning was considerably enhanced. This impressive paper shows that the synaptic plasticity of the brain can be influenced by genetic interventions. Whether this design of a 'smart' mouse can be transferred to the human situation remains to be seen. Nevertheless, these and other experiments open the door for what we would call genetic enhancement, that is, an improvement of a particular trait by genetic means. Some journalists were tempted to remind their readers of *Aldous Huxley*'s *Brave New World*, which begins with the description of a laboratory which can produce identical twins. *'Ninety-six seemed to be the limit, seventy-two a good average. From the same ovary and with gametes of the same male to manufacture as many batches of identical twins as possible, that was the best that they could do. And even that was difficult'* [2]. What was utopia in 1931 has entered a phase in 1999 that adds some reality to the science fiction of 68 years ago. What have we learned, and into which direction are we moving?

The present revolution is based on developments that started in the late sixties of the 20th century. What have we achieved so far, and where are we going? The following objectives have been reached so far:

1) Genes can be isolated through recombinant DNA technologies without technical limitations. Products of such genes have long been brought onto the drugmarket. The top 10 selling products alone had worldwide sales in 1997 of 6.578 million Euro.

2) The sequences of entire genomes are now available and are being analyzed rapidly from many model organism, including bacteria, yeast, multicellular organism, like the nematode *C. elegans*, the fruitfly, and the mouse. Recent milestones were the completion of the sequences the genomes of the roundworm *Caenorhabditis elegans* and the fruit fly *Drosophila melanogaster* with a genome sizes of 97 and 130 Megabases, respectively [3][4].

3) In 1985, the vision was expounded to sequence the entire human genome, providing a complete catalogue of all human genes [5]. The project appeared technically unfeasable at a time at which DNA sequencing technology had advanced only to such a limited extent that the time required to complete such a project was extrapolated to more the 10000 years. Not unexpectedly, technology advanced so rapidly that the human genome is expected to be completed in a preliminary form by the end of February 2001. Due to the close relationship between genomes, conclusions from the study of model organisms can often be transferred and applied to our species. A spectacular example for these surprising connections are some of the genes which predispose to *Alzheimer*'s disease, the presenilins [6]. These genes have counterparts in the worm *C. elegans*, which has only a primitive sensory nervous system. Dysfunction of preseniline genes affects the two neurons controlling egg-laying such that presenelin-defective worms cannot lay eggs anymore. The defect can be rescued not only by replacement of the defective gene with a functional copy of the worm gene but, much more surprisingly, with a functional copy of the human preseniline gene. Although man and worm have separated in evolutionary development at least 220 million years ago, the functional identity of these genes has nevertheless been fully conserved over this vast amount of time.

4) Genes are constantly being identified as the origin of various diseases. More than 100 years ago, it was discovered that red-green blindness is inherited as a Mendelian trait, proving that inheritance in *Homo sapiens* follows the same rules as found previously for plants and animals. In the meantime, the defective genes responsible for most of the 4000 plus monogenic disorders have been identified. In addition, the genetic origin of diseases other than the rare monogenic disorders is becoming more evident from day to day. In 1999, it was published that three defined genes are necessary to transform a healthy human cell in culture into a tumor cell [7]. Colon cancer is supposed to be triggered by six genetic changes in a normal colon epitheliel cells until it can develop into a full-blown and metastatic tumor. Impressive advances are made in unravelling the genetic origins of neurological disorders, like epilepsy and schizophrenia. Most recently, a mouse was constructed by genetic intervention, which had all the symptoms of a drug induced schizophrenia. No doubt, that, in time, even these multifactorial diseases will lend themselves to genetic analyses to an extent that we never thought possible only recently.

5) Genes can be transferred into cells, albeit with low efficiency. There are two principle approaches to genetic interventions, namely, somatic gene therapy and germ-line gene therapy. Germ-line interventions modify germ cells or their precursors, thereby transferring the genetic changes, unlike in the modification of somatic cells, into the progeny, *i.e.*, into subsequent generations. Technically speaking, the efficiency of gene transfer into either somatic or into germ cells is quite low; even so, considerable efforts have been spent in the past 15 years or so to improve gene transfer in the mouse. Thus, at present we are far away from the efficiency of oral vaccinations, known for example from the polio vaccination.

6) Three years ago, sheep was cloned by way of nuclear transplantation [8]. Clones are not a rare species in nature. Even in men, monozygotic twins are not uncommon and appear at a rate of 1 in 500 births. What was novel in the case of the cloned sheep *'Dolly'* was that it was produced by transplanting a nucleus from an adult, differentiated cell into an enucleated egg cells. Apparently, the genomes in somatic cells are not irreversibly modified but can be reprogrammed to the status of an embryo cell and instruct the formation of an entire organism. It is thus possible to produce clones not only by embryo splitting as known previously, but at a certain time at which the particular and potentially useful traits of an organism are already well-established. In the meantime, cloning by nuclear transplantation from somatic cells has been achieved in other species as well, including cattle, mouse, and pig. Although the intention has been spelled out repeatedly, no reports have appeared on the cloning of people by the *'Dolly'* procedure.

7) Several months ago, pluripotent stem cells have been produced and cultured from aborted human fetuses [9] and from spare embryos left over from *in-vitro* fertilization procedures [10]. By definition, such cells have the potential to differentiate into all of the 300 plus cell types that constitute a mammalian organism. Unlike an embryo, they are only pluripotent, *i.e.*, they cannot develop as such into an entire organism but only in the context of an early-stage embryo. Nevertheless, they can be genetically modified and returned as such into an early embryo. The resulting organisms would be heterozygotes. Homozygotes could only be obtained by backcrosses. By this approach, hundreds of lines of genetically modified mice have been produced in order to study the effects of gene disruptions or gene insertions on the phenotype of these animals. Some of them have even been produced as models for human diseases, with considerable success.

It has been proposed that embryonic stem cells should be used for tissue-replacement therapies by inducing them to differentiate into particular cell types, *i.e.*, heart muscle cells or neurons. At present, the factors required to differentiate embryonic cells into specific cell lineages are only partly known. In addition, the use of human embryonic stem cells is burdened with ethical and legal concerns. Fortunately, there is increasing evidence that adult stem cells may be more abundant than expected, and thus may present an alternative to human embryonic stem cells in cell therapies. Such cells had been previously known in the hemapoietic system and are used in bone-marrow transplantation for the treatment of leukemias. Only recently, they have been identified in many other organs as well, including brain [11]. More importantly even, some of these stem cells display the capacity to transdifferentiate into each other [12]. Thus, muscle stem cells have been shown to repopulate the blood stream and hemapoietic stem cells to develop into neurons. If this plasticity could be demonstrated on a broader scale, adult stem cells could replace embryonic stem cells as potential tools for cell-replacement therapies.

With all these technologies at hand, it is certainly legitimate to ponder the question, which particular goals could be realized or accomplished by applying them to human beings. Technically, there is no reason to assume that genomic technologies cannot be applied to people since human reproductive and stem cell biology are very similar to the biology of other species. The impact of genomic technologies thus will be limited only by the way we view ourselves and each other as well as by normative limits imposed by legislation of one kind or another. In a multidisciplinary study at the Institute *TTN* (*Technik-Theologie-Naturwissenschaften*) in Munich, the ethical dimensions of these various interventions were explored and analyzed in a novel approach [13]. The approach seeks to present the various kinds of genetic interventions in steps of increasing depth of intervention and to measure these against certain standards or guiding principles, which appeared relevant to the subject, namely, the concept of human dignity, the notion and definition of a disease, ethical standards of the medical profession, and the professional ethics of a scientist. These parameters speak for themselves. The principle of human dignity, for example, must guarantee the autonomy of patients and, therefore, implies that every physical intervention requires an informed consent by the patient or a legal representative if the patient is not able to consent. The constitutional right to freedom of research must, when in doubt, weigh less than the constitutional rights to life or freedom from injury.

Medical professional ethics requires the art of practicing medicine to be aimed at the benefit of the sick. Neither his or her scientific or financial interests nor the ever increasing expectations of patients must be permitted to overrule this principle. It appears to be at stake in the new field of gene therapy. Following a tragic death last fall due to a considerable lack of clinical and scientific judgement, the field would be well advised to follow proper medical practice and not to promise too much at this early time.

These parameters in mind, we analyzed a number of genetic therapies according to the extent of intervention to which they would subject a human being. A preliminary list identified three such possibilities, from substitution therapies with recombinant proteins *via* somatic gene therapy to germ-line therapy. The latter case was subdivided into four different categories of objectives extending from the therapy of monogenic disorders *via* the introduction of genes coding for disease resistance towards a number of genetic improvements that do not feature any pathological correlate whatsoever.

Step 1: The case of substitution therapies with recombinant proteins could be handled quite easily. Substitution therapy is defined as the application of proteins for the treatment of disease caused by the deficiency of single proteins, *i.e.*, insulin in diabetes, growth hormone in dwarfism, factor VIII in hemophilia B, and erythropoietin in various forms of anemia. Some of these could be produced in the past from blood or from animal organs. The majority of them, however, has become accessible only through recombinant DNA production. Its advantages are obvious: production is inexpensive. It provides material in unlimited amounts, and it is safe, as compared with blood and its possible contamination with viruses, *i.e.*, the HIV virus. When measured against the four guiding principles, substitution therapy did not violate any ethical or moral principles, granted, of course, that these proteins are handled as drugs according to accepted medical practice, including regulations for the licensing of new drugs.

Step 2 and *3* cover somatic gene therapy. Ultimately, somatic gene therapy is a logical extension of substitution therapies with recombinant proteins. Instead of substituting a gene product, a protein, the corresponding gene is replaced. In some cases, novel genes are brought into somatic cells, *i.e.*, cytokine genes. It is estimated that *ca.* 3500 patients are currently enlisted in *ca.* 200 different clinical trials. Half of these are focussed on various forms of cancer, one-third on HIV, and the rest on monogenetic disorders. As long as such studies are performed with the patients' consent, they appeared to us to belong into similar ethical categories as traditional medical interventions [14]. Their ways and means did not appear improper to us as long as strategies are optimized and tested in experimental animals, when and whereever possible. Proper gene expression, of course, has to be guaranteed whether treatment occurs *ex vivo* or *in-vivo*. It is of paramount importance that, during such experiments, the gene transfer occurs exclusively into target cells and not into other cell types of the patient's organism. A significant enhancement of vector tropism will have to be achieved for most vector systems until sufficient specificity can be guaranteed. Nevertheless, once these difficulties in gene expression and target cell specificity have been overcome, somatic gene therapy will not be inconsistent with any of the four guiding principles against which we measure ethical limits of genetic interventions.

The concept of germ-line therapies comprises all of genetic interventions which result in a targeted and permanent change in the genetic constitution of the germ-line, *i.e.*, of the genomes of egg and sperm cells and their precursors. In contrast to somatic gene therapy, genetic changes are thus not restricted to a single patient but also to individuals of subsequent generations. In the analysis two levels should be considered, *i.e.*, the means and the goals of such a therapeutic intervention. Extrapolation from mouse to man, which is certainly acceptable on this biological level, there are two sources for toti-

or pluripotent cells to target a therapeutic intervention, zygotes and embryonic stem cells. A healthy woman produces up to two egg cells per cycle. Through superovulation, this number can be raised to four to six cells per cycle. Considering the low yield of mouse embryos, which develop into transgenic animals, at best *ca.* 20%, but, in general, not more than 10%, it becomes obvious that such a procedure cannot, by any means, be extended to people. To optimize it to such an extent that six to twelve embryos of a single woman can te treated with a reasonable chance of success would require a test period with hundreds of embryos. Apart from the fact that this is prohibited in some countries, including Germany, the creation of embryos for the purpose only of optimizing the procedure would contradict the principle of human dignity in two ways: on one hand, the use of embryos as mere research tools interferes with human dignity in the sense that these embryos, while having the potential to become a human being, are spent instead for research purposes. On the other hand, the hormone treatment used to induce superovulation in women has significant side effects. It thus appears imcompatible with medical professional ethics to expose healthy women to a risky procedure that does not serve their own immediate interests. As to embryonic stem cells, they are only pluripotent, thus leading only to chimeric embryos. The *'Dolly'* procedure is morally unfeasable because of its low yields. In the case of *'Dolly'*, 277 chimeric embryos had to be produced and transferred into pseudopregnant foster mother animals to bring a single embryo to term. Several other embryos were aborted prematurely. It appears ethically unacceptable to apply such a procedure to human beings.

Assuming that technical problems had been resolved, some ot the putative goals of germ-line therapy seem to pose few if any ethical problems, for example, germ-line therapy for monogenic disorders. Replacement of a mutated gene by its natural counterpart, thereby eliminating a severe disorder, must be regarded as a noble objective. The procedure would only involve correction or replacement of a gene, not the addition of a new gene, such that the context of the genome would remain undisturbed. The procedure would not contradict professional standards, and, if restricted to severe diseases would not violate the notion or definition of a disease. Nevertheless, this 'yes' to germ-line therapy is only a conditional one. It presupposes that the development of the procedure has already been accomplished. The problem thus remains whether it is justifiable to apply a therapeutic procedure the development of which is actually illegal or immoral. At a minimum, the late vindication of a therapy freed from the risks of research leaves a certain bad aftertaste, which is ethically unsatisfactory.

The chain of arguments was much more simpler in the case of germ-line therapy for disease prevention through the introduction of genes conferring resistance towards viral disease, for example, such as HIV or influenza infections. Novel genes can change the status of other genes as much as novel proteins can influence the proteome. There are many well-known examples for this phenomenon, for example, from the case of the human papilloma viruses. Human papilloma virus type 16 codes for a protein that can bind to certain intracellular proteins. Among those is a tumor suppressor protein that is responsible for growth or cell-cycle control. If such a protein is removed through interaction with a viral protein, it will not be available anymore for its genuine task. The infected cells starts to divide; the organism will soon be confronted with an ovarian tumor. Since the actual consequences of the introduction of a novel protein into the cytoplasm can hardly be predicted, such interventions must be regarded as too risky.

The third step concerns preventive measures against deviations form behavioral standards, interventions that serve to reduce the range of variations of certain traits, which, in extreme cases, can amount to considerable risks to life, but which really do not represent dispositions to diseases, like sexual preferences, bad memory, or obesity. As to the subject of homosexuality, the geneticist *Dean Hamer* in 1993 surprised the scientific community and society at large with the identification of a particular polymorphisms on a small region of the human X-chromosome, which was supposed to occur only in homosexual brothers [15]. Although the gene in question could not be identified even up till now, the news immediately travelled around the world that the gene for homosexuality had been found. Even if this were true, the question arises whether homosexuality represents a disease in the true medical sense, or whether it is only regarded as a disease because we live in a homophobic society. In this case, the pseudodisease homosexuality should not be treated by biological means but by increased societal tolerance und other social manners.

This example cleared the way to our final stage, germ-line therapy for changes of multifactorial traits, for changes of cognitive determinants that ultimately could change the nature of our species. Examples for this type of enhancement therapy would be an increase of our intelligence, a drastic increase of our life span, or presumably the elimination of aggressive behavior. Aggression is part of the basic make-up of our species. It has to be controlled, of course, but it is indispensible for a human realization of our life. There are indications from experiments with mice that aggressive behavior can be influenced through simple genetic interventions. It is not clear whether such interventions would work in people, and whether human aggressive behavior is more a cognitive than a behavioral parameter.

There is little doubt that intelligence holds a genetic component although little, very little is known about the actual heritability of IQ, in a technical sense. A neuron contains 5000–10000 different proteins, and any relationship between the state of the corresponding genes and a given IQ must, if it existed in first place, be multi- or polyfactorial. To identify such genes and to ascribe particular effects to them appears hopeless. Some of the twin studies in which some people tried to quantify the portion of the IQ phenotype, which is accounted for by genetic and by environmental variance, are seriously flawed. They are faked, or they lack the necessary scientific quality. The construction of a mouse carrying an NMDA receptor has recently supported the *Galtonians* among us, since this genetic change clearly resulted in considerable improvements of associative learning and in memory [1]. I doubt whether this observed increase in long-term potentiation really has much to do with human intelligence. I thus totally disagree with the authors when they claim in their last sentence: *'This study also reveals a promising strategy for the creation of other genetically modified mammals with enhanced intelligence and memory'*. In matters of the cerebral cortex, the science of genetics approaches its limits.

I thus conclude that, with this step, we leave the safe ground of reality and move into utopia. The question whether such interventions can be justified on moral ground does not pose itself anymore. It becomes irrelevant to a medical doctor and/or a scientist who cannot and should not draw solid conclusions from mere speculation or science fiction.

An additional and serious ethical challenge is provided by the modern technologies of genome analyses. They permit the rapid analysis of all kind of differences in genome sequences. This can lead to a variety of ethical problems, among them:

1) The identification of disease genes that lead to disease which cannot be treated. The question raised is how serious we will be able to protect the *'right of not to know'*.

2) The identification of disease genes that only predispose to genetic disease, thereby leading to life-long uncertainties.

3) The fact that, because of the genetic generation contract, an identification of genome defects must immediately be relevant for all other family members as well. A diagnosis for a defect in a gene predisposing for breast cancer would, for example, concern all female members within this particular family.

4) The request of life- and health-insurance companies to become informed about genome defects with the possible consequence that people carrying particular mutation in disease genes will be charged different premiums or will be excluded from insurance coverage.

5) The specter of eugenics. The philosopher *Peter Sloterdijk* recently argued that aggression in man will be a serious problem in the future, and that it should be dealt with by genetic intervention, in particular by selection and by breeding. I consider these and previous proposals as extremely naive. I realize that the breeding of people is not pure science fiction anymore, because the biology of mice and man is not that much different. Such proposals are not new, if we consider the famous *CIBA* conference during which *Joshua Lederberg* argued that he wanted to increase the size of the human brain [16]. But they do not become better by repetition. I strongly feel that the delusion of eugenics will remain without a scientific basis even more today than prior to the recent advances in 'Molecular Medicine'.

There are many driving forces behind the genomic revolution, among them the scientific interest in the function of the gene and the relationship between genotype and phenotype, progress in medicine, commercial interests, and the putative promises of creating a better human being. Public acceptance of science and the future of science itself will depend on how the scientific community will be able to communicate with the public. The 'escalation model' presented in this paper tries to induce scientists and doctors as well as members of the interested public to come to grips with this field and the many possible application directed towards human beings. It is hoped that such an approach can be used to structure public discussions about ethics and risks in other situations as well.

I am grateful to Profs. *Herrmann Hepp*, *Peter Hans Hofschneider*, *Wilhelm Korff*, and *Trutz Rendtorff* for having participated in intensive discussions about the subject that led to the development of our escalation model.

REFERENCES

[1] Ya-Ping Tang *et al.*, *'Genetic enhancement of learning and memory in mice'*. *Nature* **1999**, *401*, 63–69.
[2] A. Huxley, *Brave New World*, Perennial Classics edn. 1998, p. 8.
[3] The *C. elegans Sequencing Consortium*, *'Genome sequence of the nematode C. elegans: a platform for investigating biology'*, *Science* **1998**, *282*, 2012–2018.
[4] M. D. Adams *et al.*, *'The Genome Sequence of Drosophila melanogaster'*, *Science* **2000**, *287*, 2185–2195.
[5] R. A. Dulbecco, *'Turning Point in Cancer Research: Sequencing the Human Genome'*, *Science* **1986**, *231*, 1055–1056.
[6] C. Haass, *'Presenilins: genes for life and death'*, *Neuron* **1997**, *18*, 687–690.
[7] W. C. Hahn *et al.*, *'Creation of human tumour cells with defined genetic elements'*, *Nature* **1999**, *400*, 464–468.
[8] I. Wilmut *et al.*, *'Viable offspring derived from fetal and adult mammalian cells'*, *Nature* **1997**, *385*, 810–813.
[9] K. J. Shamlott *et al.*, *'Derivation of pluripotent stem cells from cultured human primordial germ cells'*, *Proc. Natl. Acad. Sci. U.S.A.* **1998**, *95*, 13726–13731.
[10] J. A. Thomson *et al.*, *'Embryonic stem cells derived from human blastocysts'*, *Science* **1998**, *282*, 1145–1147.
[11] F. H. Gage, *'Mammalian neural stem cells'*, *Science* **2000**, *287*, 1433–1438.
[12] F. M. Watt, B. L. M. Hogan, *'Out of Eden: Stem cells and their niches'*, *Science* **2000**, *287*, 1427–1430.
[13] E. L. Winnacker, *'Gentechnik: Eingriffe am Menschen – Ein Eskalationsmodell zur ethischen Bewertung'*, Herbert-Utz-Verlag, München, 1999.

[14] T. Friedmann, *'Principles for human gene therapy studies', Science* **2000**, *287*, 2163–2165.
[15] D. Hamer, *'A linkage between DNA markers on the X-chromosome and male sexual orientation', Science* **1999**, *261*, 321–327.
[16] *Man and his future*, A Ciba Foundation Volume, The Ciba Foundation, 41 Portland Place, London W1N 4BN, 1962.

Epilogue

Synthesis of Coenzyme B_{12}: A Vehicle for the Teaching of Organic Synthesis

(Facsimile of handouts to the students of an undergraduate course on organic synthesis, given by **Albert Eschenmoser** during the summer semester of 1973 at the ETH-Zürich)

References

Total Synthesis of Vitamin B_{12}

R. B. Woodward, *Pure Appl. Chem.* **1968**, *17*, 519–547.
R. B. Woodward, *Pure Appl. Chem.* **1971**, *25*, 283–304.
R. B. Woodward, *Pure Appl. Chem.* **1973**, *33*, 145–177; R. B. Woodward, *Chem. Soc. Spec. Publ.* **1967**, *21*, 217–249.
R. B. Woodward, in'Vitamin B_{12}', Eds. B. Zagalak, W. Friedrich, Walter de Gruyter, Berlin 1979, 37–87.

A. Eschenmoser, *Pure Appl. Chem.* **1963,** *7,* 297–316.
A. Eschenmoser, *Pure Appl. Chem.* **1969,** *20,* 1–23.
A. Eschenmoser, *Quart. Rev.* **1970,** *24,* 366–415.
A. Eschenmoser, *Pure Appl. Chem.* **1971,** Suppl. Special Lectures, *23,* IUPAC Congress, Boston, Vol. *2*, 69–106.
A. Eschenmoser, *Naturwissenschaften* **1974,** *61,* 513–525; A. Eschenmoser, C. E. Wintner, *Science* **1977,** *196,* 1410–1420;

Exploratory Corrin Syntheses

E. Bertele, H. Boos, J. D. Dunitz, F. Elsinger, A. Eschenmoser, I. Felner, H. P. Gribi, H. Gschwend, E. F. Meyer, M. Pesaro, R. Scheffold, *Angew. Chem., Int. Ed.* 1964, *3*, 490–496
A. Eschenmoser, R. Scheffold, E. Bertele, M. Pesaro, H. Gschwend, *Proc. R. Soc. London, Ser. A* **1965**, *288*, 306–323.
I. Felner, A. Fischli, A. Wick, M. Pesaro, D. Bormann, E. L. Winnacker, A. Eschenmoser, *Angew. Chem., Int. Ed.* **1967**, 864–866.
A. Fischli, A. Eschenmoser, *Angew. Chem., Int. Ed.* **1967**, *6*, 866–868.
Y. Yamada, D. Miljkovic, P. Wehrli, B. Golding, P. Löliger, R. Keese, K. Müller, A. Eschenmoser, *Angew. Chem., Int. Ed.* **1969**, *8*, 343–348.
E. Götschi, W. Hunkeler, H.-J. Wild, P. Schneider, W. Fuhrer, J. Gleason, A. Eschenmoser, *Angew. Chem., Int. Ed.* **1973**, *12*, 910–912.

Conversion of Cobyric Acid to Vitamin B_{12}

W. Friedrich, G. Gross, K. Bernhauer, P. Zeller, *Helv. Chim. Acta* **1960**, *43*, 704–712.

Conversion of Vitamin B_{12} to Coenzyme B_{12}

A. W. Johnson, L. Mervyn, N. Shaw, E. L. Smith, *J. Chem. Soc.* **1963**, 4146.
K. Bernhauer, O. Müller, G. Müller, *Biochem. Z.* **1962**, 336, 102.

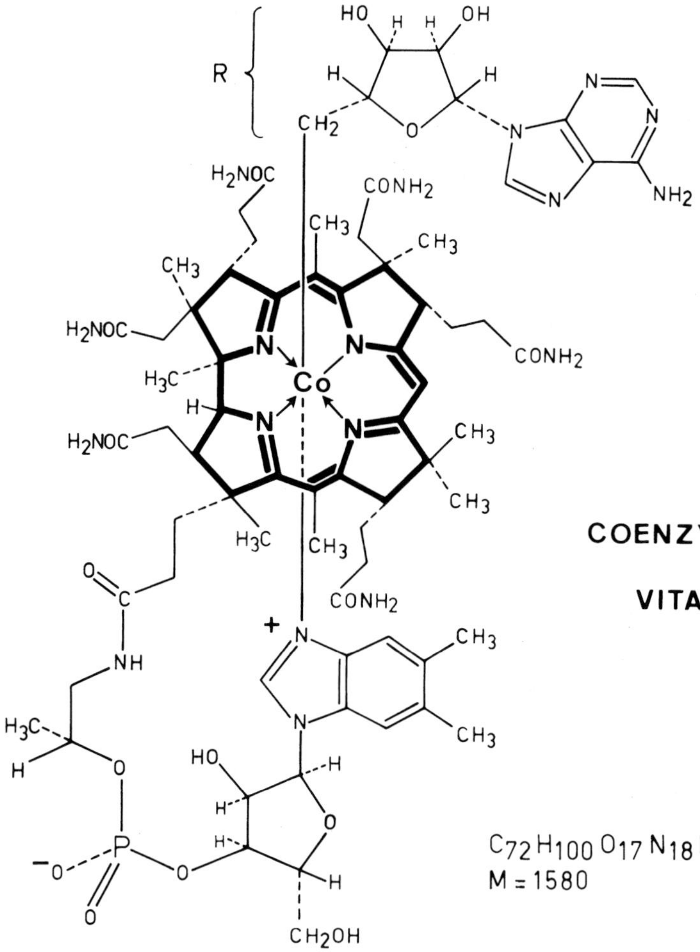

COENZYME B_{12}

VITAMIN B_{12} : R = CN

$C_{72}H_{100}O_{17}N_{18}PCo$
M = 1580

1948 Isolierung von kristallisiertem Vitamin B_{12} ("Anti-pernicious anaemia factor", Cyano-cobalamin) aus Leber.
Beginn der Entwicklung mikrobiologischer Produktionsverfahren
1956 Struktur von Vitamin B_{12} durch Röntgenstrukturanalyse
1958 Entdeckung des Coenzyms B_{12}
1960 Isolierung und Struktur der Cobyrsäure
Partialsynthese von Vitamin B_{12} aus Cobyrsäure
1961 Struktur von Coenzym B_{12} durch Röntgenstrukturanalyse
1962 Partialsynthese von Coenzym B_{12} aus Vitamin B_{12}
1972 Totalsynthese von Cobyrsäure

D-Ribose

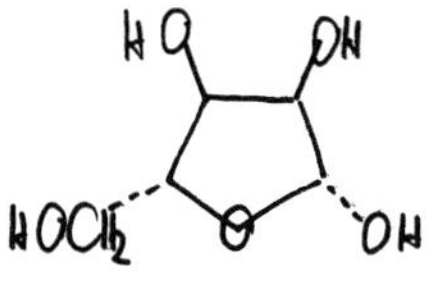

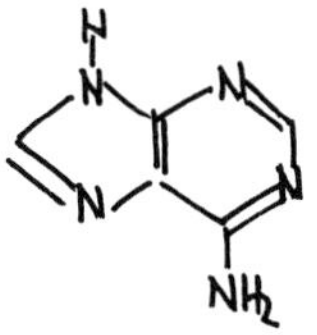

Adenin

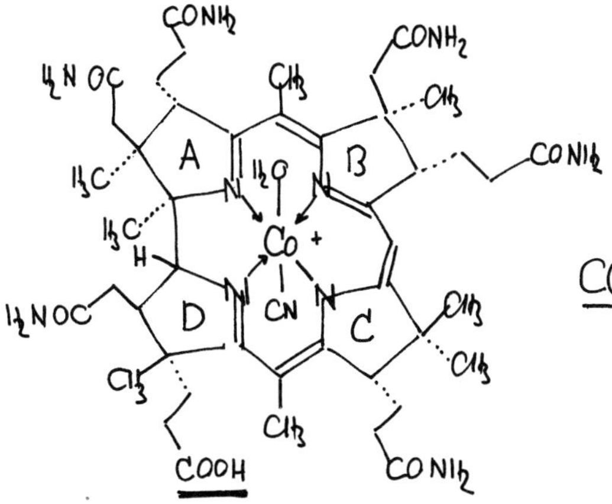

COBYRSÄURE

2R-1-Amino-2-propanol

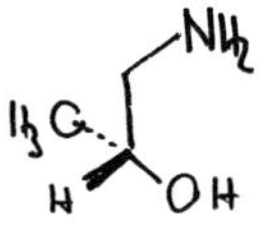

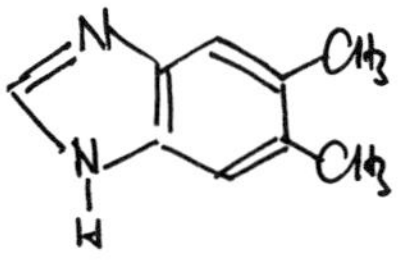

5,6-Dimethyl-benzimidazol

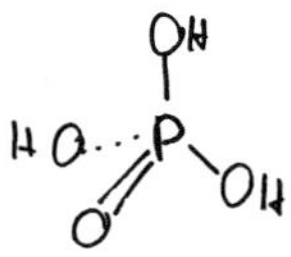

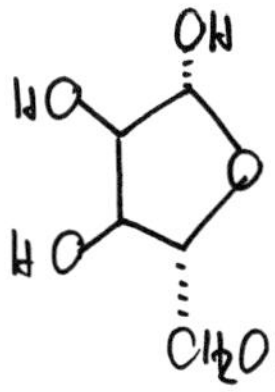

D-Ribose

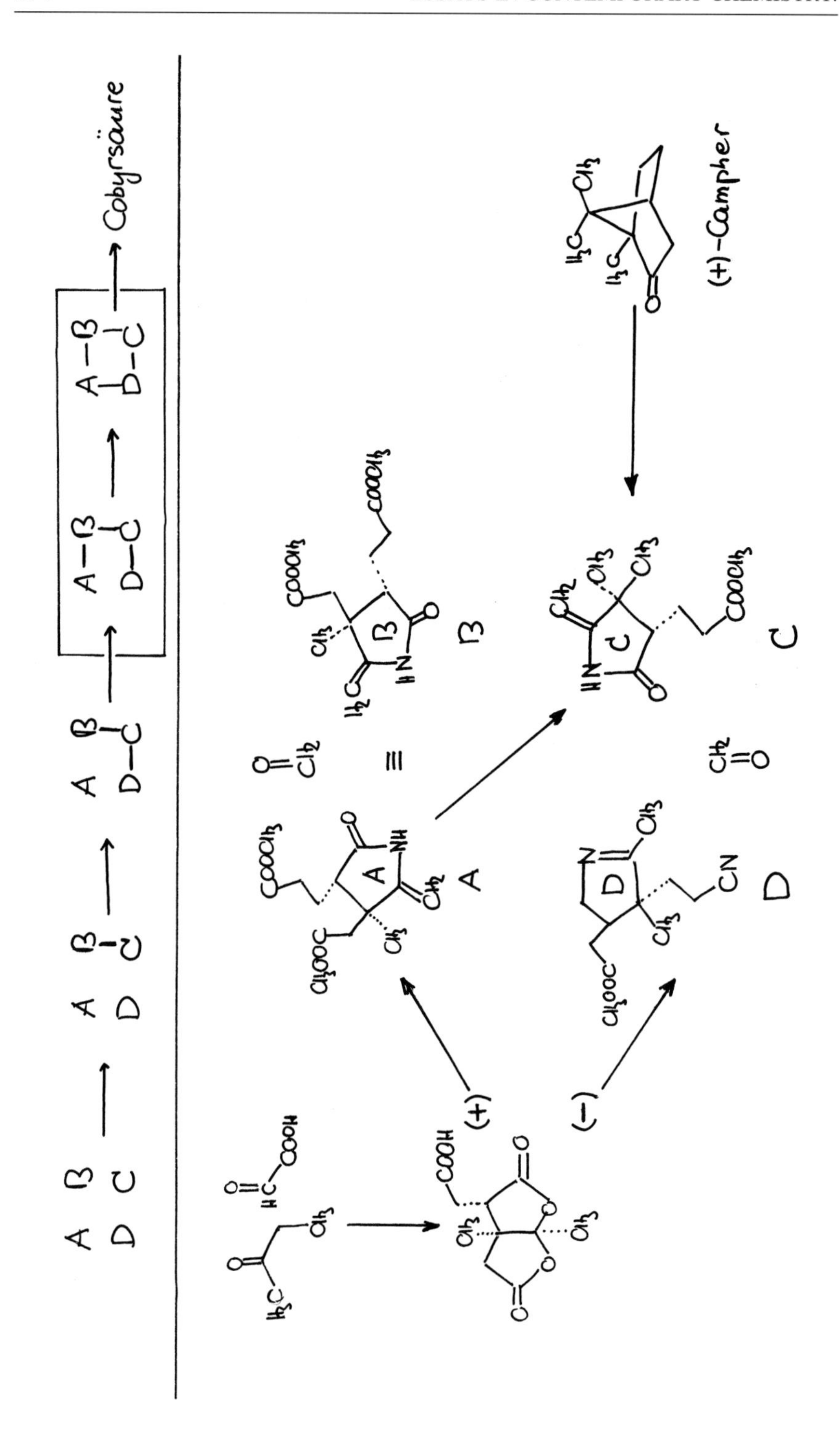

Cobyrsäure
(+)-Campher
(+)
(−)
A
B
C
D

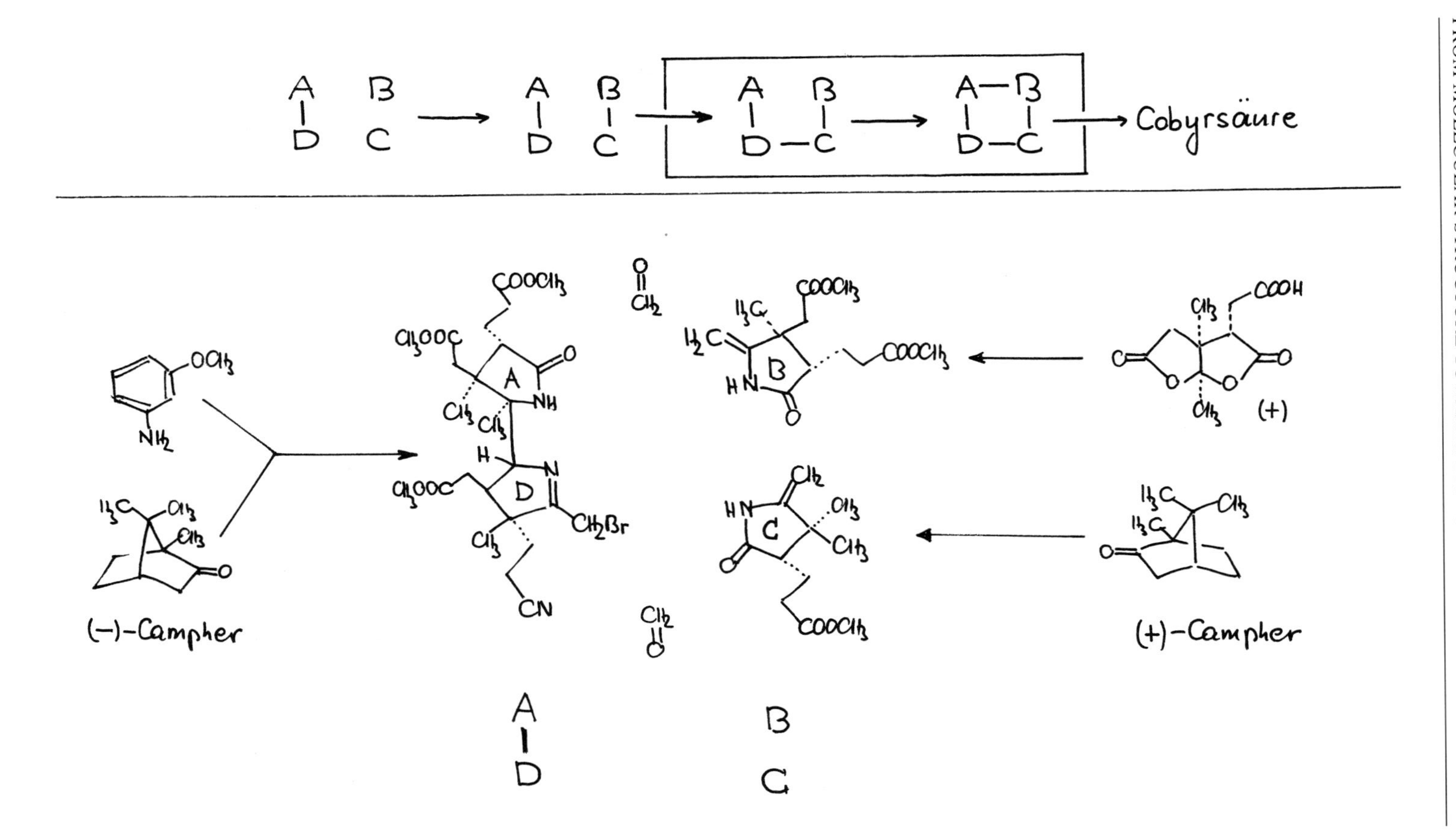
Cobyrsäure
COOCH3
COOH
(+)
CH2Br
CN
(−)-Campher
(+)-Campher

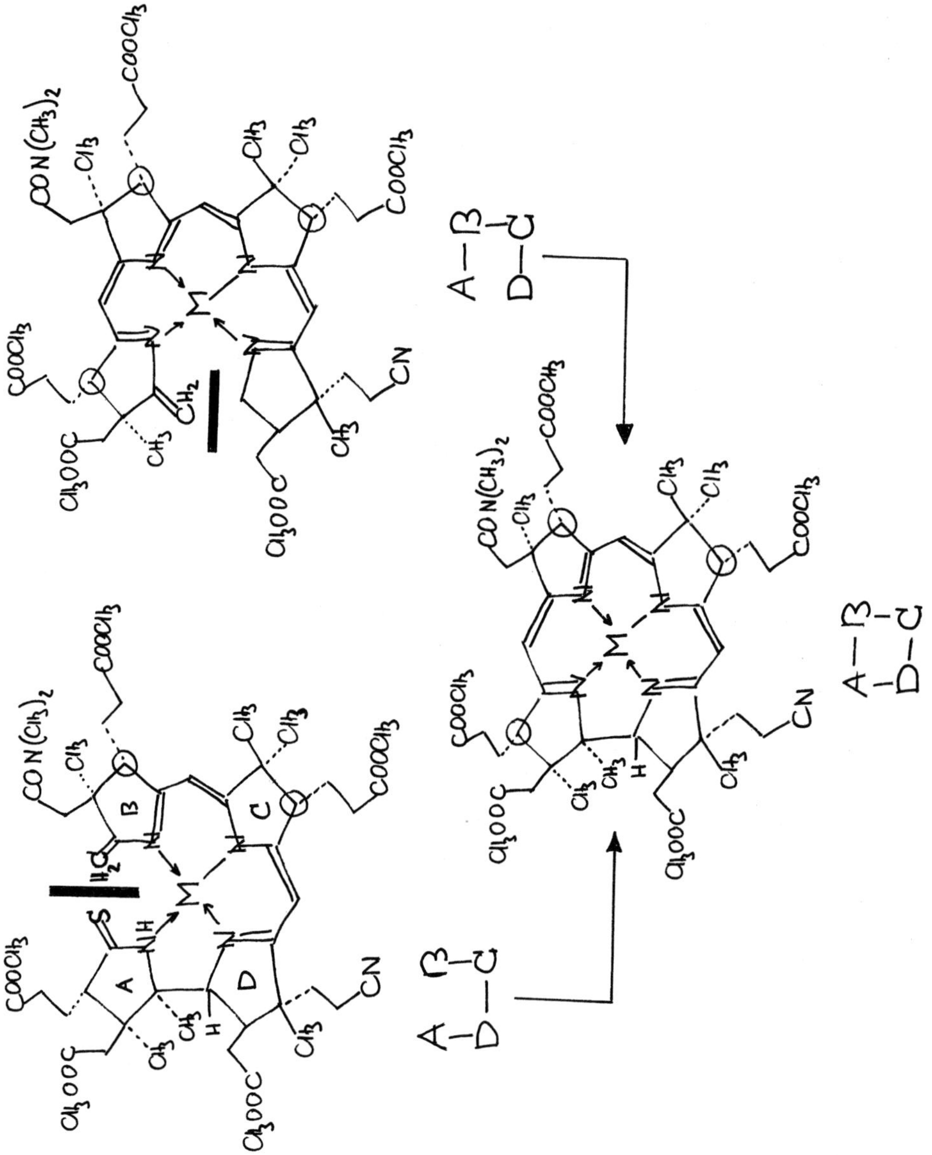

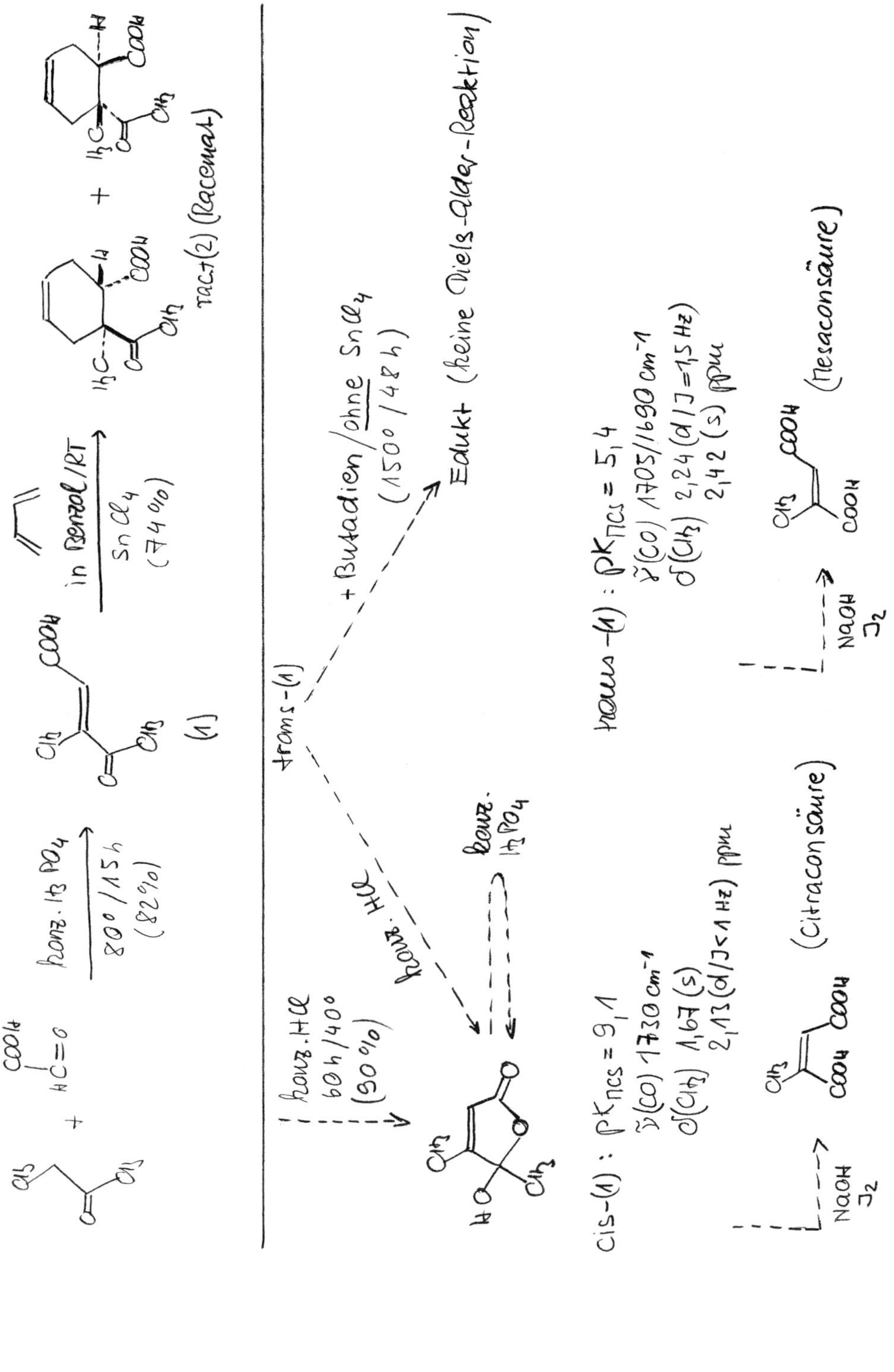

CH_3COCH_3 + $HC(=O)COOH$ —(konz. H_3PO_4, 80°/15 h, (82%))→ (1) —(Butadien, in Benzol/RT, $SnCl_4$, (74%))→ rac-(2) (Racemat)

trans-(1)

konz. HCl, 60 h/40°, (90%)

konz. HCl

konz. H_3PO_4

+ Butadien/ohne $SnCl_4$ (150°/48 h) → Edukt (keine Diels-Alder-Reaktion)

cis-(1) : pK_{MCS} = 9,1
$\tilde{\nu}$(CO) 1730 cm^{-1}
$\delta(CH_3)$ 1,67 (s)
2,13 (d/J<1 Hz) ppm

trans-(1) : pK_{MCS} = 5,4
$\tilde{\nu}$(CO) 1705/1690 cm^{-1}
$\delta(CH_3)$ 2,24 (d/J=1,5 Hz)
2,42 (s) ppm

NaOH, I_2 → (Citraconsäure)

NaOH, I_2 → (Mesaconsäure)

Trennung des Racemats rac.-(2) in die Enantiomeren:

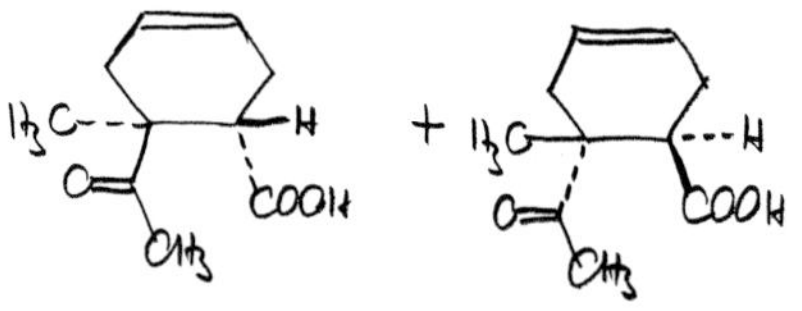

Racemat rac.-(2): Smp. 105°
$[\alpha]_D = 0°$

(−)-α-Phenyl-äthylamin in $CHCl_3$/Hexan

↓

Schwerer lösliches Salz
Smp. 147°; $[\alpha]_D = +49°$

verd. HCl →

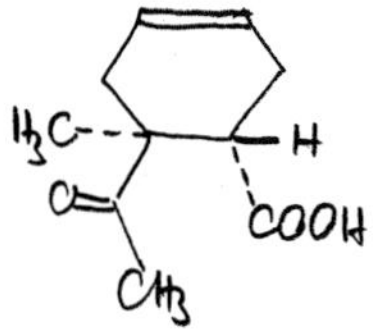

(+)-(2): Smp. 52°
$[\alpha]_D = +89°$

+ "Mutterlauge"

↓

Spaltung mit verd. HCl und Umsetzung der Säure mit (+)-α-Phenyl-äthylamin

↓

Schwerer lösliches Salz
Smp. 147°; $[\alpha]_D = -49°$

verd. HCl →

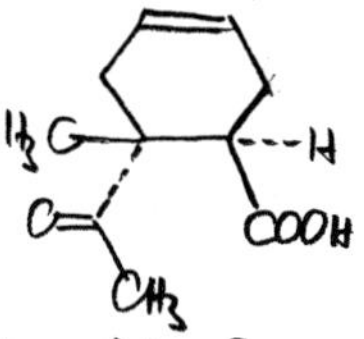

(−)-(2): Smp. 52°
$[\alpha]_D = -89°$

(Chiralitäts-Zuordnung vorweggenommen)

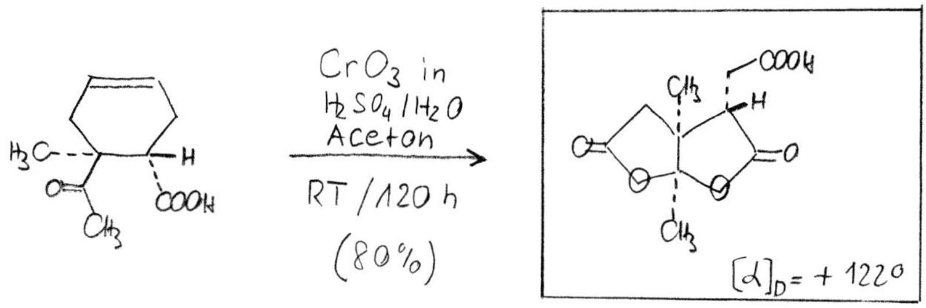

CrO_3 in H_2SO_4/H_2O Aceton

RT / 120 h

(80%)

$[\alpha]_D = +122°$

(+)-(2) (+)-(3)

C_{10}-Dilacton-carbonsäure)
Ausgangsprodukt für
B_{12}-Ringvorläufer A, B, C

(–)-(2) ⟶ (–)-(3)

Ausgangsprodukt für
B_{12}-Ringvorläufer D

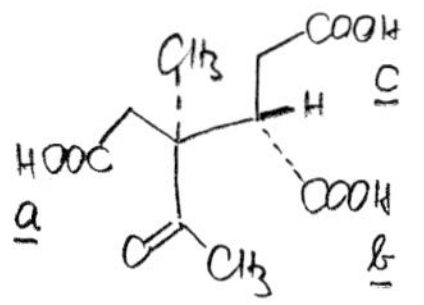

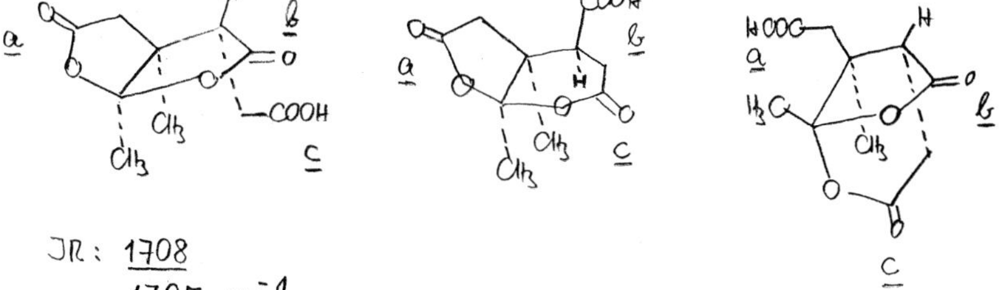

JR: 1708
1795 cm^{-1}

(Beweis: vgl. unabhängige Synthese
der C_{11}-Dilacton-carbonsäure (5))

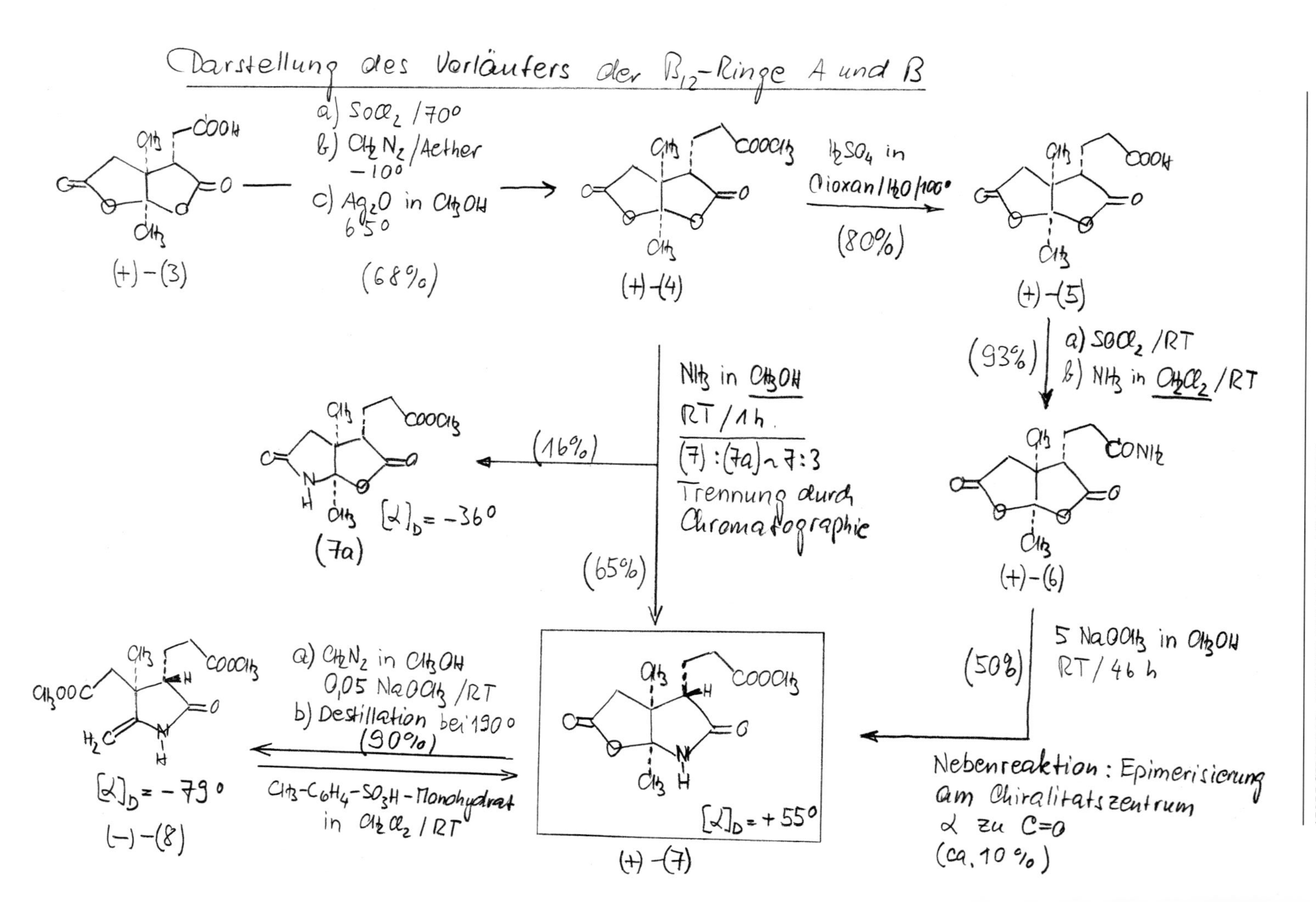
Darstellung des Vorläufers der B12-Ringe A und B
(+)-(3)
a) SOCl2 /70°
b) CH2N2/Aether −10°
c) Ag2O in CH3OH 65°
(68%)
(+)-(4)
H2SO4 in Dioxan/H2O/100°
(80%)
(+)-(5)
a) SOCl2 /RT
b) NH3 in CH2Cl2 /RT
(93%)
CONH2
(+)-(6)
5 NaOCH3 in CH3OH
RT / 46 h
(50%)
Nebenreaktion: Epimerisierung
am Chiralitätszentrum
α zu C=O
(ca. 10%)
NH3 in CH3OH
RT / 1 h
(7) : (7a) ~ 7:3
Trennung durch
Chromatographie
(16%)
(65%)
[α]D = −36°
(7a)
[α]D = +55°
(+)-(7)
a) CH2N2 in CH3OH
0,05 NaOCH3 /RT
b) Destillation bei 190°
(90%)
CH3-C6H4-SO3H-Monohydrat
in CH2Cl2 /RT
[α]D = −79°
(−)-(8)

Kommentare zur Reaktionsfolge (+)-(3)⟶(-)-(8)

(3)⟶(4):

= Arndt-Eistert Reaktion (Kettenverlängerung einer Carbonsäure um eine CH_2-Gruppe)

Diazo-keton

Wolff'sche Umlagerung

Keto-carben, vermutlich mit Ag^+ komplexiert

- Behandlung des Säurechlorids mit CH_3OH/Pyridin bei RT gibt den gleichen Methylester von (3), wie er aus (3) durch Veresterung mit Diazomethan erhalten wird; d.h. bei der Darstellung des Säurechlorids findet keine andere Strukturänderung (z.B. Epimerisierung) statt.

(4)⟶(5):

Zweifellos wird unter diesen Reaktionsbedingungen auch die Dilactongruppierung angegriffen, aber offenbar reversibel. Die Veresterung von (5) gibt den Methylester (4) zurück, d.h. es findet keine Epimerisierung am potentiell labilen Chiralitätszentrum statt.

(6)⟶(7):

Hypothetischer Mechanismus der intramolekularen Amidstickstoff-Verschiebung:

Methanolyse sterisch ungehindert

beim Ansäuern ⟶ (7)

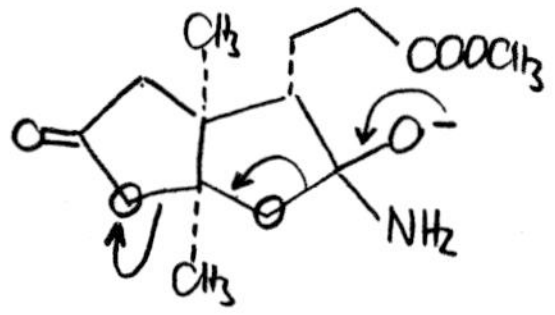

(4)→(7) + (7a): Reaktionstyp:

- Die Carbonylgruppen der Dilactongruppierung sind gegenüber NH_3 viel nukleophiler als normale Lacton- oder Estergruppen.
- "Ammonolysen" dieser Art verlaufen in CH_3OH viel rascher als in nicht-hydroxyl-haltigen Lösungsmitteln, wie z.B. CH_2Cl_2 (vgl. (5)→(6), Ammonolyse des Säurechlorids).
- Die Konstitutionszuordnung für die isomeren Lacton-lactame (7) und (7a) ergab sich aus der Identität von (7) mit dem Hauptprodukt der intramolekularen NH-Uebertragung.
- Folgendes Experiment beweist, dass das potentiell labile Chiralitätszentrum in (7) nicht epimerisiert ist:

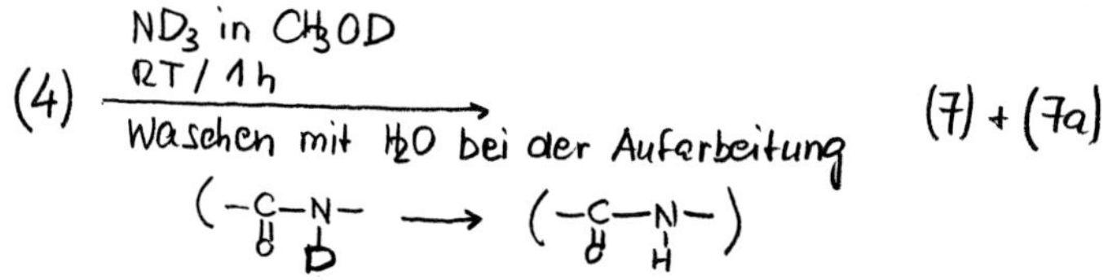

(4) —[ND_3 in CH_3OD, RT / 1 h; Waschen mit H_2O bei der Aufarbeitung]→ (7) + (7a)

(–C(=O)–N(D)–) → (–C(=O)–N(H)–)

Wenn unter diesen Reaktionsbedingungen eine Epimerisierung des in Frage stehenden Chiralitätszentrums stattgefunden hätte, müssten die Lacton-lactame an diesem Zentrum ein Deuterium aufweisen. Experimentell: sowohl nach NMR, als auch nach MS praktisch kein Einbau von Deuterium nachweisbar.
- Trotz der Notwendigkeit der chromatographischen Trennung von (7) und (7a) ist für die Produktion von (7) die direkte Ammonolyse von (4) einfacher als (4)→(5)→(6)→(7). ((7a) kann übrigens nachträglich durch Behandlung mit ver. Säure mit (7) äquilibriert werden).

(7)→(8):

Normalerweise werden bei der Veresterung mit Diazomethan die freien Carbonsäuren mit einer ätherischen Lösung von CH_2N_2 umgesetzt:

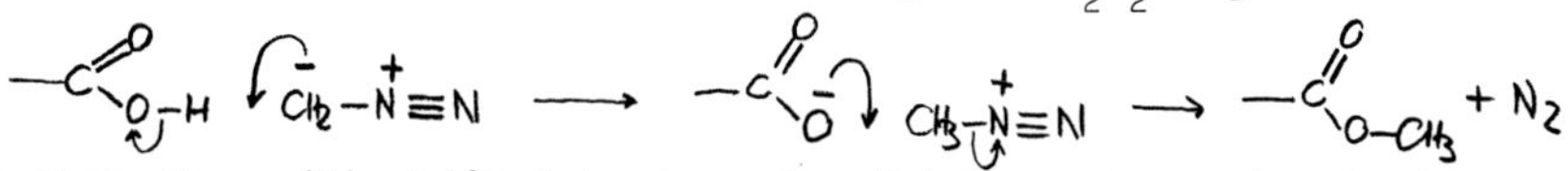

$$-C(=O)-O-H + \bar{C}H_2-\overset{+}{N}\equiv N \longrightarrow -C(=O)-\bar{O} \; CH_3-\overset{+}{N}\equiv N \longrightarrow -C(=O)-O-CH_3 + N_2$$

Die Umwandlung (7)→(8) ist eines der (bisher seltenen) Beispiele einer Veresterung in basischem Medium:

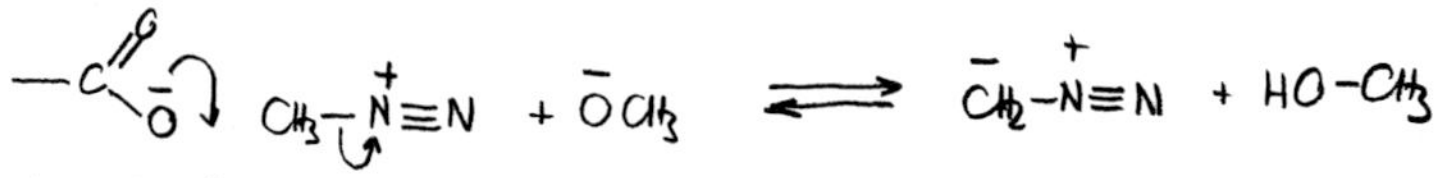

$$-C(=O)-\bar{O} \; CH_3-\overset{+}{N}\equiv N + \bar{O}CH_3 \rightleftharpoons \bar{C}H_2-\overset{+}{N}\equiv N + HO-CH_3$$

Die katalytische Menge $NaOCH_3$ dient der Einstellung des folgenden Gleichgewichtes:

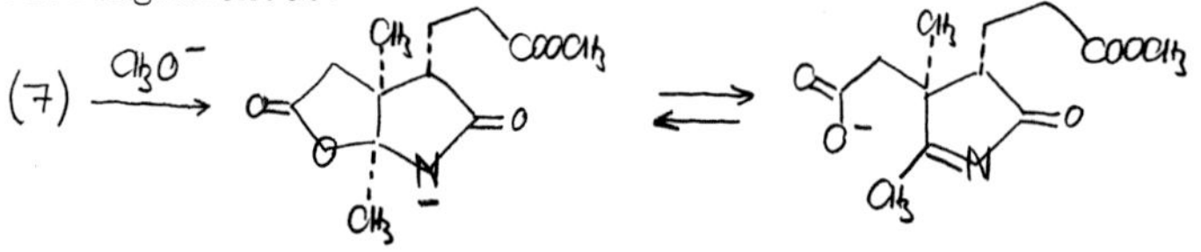

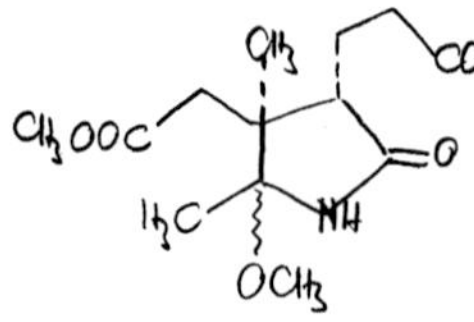

(7) —CH_3O^-→ [Strukturformeln]

(8a)

Primär wird das Methanol-Additionsprodukt (8a) isoliert; dieses liefert bei der Destillation unter Abspaltung von CH_3OH das Enamid-Derivat (8).

- Erfahrungsgemäss liegen Tautomerie-Gleichgewichte des folgenden Typs vollständig auf der Seite der Enamidform

"isoliertes", nicht-bindendes sp^2-Elektronenpaar

in π-Delokalisation involviertes p-Elektronenpaar (vgl. hohe Stabilisierungsenergie von Amidgruppen)

Trägt der Stickstoff jedoch keine Acyl-, sondern eine Alkylgruppe, so liegt das Gleichgewicht auf der Seite der Ketiminform (vgl. Ring D Vorläufer).

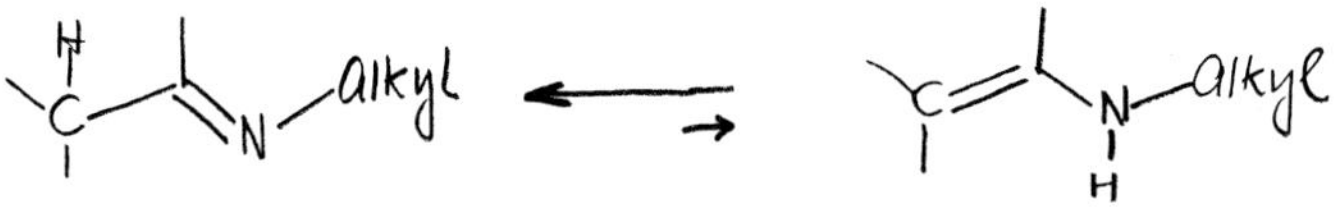

alkyl alkyl

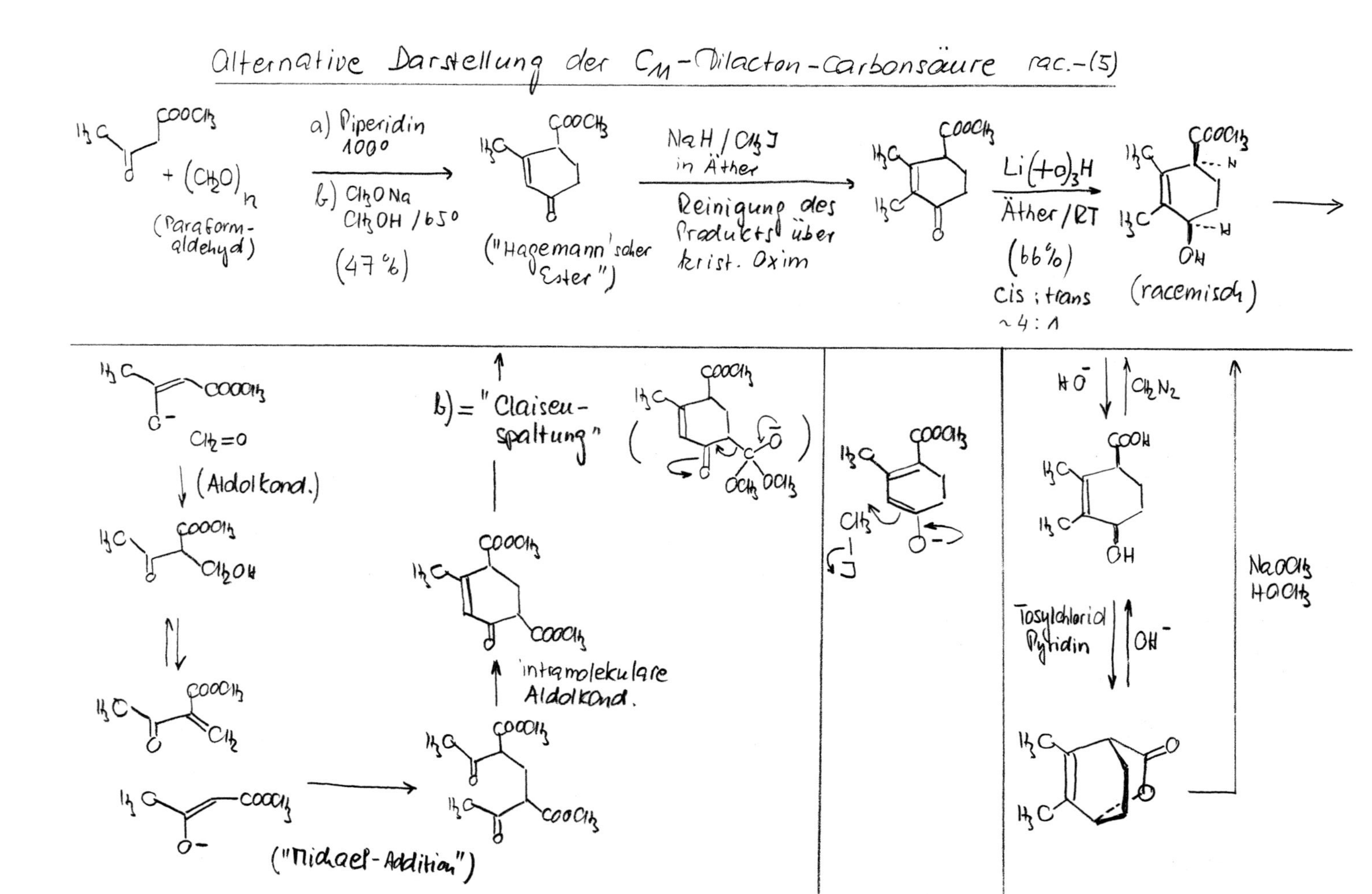
Alternative Darstellung der C11-Dilacton-Carbonsäure rac.-(5)
a) Piperidin 100°
b) CH3ONa CH3OH / 65°
(47 %)
(Paraform-aldehyd)
("Hagemann'scher Ester")
NaH / CH3J in Äther
Reinigung des Produkts über krist. Oxim
Li(+O)3H Äther/RT
(66%)
cis : trans ~4 : 1
(racemisch)
b) = "Claisen-spaltung"
(Aldolkond.)
intramolekulare Aldolkond.
("Michael-Addition")
Tosylchlorid Pyridin
NaOCH3 HOCH3

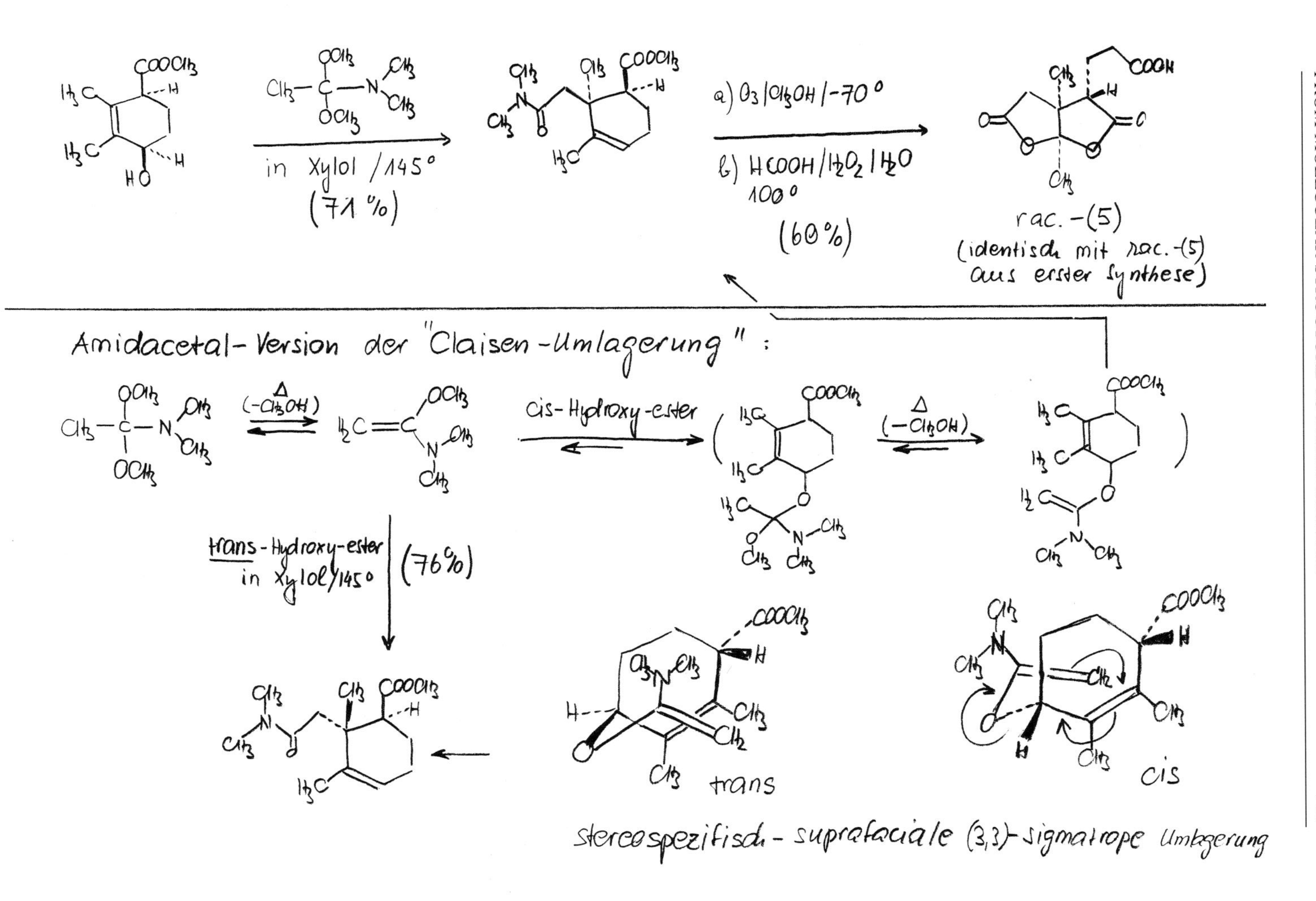
in Xylol / 145°
(71 %)
a) O3/CH3OH/-70°
b) HCOOH/H2O2/H2O
100°
(60 %)
rac.-(5)
(identisch mit rac.-(5)
aus erster Synthese)
Amidacetal-Version der "Claisen-Umlagerung":
cis-Hydroxy-ester
trans-Hydroxy-ester
in Xylol/145°
(76 %)
trans
cis
stereospezifisch - suprafaciale (3,3)-sigmatrope Umlagerung

Darstellung des Vorläufers von B₁₂-Ring D

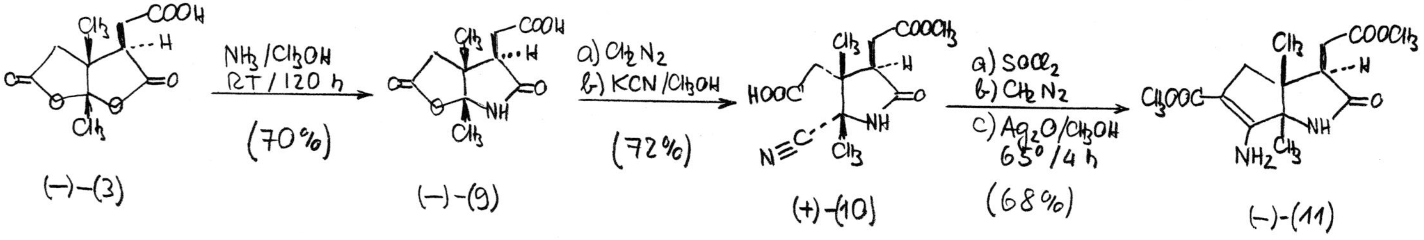

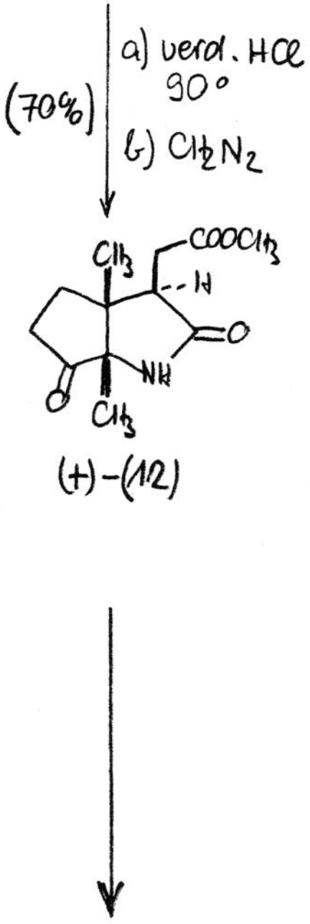

Kommentare:

(3) ⟶ (9): Nebenprodukt:
(Isomerenverhältnis ~ 9 : 1)

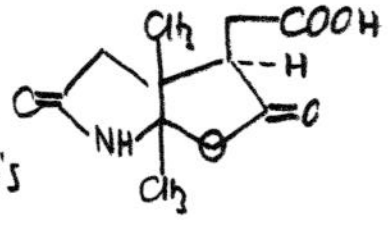

Strukturbeweis für Hauptprod. (−)-(9):
(−)-(9) ⟶ Arndt-Eistert ⟶ (−)-(7)

(9) ⟶ (10): Nebenprodukt:
(Isomerenverhältnis ~ 4 : 1)

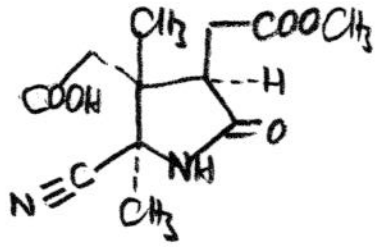

Konfiguration des Hauptprod. (+)-(10):
aus Reaktion (10) ⟶ (11) ersichtlich

(10) ⟶ (11): "Fehlgeschlagene" Arndt-Eistert-Reaktion!
Vermutlich:

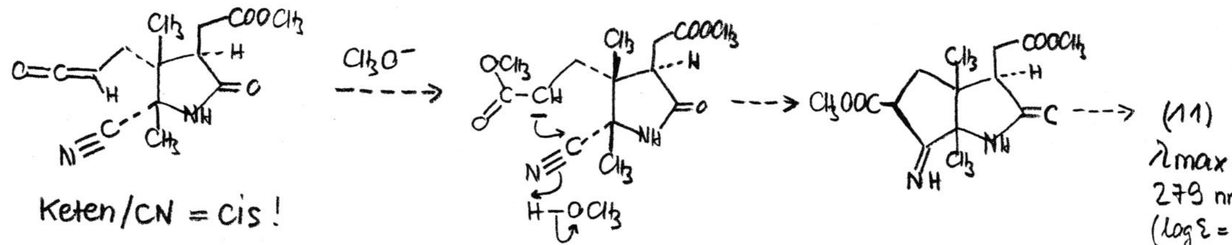

Keten/CN = cis!

λmax = 279 nm
(log ε = 4,2)!

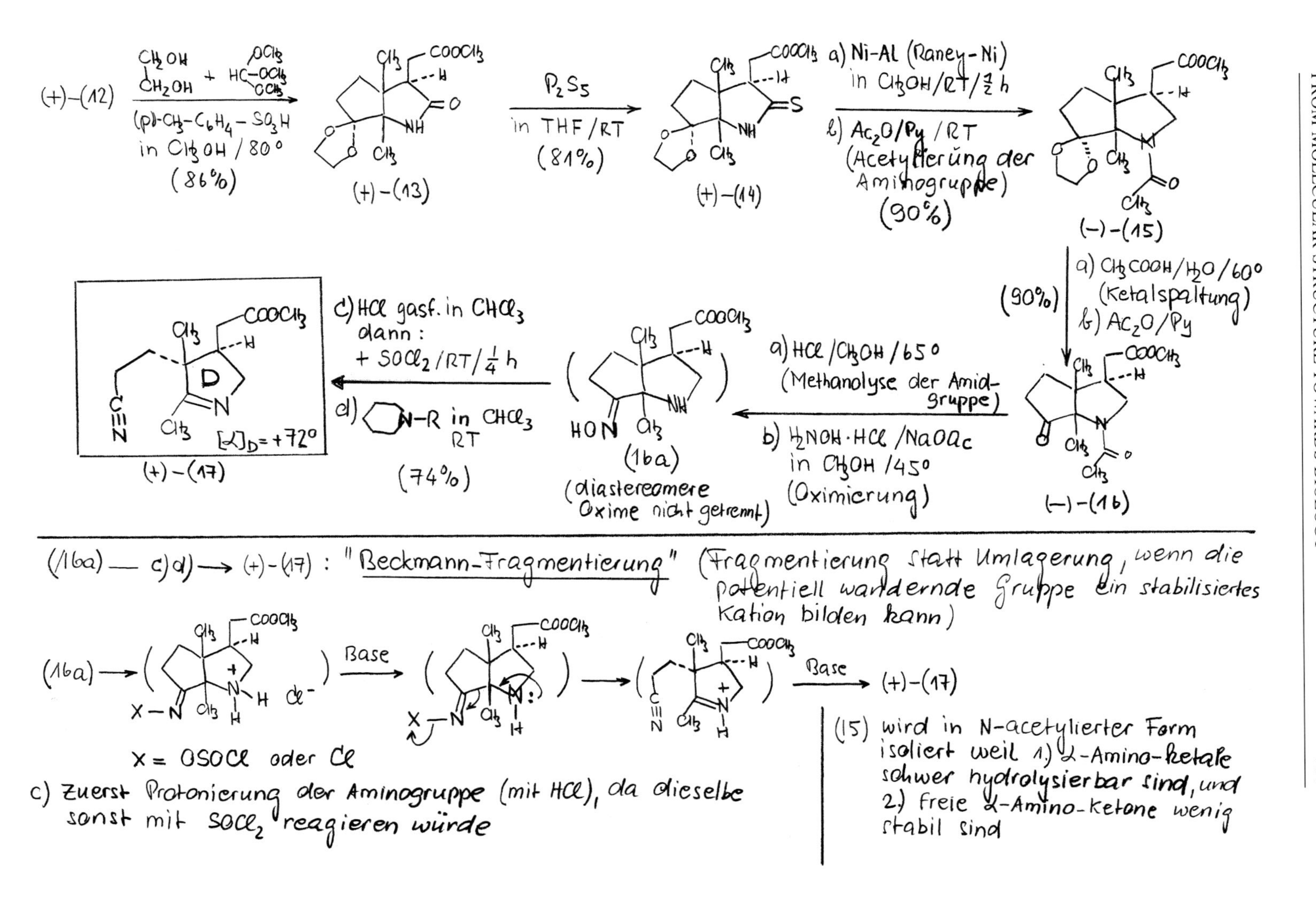
(+)-(12)
(p)-CH3-C6H4-SO3H
in CH3OH / 80°
(86%)
(+)-(13)
P2S5
in THF/RT
(81%)
(+)-(14)
a) Ni-Al (Raney-Ni)
in CH3OH/RT/½ h
b) Ac2O/Py/RT
(Acetylierung der Aminogruppe)
(90%)
(–)-(15)
a) CH3COOH/H2O/60°
(Ketalspaltung)
b) Ac2O/Py
(90%)
(–)-(16)
a) HCl/CH3OH/65°
(Methanolyse der Amid-gruppe)
b) H2NOH·HCl/NaOAc
in CH3OH/45°
(Oximierung)
(16a)
(diastereomere Oxime nicht getrennt)
c) HCl gasf. in CHCl3
dann:
+ SOCl2/RT/¼ h
d) N-R in CHCl3
RT
(74%)
[α]D = +72°
(+)-(17)
(16a) — c) d) → (+)-(17) : "Beckmann-Fragmentierung" (Fragmentierung statt Umlagerung, wenn die potentiell wandernde Gruppe ein stabilisiertes Kation bilden kann)
Base
Base
X = OSOCl oder Cl
c) Zuerst Protonierung der Aminogruppe (mit HCl), da dieselbe sonst mit SOCl2 reagieren würde
(15) wird in N-acetylierter Form isoliert weil 1.) α-Amino-Ketale schwer hydrolysierbar sind, und 2.) Freie α-Amino-Ketone wenig stabil sind

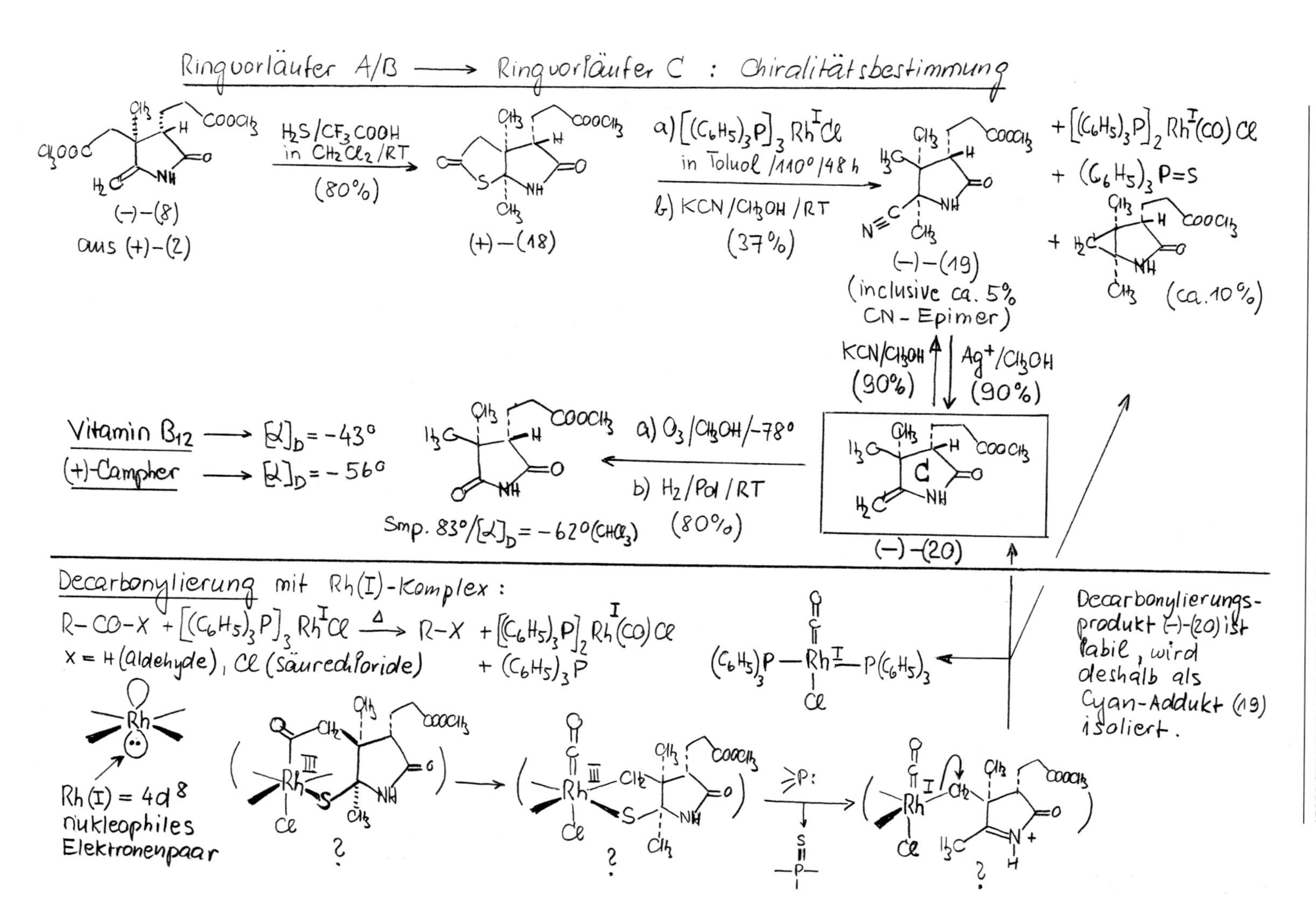
Ringvorläufer A/B ⟶ Ringvorläufer C : Chiralitätsbestimmung
(–)-(8)
aus (+)-(2)
H2S/CF3COOH
in CH2Cl2/RT
(80%)
(+)-(18)
a) [(C6H5)3P]3 RhICl
in Toluol /110°/48 h
b) KCN/CH3OH/RT
(37%)
(–)-(19)
(inclusive ca. 5% CN-Epimer)
+ [(C6H5)3P]2 RhI(CO)Cl
+ (C6H5)3 P=S
(ca. 10%)
KCN/CH3OH (90%)
Ag+/CH3OH (90%)
Vitamin B12 ⟶ [α]D = –43°
(+)-Campher ⟶ [α]D = –56°
a) O3/CH3OH/–78°
b) H2/Pd/RT
(80%)
Smp. 83°/[α]D = –62°(CHCl3)
(–)-(20)
Decarbonylierung mit Rh(I)-Komplex :
R–CO–X + [(C6H5)3P]3 RhICl ⟶ R–X + [(C6H5)3P]2 RhI(CO)Cl
X = H (Aldehyde), Cl (Säurechloride) + (C6H5)3P
Decarbonylierungs-produkt (–)-(20) ist labil, wird deshalb als Cyan-Addukt (19) isoliert.
Rh(I) = 4d8
nukleophiles Elektronenpaar

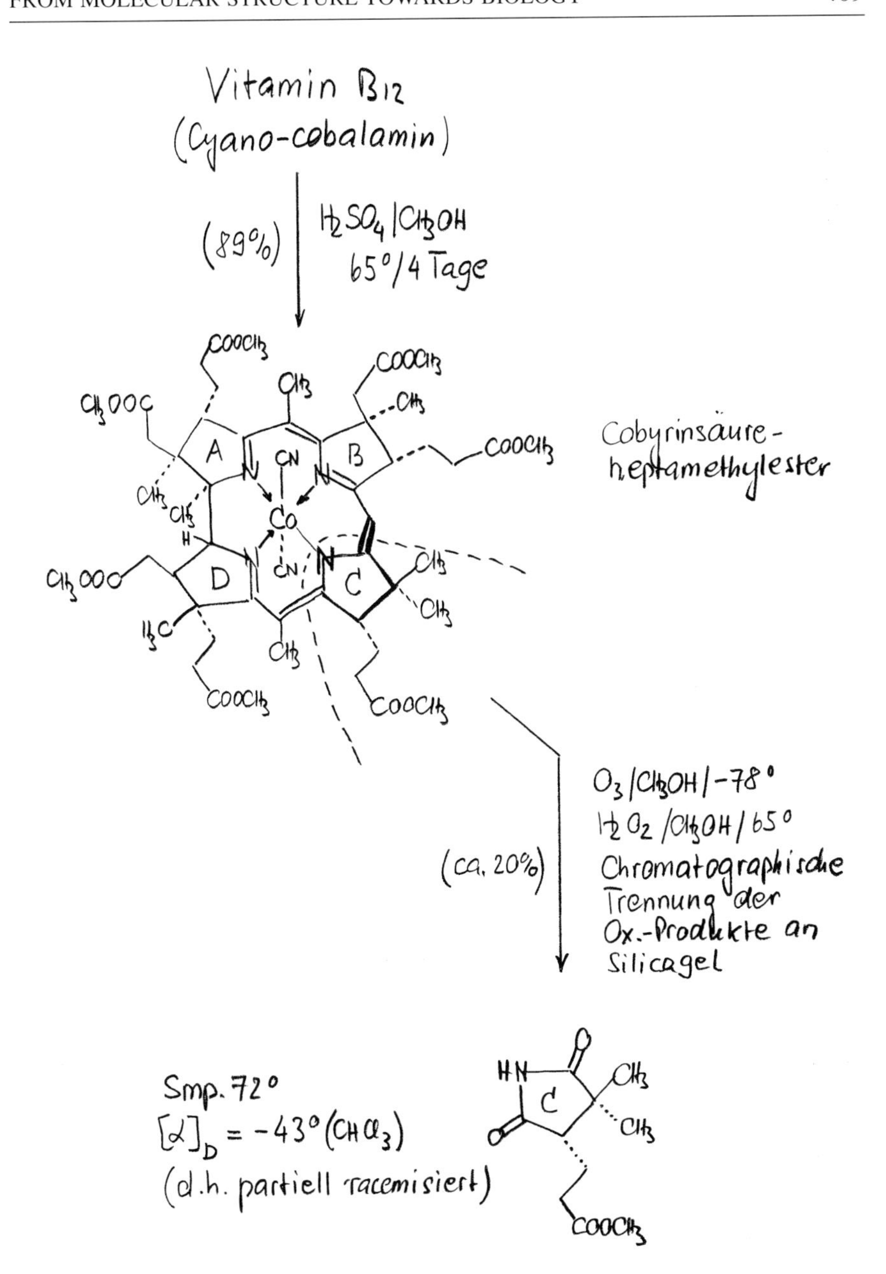
Vitamin B12
(Cyano-cobalamin)
(89%)
H2SO4/CH3OH
65°/4 Tage
COOCH3
CH3OOC
CH3
A
B
CN
Co
CN
D
C
H
CH3
COOCH3
Cobyrinsäure-heptamethylester
O3/CH3OH/−78°
H2O2/CH3OH/65°
Chromatographische Trennung der Ox.-Produkte an Silicagel
(ca. 20%)
Smp. 72°
[α]D = −43° (CHCl3)
(d.h. partiell racemisiert)
HN
O
C
CH3
COOCH3

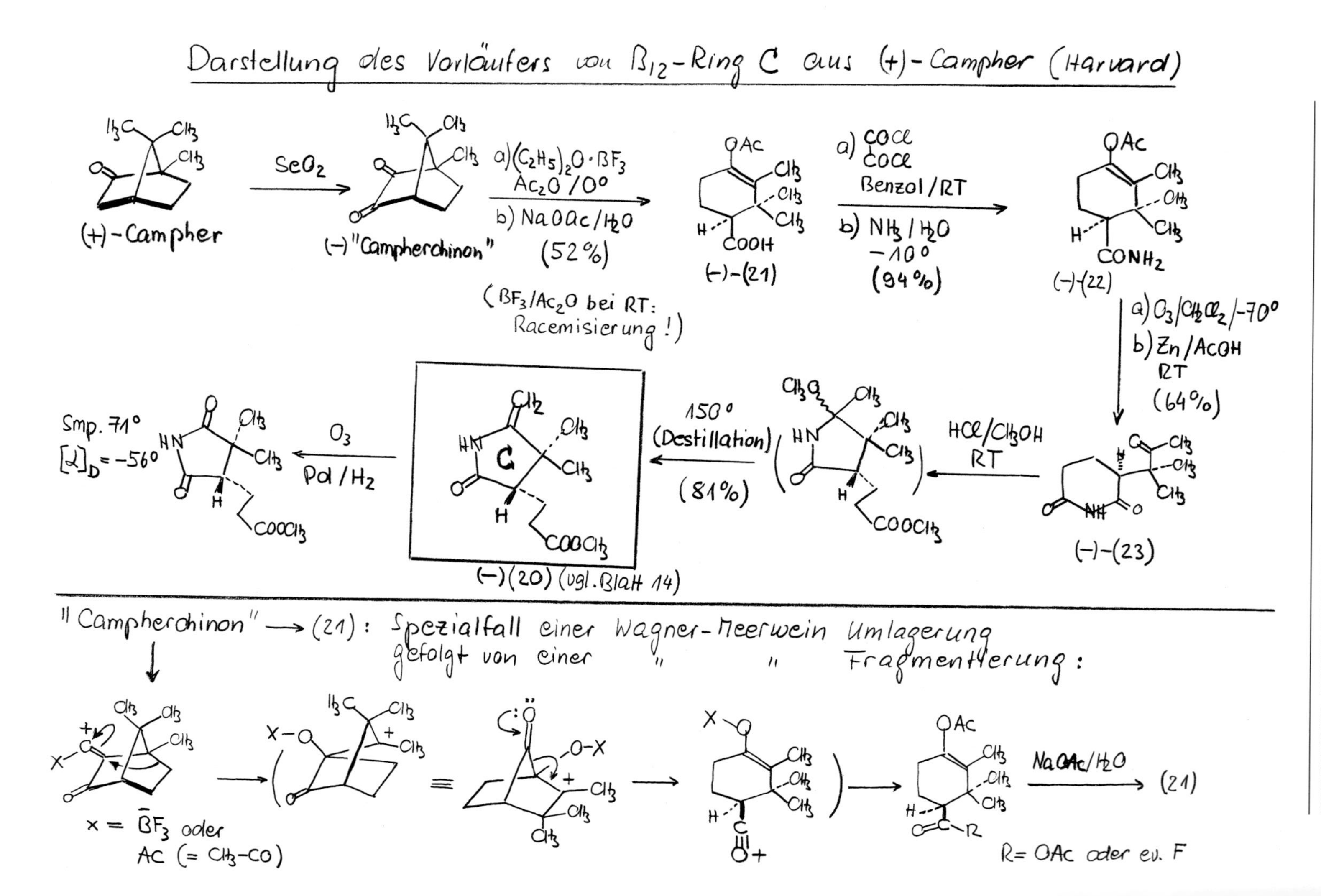
Darstellung des "Vorläufers" von B12-Ring C aus (+)-Campher (Harvard)
(+)-Campher
SeO2
(-)-"Campherchinon"
a) (C2H5)2O·BF3 Ac2O / 0°
b) NaOAc/H2O
(52%)
(BF3/Ac2O bei RT: Racemisierung!)
(-)-(21)
a) COCl COCl Benzol/RT
b) NH3/H2O −10°
(94%)
(-)-(22)
a) O3/CH2Cl2/−70°
b) Zn/AcOH RT
(64%)
(-)-(23)
HCl/CH3OH RT
150° (Destillation)
(81%)
(-)(20) (vgl. Blatt 14)
O3
Pd/H2
Smp. 71°
[α]D = −56°
"Campherchinon" → (21): Spezialfall einer Wagner-Meerwein Umlagerung
gefolgt von einer " " Fragmentierung:
X = BF3 oder Ac (= CH3-CO)
NaOAc/H2O
R = OAc oder ev. F

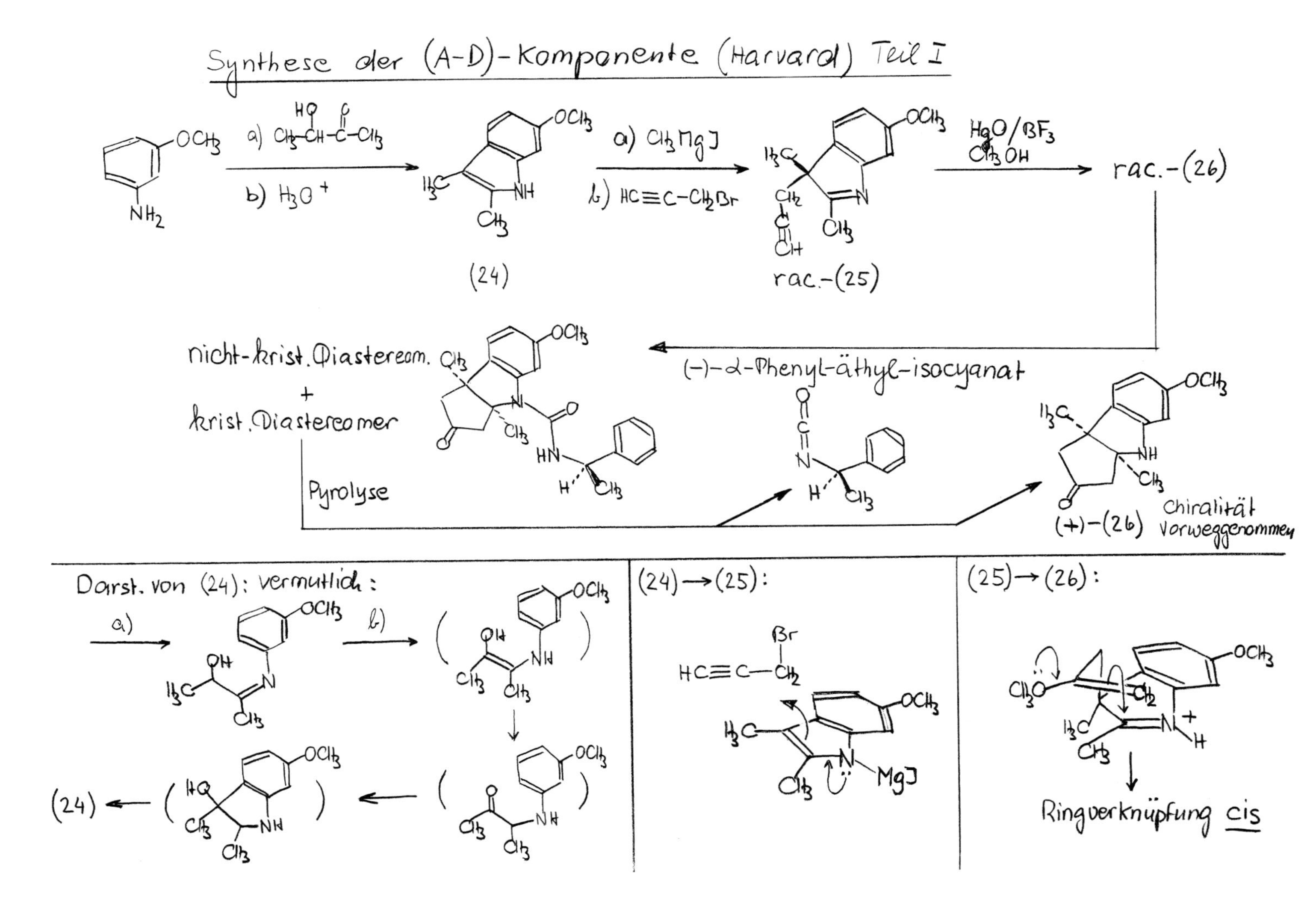
Synthese der (A-D)-Komponente (Harvard) Teil I
a) CH3-CH(OH)-C(=O)-CH3
b) H3O+
(24)
a) CH3MgJ
b) HC≡C-CH2Br
rac.-(25)
HgO/BF3
CH3OH
rac.-(26)
nicht-krist. Diastereom.
+
krist. Diastereomer
(−)-α-Phenyl-äthyl-isocyanat
Pyrolyse
(+)-(26)
Chiralität vorweggenommen
Darst. von (24): vermutlich:
(24)→(25):
(25)→(26):
Ringverknüpfung cis

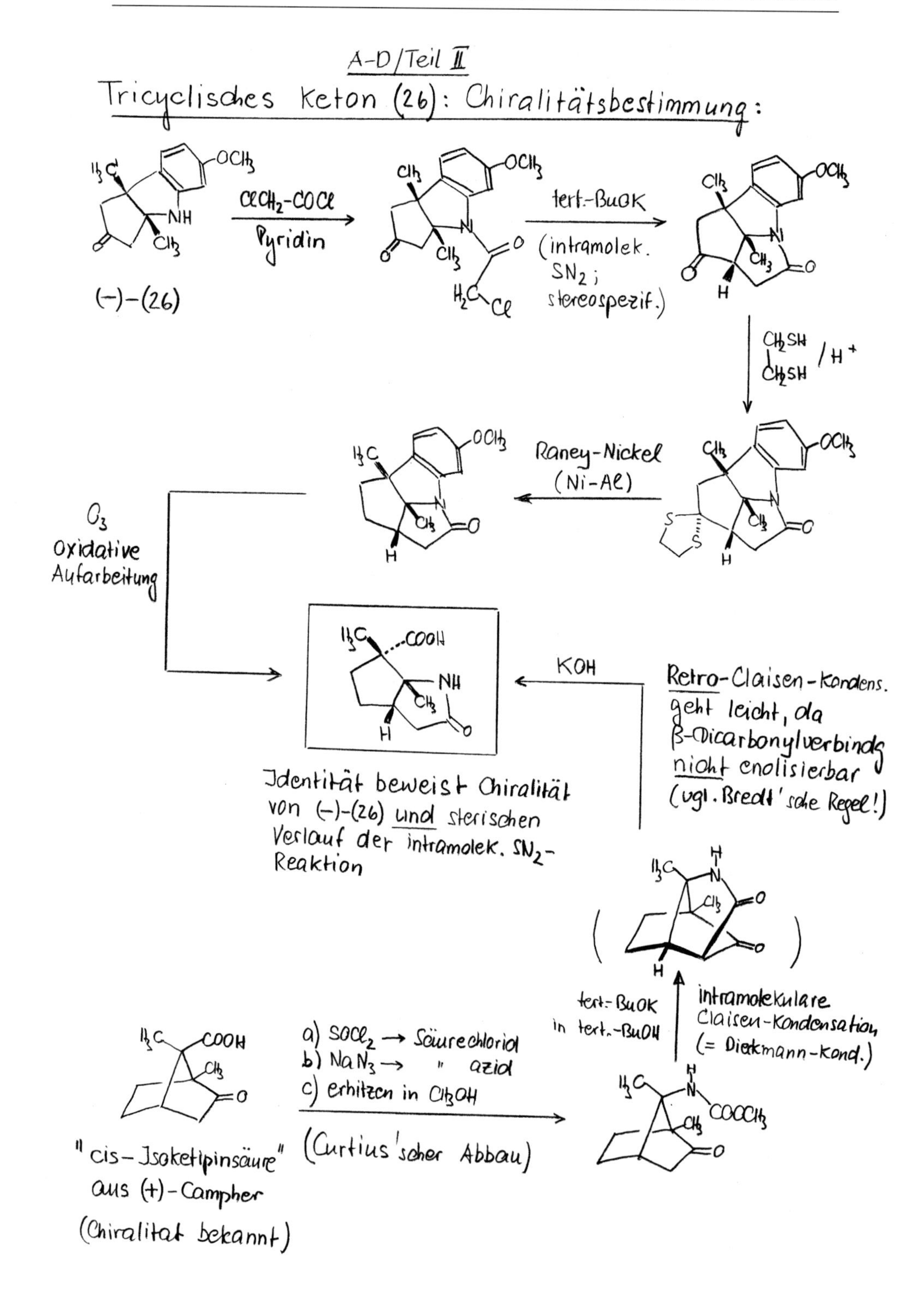
A-D/Teil II
Tricyclisches Keton (26): Chiralitätsbestimmung:
(–)-(26)
ClCH2-COCl
Pyridin
tert.-BuOK
(intramolek. SN2; stereospezif.)
CH2SH / CH2SH /H+
Raney-Nickel
(Ni-Al)
O3
oxidative Aufarbeitung
KOH
Retro-Claisen-Kondens. geht leicht, da β-Dicarbonylverbindg nicht enolisierbar (vgl. Bredt'sche Regel!)
Identität beweist Chiralität von (–)-(26) und sterischen Verlauf der intramolek. SN2-Reaktion
tert.-BuOK in tert.-BuOH
intramolekulare Claisen-Kondensation (= Dieckmann-Kond.)
a) SOCl2 → Säurechlorid
b) NaN3 → " azid
c) erhitzen in CH3OH
(Curtius'scher Abbau)
"cis-Isoketipinsäure" aus (+)-Campher
(Chiralität bekannt)

OC/B (1973)

Übung 4

Tricyclisches Keton (+)-(26): Chiralitätsbestimmung

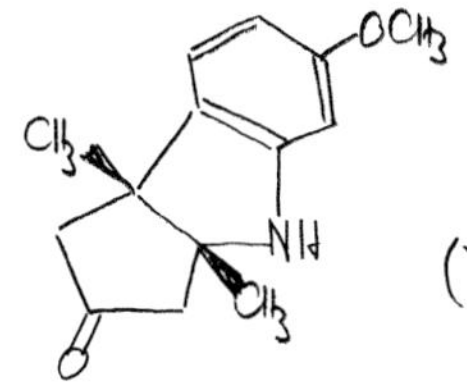

OCH_3 CH_3 NH CH_3 O

(–)-(26)

5 Stufen

(Tip: CH_2Cl–COCl !)

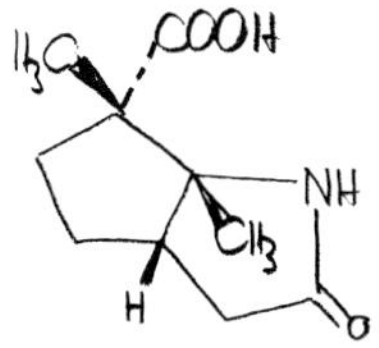

H_3C COOH NH CH_3 H O

identisch, inclusive opt. Drehung

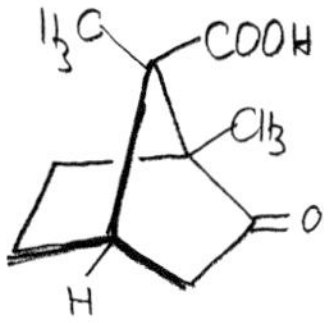

H_3C COOH CH_3 O H

"cis-Isoketipinsäure" aus (+)-Campher (Chiralitätssinn bekannt)

5 Stufen

(Tip: Retro-Claisen-Kondensation !)

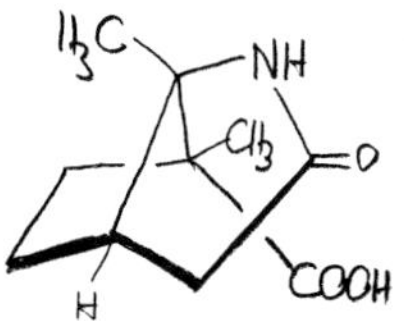

H_3C NH CH_3 O H COOH

deshalb:

(+)-(26) ≡ CH_3 OCH_3 NH CH_3 O

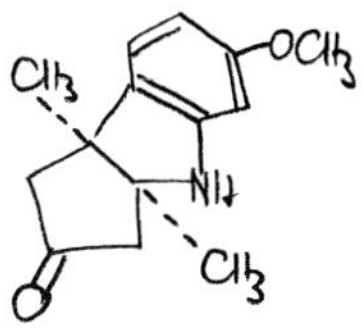

Rekonstruieren Sie Art und Reihenfolge der verwendeten Reaktionsstufen

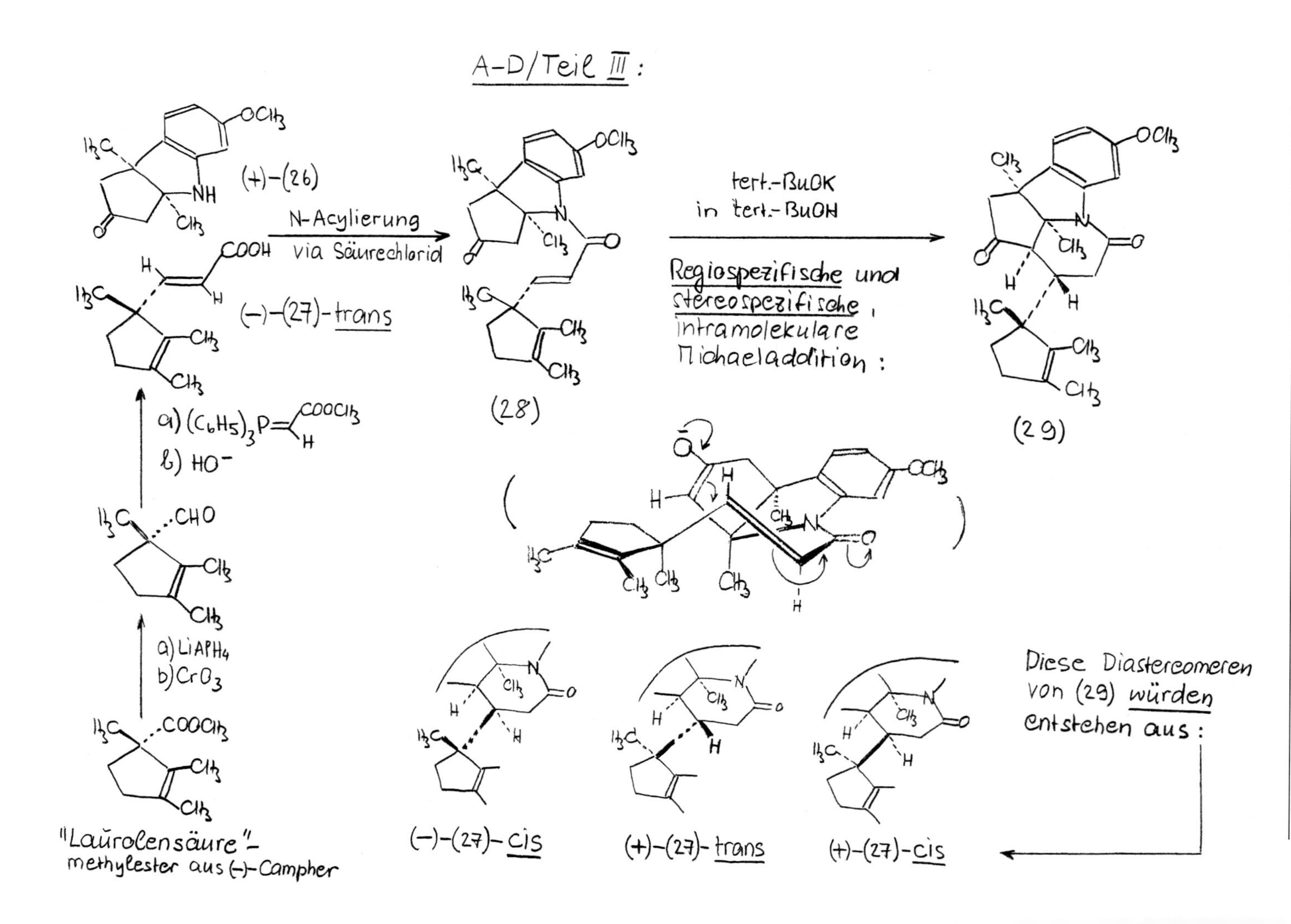
A–D/Teil III:
(+)-(26)
N-Acylierung
via Säurechlorid
(−)-(27)-trans
(28)
tert.-BuOK
in tert.-BuOH
Regiospezifische und
stereospezifische,
intramolekulare
Michaeladdition:
(29)
a) $(C_6H_5)_3P=C(COOCH_3)H$
b) HO^-
a) $LiAlH_4$
b) CrO_3
"Laurolensäure"-
methylester aus (−)-Campher
Diese Diastereomeren
von (29) würden
entstehen aus:
(−)-(27)-cis
(+)-(27)-trans
(+)-(27)-cis

Entdeckung des Prinzips der Erhaltung der Orbitalsymmetrie (R.B. Woodward, 1965)

Syntheseplan:

Ketalisierung

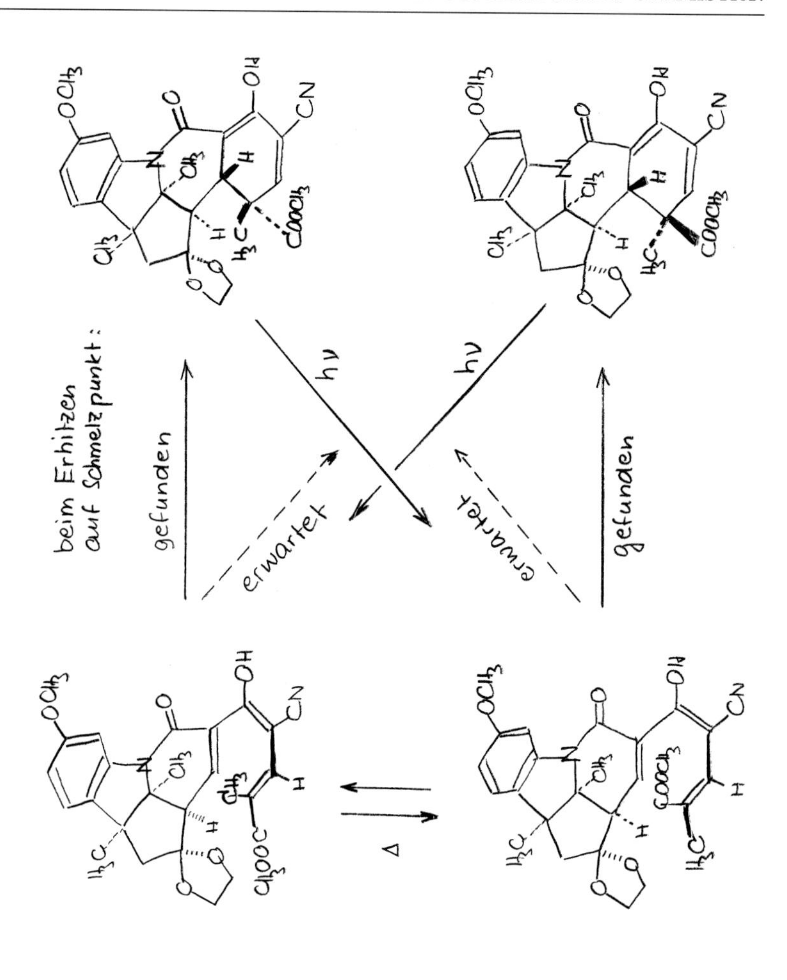
Experimentelle Feststellungen :
beim Erhitzen auf Schmelzpunkt :
gefunden
erwartet
hν
hν
erwartet
gefunden
Δ

Interpretation:

1) Ringschluss von unten sterisch günstiger

2) Zwei Möglichkeiten für Ringschluss-Konformation:

A

erwartet, da geringere π-Entkopplung als bei B

B

Konformation B entspricht der experimentell beobachteten Konfiguration des Cyclisationsproduktes, offenbar durch (unbekannten) Reaktivitätsfaktor begünstigt ⟶ Orbitalsymmetrie.

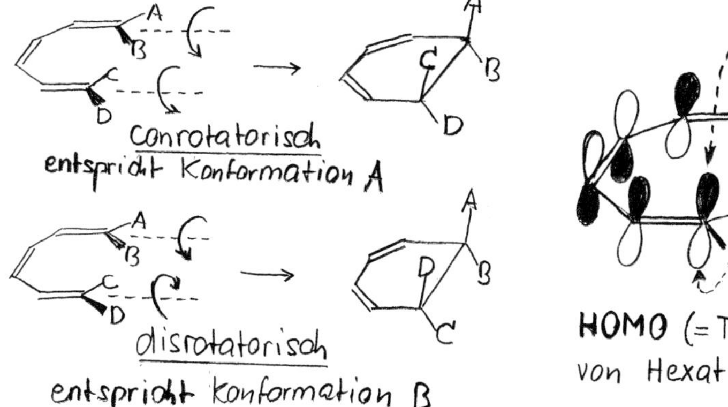

disrotatorisch

conrotatorisch

HOMO (= π_3) von Hexatrien

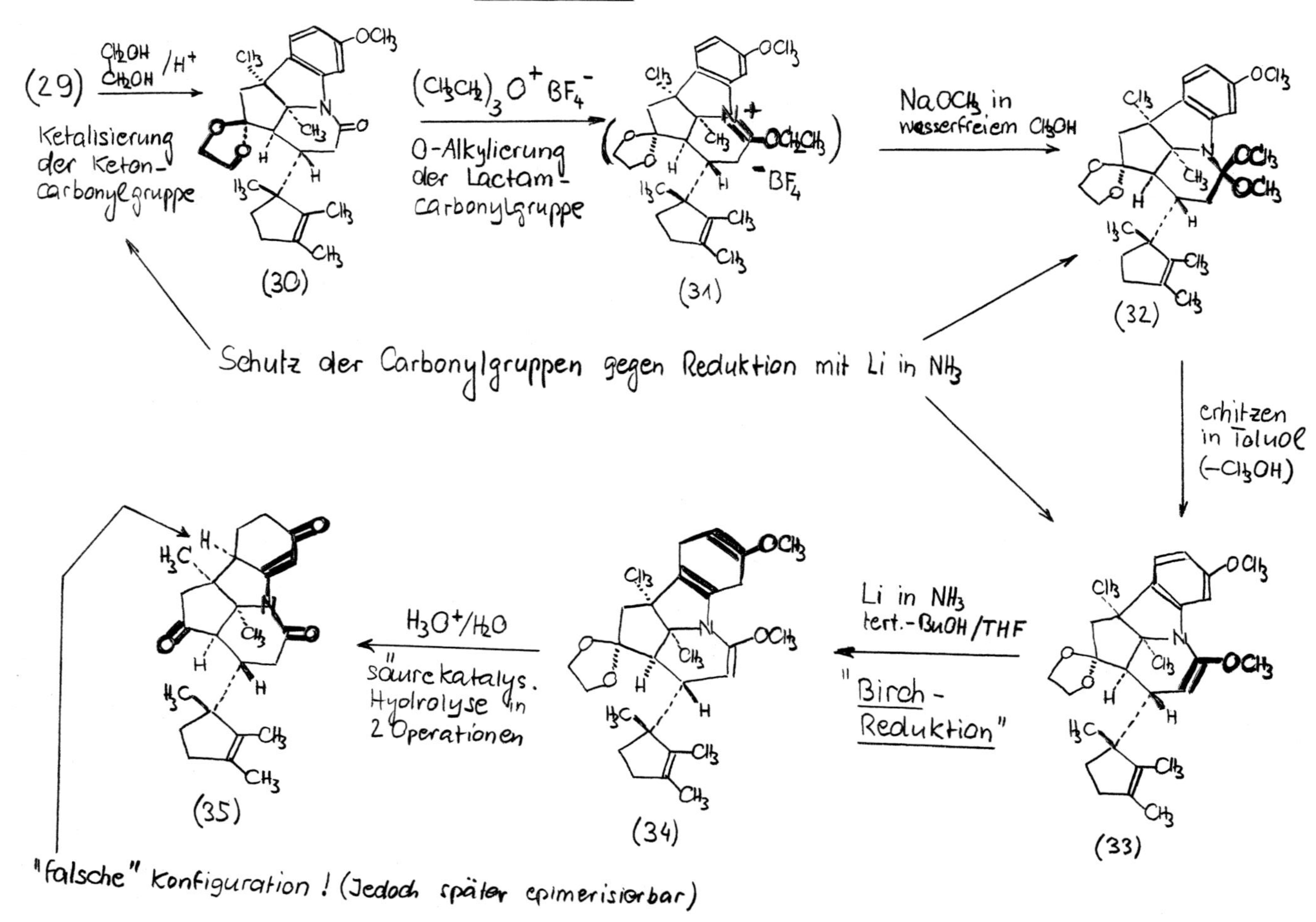
A-D/Teil IV: Birch-Reduktion
(29)
CH2OH / CH2OH /H+
Ketalisierung der Keton-carbonylgruppe
(30)
(CH3CH2)3O+ BF4-
O-Alkylierung der Lactam-Carbonylgruppe
(31)
NaOCH3 in wasserfreiem CH3OH
(32)
Schutz der Carbonylgruppen gegen Reduktion mit Li in NH3
erhitzen in Toluol (−CH3OH)
(33)
Li in NH3 tert.-BuOH/THF
"Birch-Reduktion"
(34)
H3O+/H2O
"saure katalys. Hydrolyse" in 2 Operationen
(35)
"falsche" Konfiguration! (Jedoch später epimerisierbar)

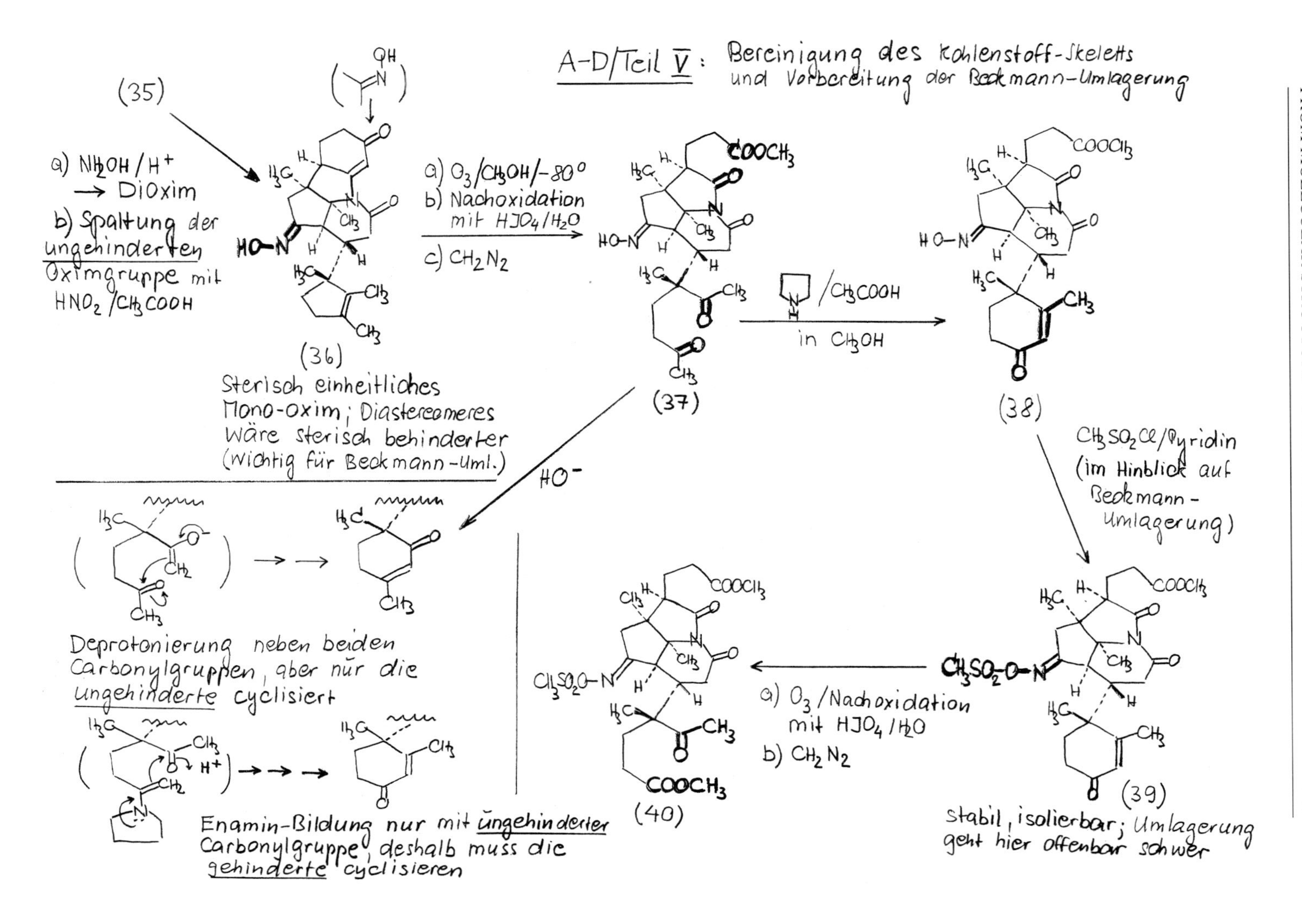
A–D/Teil V: Bereinigung des Kohlenstoff-Skeletts und Vorbereitung der Beckmann-Umlagerung
(35)
a) NH_2OH/H^+ → Dioxim
b) Spaltung der ungehinderten Oximgruppe mit HNO_2/CH_3COOH
(36)
Sterisch einheitliches Mono-oxim; Diastereomeres wäre sterisch behinderter (wichtig für Beckmann-Uml.)
a) $O_3/CH_3OH/-80°$
b) Nachoxidation mit HJO_4/H_2O
c) CH_2N_2
(37)
$/CH_3COOH$ in CH_3OH
(38)
CH_3SO_2Cl/Pyridin (im Hinblick auf Beckmann-Umlagerung)
(39)
stabil, isolierbar; Umlagerung geht hier offenbar schwer
a) O_3/Nachoxidation mit HJO_4/H_2O
b) CH_2N_2
(40)
HO^-
Deprotonierung neben beiden Carbonylgruppen, aber nur die ungehinderte cyclisiert
Enamin-Bildung nur mit ungehinderter Carbonylgruppe, deshalb muss die gehinderte cyclisieren

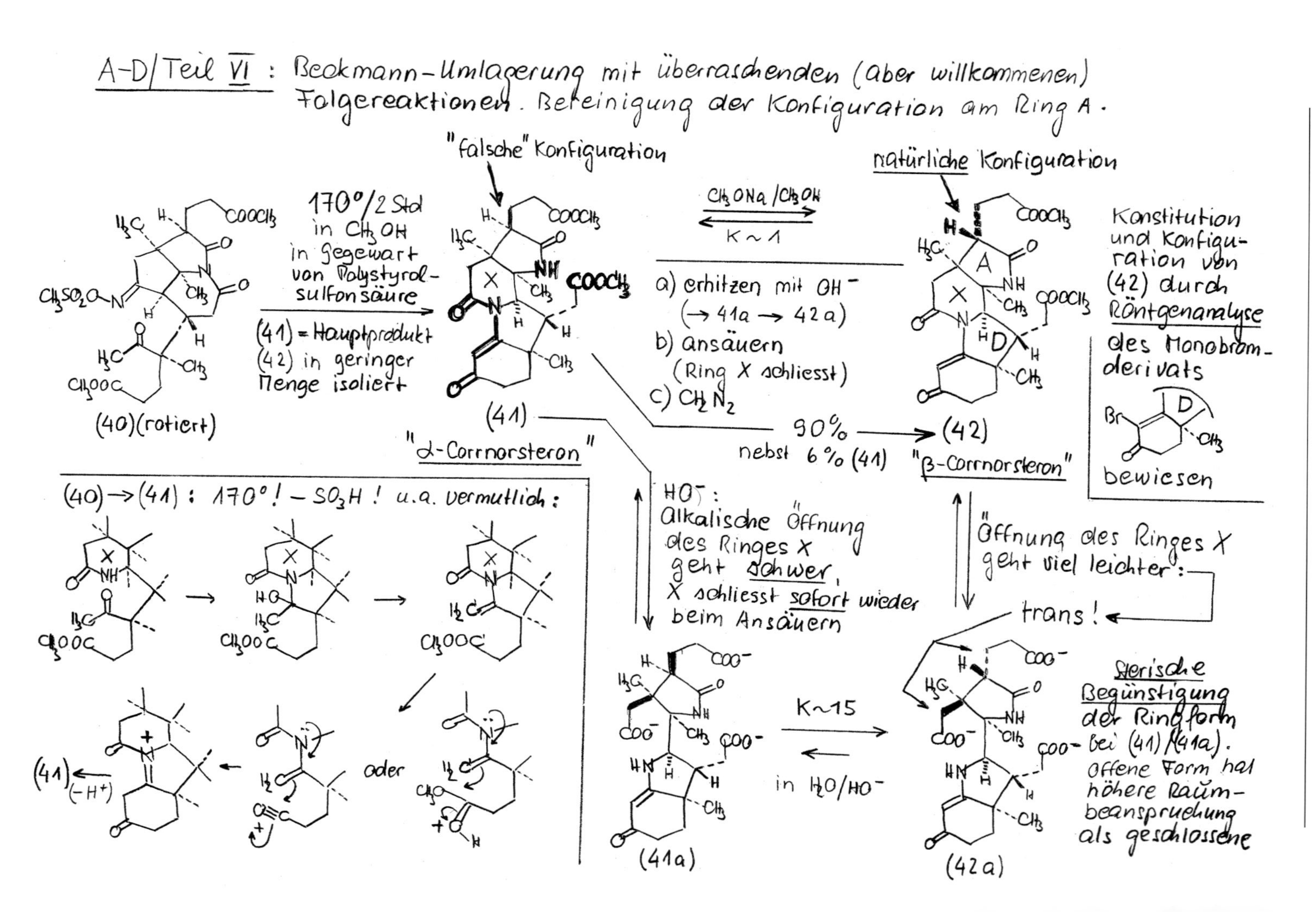

A-D/Teil VI : Beckmann-Umlagerung mit überraschenden (aber willkommenen) Folgereaktionen. Bereinigung der Konfiguration am Ring A.
"falsche" Konfiguration
natürliche Konfiguration
170°/2 Std in CH3OH in Gegenwart von Polystyrol-sulfonsäure
(41) = Hauptprodukt
(42) in geringer Menge isoliert
(40)(rotiert)
(41)
"α-Corrnorsteron"
CH3ONa/CH3OH
K ~ 1
a) erhitzen mit OH⁻ (→ 41a → 42a)
b) ansäuern (Ring X schliesst)
c) CH2N2
90%
nebst 6% (41)
(42)
"β-Corrnorsteron"
Konstitution und Konfiguration von (42) durch Röntgenanalyse des Monobromderivats
bewiesen
HO⁻:
Alkalische Öffnung des Ringes X geht schwer, X schliesst sofort wieder beim Ansäuern
Öffnung des Ringes X geht viel leichter:
trans!
K~15
in H2O/HO⁻
(41a)
(42a)
Sterische Begünstigung der Ringform bei (41)/(41a). Offene Form hat höhere Raumbeanspruchung als geschlossene
(40)→(41) : 170°! –SO3H ! u.a. vermutlich:
oder
(41)
(–H⁺)

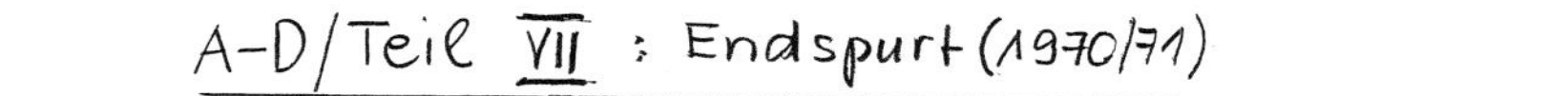
A–D/Teil VII: Endspurt (1970/71)

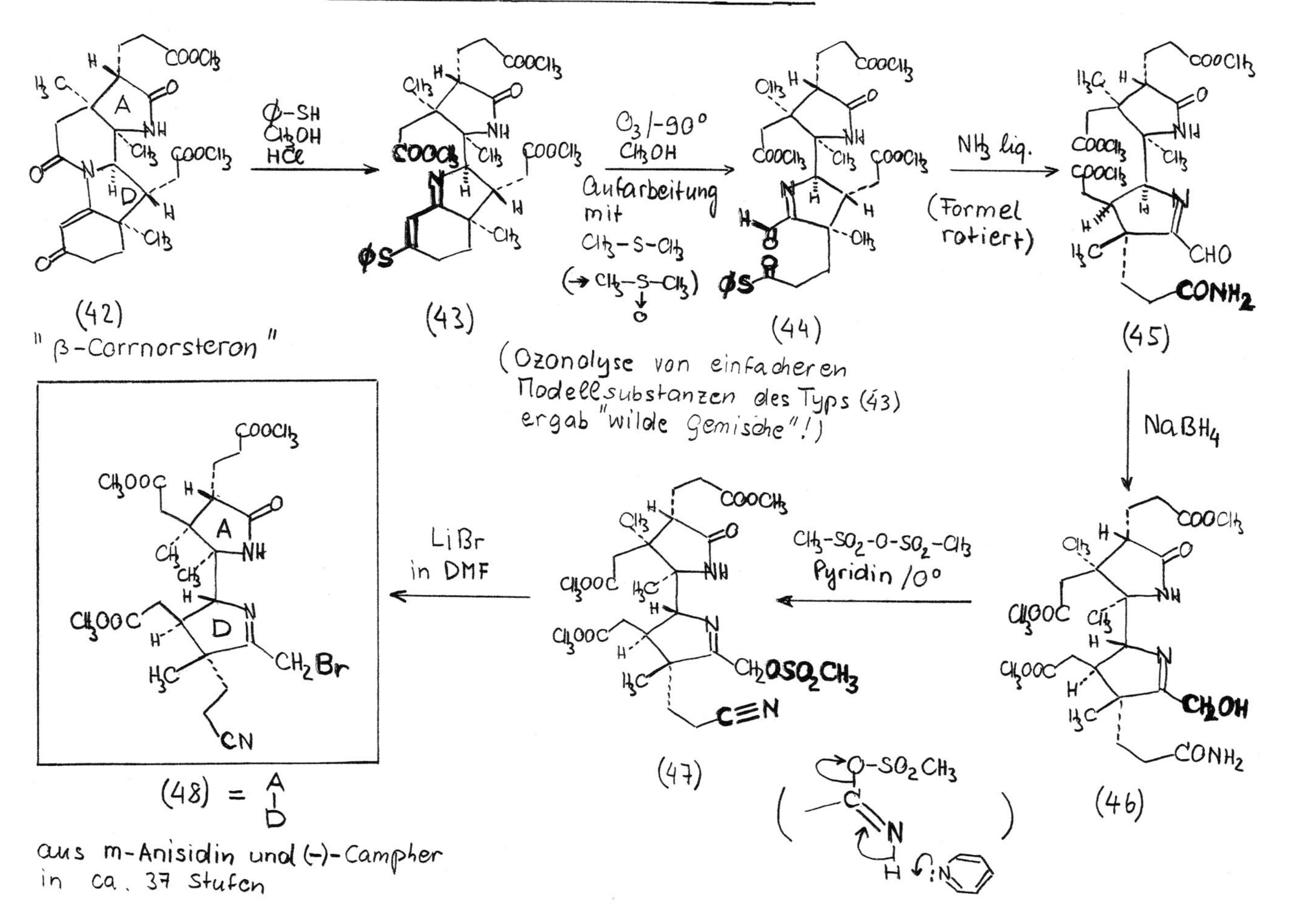
(42)
"β-Corrnorsteron"
φ-SH
CH3OH
HCl
(43)
O3/-90°
CH3OH
Aufarbeitung mit
CH3-S-CH3
(44)
(Ozonolyse von einfacheren Modellsubstanzen des Typs (43) ergab "wilde Gemische"!)
NH3 liq.
(Formel rotiert)
(45)
NaBH4
(46)
CH3-SO2-O-SO2-CH3
Pyridin/0°
(47)
LiBr
in DMF
(48) = A–D
aus m-Anisidin und (-)-Campher
in ca. 37 Stufen

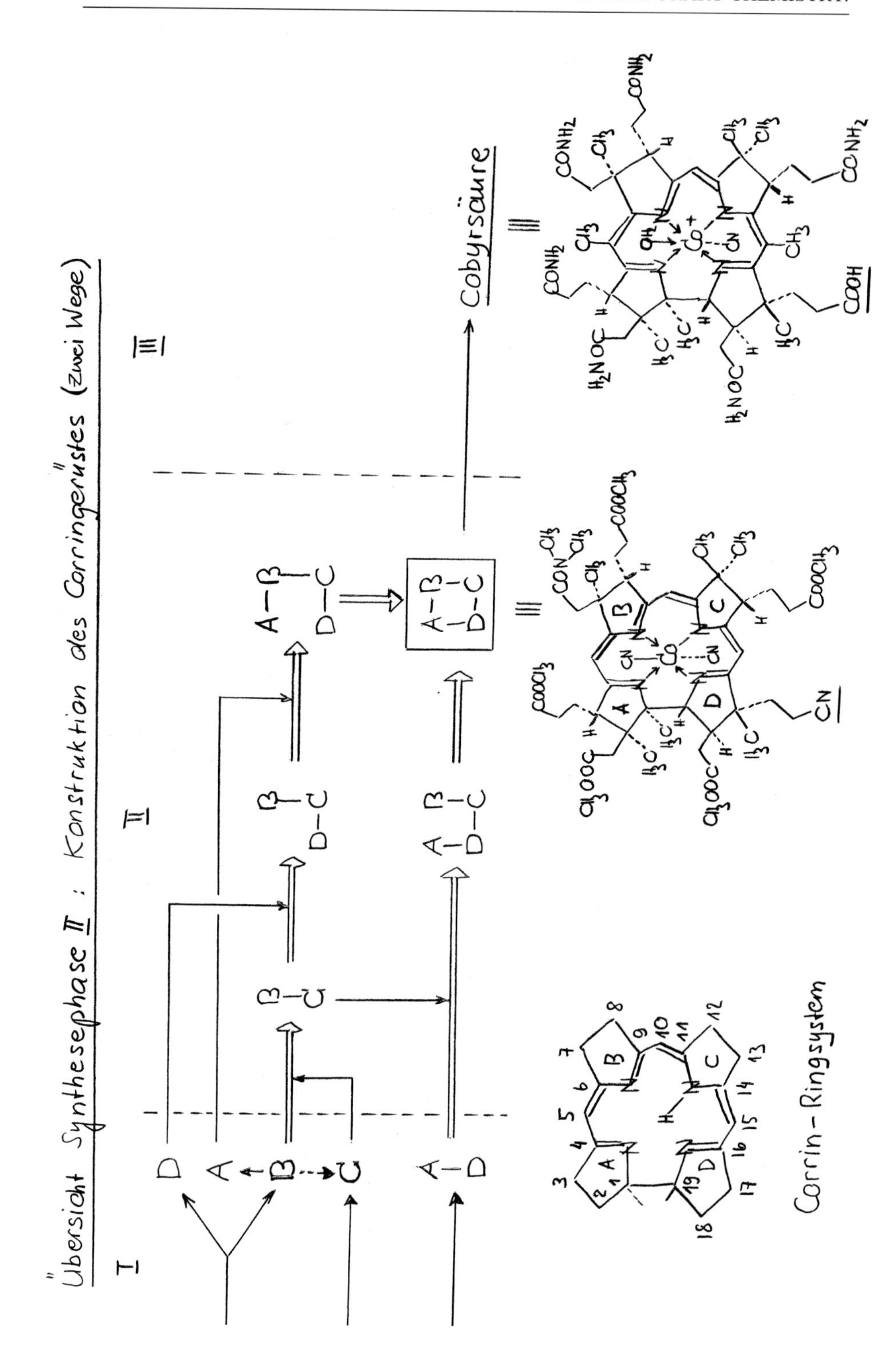
"Übersicht Synthesephase II : Konstruktion des Corringerüstes" (zwei Wege)
I
II
III
Cobyrsäure
Corrin-Ringsystem

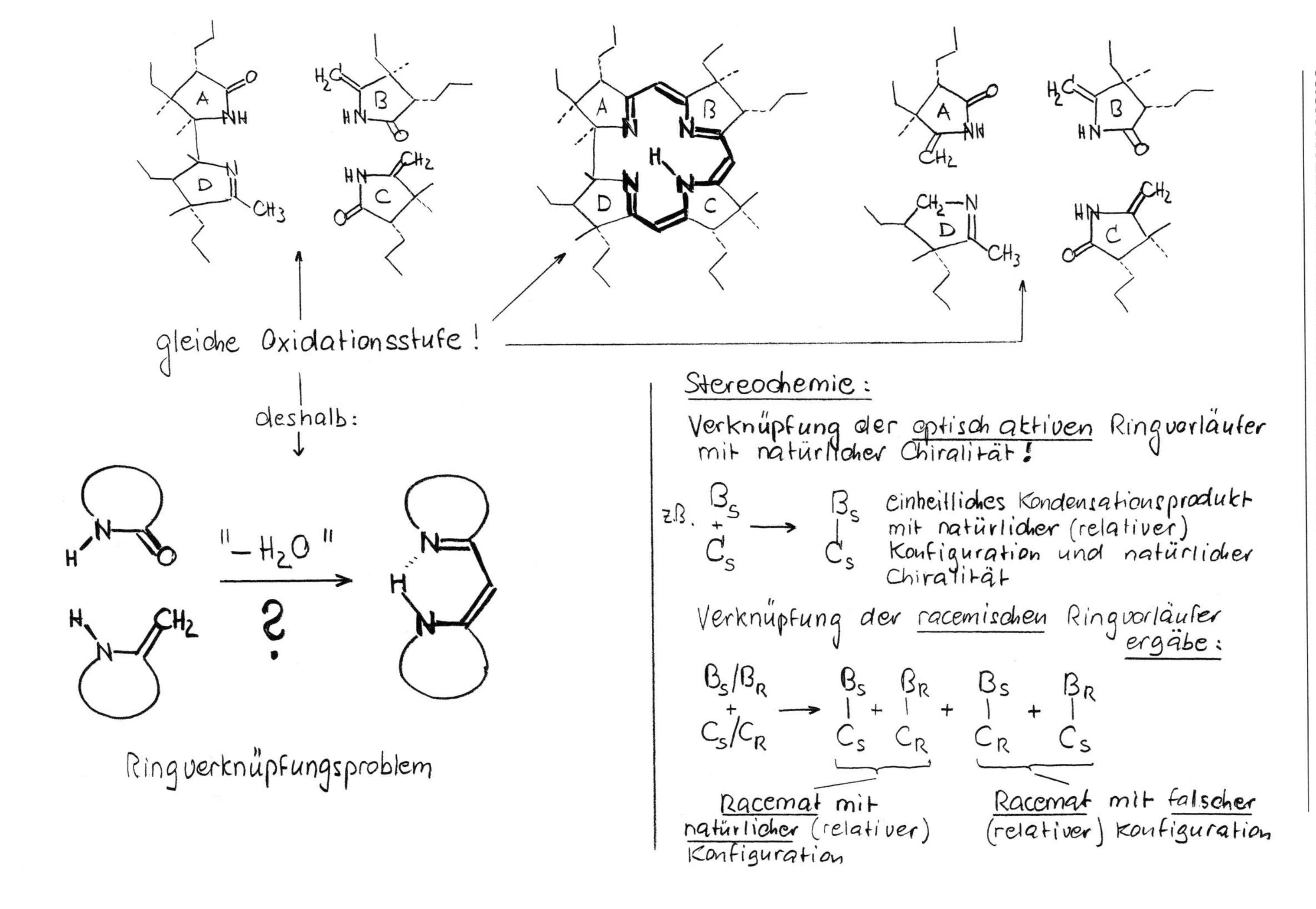
gleiche Oxidationsstufe !
deshalb:
"– H2O"
?
Ringverknüpfungsproblem
Stereochemie:
Verknüpfung der optisch aktiven Ringvorläufer mit natürlicher Chiralität!
z.B.
einheitliches Kondensationsprodukt mit natürlicher (relativer) Konfiguration und natürlicher Chiralität
Verknüpfung der racemischen Ringvorläufer ergäbe:
Racemat mit natürlicher (relativer) Konfiguration
Racemat mit falscher (relativer) Konfiguration

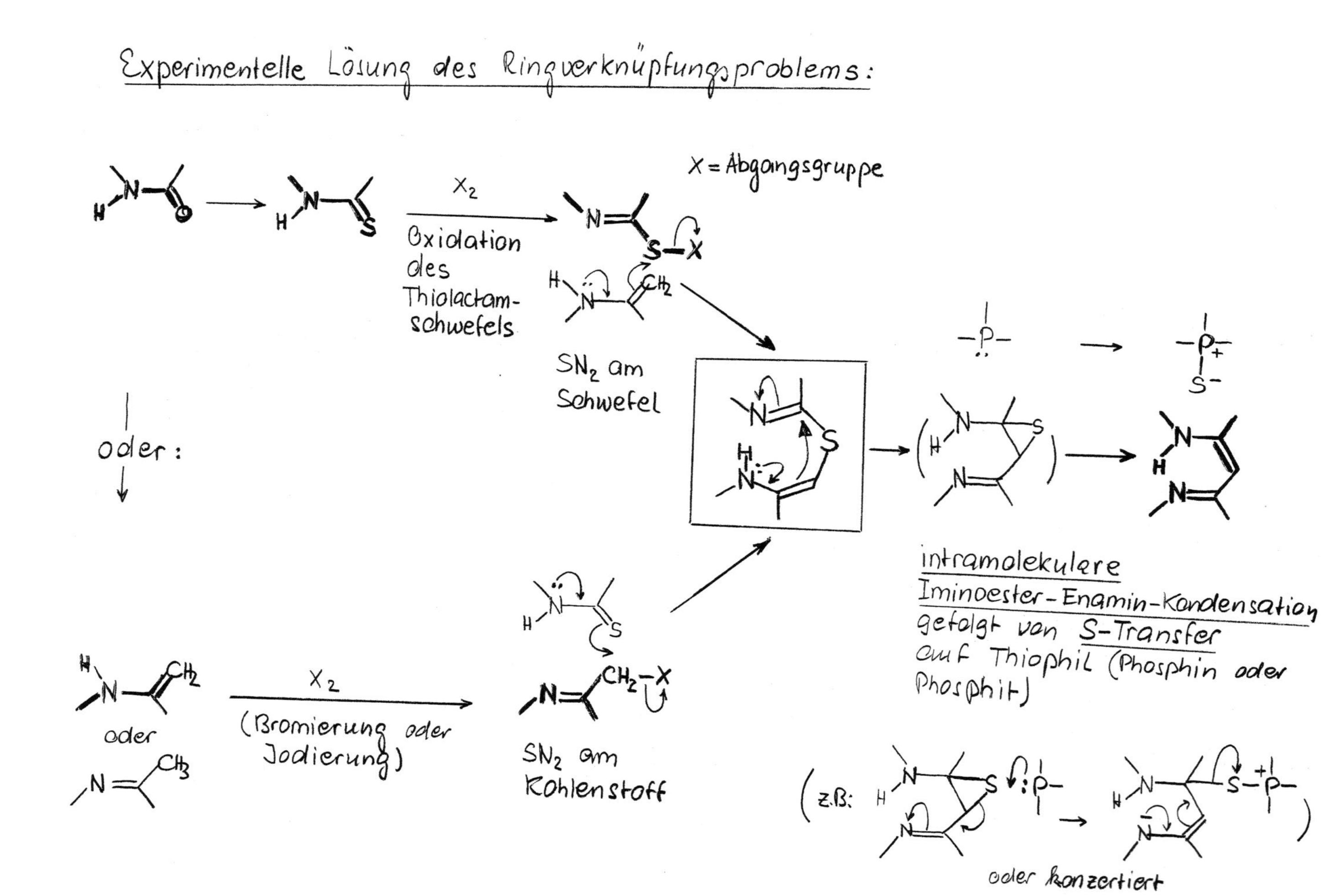
Experimentelle Lösung des "Ringverknüpfungsproblems":
X = Abgangsgruppe
X2
Oxidation des Thiolactam-schwefels
SN2 am Schwefel
oder:
oder
X2
(Bromierung oder Jodierung)
SN2 am Kohlenstoff
intramolekulare Iminoester-Enamin-Kondensation gefolgt von S-Transfer auf Thiophil (Phosphin oder Phosphit)
(z.B:
oder konzertiert)

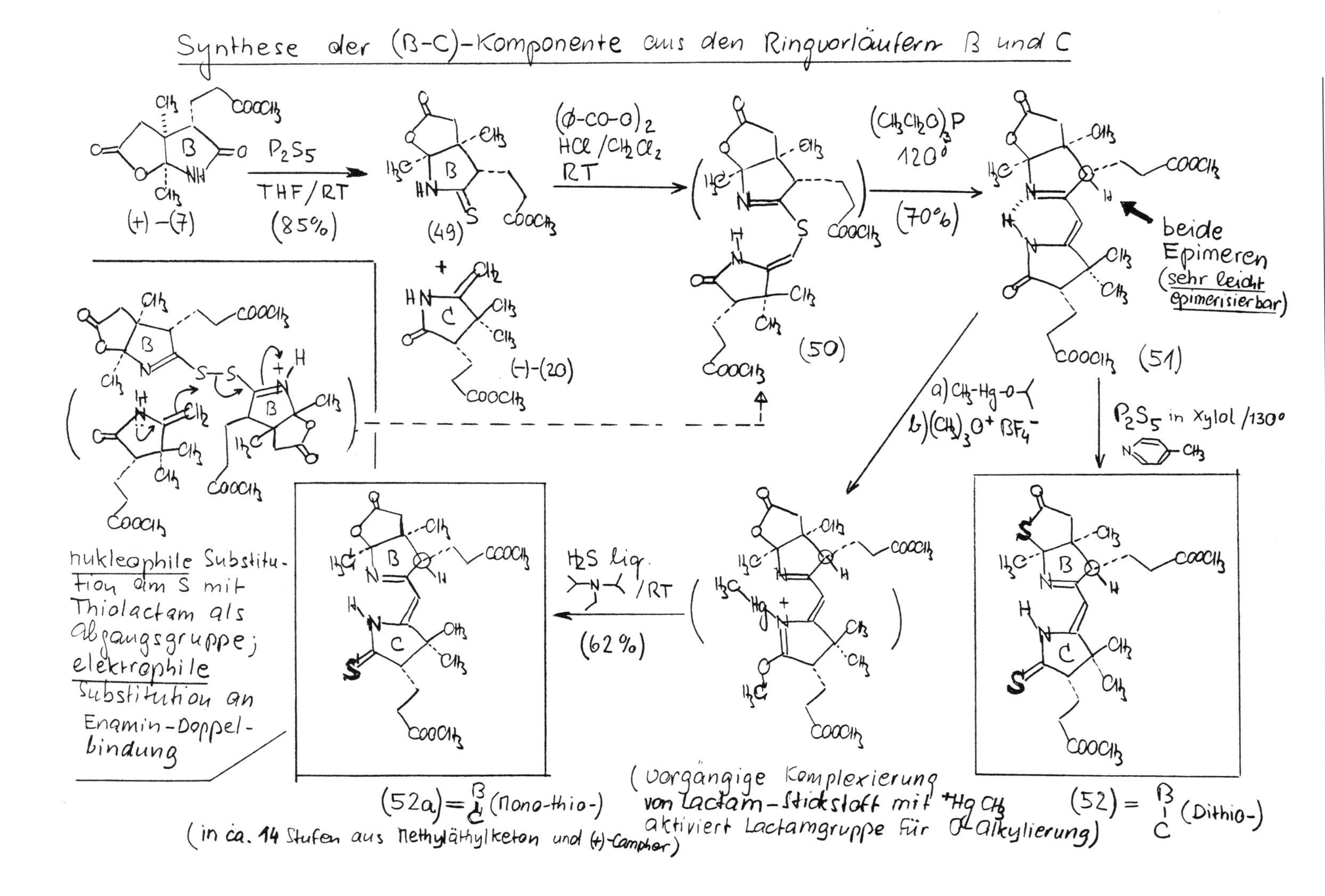
Synthese der (B-C)-Komponente aus den Ringvorläufern B und C
(+)-(7)
P2S5
THF/RT
(85%)
(49)
(−)-(20)
(φ-CO-O)2
HCl/CH2Cl2
RT
(50)
(CH3CH2O)3P
120°
(70%)
beide Epimeren
(sehr leicht epimerisierbar)
(51)
P2S5 in Xylol/130°
a) CH3-Hg-O-
b) (CH3)3O+ BF4−
H2S liq.
/RT
(62%)
nukleophile Substitution am S mit Thiolactam als Abgangsgruppe; elektrophile Substitution an Enamin-Doppelbindung
(52a) = B/C (Mono-thio-)
(in ca. 14 Stufen aus Methyläthylketon und (+)-Campher)
(vorgängige Komplexierung von Lactam-Stickstoff mit +HgCH3 aktiviert Lactamgruppe für O-Alkylierung)
(52) = B/C (Dithio-)

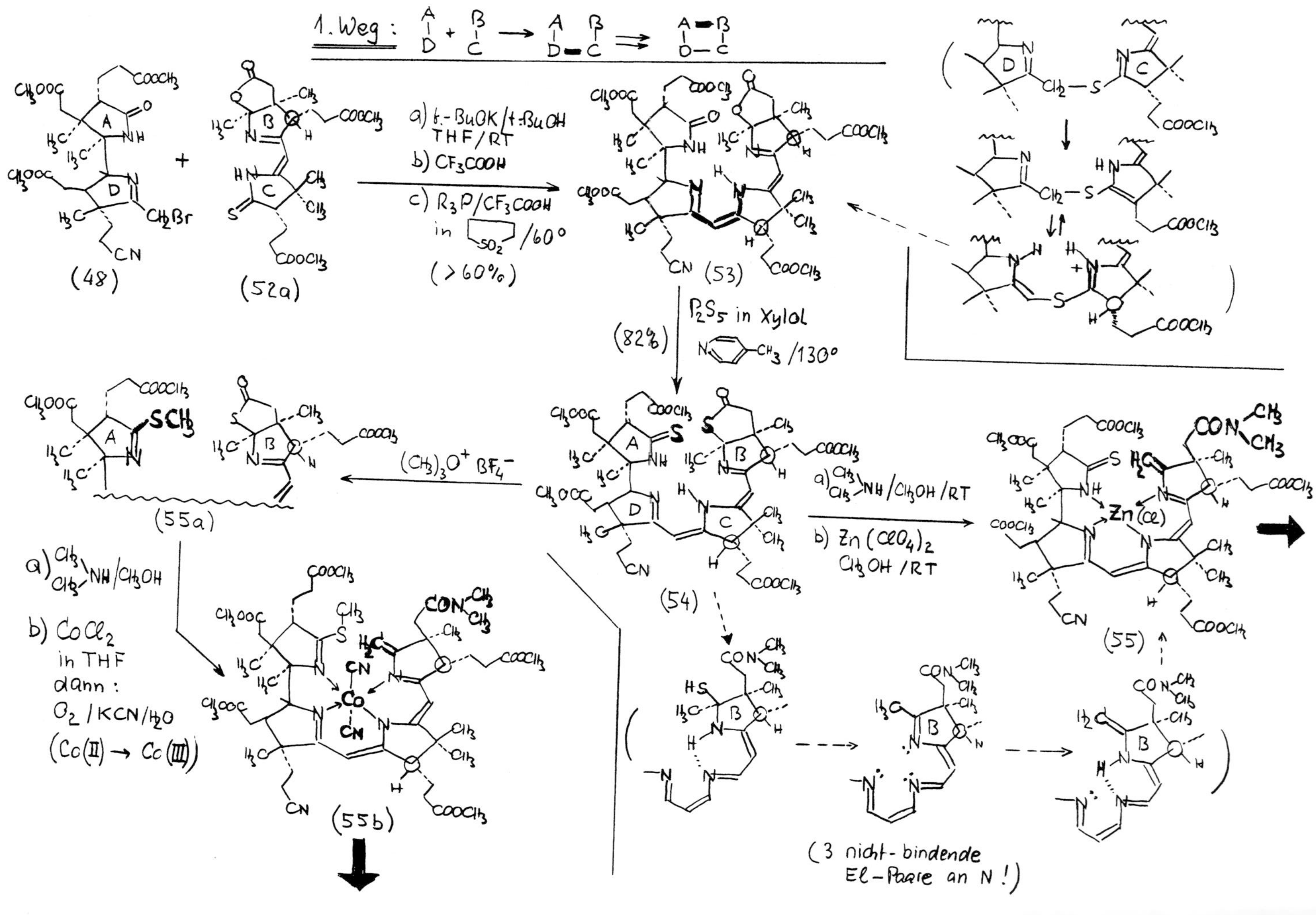

1. Weg: A/D + B/C → A–B/D=C ⇉ A=B/D–C

(48) + (52a)

a) t-BuOK / t-BuOH THF/RT
b) CF_3COOH
c) R_3P/CF_3COOH in sulfolane (SO_2) /60°
(> 60%)

(53)

P_2S_5 in Xylol, 4-CH_3-pyridine /130°
(82%)

(54)

a) $(CH_3)_2NH/CH_3OH/RT$
b) $Zn(ClO_4)_2$ CH_3OH/RT

(55)

(3 nicht-bindende El-Paare an N!)

$(CH_3)_3O^+ BF_4^-$

(55a)

a) $(CH_3)_2NH/CH_3OH$
b) $CoCl_2$ in THF
dann: $O_2/KCN/H_2O$
(Co(II) → Co(III))

(55b)

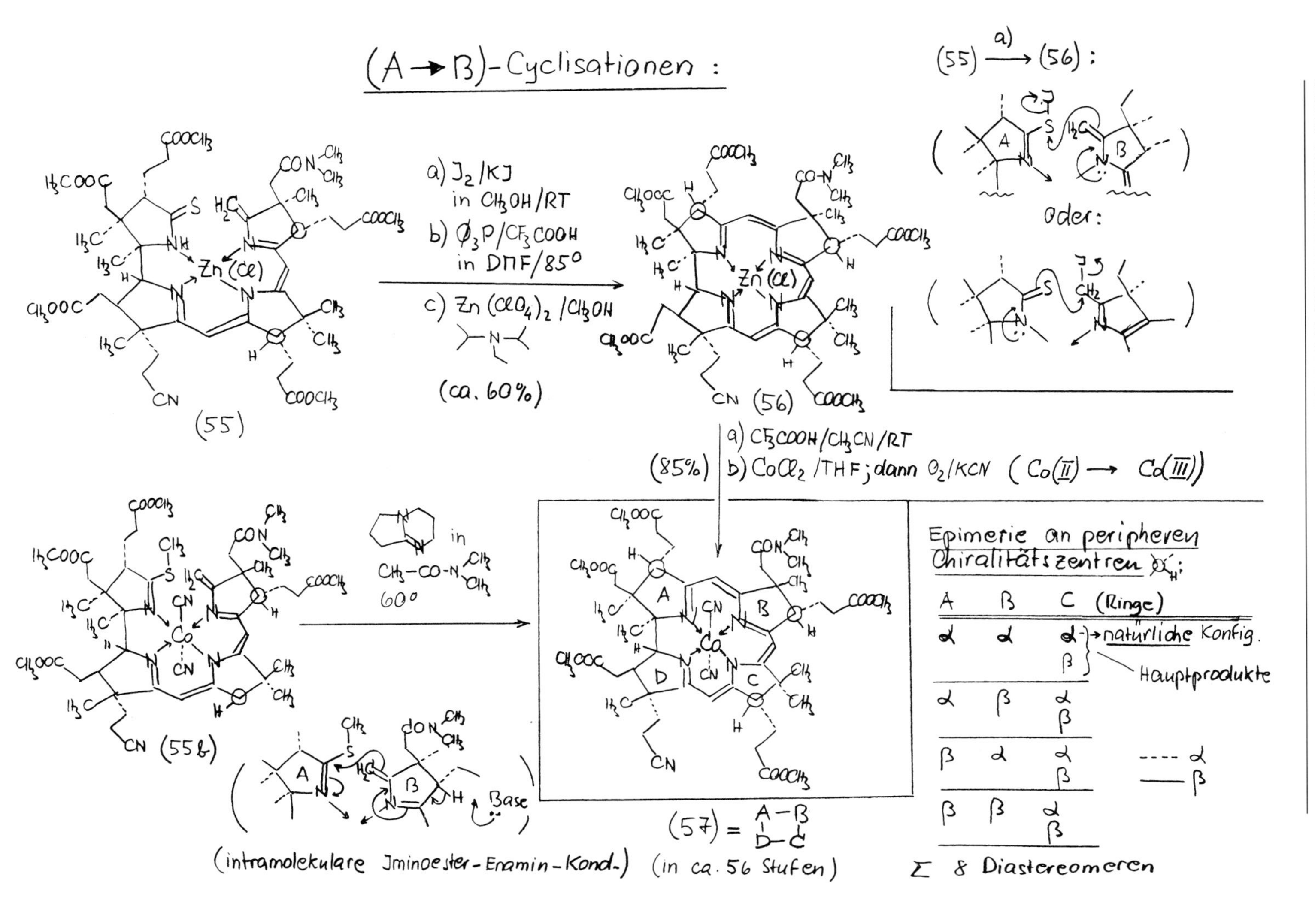
(A→B)-Cyclisationen:
a) J2/KJ in CH3OH/RT
b) Ø3P/CF3COOH in DMF/85°
c) Zn(ClO4)2/CH3OH
(ca. 60%)
(55)
(56)
(55) a) → (56):
oder:
a) CF3COOH/CH3CN/RT
b) CoCl2/THF; dann O2/KCN (Co(II) → Co(III))
(85%)
in CH3-CO-N(CH3)2 60°
(55b)
(intramolekulare Iminoester-Enamin-Kond.)
(57) = A−B D−C
(in ca. 56 Stufen)
Epimerie an peripheren Chiralitätszentren
A B C (Ringe)
natürliche Konfig.
Hauptprodukte
Σ 8 Diastereomeren

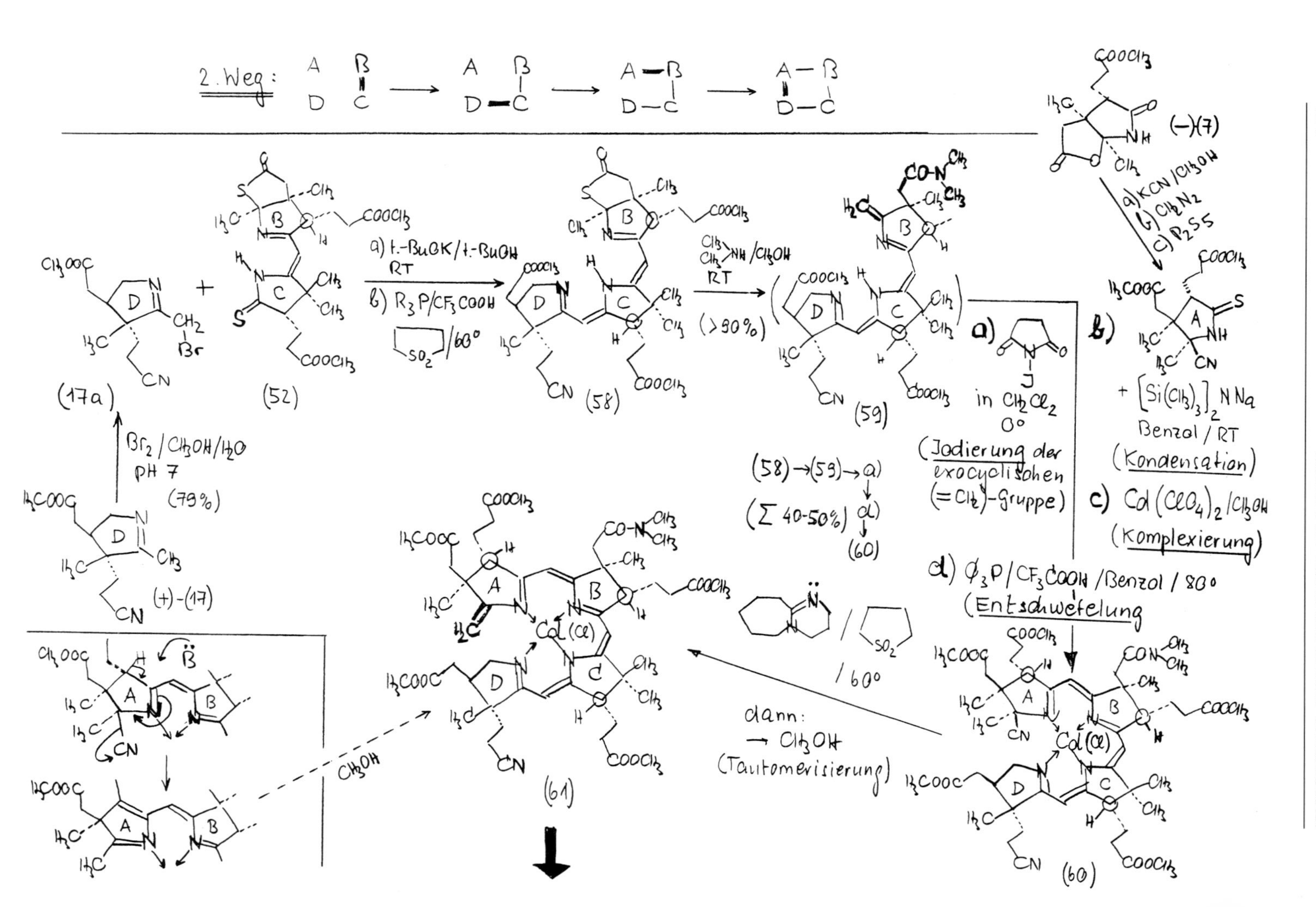
2. Weg:
a) t.-BuOK/t.-BuOH RT
b) R3P/CF3COOH / 60°
(> 90%)
(−)-(7)
a) KCN/CH3OH
b) CH2N2
c) P2S5
(17a)
(52)
(58)
(59)
Br2/CH3OH/H2O pH 7 (79%)
(+)-(17)
a) in CH2Cl2 0°
(Jodierung der exocyclischen (=CH2)-Gruppe)
b) + [Si(CH3)3]2NNa Benzol/RT (Kondensation)
c) Cd(ClO4)2/CH3OH (Komplexierung)
d) Ø3P/CF3COOH/Benzol/80° (Entschwefelung)
(58) → (59) → a) → d) → (60)
(Σ 40-50%)
/ 60°
dann: → CH3OH (Tautomerisierung)
CH3OH
(61)
(60)

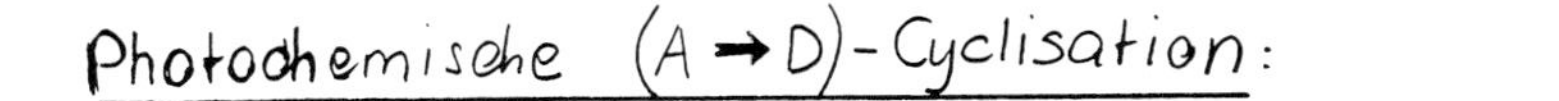
Photochemische (A→D)-Cyclisation:

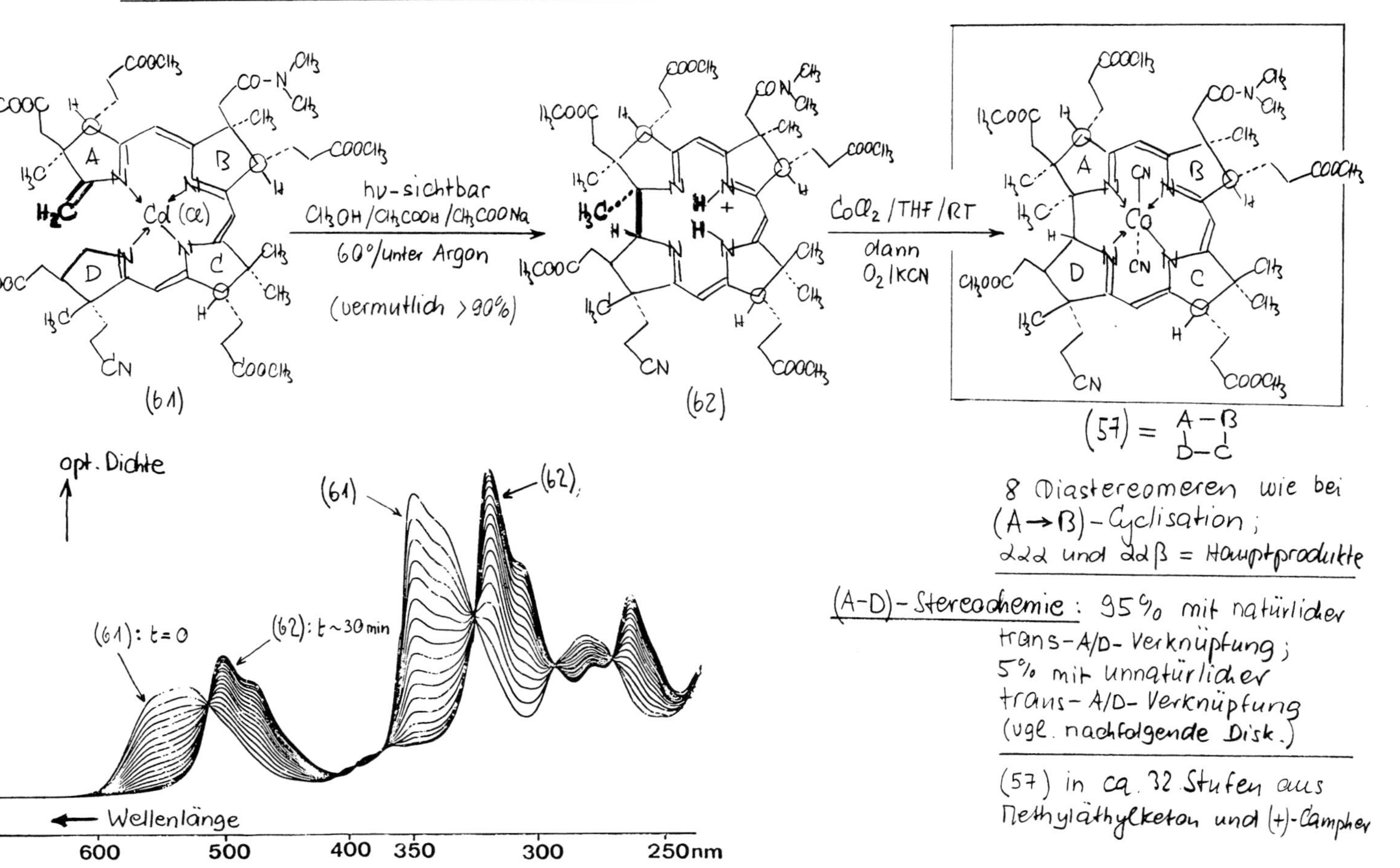
(61)
(62)
hν-sichtbar
CH3OH/CH3COOH/CH3COONa
60°/unter Argon
(vermutlich >90%)
CoCl2/THF/RT
dann
O2/KCN
(57) = A—B, D—C
8 Diastereomeren wie bei
(A→B)-Cyclisation;
ααα und ααβ = Hauptprodukte
(A-D)-Stereochemie: 95% mit natürlicher
trans-A/D-Verknüpfung;
5% mit unnatürlicher
trans-A/D-Verknüpfung
(vgl. nachfolgende Disk.)
(57) in ca. 32 Stufen aus
Methyläthylketon und (+)-Campher
opt. Dichte
(61): t=0
(62): t~30 min
(61)
(62)
Wellenlänge
600
500
400
350
300
250nm

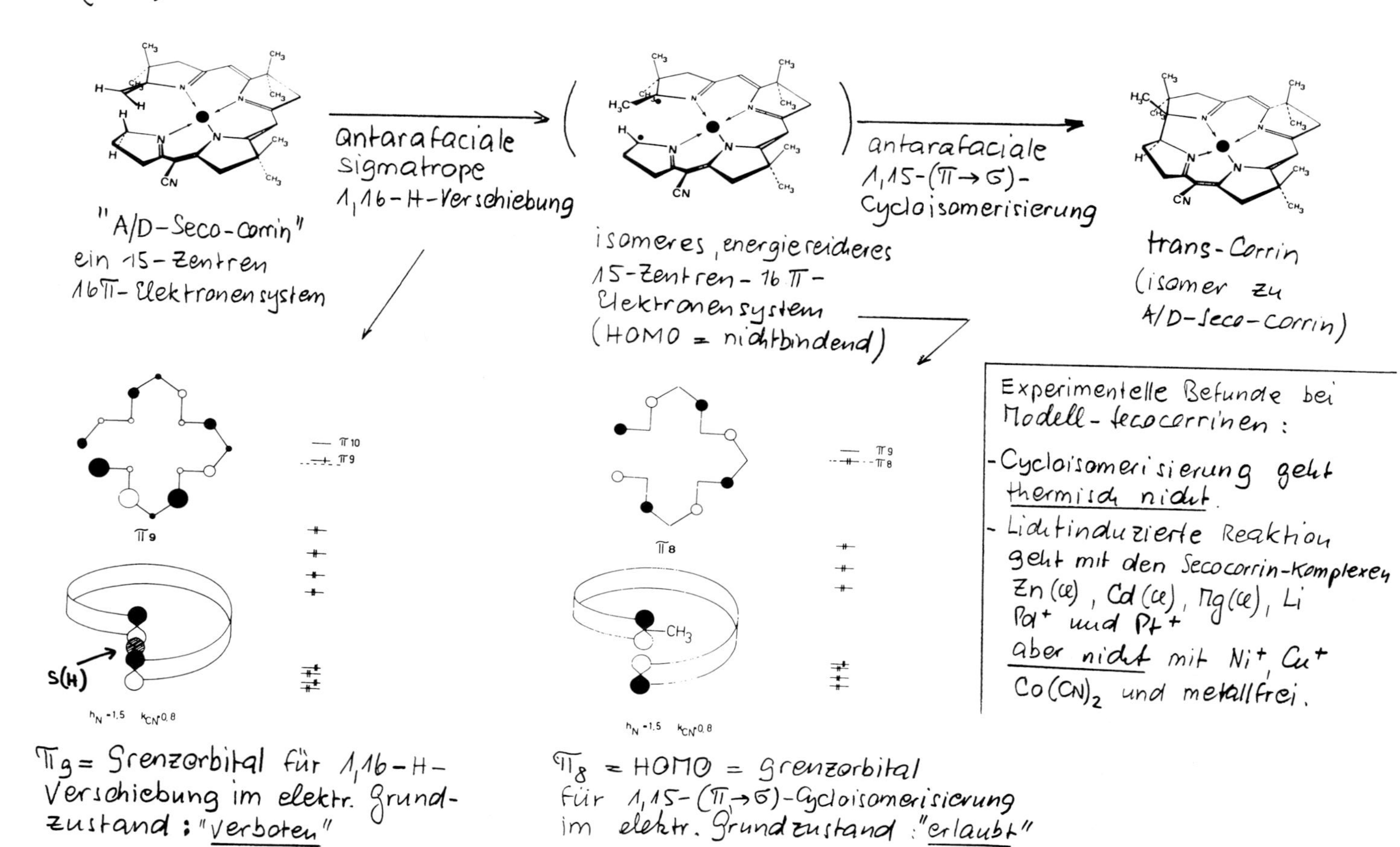
(A→D)-Cycloisomerisierung: Mechanismus / Grenzorbitalanalyse
"A/D-Seco-Corrin"
ein 15-Zentren 16π-Elektronensystem
antarafaciale sigmatrope 1,16-H-Verschiebung
isomeres, energiereicheres 15-Zentren-16π-Elektronensystem (HOMO = nichtbindend)
antarafaciale 1,15-(π→σ)-Cycloisomerisierung
trans-Corrin (isomer zu A/D-Seco-Corrin)
π9
π10
π9
S(H)
π8
π9
π8
CH3
Experimentelle Befunde bei Modell-Secocorrinen:
-Cycloisomerisierung geht thermisch nicht.
-Lichtinduzierte Reaktion geht mit den Secocorrin-Komplexen Zn(Cl), Cd(Cl), Mg(Cl), Li, Pd+ und Pt+
aber nicht mit Ni+, Cu+, Co(CN)2 und metallfrei.
π9 = Grenzorbital für 1,16-H-Verschiebung im elektr. Grundzustand: "verboten"
π8 = HOMO = Grenzorbital für 1,15-(π→σ)-Cycloisomerisierung im elektr. Grundzustand: "erlaubt"

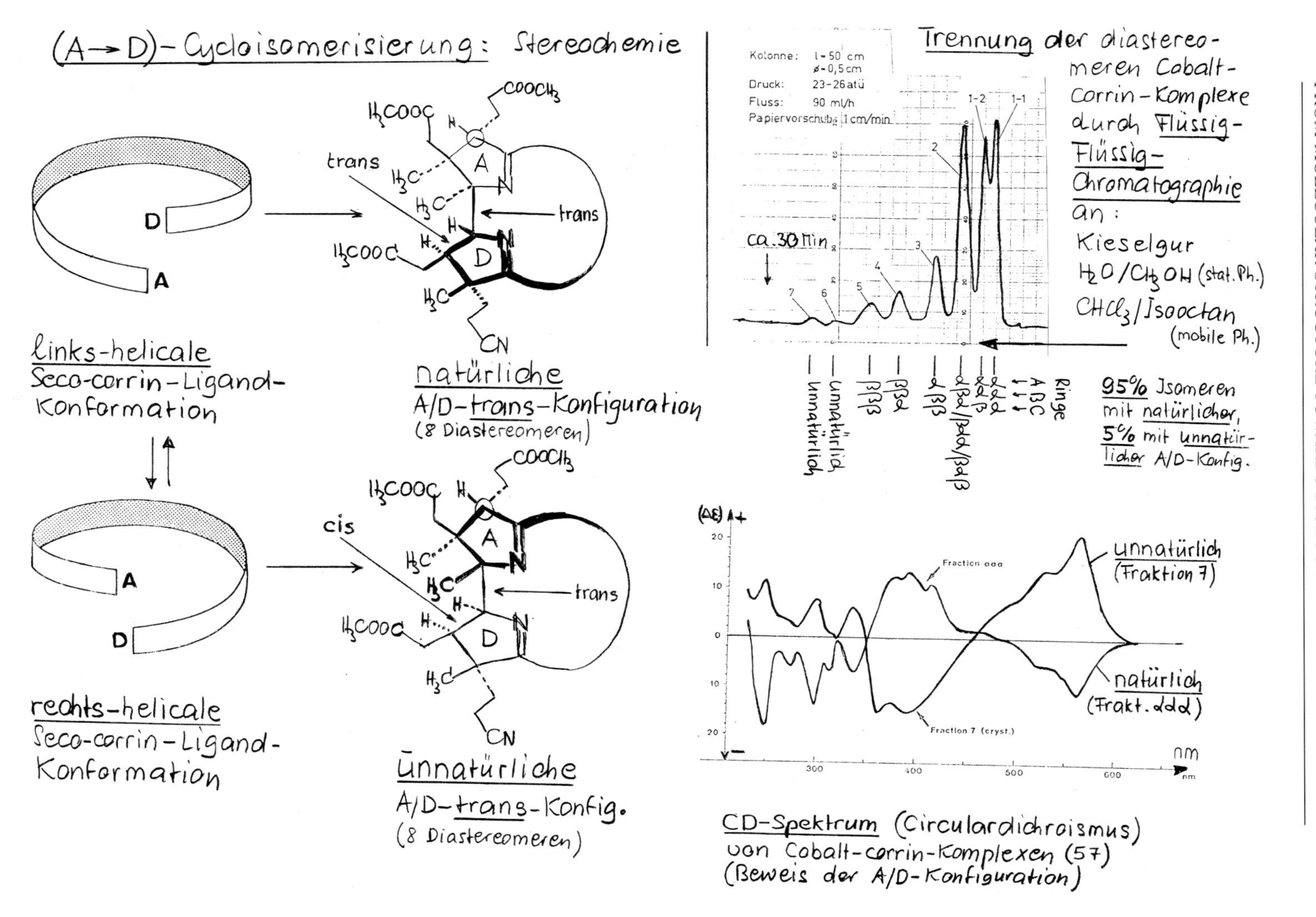

(A→D)-Cycloisomerisierung: Stereochemie
links-helicale Seco-corrin-Ligand-Konformation
rechts-helicale Seco-corrin-Ligand-Konformation
natürliche A/D-trans-Konfiguration (8 Diastereomeren)
unnatürliche A/D-trans-Konfig. (8 Diastereomeren)
trans
cis
Trennung der diastereomeren Cobalt-Corrin-Komplexe durch Flüssig-Flüssig-Chromatographie
an: Kieselgur
H2O/CH3OH (stat. Ph.)
CHCl3/Isooctan (mobile Ph.)
Kolonne: l = 50 cm ø = 0,5 cm
Druck: 23-26 atü
Fluss: 90 ml/h
Papiervorschub: 1 cm/min
ca. 30 Min
Ringe A B C
ααα
ααβ
αβα/βαα/βαβ
αββ
ββα
βββ
unnatürlich
unnatürlich
95% Isomeren mit natürlicher, 5% mit unnatürlicher A/D-Konfig.
unnatürlich (Fraktion 7)
natürlich (Frakt. ααα)
Fraction ααα
Fraction 7 (cryst.)
(Δε)
nm
CD-Spektrum (Circulardichroismus) von Cobalt-corrin-Komplexen (57) (Beweis der A/D-Konfiguration)

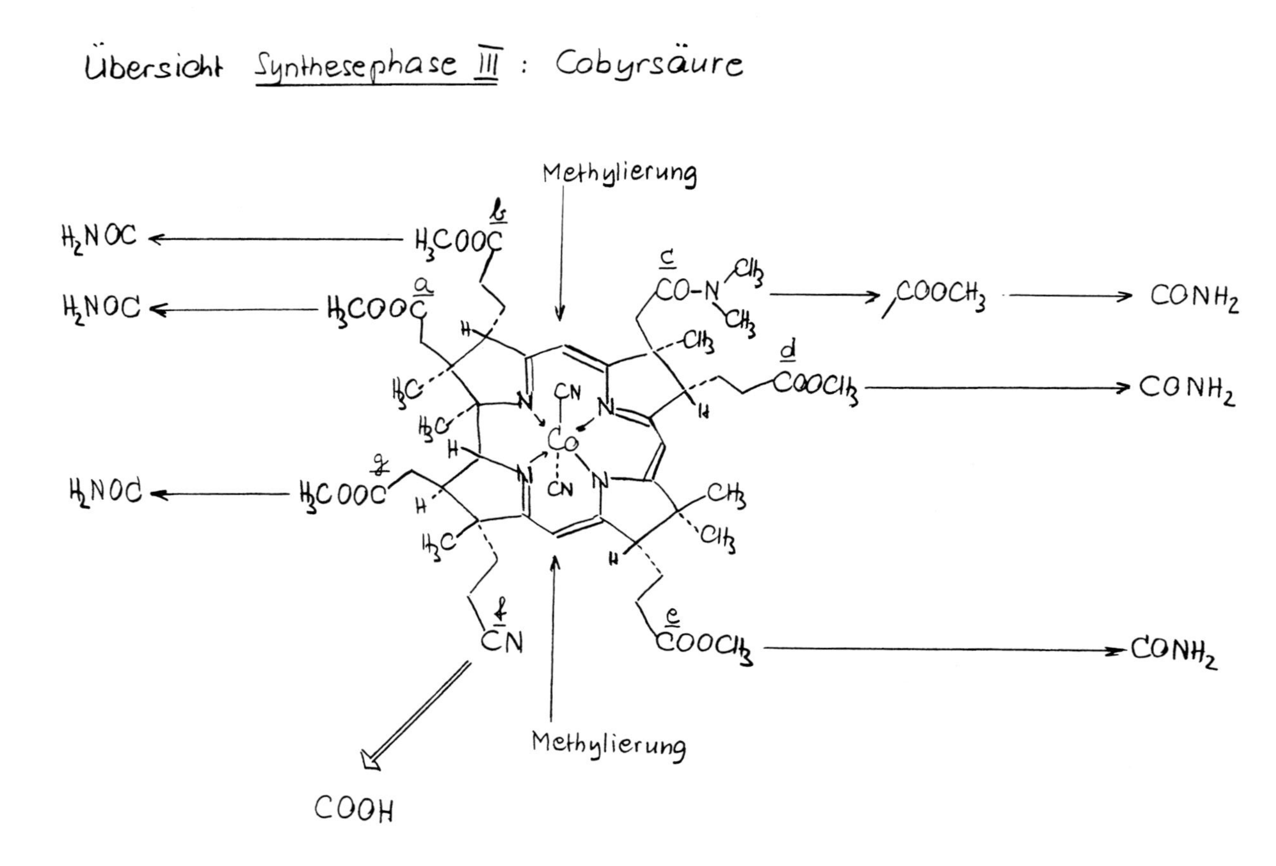
Übersicht Synthesephase III : Cobyrsäure
Methylierung
H2NOC
H3COOC
b
H2NOC
H3COOC
a
c
CO-N
CH3
CH3
COOCH3
CONH2
d
COOCH3
CONH2
CN
Co
CN
H2NOC
H3COOC
g
f
CN
COOH
e
COOCH3
CONH2
Methylierung

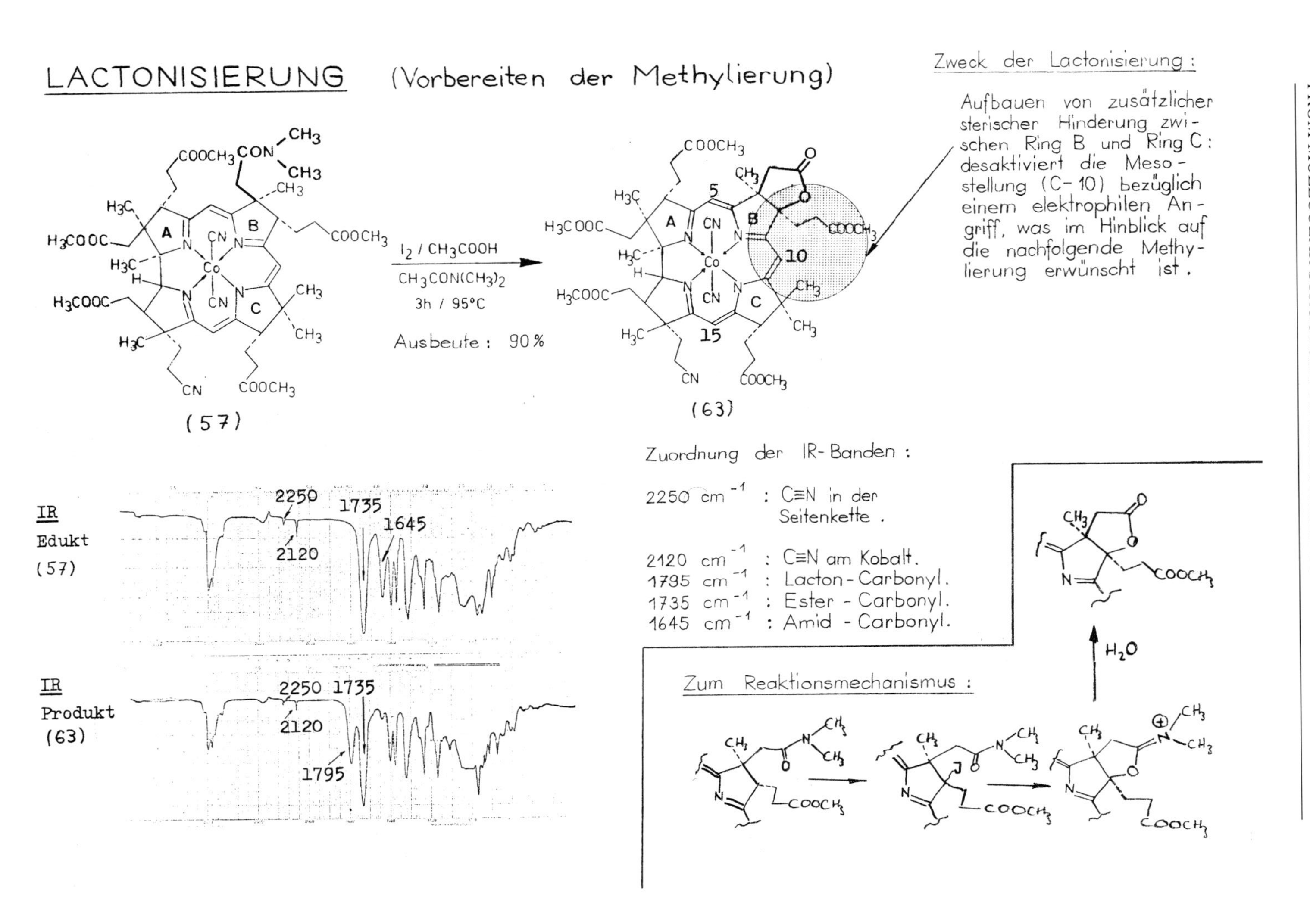
LACTONISIERUNG (Vorbereiten der Methylierung)
(57)
I2 / CH3COOH
CH3CON(CH3)2
3h / 95°C
Ausbeute : 90 %
(63)
Zweck der Lactonisierung :
Aufbauen von zusätzlicher sterischer Hinderung zwischen Ring B und Ring C: desaktiviert die Mesostellung (C-10) bezüglich einem elektrophilen Angriff, was im Hinblick auf die nachfolgende Methylierung erwünscht ist.
IR Edukt (57)
IR Produkt (63)
Zuordnung der IR-Banden :
2250 cm-1 : C≡N in der Seitenkette.
2120 cm-1 : C≡N am Kobalt.
1795 cm-1 : Lacton-Carbonyl.
1735 cm-1 : Ester - Carbonyl.
1645 cm-1 : Amid - Carbonyl.
Zum Reaktionsmechanismus :
H2O

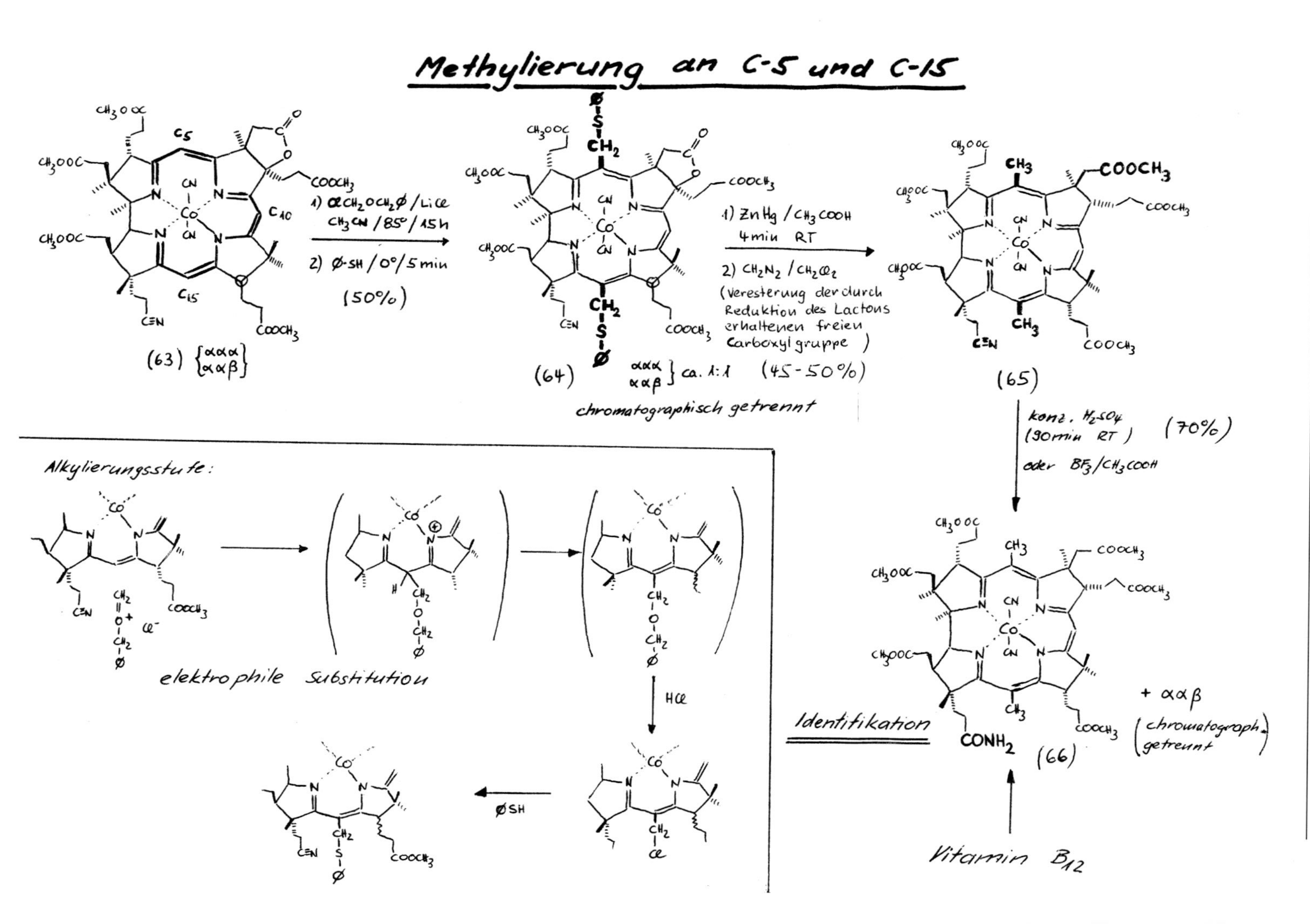
Methylierung an C-5 und C-15
(63)
1) ClCH2OCH2Ø / LiCl CH3CN / 85° / 15h
2) Ø-SH / 0° / 5 min
(50%)
(64)
ca. 1:1
chromatographisch getrennt
1) ZnHg / CH3COOH 4min RT
2) CH2N2 / CH2Cl2
(Veresterung der durch Reduktion des Lactons erhaltenen freien Carboxylgruppe)
(45-50%)
(65)
konz. H2SO4 (90min RT)
oder BF3/CH3COOH
(70%)
(66)
+ ααβ (chromatograph. getrennt)
Identifikation
Vitamin B12
Alkylierungsstufe:
elektrophile Substitution

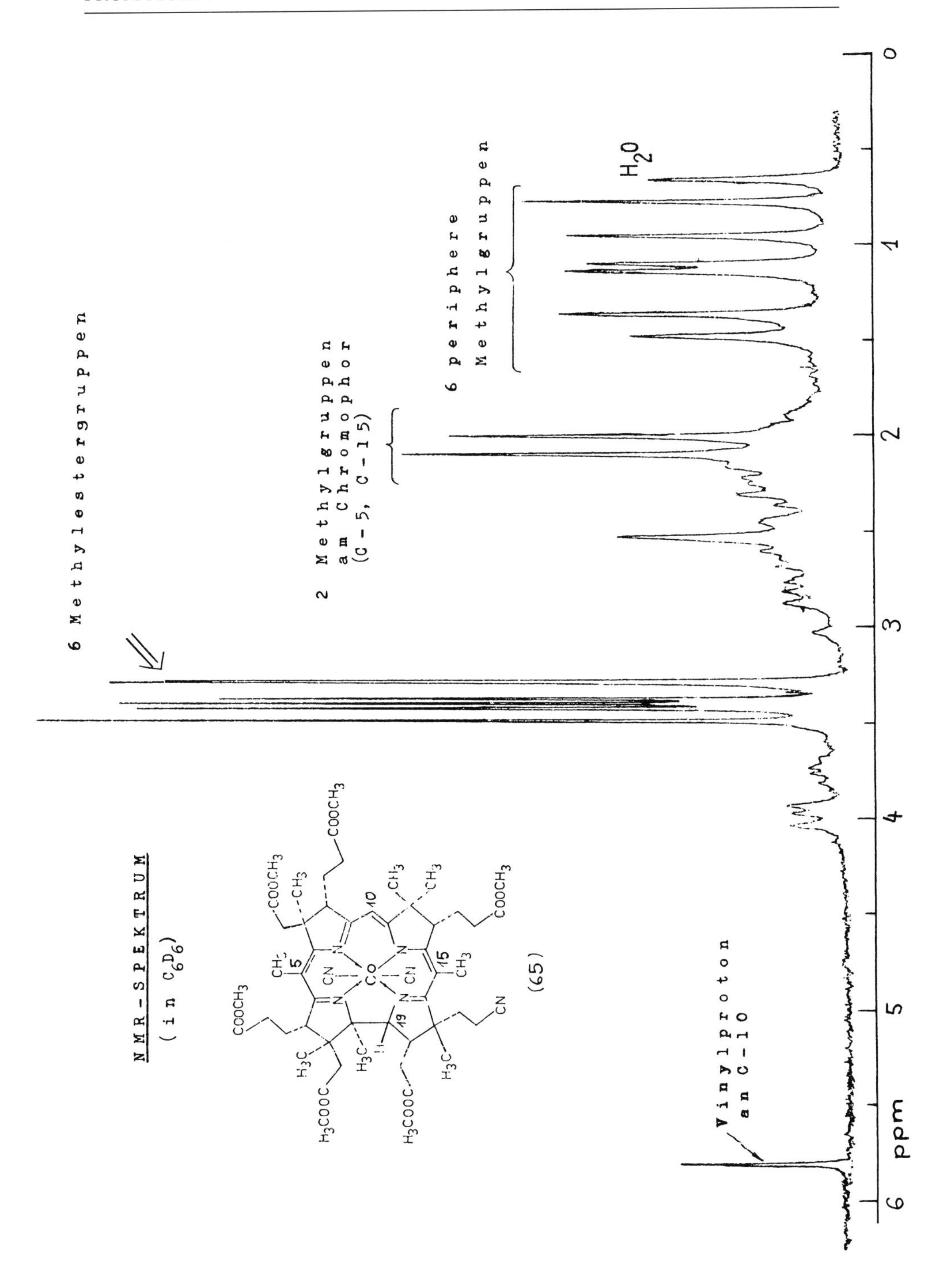
NMR-SPEKTRUM
(in C_6D_6)
6 Methylestergruppen
2 Methylgruppen am Chromophor (C-5, C-15)
6 periphere Methylgruppen
H_2O
Vinylproton an C-10
(65)
0
1
2
3
4
5
6
ppm

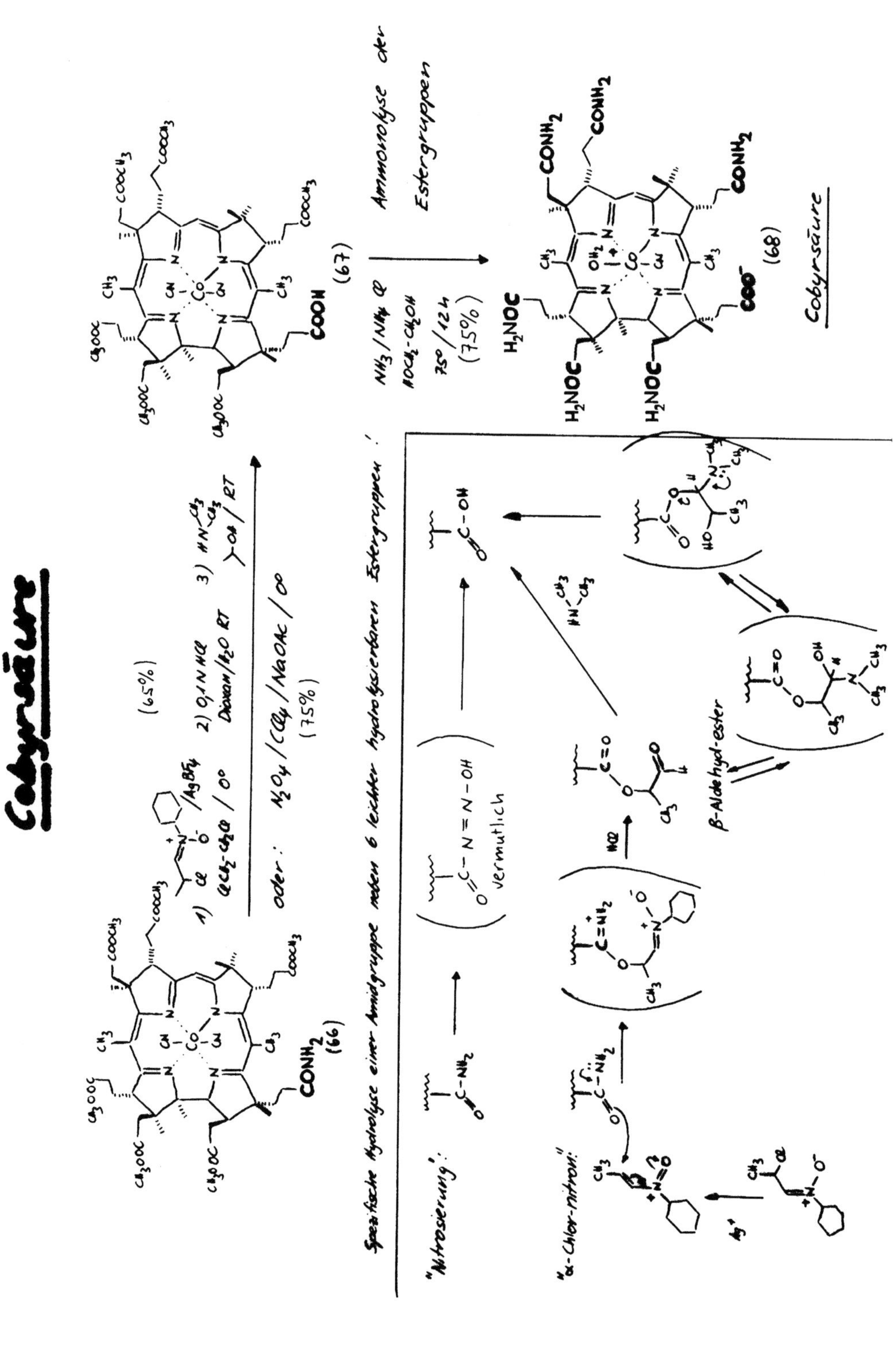

Cobyrsäure

(66) → (67)

1) Chlor-nitron / $AgBF_4$, $ClCH_2$-CH_2Cl / 0°
2) 0,1 N HCl, Dioxan/H_2O RT
3) $HN(CH_3)_2$, Isopropanol / RT
(65%)

oder: N_2O_4 / CCl_4 / NaOAc / 0°
(75%)

Spezifische Hydrolyse einer Amidgruppe neben 6 leichter hydrolysierbaren Estergruppen!

(67) → (68): NH_3 / NH_4Cl, $HOCH_2$-CH_2OH, 75° / 12 h (75%)

Ammonolyse der Estergruppen

(68) Cobyrsäure

"Nitrosierung": —C(=O)-NH_2 → (—C(=O)-N=N-OH, vermutlich) → —C(=O)-OH

"α-Chlor-nitron": Ag^+; HCl; $HN(CH_3)_2$

β-Aldehyd-ester

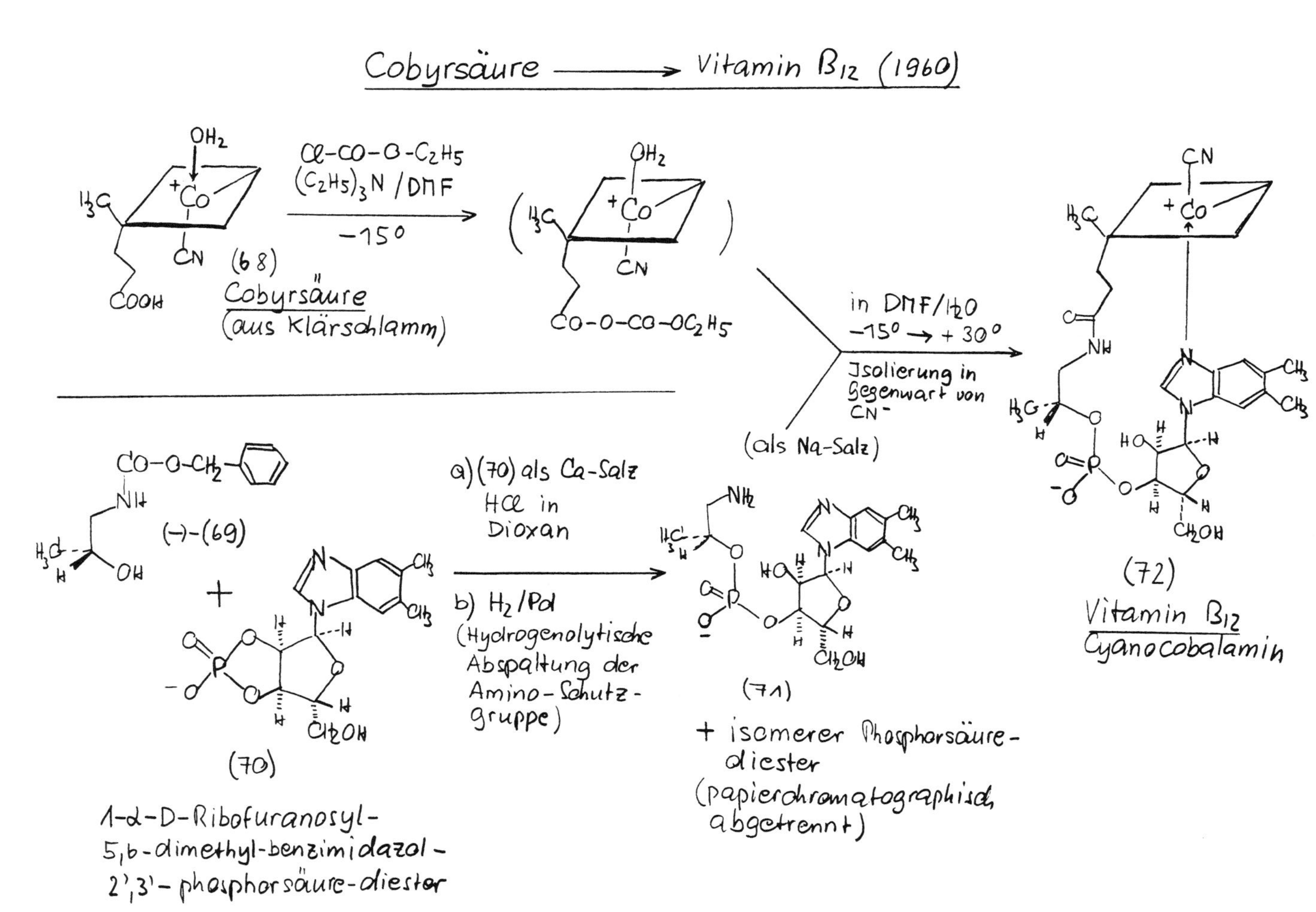
Cobyrsäure ⟶ Vitamin B12 (1960)
Cl-CO-O-C2H5
(C2H5)3N / DMF
−15°
(68)
Cobyrsäure
(aus Klärschlamm)
CO-O-CO-OC2H5
in DMF/H2O
−15° → + 30°
Isolierung in Gegenwart von CN⁻
(als Na-Salz)
(−)-(69)
a) (70) als Ca-Salz
HCl in
Dioxan
b) H2/Pd
(Hydrogenolytische
Abspaltung der
Amino-Schutz-
gruppe)
(70)
1-α-D-Ribofuranosyl-
5,6-dimethyl-benzimidazol-
2',3'-phosphorsäure-diester
(71)
+ isomerer Phosphorsäure-
diester
(papierchromatographisch
abgetrennt)
(72)
Vitamin B12
Cyanocobalamin

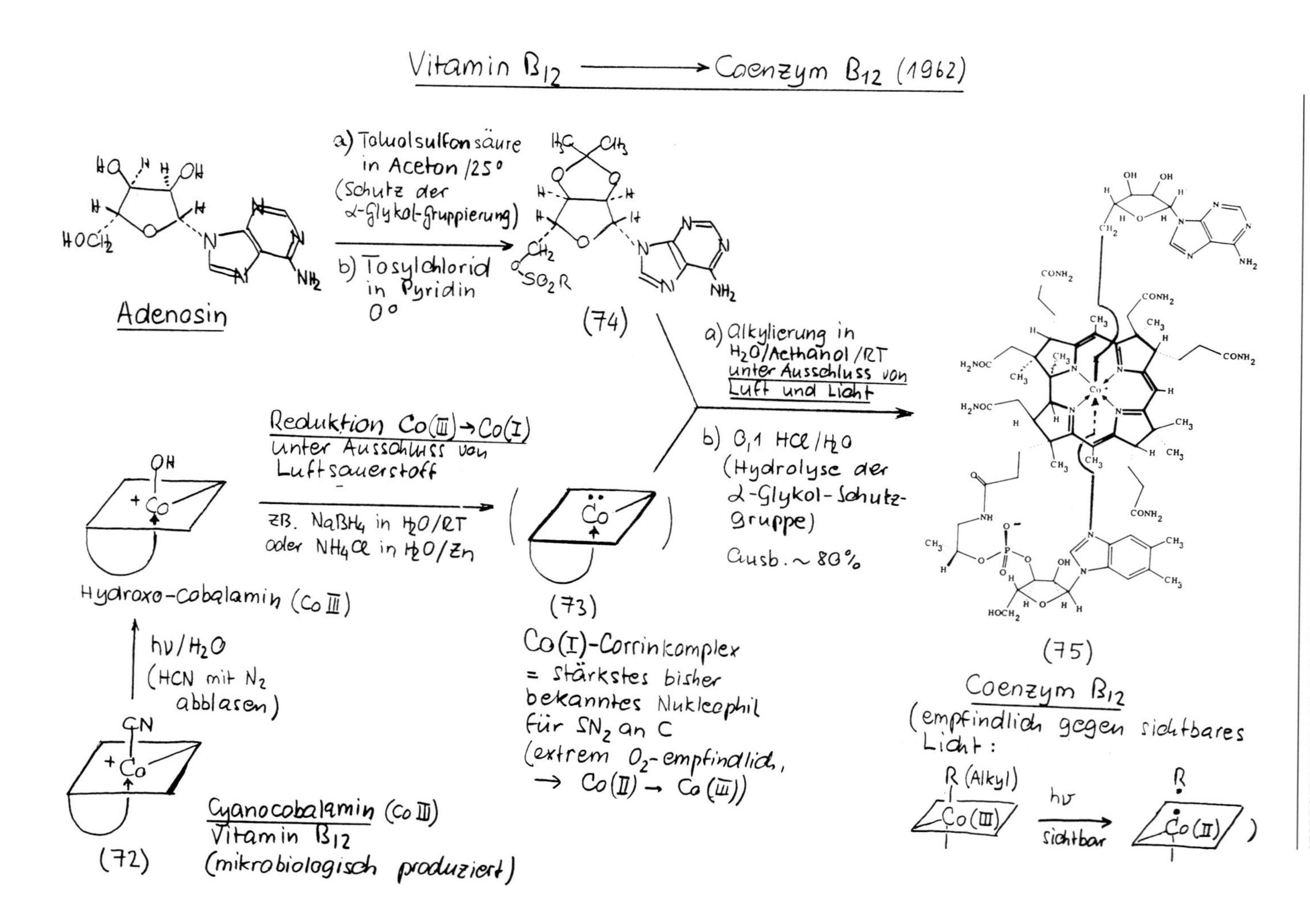
Vitamin B12 → Coenzym B12 (1962)
Adenosin
a) Toluolsulfonsäure in Aceton /25°
(Schutz der α-Glykol-Gruppierung)
b) Tosylchlorid in Pyridin 0°
(74)
Reduktion Co(III) → Co(I)
unter Ausschluss von Luftsauerstoff
zB. NaBH4 in H2O/RT
oder NH4Cl in H2O/Zn
Hydroxo-Cobalamin (Co III)
hν/H2O
(HCN mit N2 abblasen)
(72)
Cyanocobalamin (Co III)
Vitamin B12
(mikrobiologisch produziert)
(73)
Co(I)-Corrinkomplex
= stärkstes bisher bekanntes Nukleophil für SN2 an C
(extrem O2-empfindlich, → Co(II) → Co(III))
a) Alkylierung in H2O/Aethanol /RT
unter Ausschluss von Luft und Licht
b) 0,1 HCl/H2O
(Hydrolyse der α-Glykol-Schutz-gruppe)
Ausb. ~ 80%
(75)
Coenzym B12
(empfindlich gegen sichtbares Licht:
R (Alkyl)
Co(III)
hν
sichtbar
Co(II))

PRO MEMORIA OC/B 1973 Organische Synthese

Aufbau von Kohlenstoffgerüsten

- Aldolkondensation (Essay Nr.4)
- Wittig-Reaktion (Essay Nr. 4)
- Mannich-Kondensation (Essay Nr.4)
- Claisen'sche Esterkondensation und Claisen-Spaltung
- Iminoester-Enamin-Kondensation
- Michael-Addition (Essay Nr.5)
- α-Alkylierung und α-Acylierung von Carbonylsystemen
- Claisen'sche Umlagerungen (Essay Nr.10)
- Fischer'sche Indolsynthese (Essay Nr.10)
- Diels-Alder Reaktion (Essay Nr.2) und verwandte Cycloadditionen
- Elektrocyclische ($\pi \leftrightarrows \sigma$)-Isomerisierungen
- Elektrophile Substitution an cyclischen π-Systemen

Modifikation von Kohlenstoffgerüsten

- Oxidative Spaltung von Doppelbindungen (Essay Nr.1)
- Oxidative Spaltung von α-Dicarbonylsystemen
- Arndt-Eistert Reaktion (Essay Nr.3)
- Reaktionen mit Diazamethan (Essay Nr.3)
- Beckmann'sche Umlagerung (Essay Nr.8)
- Beckmann'sche Fragmentierung
- Wagner-Meerwein'sche Umlagerung (Essay Nr.8)
- Curtius'scher Abbau (Essay Nr.8)
- Sigmatrope H-Verschiebungen
- Decarbonylierung von Aldehyden und Säurechloriden
- Enaminreaktionen (Essay Nr.11)

Umwandlung von funktionellen Gruppen

- Bildung von Estern, Lactonen, Amiden, Lactamen etc. aus Carbonsäuren
- Hydrolyse, Alkoholyse, Aminolyse von Carbonsäurederivaten
- Reduktion von Carbonylverbindungen mit Metallhydriden (Essay Nr.6)
- Schutz von Carbonylgruppen, Hydroxylgruppen, α-Glykolgruppen Aminogruppen
- Reduktive Entschwefelung von Thioketalen, Thioamiden und Sulfiden
- Hydrierung, Halogenierung, Hydroxylierung von Doppelbindungen
- Reaktionen mit Singlett-Sauerstoff (Essay Nr.7)
- Hydratisierung von Dreifachbindungen
- Birch-Reduktion und Folgereaktionen (Essay Nr.9)
- Alkylierung von Amiden mit Oxoniumsalzen

Stereochemische Aspekte

- Prinzip der "Erhaltung der Orbitalsymmetrie"
- Grenzorbitalanalyse
- Chiralität-Helizität
- Uebergangszustands-Konformation → Produkt-Konfiguration
- Vermeidung der Bildung von Diastereomeren bei Verknüpfung chiraler Edukte durch Einsatz derselben in optisch aktiver Form mit korrektem Chiralitätssinn
- Kontrolle des sterischen Reaktionsverlaufs durch Intramolekularität von Syntheseschritten
- Intramolekular induzierte Stereospezifität bei bimolekularen Syntheseschritten
- Konfigurationsbeweise durch konfigurationsspezifische Ringschlussreaktionen
- Epimerisierbarkeit von Carbonyl-α-Stellungen
- Spaltung von Racematen in Enantiomeren

Einige Aspekte der Syntheseplanung

- Retrosynthetische Analyse:
  schrittweise Zerlegung der Produktstruktur in Zwischenprodukt- und Eduktstruktur-Serien durch Formalumkehr bekannter Syntheseschritte
- Potentielle Gleichwertigkeit transformierbarer funktioneller Gruppen (Potentielle Funktionalität)
- Assoziation von Strukturtyp mit Reaktionstyp
  zum Beispiel:
  Cyclohexene → Diels-Alder
  1,n-Dicarbonylverb. → Oxidative Spaltung von n-gliedrigen Cycloalkenen
  1,3-Dicarbonylverb. → Claisen'sche Esterkondensation und verwandte Reaktionen
  1,5-Dicarbonylverb. → Michaeladdition
  α,β-ungesättigte Carbonylverb. → Aldolkondensation und verwandte Reaktionen
  Lactame → Beckmann'sche Umlagerung von cyclischen Oximen
  usw. usw.
- Erzielung von Regio- und Stereospezifizität durch Kontrollgruppen bzw. Kontrollelemente
  z.B. Schutzgruppen
  Aktivierungsgruppen
  Intermediär-Ringe
  Intermediär-Verbrückungen
  Metallionen als Koordinationszentren
  usw.

Fragekatalog zur Vorlesung von Prof. Eschenmoser (Totâlsynthese des Vitamin B 12)

---

1. Welches Interesse bestimmte zu Beginn die Syntheseplanung - für das Vitamin B 12
   - Vitamin B 12 synthetisch in grossen Mengen herzustellen, für industriell-kommerzielle Verwendungszwecke?
   - Den Mechanismus der Natursynthese des Vitamin B 12 aufzudecken
   - Auch einen höchst komplizierten Naturstoff "machen zu können" (Beherrschen der Natur?)
2. Gab es spezifische oder grundsätzliche Probleme, die man durch das Studium des Vitamin B 12 studiert werden wollten - war die Orbitalsymetrieerhaltung ein Zufallsprodukt?
3. Warum wurde die Synthese des Vitamin B 12 in Angriff genommen und nicht ein ebenso kompliziertes Molekül, was gab den Ausschlag dafür, dass die grossen finanziellen Aufwendungen gemacht wurden?
4. Wer bezahlte die Vitamin B 12 Forschung, welche Industrie hatte zum vornherein ein Interesse an den Resultaten angemeldet?
5. Wozu wird heute organische Synthese in erster Linie von der chemischen Industrie eingesetzt,
   welche Stoffklassen müssen - sollen noch synthetisiert werden, aus welchen Gründen?
6. Wozu wird das Grundlagenwissen von der Industrie benützt, das heisst, für welche Aufgaben bilden wir uns aus (Diplom, Dissertation)

Index

A

B

C

D

I

J

K

N

O

P

T

U

V

W